Nonlinear Optics in Signal Processing

Engineering Aspects of Lasers Series

Series Editor

Dr T. A. Hall

Reader in Physics, University of Essex

SERIES EDITOR'S PREFACE

In the late 1960s and early 1970s the laser was still something of a scientific curiosity with only a limited practical use. The extent of the four volumes in this series shows the enormous change that has happened since that time. The laser is now an indispensable addition to the toolbox of the engineer and scientist. The progress from the time when the laser was often dubbed 'a solution in search of a problem' to today, when engineers of all disciplines frequently use lasers as a matter of course, is a remarkable transformation. Even so, the use of lasers in engineering and other walks of life is still in its infancy and has been held back partly by their relatively high cost and in some cases by their inconvenience in use. The cost of many laser systems has been falling for several years, they are becoming much more convenient to use and no longer need trained personnel to operate them. As these problems are overcome lasers will find wider and wider applications and there is an ever increasing need for engineers and scientists, who perhaps have little interest in lasers themselves to have access to an authoritative source which not only acts as an introduction but also takes the reader up to the latest developments in laser applications.

The four books of the series 'Engineering Aspects of Lasers' arose from a series of laser workshop courses which have been held annually at the University of Essex since 1979. These courses have evolved very considerably since their inception but aspects of their organization have not changed – the contents of the courses have always been coordinated by the recognized international authority in each subject area and the lectures given by experts in the particular field from industry, government laboratories or universities. When the idea of publishing a series of books based upon the contents of these courses was first suggested, the course coordinators at that time became the editors of each volume and the lecturers were asked to contribute.

The workshop courses are self-supporting courses which also form part of the MSc degree in Lasers and their Applications. There are many people who have contributed much to these courses over the years and have made them the success that they have been. I would like to express my gratitude to them all. The courses and the MSc were the brainchild of T. P. Hughes who was then Reader in Physics at Essex University. Without his foresight, hard work and determination in setting up the courses, this series of books would not have been written.

Other titles in series

Laser Processing in Manufacturing
Edited by R. C. Crafer and P. J. Oakley

Optical Methods in Engineering Metrology
Edited by D. C. Williams

Advances in Optical Communications
Edited by N. Doran and I. Garrett

Nonlinear Optics in Signal Processing

Edited by

Robert W. Eason
Senior Lecturer in Optics
University of Southampton
UK

and

Alan Miller
Professor of Physics and Electrical Engineering
Center for Research in Electro-Optics and Lasers
University of Central Florida
Orlando
USA

CHAPMAN & HALL
London · Glasgow · New York · Tokyo · Melbourne · Madras

Published by Chapman & Hall, 2–6 Boundary Row, London SEI 8HN

Chapman & Hall, 2–6 Boundary Row, London SEI 8HN, UK

Blackie Academic & Professional, Wester Cleddens Road, Bishopbriggs. Glasgow G64 2NZ, UK

Van Nostrand Reinhold Inc, 115 5th Avenue, New York, NY10003, USA

Chapman & Hall Japan, Thomson Publishing Japan, Hirakawacho Nemoto Building, 7F, 1-7-11 Hirakawa-cho, Chiyoda-ku, Tokyo 102, Japan

Chapman & Hall Australia, Thomas Nelson Australia, 102 Dodds Street, South Melbourne, Victoria 3205, Australia

Chapman & Hall India, R. Seshadri, 32 Second Main Road, CIT East, Madras 600 035, India

First edition 1993

Typeset in 10/12 Times by Thomson Press (India) Ltd, New Delhi
Printed in Great Britain at the University Press, Cambridge

ISBN 0 412 39560 6 0 442 31658 5 (USA)

A catalogue record for this book is available from the British Library

Library of Congress Cataloging-in-Publication data

Nonlinear optics in signal processing / edited by R. W. Eason and A. Miller. — 1st ed.
p. cm.
Includes bibliographical references and index.
ISBN 0–442–31658–5
1. Signal processing. 2. Nonlinear optics. I. Eason, R. W. II. Miller, A.
TK5102.5.N564 1993
621.382′2—dc20 92–31149
CIP

Printed on permanent acid-free text paper, manufactured in accordance with the proposed ANSI/NISO Z 39.48–199X and ANSI Z 39.4–1984

Contents

Preface

Recent years have witnessed what may be described as a 'second wave' of research activity in the field of nonlinear optics. The first wave began in the early 1960's soon after the invention of the laser. Anharmonic displacements of bound electrons could be readily accessed for the first time with the availability of large optical field strengths of these new coherent light sources. The practical consequence was the provision of an extended range of laser wavelengths via phenomena such as second (and higher) harmonic generation, parametric amplification, stimulated Raman scattering, etc. The second wave began in the late 1970's, and has grown rapidly over the last decade. This was driven by the discovery of optical nonlinearities which were many orders of magnitude more sensitive than had been previously available. Nonlinear optical effects became available at the milliwatt optical power level in semiconductors, organics, photorefractives, liquid crystals and glass fibres. These have opened new possibilities for exploiting laser light in communications, image processing and general purpose computing.

Laser light is currently finding its way into many areas of information technology. Fibre communications, optical discs, bar code readers and laser printers are a few well known examples. Optics is also finding application providing board to board interconnects within high performance computers. Optical signal processing employing linear optics has been available for some time. This makes use of the coherence and interconnect properties of laser light to perform mathematical functions which are computationally time consuming for electronic computers. However, linear signal processing has yet to establish any major applications. An incentive for developing our knowledge and understanding of nonlinear optical materials processes is the ability it gives in controlling one laser beam by another. The challenge for nonlinear optics is therefore to extend the role of light to the processing stage, within signal processing systems. The question is whether new phenomena such as optical bistability, phase conjugation and optical solitons are capable of exploiting the new types of material nonlinearity for digital processing, image processing, data storage, signal routing and computing.

Nonlinear optics is now a vast and often bewildering cross-disciplinary subject involving a multitude of different materials, phenomena and applications. The days when a complete coverage of nonlinear optics in one volume to any reasonable depth are long gone. Previous books on nonlinear optics, notably those by Bloembergen (1964), Akhamanov and Khokhlov (1972), Zerike and Midwinter (1973), Shen (1984), Schubert and Wilhelmi (1986), Butcher and Cotter (1990) and Boyd (1991), lay the foundations for the principles of nonlinear optics. Introductory treatments are contained in texts by Yariv (1985) and others. Conference and summer school proceedings, e.g. Harper and Wherrett (1977), Oudar (1986) provide more detailed discussions, but are typically advanced and very specialized. This book concentrates on providing a survey of advances in nonlinear optics relevant to signal processing, while also covering current research trends towards applications. The aim has been to establish a tutorial style introduction of the topic, to provide an illustration of the principles and concepts through examples, and to highlight key advances in this rapidly developing field.

This book is based on some of the lectures presented at a one week workshop entitled Nonlinear Optics in Signal Processing held at the University of Essex, Colchester, UK. This annual workshop forms part of a full-time MSc course in Lasers and Their Applications, but is also attended by participants from a broad cross-section of industry. The authors are all prominent scientists from university, government and industrial research laboratories, providing authoritative survey of the field from basics through recent advances to future possibilities. As in the workshop course, this book is prepared for physics or electrical engineering graduate level students who wish to bridge the gap between introductory treatments of nonlinear optics and advanced concepts applicable to signal processing. As such, the book should also appeal to those researchers already working in nonlinear optics or signal processing, who require a broad overview of the topic.

The contributors of this volume would like to thank the staff of the University of Essex for the opportunity to produce this book and for the excellent hospitality and congenial surroundings provided at Wivenhoe House, University of Essex which have made the workshops courses so successful and enjoyable.

Robert Eason, Southampton
Alan Miller, Orlando

Contributors

Alan Miller
Center for Research in Electro-Optics and Lasers, University of Central Florida, 12424 Research Parkway, Orlando, Florida 32826, USA

Simon Allen
Wilton Materials Research Centre, ICI Chemicals and Polymers Ltd, PO Box No 90, Wilton Middlesbrough, Cleveland, TS6 8JE, UK

Trevor Hall et al.
Department of Physics, King's College London, Strand, London, WC2R 2LS, UK

Brian Wherrett et al.
Department of Physics, Heriot-Watt University, Riccarton, Edinburgh, EH14 4AS, UK

Robert Eason
Department of Physics and Optoelectronic Research Centre, Southampton University, Southampton, SO9 5NH, UK

Ian Bennion et al.
Department of Electronic Engineering and Applied Physics, Aston University, Aston Triangle, Birmingham, B4 7ET, UK

David Cotten
BT Laboratories, Martlesham Heath, Ipswich, IP5 7RE, UK

Wilson Sibbett
Department of Physics and Astronomy, University of St Andrews, North Haugh, St Andrews, Fife KY16 9SS, Scotland

Introduction 1

Alan Miller

1.1. INITIAL COMMENTS

Recent years have witnessed a surge of research activity in the field of nonlinear optics. This has occurred for several reasons. Developments in material and laser technology have triggered the discovery of a wealth of new nonlinear optical effects in a wide range of materials. New phenomena such as optical bistability and other all-optical switching phenomena, phase conjugation and real time holography, solitons, etc., provide mechanisms for exploiting these new nonlinear optical interactions. Simultaneously, the increasing use of laser light in storage, display, communications and computing has given the incentive for utilizing nonlinear optical phenomena in information processing to extend systems capabilities.

The subject of nonlinear optics underwent an initial rapid growth in the early 1960s soon after the invention of the laser. It was the laser which provided electric field strengths comparable with interatomic fields for the first time, so that spectacular effects such as the second harmonic up-conversion of infrared radiation to visible light could be demonstrated. Using crystals, vapours and liquids and just a few fixed laser output lines, new wavelengths extending from the far infrared to the ultraviolet were soon being created by second harmonic generation, sum and difference frequency mixing and stimulated Raman scattering. In addition, parametric gain in nonlinear optical crystals allowed coherent oscillation and tunable outputs (Bloemberger, 1965; Zernike and Midwinter, 1973; Harper and Wherrett, 1977; Shen, 1984; Hopf and Stegeman, 1986; Butcher and Cotter, 1990).

The standard materials for the generation of new frequencies are high optical quality single crystals which utilize nonresonant 'bound electron' interactions with input and output light in the transparency range of the material. An essentially instantaneous response is necessary for wavelength conversion since the polarization of the crystal needs to be modulated at the new optical frequency. A classical description is normally adequate to describe these interactions whereby the nonlinearity is related to the anharmonicity in the potential of the bound electrons when driven to larger displacements.

Although concepts for optical signal processing using nonlinear optics were envisaged at an early stage, these were generally ruled out for most practical purposes because of the limited magnitude of the nonlinear optical coefficients and therefore requirements for very high optical intensities.

More recently, *resonant* interactions have been providing very large optical nonlinearities in many types of materials, e.g. organics, semiconductors, photorefractive crystals and liquid crystals, thus offering new opportunities for utilizing the unique properties of laser beams for signal processing. Even nonresonant nonlinearities are now finding applications in cases where the interaction length can be very long, such as in glass fibres developed for optical communications. Because mode-locked lasers can produce ultrashort pulses in the femtosecond regime, the required intensities can be achieved at moderate pulse energies. Nonlinear interactions have also yielded the means to monitor and unravel the physics of events that take place on ultrashort timescales. Thus, the development of femtosecond pulse lasers has given access to timescales not attainable in electronic devices by making use of nonlinear optical interactions in pump and probe laser measurements in which a probe pulse is time delayed by employing a variable spatial delay. Other fundamental aspects of nonlinear optics which have developed and can have bearing on nonlinear optical device performance include quantum statistics, chaos, coherence phenomena and squeezed light.

Research on *dynamical* optical nonlinearities began in earnest in the early 1980s. There now exists a bewildering variety of transient nonlinear optical phenomena to choose from, whether they be passive or active, resonant or nonresonant, purely optical or hybrid with electronics. Simultaneously, new phenomena such as optical bistability and phase conjugation have appeared which can utilize these large nonlinearities in novel ways (Gibbs, 1985; Fisher, 1983). Parallel advances in lasers in terms of size, power, threshold, pulse length and reliability makes this topic all the more interesting. Thus, nonlinear optics, after a period of relatively low activity during the 1970s, has blossomed over the last decade into a very active research area in many material systems, with the potential for exploitation in various areas of lasers, optoelectronics, signal processing, image processing and computing.

1.2. NONLINEAR OPTICAL INTERACTIONS

A **linear system** is defined as one which has a response proportional to external influence and has a well-known property, i.e. if influences, $F_1, F_2 \ldots, F_n$ are applied simultaneously, the response produced is the sum of the responses that would be produced if the influence were applied seperately.

A **nonlinear system** is one in which the response is not strictly proportional to the influence and the transfer of energy from one influence to another can occur.

If the influences are periodic in time, the response of a nonlinear system can contain frequencies different from those present in the influences. However, the point to emphasize here is that, as well as the generation of new frequencies, nonlinear optics provides the ability to control light with light and so to transfer information directly from one beam to another without the need to resort to electronics.

Traditionally, nonlinear optics has been described phenomenologically in terms of the effect of an electric field on the polarization within a material. Electromagnetic processes are described by Maxwell's equations which constitute a set of linear equations. In SI units,

$$\nabla \cdot \mathbf{D} = \rho \tag{1.1}$$

$$\nabla \times \mathbf{E} = -\frac{\partial \mathbf{B}}{\partial t} \tag{1.2}$$

$$\nabla \cdot \mathbf{B} = 0 \tag{1.3}$$

$$\nabla \times \mathbf{H} = \frac{\partial \mathbf{D}}{\partial t} + \mathbf{J} \tag{1.4}$$

where the electric and magnetic fields are **E** and **B** respectively. The displacement fields **D** and **H** arise from the external charge and conduction current densities ρ and **J**. In most cases of interest in nonlinear optices, $\rho = 0$ and $\mathbf{J} = \mathbf{0}$.

'Constitutive relations' connect the charge and current distributions within the medium and the displacement fields to the electric and magnetic fields. The relation of interest for nonlinear optics is

$$\mathbf{D} = \mathbf{P} + \varepsilon_0 \mathbf{E} = \varepsilon \mathbf{E} \tag{1.5}$$

where **P** is the induced polarization in the medium resulting from the field **E**, ε is defined as a dielectric constant and ε_0 is the permittivity of free space ($8.85 \times 10^{-12}\,\mathrm{F\,m^{-1}}$ in MKS units).

In the usual description of electromagnetic interactions, the constitutive relations are linearized and the principle of superposition of waves can be applied. However, this linear approximation is only applicable to low field strengths so that the existence of nonlinear optical phenomena in materials should come as no surprise. Indeed, the concept of harmonic generation was quite familiar at radio frequencies long before frequency mixing at optical frequencies became possible. The conventional way of describing optical nonlinearity is in terms of susceptibilities, χ, which are defined by expanding the induced polarization, **P**, in powers of electric field, **E**. Thus,

$$\mathbf{P} = \varepsilon_0 [\chi^{(1)} + \chi^{(2)}\mathbf{E} + \chi^{(3)}\mathbf{E}^2 + \cdots]\mathbf{E} \tag{1.6}$$

Table 1.1 lists passive nonlinear optical phenomena in relation to their order of nonlinear susceptibility. The susceptibilities are tensor quantities with the

Table 1.1 Passive nonlinear uptical effects

Process	*Susceptibility*	*Input*	*Output*
Linear dispersion	$\chi^{(1)}(\omega_1;\omega_1)$	ω_1	ω_1
Linear electro-optic (Pockels effect)	$\chi^{(2)}(\omega_1;\omega_1,0)$	$\omega_1,0$	ω_1
Sum frequency mixing	$\chi^{(2)}(\omega_3;\omega_1,\omega_2)$	ω_1,ω_2	$\omega_3=\omega_1+\omega_2$
Second harmonic generation	$\chi^{(2)}(\omega_3;\omega_1,\omega_1)$	ω_1	$\omega_3=2\omega_1$
D.C. rectification	$\chi^{(2)}(0;-\omega_1,\omega_1)$	ω_1	0
Difference frequency mixing	$\chi^{(2)}(\omega_2;-\omega_3,\omega_1)$	ω_1	$\omega_2,\omega_3(\omega_1=\omega_2+\omega_3)$
Quadratic electro-optic (Kerr effect)	$\chi^{(3)}(\omega_1;0,0,\omega_1)$	$\omega_1,0$	ω_1
Sum mixing	$\chi^{(3)}(\omega_4;\omega_1,\omega_2,\omega_3)$	$\omega_1,\omega_2,\omega_3$	$\omega_4=\omega_1+\omega_2+\omega_3$
Third harmonic	$\chi^{(3)}(\omega_4;\omega_1,\omega_1,\omega_1)$	ω_1	$\omega_4=3\omega_1$
Optical Kerr effect	$\chi^{(3)}(\omega_1;\omega_1,-\omega_1,\omega_1)$	ω_1	ω_1
Four-wave difference frequency mixing	$\left\{\begin{matrix}\chi^{(3)}(\omega_3;-\omega_4,\omega_1,\omega_2)\\ \chi^{(3)}(\omega_4;-\omega_3,\omega_1,\omega_2)\end{matrix}\right\}$	ω_1,ω_2	$\omega_3,\omega_4(\omega_1+\omega_2=\omega_3+\omega_4)$
	$\chi^{(3)}(\omega_2;-\omega_3,-\omega_4,\omega_1)$	ω_1	$\omega_2,\omega_3,\omega_4(\omega_1=\omega_2+\omega_3+\omega_4)$

components giving the response for a given orientation of the crystallographic axes relative to the incident field. For linear interactions in isotropic crystals, χ and ε are scalars but the tensor components must always be taken into account in higher orders.

Linear optical properties are described by $\chi^{(1)}$ which is related to the complex index, η, and dielectric constant, ε, by

$$\eta^2 = \varepsilon = \varepsilon_0(1 + \chi^{(1)}) \tag{1.7}$$

The real part of $\chi^{(1)}$ describes refractive properties while the imaginary part describes absorption.

A classical description is usually adequate for the optical properties of dielectric materials. Thus a harmonic oscillator with a resonance frequency, ω_0, and damping constant, γ, gives a frequency-dependent linear susceptibility,

$$\chi^{(1)}(\omega) = \frac{Ne^2}{\varepsilon_0 m} \frac{1}{(\omega_0^2 - \omega^2 - 2i\gamma\omega)} \tag{1.8}$$

where N is the density of electrons. Real and imaginary parts of the susceptibility are illustrated in Figure 1.1 for more than one oscillator. Causality between the real and imaginary parts is usefully expressed using the Kramers–Kronig relations, for example, for the refractive index, n (real part of η) and the Beer's law absorption coefficient, α,

$$n(\omega) - 1 = \frac{c}{\pi}\mathscr{P}\int_0^\infty \frac{\alpha(\Omega)\mathrm{d}(\Omega)}{\Omega^2 - \omega^2} \tag{1.9}$$

where $\mathscr{P}$ denotes the Cauchy principal value and c is the speed of light. Thus, the absorption coefficient at all frequencies, Ω, contributes to the refractive index at the observation frequency, ω.

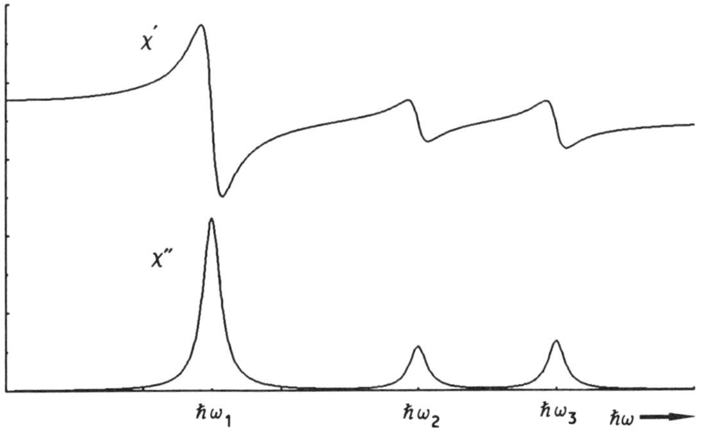

Figure 1.1 Real and imaginary parts of the susceptibility for more than one oscillator.

The second-order susceptibility, $\chi^{(2)}$ describes three-wave mixing processes such as second harmonic generation (SHG), sum and difference frequency mixing (Table 1.1). For instance, introducing a quadratic displacement term proportional to a constant, b, into the equation of motion for electrons predicts terms which oscillate at 2ω and a relation for the second-order susceptibility given by,

$$\chi^{(2)} = \frac{\varepsilon_0^2 mb}{N^2 e^3}[\chi^{(1)}(\omega)]^2[\chi^{(1)}(2\omega)] \tag{1.10}$$

Second-order susceptibilities vary between 10^{-7} and 10^{-15} esu, the conversion factor to SI units being $\chi^{(2)}(\text{esu}) = 2.387 \times 10^3 \chi^{(2)}(\text{m V}^{-1})$.

The third-order term, $\chi^{(3)}$, describes four-wave mixing processes. Thus, third harmonic generation and the various sum and difference combinations produced by superimposing two or three frequencies are given by the third-order susceptibility, (Table 1.1). The real part of $\chi^{(3)}$ also accounts for nonlinear refraction (or optical Kerr effect) employed in a number of nonlinear optical devices. The magnitude of $\chi^{(3)}$ for nonresonant interactions is typically from 10^{-11} to about 10^{-14} esu. (The conversion factor from esu to SI units is $\chi^{(3)}(\text{esu}) = 7.162 \times 10^7\, \chi^{(3)}(\text{m}^2\,\text{V}^{-2})$.) An alternative parameter which describes nonlinear refraction, n_2, when taken as a product with the laser beam intensity, I, is given by,

$$n = n_0 + n_2 I \tag{1.11}$$

where n_0 is the linear refractive index.

This definition of nonlinear refraction allows a more comfortable depiction of resonant phenomena when a change in linear susceptibility, $\Delta\chi^{(1)}$, may be a more appropriate description. Since the intensity I is proportional to $\langle|\mathbf{E}|^2\rangle$, we can relate n_2 to $\chi^{(3)}$ or an effective third-order susceptibility, $\chi^{(3)}_{\text{eff}}$, from,

$$\mathbf{P}(\omega) = \varepsilon_0(\chi^{(1)} + \Delta\chi^{(1)})\mathbf{E}(\omega) = \varepsilon_0(\chi^{(1)} + \chi^{(3)}_{\text{eff}}\langle|\mathbf{E}(t)|^2\rangle)\mathbf{E}(\omega) \tag{1.12}$$

to give,

$$n_2 = \frac{\text{Re}[\chi^{(3)}_{\text{eff}}]}{2\varepsilon_0 c n_0^2} \tag{1.13}$$

An alternative definition of n_2, for bound electron interactions, gives the refractive index change in terms of electric field strength rather than intensity.

Just as the imaginary part of $\chi^{(1)}$ accounts for absorption, the imaginary part of $\chi^{(3)}$ describes the dissipative processes two-photon absorption, saturation, a.c. Stark effect and Raman and Brillouin scattering (Figure 1.2 and Table 1.2). Causality again links the real and imaginary parts of $\chi^{(3)}$. This is illustrated in Figure 1.3 for saturation of a two-level atomic system. Saturation of the Lorentzian absorption lineshape predicts a decrease in refractive index below the resonance frequency and an increase above it; however, the refractive

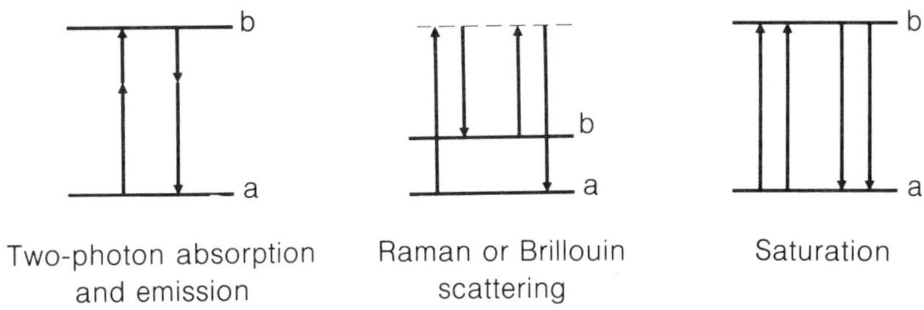

Figure 1.2 Examples of active nonlinearities.

Table 1.2 Resonant optical nonlinearities

Process	*Susceptibility*	
Linear absorption	$\chi^{(1)}(\omega_1;\omega_1)$	$\omega_1 \simeq \omega_0$
Saturation	$\chi^{(3)}(\omega_1;\omega_1,-\omega_1,\omega_1)$	$\omega_1 \simeq \omega_0$
Two-photon absorption	$\chi^{(3)}(\omega_1;\omega_1,-\omega_1,\omega_1)$	$2\omega_1 \simeq \omega_0$
Raman scattering Brillouin scattering	$\chi^{(3)}(\omega_2;\omega_1,-\omega_1,\omega_2)$	$\omega_2 = \omega_1 \mp \omega_0$

index change is zero exactly on resonance. This simple atomic picture translates well to more complicated systems consisting of many states. For example, band filling saturation in semiconductors produces negative changes in index at frequencies below the band gap energy which can be understood in terms of a set of optically coupled two-level states because of the restriction to vertical transitions in reciprocal space. Figure 1.4 shows how a strong pump beam can alter the real and imaginary absorptive and refractive spectral properties of a medium via two-photon absorption to a band at frequency ω_g and Raman scattering from a lattice vibration at ω_v.

An extension of the Kramers–Kronig relation to the nonlinear regime gives insight into the relationship between the real and imaginary parts. It can be shown that changes in refractive index, Δn, can be related to changes in absorption, $\Delta\alpha$, in a general way by the relation,

$$\Delta n(\omega;\zeta(Xi)) = \frac{c}{\pi}\mathscr{P}\int_0^\infty \frac{\Delta\alpha(\Omega;\zeta(Xi))}{\Omega^2-\omega^2}\,\mathrm{d}\Omega \tag{1.14}$$

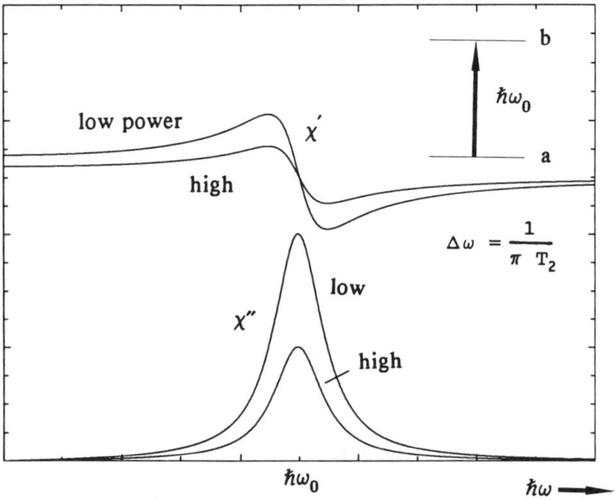

Figure 1.3 Saturation of a two-level atomic system.

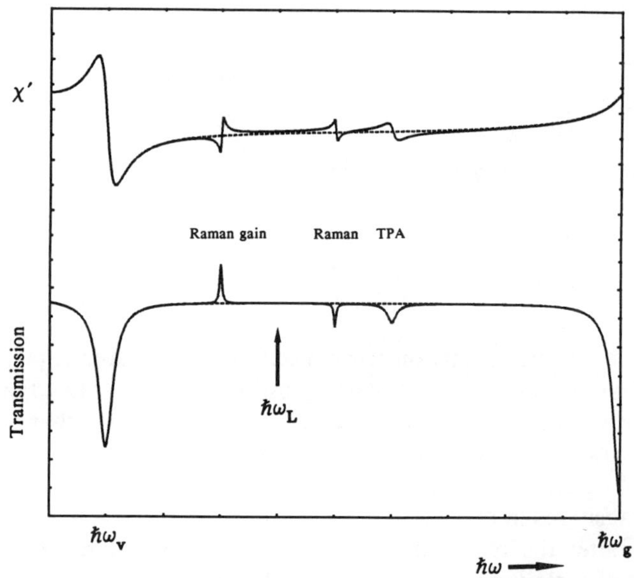

Figure 1.4 Effect of a pump beam on spectral properties.

where $\zeta(Xi)$ denotes some perturbation of the material caused by the light. This perturbation may be due to generated carriers, thermo-optically induced effects, an electric field or a 'dressed' state produced by the light–material interaction. It is important to note, particularly for the latter case, that the perturbation must be independent of the frequency of observation, ω, thus allowing the integration to proceed. The integration given in equation 1.14

is often easier to perform than the linear case of equation 1.9 since *changes* in absorption typically occur only over a limited spectral region.

Any influence leading to light modulation can be considered as nonlinear optical if it depends on the nonlinearity described by the constitutive relations. Thus, the Pockels, Kerr and Faraday effects are now included as nonlinear optical phenomena, but the frequency of the applied field inducing the nonlinearity is low (d.c. to microwave) compared with optical frequencies. Resonant optical nonlinearities are often more appropriately described by quantum mechanical transition rates for the excitation combined with rate equations to take account of the time constants and the density matrix formalism employed.

Under the above definition of a nonlinear system, we can choose to define our nonlinear system very broadly. For instance, the 'system' might consist of a detector, some electronics and a light emitter to produce an appropriate nonlinear input–output response. Alternatively, we might combine optical excitation, motion of generated free charge carriers and optical modulation. An example of an optical nonlinearity which depends on carrier motion is the photorefractive effect which combines photogeneration of carriers and electro-optic modulation. In this case, free carriers optically generated from trap states move distances on the scale of a wavelength to build up space charge fields which in turn induce changes in refractive index via the Pockels electro-optic effect. Four-wave mixing in photorefractives is thus achieved by combining the imaginary part of $\chi^{(1)}$ (linear absorption) with the real part of $\chi^{(2)}$ (Pockels effect) to give four-wave mixing normally associated with $\chi^{(3)}$. This type of nonlinear optical interaction extends the traditional description of a microscopically 'local' nonlinear response to spatially extended interactions and carrier transport. The result is some of the most sensitive nonlinear optical interactions. Self-electro-optic effect devices (SEEDs), which at present provide the most attractive option for logic elements for testing optical computing concepts, are an extension of this concept. These devices combine photodetection and modulation within a multiple quantum well (MQW) semiconductor with feedback provided by external circuitry. (Several different variations of this basic idea have been demonstrated.) The high sensitivity of these devices comes from the separation of planes of charge. The effective $\chi^{(3)}$ can be on the order of 10^7 esu! Optical switching has been demonstrated in SEED devices with no external bias. Thus, the distinction between an all-optical nonlinear process and a hybrid optical–electrical interaction which creates the desired response becomes fuzzy. Indeed, the limiting cases of a combination of very highly integrated optoelectronic devices producing a nonlinear input–output response, and a nonlinear optical process which allows some charge transport, become one and the same thing.

A distinct class of nonlinear optical phenomenon arises from thermo-optic processes. Resonant excitation always produces some heating due to absorption of energy from the light beam. Absorptive and refractive properties of most materials are sensitive to temperature such that, for continuous wave excitation at least, they dominate over electronic optical nonlinearities. Although these optothermal effects are for the most part slow and a nuisance, they have also been put to good use. For instance, thermo-optic effects in semiconductors and liquid crystals have rendered useful low power optically bistable devices. These have been fabricated into arrays and used to demonstrate prototype digital optical circuits and optical computing concepts.

The majority of this volume deals with light–material interactions which are resonant with a vibration, rotation or electronic transition of a material, in which case the process is dissipative and described by $\chi^{(3)}_{\rm eff}$. This also implies that the material response time is long compared with the reciprocal of the optical frequency, so that the harmonic generation of new frequencies is not possible. Although the frequency mixing susceptibility formalism may not be the most appropriate description, effective susceptibilities are often deduced for resonant interactions in order to compare magnitudes.

A useful categorization of third-order optical nonlinearities is by response time. The fastest processes involve bound electrons giving ultrafast but relatively insensitive interactions. Resonance with material energy levels enhances the magnitude of these effects to provide much increased sensitivities for the all-optical control of light beams, but generally at the expense of speed. Charge transport optical nonlinearities can further enhance the size of the interaction with the response time determined by the details of the charge transport mechanisms. For small dimensions, e.g. in semiconductors, charge transfer can be very fast. Thermal effects can be very sensitive but are usually associated with a slow recovery time, however thermal engineering can reduce this to a few nanoseconds. Indeed, most resonant optical nonlinearities can offer relatively fast responses if conditions are appropriately chosen.

In general, there is a trade-off between sensitivity and speed. The speed of response sets the application. For instance, ultrafast responses are most appropriate for applications which employ a waveguide format supporting relatively few channels whereas slower, sensitive responses can achieve fast systems speeds overall by exploiting the massive parallel processing capabilities offered by optics.

1.3. A SKETCH HISTORY OF NONLINEAR OPTICS IN DIELECTRIC MATERIALS

The invention of the laser launched the subject called nonlinear optics as a research field, but nonlinear optical effects had been predicted and observed previously. Phenomena such as the Pockels and Kerr effects, Raman scattering, saturation and two-photon absorption were known before high brightness sources were available.

The Kerr effect was discovered as early as 1875. Kerr found that when d.c. electric fields were applied to certain transparent liquids, the refractive index changed in proportion to the square of the field. Later, Pockels and others observed refractive index changes proportional to applied field in certain crystals. The Raman effect was discovered in 1927 and allowed a shift of wavelength for monochromatic light in transparent media due to inelastic scattering with atomic vibrations. Two-photon absorption processes were considered in 1931 and the implications of expanding the dipole moments of molecules in powers of the field were studied theoretically in the 1950s.

The announcement of laser action in Ruby in 1960 made high intensity optical pulses available for the first time and very quickly led to the demonstration of a host of nonlinear optical phenomena. Lasers were soon providing extremely high power densities—intensities of over $10^9\ \mathrm{W\,cm^{-2}}$ (optical field strengths of $10^6\ \mathrm{V\,cm^{-1}}$) could be applied to crystals in short optical pulses without damage. Optical harmonics were already being generated by 1961. Other nonlinear phenomena using lasers soon followed, such as sum and difference frequency mixing, the optical Kerr effect, four-wave mixing, stimulated scattering (Raman and Brillouin) and self-induced transparency. Parametric oscillation in nonlinear crystals provided enough gain in three-wave mixing configurations in 1965 to provide tunable coherent sources. A great deal of effort was expended throughout the 1960s and early 1970s on identifying inorganic crystals with the correct combination of properties to give efficient generation at new frequencies to extend the coverage of available laser wavelengths. The simultaneous requirements of nonlinearity, phase matching, transparency, excellent crystal quality and mechanical strength limit the number of potentially useful nonlinear optical crystals to relatively few out of the many thousands of natural and synthesized crystals.

Although *ab initio* calculation of nonlinear optical coefficients is a complicated procedure for most materials, there are a few guiding principles which can be used. For instance, centrosymmetric crystal structures have an identically zero $\chi^{(2)}$ because of symmetry. A useful empirical rule suggested by equation 1.10, first proposed by R.C. Miller, relates $\chi^{(2)}$ to $\chi^{(1)}$ according to

$$\delta = \frac{\chi^{(2)}}{[\chi^{(1)}(\omega)]^2[\chi^{(1)}(2\omega)]} \simeq \text{constant} \qquad (1.15)$$

δ is expected to be constant for materials of the same symmetry; however, it is found to vary by less than a factor of two over most materials. An interpretation of Miller's rule is that the more polarizable the material, the stronger the first-order optical nonlinearity. Additional symmetry rules for the tensor components give the simplifying result that $\chi_{ijk}{}^{(2)}$ is symmetric under any permutation of its indices.

Large optical nonlinearity alone is not enough for efficient second harmonic generation. Three-wave mixing processes such as SHG require phase matching of input and output waves over the length of the interaction, otherwise light generated at one point in the crystal will destructively interfere with light from another location. Because of dispersion, Figure 1.1, it is not possible to have the same phase velocity for both fundamental and harmonic waves in isotropic materials. A coherence length in the crystal is defined by

$$l_{coh} = \frac{\pi}{\Delta k} = \frac{\lambda_\omega}{4(n_{2\omega} - n_\omega)} \tag{1.16}$$

where $\Delta k = k^{2\omega} - 2k^\omega$ and k^ω and $k^{2\omega}$ are the wavevectors of the fundamental and harmonic waves respectively. Coherence lengths are typically in the range 10–100 μm. Phase matching, $\Delta k = 0$, may be achieved in birefringent crystals by the appropriate choice of propagation direction and polarization to give the maximum interaction length. This is important since conversion efficiencies depend on the square of the interaction length. It is necessary to know refractive indices to three significant places (a) to determine whether the birefringence has the correct magnitude (it can be too small or too large) for phase matching, and (b) to determine the orientation of the crystal. Phase matching may only be achievable over a restricted range of the transmission window of the crystal. Beam walk-off is also very important (i.e. the Poynting vectors of the fundamental and second harmonic are not necessarily co-directional in anisotropic crystals). Noncritical conditions occur when phase matching can be achieved for propagation directions along the principal axes. Other effects such as the spatial profile of the beam, focusing, nonlinear self-focusing and damage also must be taken into consideration. Because phase matching places a restriction on the propagation direction within the crystal, the appropriate component of the susceptibility tensor must be chosen. Alternative techniques for phase matching are available, for example, inversion of the crystal about its symmetry axis every coherence length or by making use of waveguide dispersion.

Because of the many restrictions placed on crystals for second harmonic generation, relatively few nonlinear optical crystals are commonly used. Those in frequent use are Te (tellurium), KDP (potassium dihydrogen phosphate), ADP (ammonium dihydrogen phosphate), $LiNbO_3$ (lithium niobate) and $LiIO_3$ (lithium iodate). Efficient frequency conversion of the more common laser sources has become well established with these crystals. For example, Te is used to frequency double CO_2 laser from 10 to 5 μm; KDP can produce second, third and fourth harmonics of a 1.06 μm Nd:YAG laser with high conversion efficiencies; and $LiNbO_3$ yields a parametric oscillator giving tunable output in the 1.4–4 μm range when pumped with 1.06 μm Nd:YAG output.

Laser developments in the areas of high power semiconductor heterostructure lasers and arrays, diode-pumped solid state lasers, tunable solid state sources such as Ti:sapphire, colour centres, alexandrite, and other vibronic lasers have extended the availability of laser wavelengths. However, this has only increased the demand for nonlinear frequency conversion crystals in order to extend further the applications of these lasers since some wavelengths, such as the visible range, are still difficult to access with solid state lasers alone. Advances in mode-locking of near-infrared lasers make femtosecond pulses much more readily available, thus providing high intensities for efficient nonlinear conversion to the visible. Resonant intracavity second harmonic generation in diode-pumped solid state lasers provides good opportunities for compact visible sources.

Efficient frequency generation has demanded large crystals of high optical quality. A great deal of effort has gone into crystal growth throughout the 1970s and 1980s. For instance, the family of chalcopyrite-structured semiconductor crystals offers over 30 different birefringent crystals with large nonlinear optical coefficients. Although visible wavelengths tend to be limited by the fundamental absorption edge, $CdGeAs_2$, $ZnGeAs_2$, $AgGaSe_2$ and $AgGaS_2$ are very useful materials for the infrared, but they are only just becoming readily available in good quantity after over 15 years of research into their growth. $AgGaS_2$ has been employed for downconversion of picosecond pulses to the 1.5–10 μm range. Thallium arsenic selenide (TAS) has also proved difficult to grow to good quality but is becoming a most attractive alternative to Te for second harmonic frequency conversion of CO_2 lasers because of the multiphoton absorption limitations of Te at high intensities.

KTP (potassium titanyl phosphate), $KNbO_3$ (potassium niobate), LBO (lithium triborate) and BBO (β-barium borate) are examples of new crystals recently showing good success. KTP is now supplied by laser manufacturers as the standard crystal for frequency doubling of Nd:YAG lasers from 1.06 to 0.53 μm. Noncritical phase matching can be achieved in KTP and $KNbO_3$. KTP has also produced tunable radiation in the 0.6–4.5 μm region by difference frequency mixing 1.06 μm and the output from a visible tunable dye laser. As an OPO, 1.8–2.4 μm tuning has been demonstrated with 1.3 and 4 μm theoretically possible.

LBO and BBO look even more attractive. LBO can achieve higher second harmonic conversion efficiencies for mode-locked YAG lasers than KTP because of damage limitations in the latter. BBO has a wide transparency between 189 nm and 3 μm. Fifth harmonic generation of 1.06 μm Nd:YAG to 213 nm has been demonstrated in BBO with 11% overall efficiency. Two crystals, Nd:MgO:$LiNbO_3$ (Nd- and MgO-doped lithium niobate) and NYAB (neodymium yttrium aluminium borate), provided simultaneous lasing and self-frequency doubling.

Organic materials offer good prospects with very large nonlinear coefficients. Urea is the best developed but many others are being pursued (Chapter 2). Organic molecules can allow optimization of their nonlinear optical response by molecular engineering. Other forms of engineered structures can provide the birefringence necessary to achieve phase matching. For instance III–V semiconductors such as GaAs offer large nonlinear optical coefficients, but crystallize in a cubic structure. Birefringence can be induced, however, by epitaxial growth of heterostructures of GaAs and AlGaAs. Although thickness will always be limited to a few microns, waveguide geometries provide sufficient interaction lengths which could provide one solution to the challenge of providing up-conversion of semiconductor laser diodes from the infrared into the visible. Further progress in inorganic and organic crystals may allow some of the previously abandoned early applications of nonlinear optics to be reconsidered. For instance, up-conversion of infrared images could take advantage of superior visible detectors for image processing.

Material requirements are not so restrictive for third-order processes. Centrosymmetric crystals can be included and phase matching can be more readily accomplished even in isotropic materials. Four-wave mixing processes can be readily phase matched by suitable choice of propagation directions. The optical Kerr effect, stimulated Raman and Brillouin scattering and two-photon absorption are automatically phase matched irrespective of the material dispersion. For nonresonant optical nonlinearities, third-order coefficients are in general smaller than those of second order; however, resonance can provide greatly enhanced interactions. A large number of solids, liquids and vapours have been investigated. Many of these interactions are discussed in this volume.

1.4. OPTICAL SIGNAL PROCESSING

The very large bandwidth of light, $\sim 10^{15}$ Hz, gives optics a potential speed for signal processing which is well beyond any electronics. Indeed, the shortest optical pulses of < 10 fs give light three orders of magnitude advantage over the shortest electrical pulses achievable. (Subpicosecond electrical pulses can be produced using mode-locked lasers.) However, exploiting this huge bandwidth for signal processing is not simple if only for the reason that this advantage turns into a problem of interfacing with conventional electronics with bandwidths typically no better than 10^9 Hz.

Various linear signal processing devices utilizing acousto-optical, electro-optical and two-dimensional properties of light have been under development for some time. Such devices as spectrum analysers, correlators, filters, and matrix–vector multipliers are very powerful and have shown early promise, but have yet to make a significant impact on signal processing applications. On the other hand, optical fibres have been an outstanding success of optics. The principal reasons for the success of fibres in telecommunication systems are low cost of fabrication and low loss propagation which translates into price competitiveness compared with copper coaxial cable. However, the wide deployment of fibres gives the opportunity for exploiting the wide bandwidths made available. For instance, the use of optical solitons is currently being considered for long haul, intercontinental submarine communications. Applications of optical fibres are being extended to sensors, gyros, microwave filters, and phased array radar. Other recent high volume sales successes of lasers have been in optical storage and bar code readers. The acceptance of the audio disc to replace the vinyl long-playing record is placing lasers in virtually every home. In this case, advantage is being taken of the high spatial resolution from the small spot size that can be achieved with focused laser light. These successes, putting more information onto beams of light, gives the main incentive for carrying out routing or signal processing functions while the information is still in the form of light.

Parallel processing is the alternative area where the use of light suggests considerable potential. Light offers a degree of parallelism unapproachable by electronics because of the huge interconnect capability of non-interfering laser beams. For example, a $1\,cm^2$ area optical lens has an inherent resolution set by diffraction of about 10^8 channels. Both digital and analogue parallel processing is being activity pursued using optics, while neural computing would seem to be well suited to the optical domain. In particular, image processing is a very taxing problem for electronic computing and would benefit enormously by some form of optical pre-processing. In parallel processing systems, the energy required to drive each element or pixel must be minimized, thus demanding more sensitive interactions. Individual optical devices cannot compete with electronic devices on energy comparisons alone; however, when interconnect and parallelism are considered, optical systems may have much to offer. Thus, optical nonlinearities when applied via processes such as optical bistability, phase conjugation, and associated phenomena, or may offer advantages over electronics if the full potential of parallelism and coherence can be exploited with low energy interactions.

An application of nonlinear optics not widely heralded has been its success in passive mode locking of lasers to produce optical pulses only a few optical cycles in duration (a few femtoseconds). Ultrashort pulses have been employed mainly for scientific purposes, but their applications could extend to optical communications and computing systems within a few years. Placing saturable dyes within laser cavities is a standard method of pulse shortening and passive mode locking. At the very high intensities available from mode-locked lasers the laser gain medium itself is invariably nonlinear and contributes to the pulse shortening. Pulses can be further compressed very conveniently by exploiting optical nonlinearities in optical fibres after leaving the laser cavity (Chapter 10).

We may thus list the potential advantages of using optics in signal processing as shown in Table 1.3. Against these advantages, we have to accept or overcome a number of problems and limitations. The largest hurdle for the implementation of optics in information processing systems is probably the phenomenal success of electronics and computing over the past few decades. It is difficult to introduce a new technology into another which is still advancing so quickly. Optics must offer very substantial advantages to outpace developments in electronics. However, electronics has some funda-

Table 1.3 Advantages and disadvantages of optics

Advantages of optics	*Disadvantages of optics*
Large bandwidth	High power requirements
Low loss propagation	Interfacing with electronics
Low crosstalk	Rayleigh scattering
High degree of parallelism	Wavefront distortions
Small dimensions	Mechanical stability
Ultrashort pulses	
Low time skew	
Coherence properties	

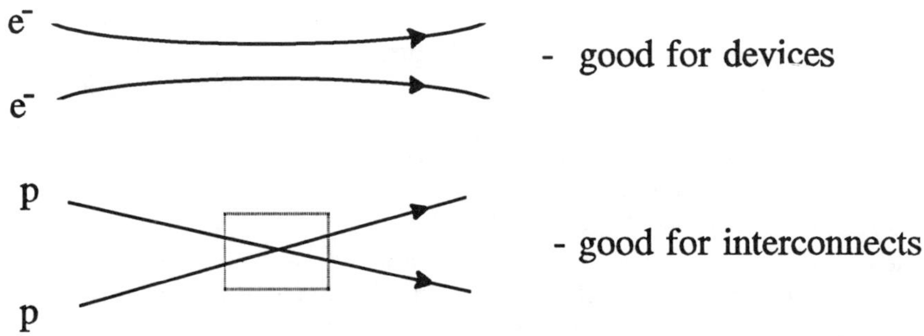

Figure 1.5 Uses of electrons and photons.

mental limitations. For instance, more space is being devoted to interconnections on a chip than for the devices. Capacitance of these interconnects will set the speed limits of integrated circuits rather than the speed of the individual devices. High frequencies resulting in excessive cross-talk and extracting data from integrated circuits limited by metal pin connectors are other major bottlenecks. The fundamental differences between electronics and photonics resides in the fact that photons pass through each other unperturbed in the absence of a nonlinear interaction, whereas electrons interact with each other even at a distance (Figure 1.5). Electrons are thus more naturally suited to active devices while photons are more appropriate for interconnection. However, once information is in the form of light, there are many attractions to taking the next step of processing data without repeated conversion to electronics and vice versa. It seems likely that electronic devices will always be more energy efficient, but the advantages of speed and/or parallelism can give optics a valuable role in signal processing.

1.5. CONCLUSIONS

To conclude this introduction, we emphasize that nonlinear optics is now a very broad subject with a wealth of phenomena accessible in a wide range of materials. Generation of new frequencies by mixing techniques is well established, but there still remains a need for better crystals. An alternative to frequency mixing is stimulated Raman scattering which can access some wavelengths more conveniently. The availability of high intensities from ultrashort pulse, mode-locked lasers now allows even very small optical nonlinearities to be accessed which has had some major implications for optical fibres. Resonant third-order optical nonlinearities can be very sensitive for applications in optical logic devices and coherent parallel processing. Advances in materials such as organics, semiconductors, multiple quantum wells, liquid crystals, and photorefractives give a great deal of scope for applications in signal processing which will be discussed in this volume.

REFERENCES

Bloemberger, N. (1965) *Nonlinear Optics*, Benjamin, New York.

Butcher, P. N. and Cotter, D. (1990) *The Elements of Nonlinear Optics*, Cambridge Studies in Modern Optics, **9**, Cambridge University Press, Cambridge.

Fisher, R. A. (1983) *Optical Phase Conjugation*, Academic Press, New York.

Gibbs, H. M. (1985) *Optical Bistability: Controlling Light with Light*, Academic Press, Orlando, FL.

Harper, P. and Wherrett, B. S. (1977) *Nonlinear Optics*, Academic Press, London.

Hopf, F. A. and Stegeman, G. I. (1986) *Applied Classical Electrodynamics*, Vol. II, *Nonlinear Optics*, Wiley, New York.

Shen, Y. R. (1984) *The Principles of Nonlinear Optics*, Wiley, New York.

Zernike, F. and Midwinter, J. E. (1973) *Applied Nonlinear Optics*, Wiley, New York.

2 Nonlinearities in organic solids

Simon Allen

Although there has been a high level of interest in nonlinear optical effects for more than 20 years now, most of the work in this field has concentrated on the development of noncentrosymmetric inorganic compounds such as $LiNbO_3$ and analogues of potassium dihydrogen phosphate (KDP). Many of these inorganic crystals have been commercialized in the form of single-crystal frequency convertors, electro-optic modulators, and more recently as simple waveguide couplers and switches. Since about 1975 there has been increasing interest in organic compounds with nonlinear optical properties as potential alternatives to the inorganics mentioned above (Williams, 1983, 1984; Chemla and Zyss, 1986). Much of the work has concentrated on second-order properties, although more recently third-order effects have also been described.

In section 2.1 the relative strengths and weaknesses of organic and inorganic materials are discussed, and then in the following section the molecular basis of the nonlinearities in organic materials is described in some detail. Experimental and theoretical methods for the determination of molecular nonlinearities are outlined. Sections 2.3, 2.4 and 2.5 cover in turn three different routes to the achievement of highly nonlinear solids, by the aggregation of the individual molecules in crystals, polymers and Langmuir–Blodgett films respectively.

The structures of molecules described within this chapter are shown in Figure 2.1, along with their commonly used acronyms. Full chemical nomenclature will not be used in this work.

As described in Chapter 1, the induced polarization **P** of a nonlinear optical medium, due to an electromagnetic field **E**, can be written

$$\mathbf{P} = \chi^{(1)}\mathbf{E} + \chi^{(2)}\mathbf{E}^2 + \chi^{(3)}\mathbf{E}^3 + \dots \qquad (2.1)$$

where it is understood that the macroscopic nonlinear susceptibilities $\chi^{(i)}$ are in fact tensors of rank $i + 1$. A similar equation can be written to relate the molecular polarization **p** induced in a single molecule by the application of the

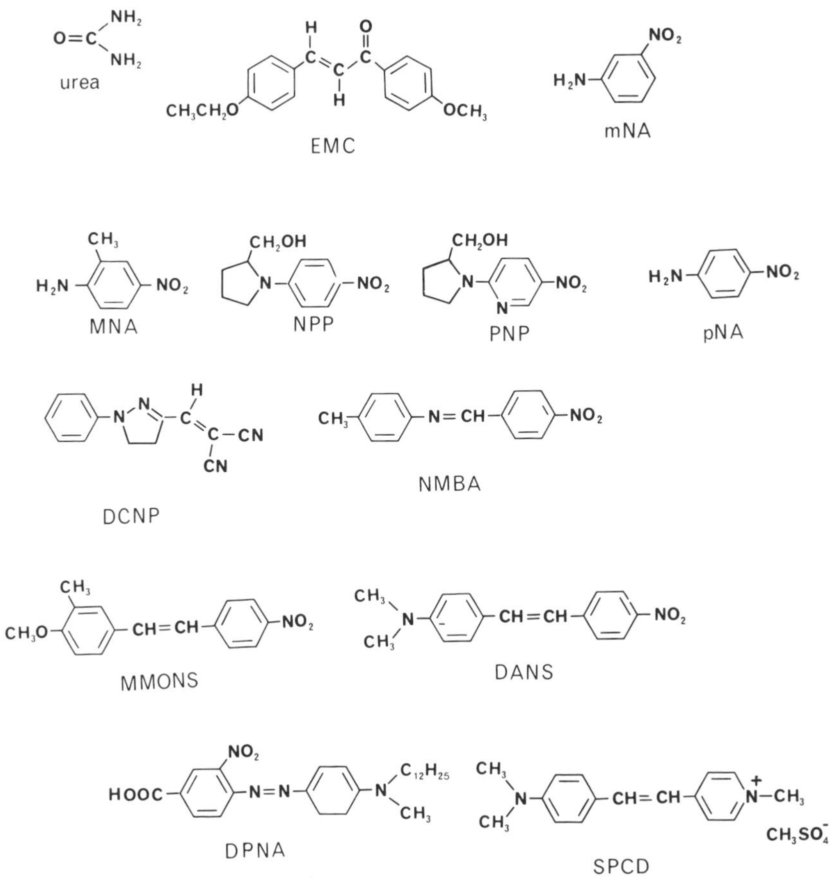

Figure 2.1 Structures and acronyms for the molecules discussed in this work.

same external field **E**:

$$\mathbf{p} = \boldsymbol{\alpha}\mathbf{E} + \boldsymbol{\beta}\mathbf{E}^2 + \boldsymbol{\gamma}\mathbf{E}^3 + \ldots \tag{2.2}$$

Here again $\boldsymbol{\alpha}, \boldsymbol{\beta}, \boldsymbol{\gamma}$ are second-, third- and fourth-rank tensors. The organic media we are considering in this chapter (such as crystals and polymers) are built up from individual molecular units. In general the intermolecular forces are only relatively weak (van der Waals forces), and the intermolecular interactions are much smaller than the intramolecular ones. Because of this the bulk nonlinear optical properties can to a very good approximation be described as the tensor sum of the individual molecular properties taking due account of orientational factors, molecular volume and the effects of the local fields. Because of the presence of neighbouring molecules (normally polar) the electric field at any point within the medium will not be the same as the external applied field. Due account of this can be taken using theories of varying sophistication (Levine and Bethea, 1975; Hurst and Munn, 1986). Thus for second-order effects the macroscopic coefficient $\chi^{(2)}$ is related to the microscopic coefficient $\boldsymbol{\beta}$ by the equation

$$\chi^{(2)}_{IJK} = (1/V)\Sigma f^{2\omega}(f^{\omega})^2 a_{Ii}a_{Jj}a_{Kk}\beta_{ijk} \tag{2.3}$$

where V is the volume (e.g. of the crystal unit cell), f are the local field factors, a_{Ii} are the direction cosines of the molecular frame of reference xyz with respect to the crystal frame XYZ (often denoted 123) and the sum is over all molecules within the volume V. A similar equation relates the third-order coefficient γ and $\chi^{(3)}$:

$$\chi^{(3)}_{IJKL} = (1/V)\Sigma f^{3\omega}(f^{\omega})^3 a_{Ii}a_{Jj}a_{Kk}a_{Ll}\gamma_{ijkl} \tag{2.4}$$

The role of the direction cosine terms in equations 2.3 and 2.4 is particularly important. In either second-order or third-order case it may happen that molecules which themselves have high intrinsic nonlinearities are arranged in the solid state so that, within the summation of coefficients over neighbouring molecules, the major components from different molecules cancel each other out. This will result in a low value for the bulk nonlinearity $\chi^{(2)}$ or $\chi^{(3)}$. In the case of the second-order terms the form of the summation is such that the bulk coefficient $\chi^{(2)}$ will be identically zero in all structures which contain a centre of symmetry. It is also common in practice that structures are formed with pseudo-centrosymmetric arrangements of molecules. In these cases the bulk nonlinearities $\chi^{(2)}$ will be very small. This severe restriction on the useful range of solid state molecular packing does not hold for the third-order terms. Materials with centrosymmetric molecular structures may have high third-order nonlinearities $\chi^{(3)}$.

The optimum arrangement of molecules in the solid state (i.e. the preferred values of a_{Ii}) depends on the application considered, as described in following sections. It can be seen from the above that the design of novel, highly active organic nonlinear optical materials can be divided into two distinct parts: the identification, design and synthesis of molecules with high molecular nonlinearities $\boldsymbol{\beta}$ or γ, and then the design of bulk materials in which the relative orientation of these active molecules is optimized for a particular application.

2.1. A COMPARISON OF ORGANIC AND INORGANIC MATERIALS

2.1.1. Second-order nonlinearities

The main advantage that organic materials have over the inorganic materials such as KDP and $LiNbO_3$ is that of significantly higher nonlinear efficiencies, as shown in Figure 2.2, where large organic molecules (such as NMBA and

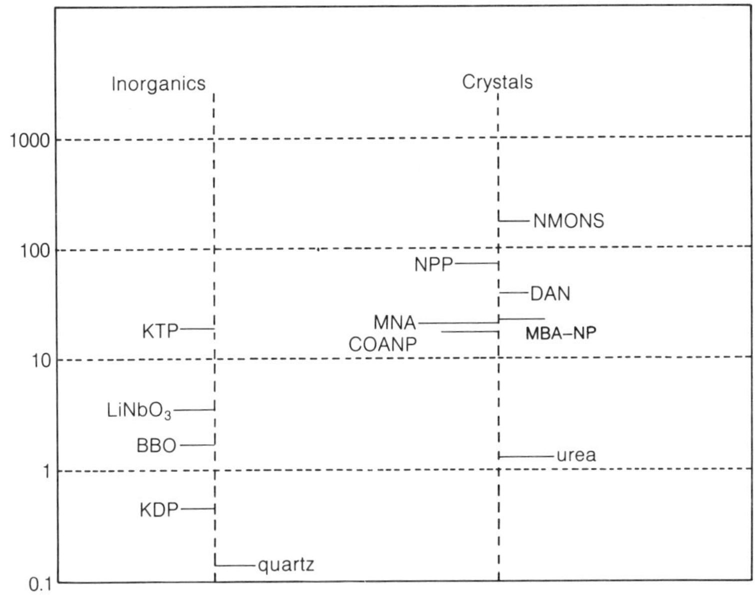

Figure 2.2 A comparison of the second-order nonlinearities $\chi^{(2)}$ for second harmonic generation in a range of inorganic and organic crystalline compounds.

DPNA) can have $\chi^{(2)}$ coefficients one or two orders of magnitude greater than that of $LiNbO_3$. In fact, as shown in Chapter 1, the efficiency of any second-order process such as second harmonic generation does not depend simply on the coefficient $\chi^{(2)}$. For this reason it is possible to define a different 'figure of merit' for each process. For second harmonic generation the figure of merit is often chosen to be d^2/n^3 (where d is the second harmonic coefficient, related to $\chi^{(2)}$ by $d=\chi^{(2)}/2$, and n is the refractive index). Amongst the organics the highest phase-matchable value for MNA is about 45 times that of $LiNbO_3$ (Levine *et al.*, 1979) (although a figure of merit of about 100 times that of $LiNbO_3$ is obtained in a non-phase-matchable direction), and the phase-matchable figures of merit for other similar compounds PNP and NPP are about 200–300 times the highest phase-matchable figure for $LiNbO_3$.

One figure of merit for the electro-optic (Pockels) effect may be taken as the reduced half-wave voltage V_π^* (which is the voltage required to switch a unit cube of the material from 100% to 0% transmission at a given wavelength, when placed between crossed polarizers). In this case clearly the aim is for as low a value for V_π^* as possible. MNA, again, has a reduced half-wave voltage 2–3 times lower than lithium niobate, and more recently a material having an order of magnitude lower V_π^* than $LiNbO_3$ has been described (Allen *et al.*, 1988a).

There is also evidence that some organic materials show significantly greater resistance to damage by intense laser radiation than do inorganics. POM, for example has a damage threshold of 50 MW cm^{-2} at 530 nm with 15 ns pulses and 270 MW cm^{-2} with 7 ps pulses (Zyss *et al.*, 1981), and that of urea is about 5 GW cm^{-2} at 1.06 μm using 10 ns pulses (Cassidy *et al.*, 1979), whereas that of $LiNbO_3$ is only about 20 MW cm^{-2} (Linford and Clendening, 1986). However, some inorganic materials described in the literature recently, such as β-barium borate, also have high damage thresholds.

Most polar organic single crystals also have very large birefringence (due to the anisotropic nature of the molecular polarizability), which can be of an advantage for the achievement of phase matching for many second-order processes such as second harmonic generation, frequency mixing and parametric amplification.

A final advantage of organic compounds is the possibility for 'molecular engineering', in which subtle chemical changes in the molecular structure can be effected, having little impact on the nonlinear optical properties of the molecule, but being designed to improve other properties such as molecular packing, melting point, stability or solubility of the molecule within a polymer host.

2.1.2. Third-order nonlinearities

As outlined in later chapters in this book, over recent years there has been a large amount of work carried out studying third-order nonlinear optical effects in semiconductor materials. Initially this work concentrated on the materials InSb and GaAs, but more recently it has been extended to a great range of semiconducting compounds. Particular emphasis has been placed on nonlinearities in GaAlAs multiple quantum well (MQW) structures. The nonlinearities can usually be described in terms of an intensity-dependent refractive index n,

$$n = n_0 + n_2 I \tag{2.5}$$

where I denotes the incident light intensity, and n_0 is the refractive index in the limit of $I = 0$.

When a material having a high n_2 term is placed in some form of optical feedback device (such as a Fabry–Pérot cavity) optical bistability can be observed. Optically bistable devices have many potential applications in signal processing and optical computing, as described elsewhere in this book. The nonlinear term $n_2(I)$ can be formally related to the third-order nonlinear optical coefficient $\chi^{(3)}$ by the equation

$$n_2 = (4\pi^2/c)\chi^{(3)}/n^2 \tag{2.6}$$

Here $\chi^{(3)}$ is in units of cm^4 esu^{-1}. In fact this relationship does not have great significance, as the definition of $\chi^{(3)}$ (as in equation 2.1) is based, as already described, on a model of anharmonic local potentials for electron motion, and relates to an 'intrinsic' nonlinearity which will be present to a greater or lesser extent at all wavelengths. The origin of the nonlinear refractive index n_2 in semiconductors lies in more complex excitation and relaxational processes involving the absorption of light within the tail of the absorption band (Wherrett and Smith, 1985). Thus the effects occur only at a specific wavelength, and this will limit the versatility of any device utilizing such materials. In compensation for this, however, the magnitude of the nonlinearities obtained in semiconducting materials is several orders of magnitude higher than the intrinsic third-order nonlinearities observed in organic and polymeric systems. Demonstrations of a number of optical signal processing functions have been achieved, particularly using hybrid electrical–optical MQW devices known as self-electro-optic devices (SEEDs). Effective $\chi^{(3)}$ values of about 10^{-6} esu have been obtained in semiconductor materials, whereas the best obtained to date in organic compounds are about 10^{-8}–10^{-9} esu. Nevertheless, there are still some attractions in the use of organic materials for third-order processes:

1. Because, unlike semiconductor compounds, the nonlinearities are not due to some real excitation of the molecule they are not specific to a particular wavelength. The materials can therefore be used at any point within the spectrum where they do not absorb a significant proportion of the incident or emitted light.
2. The absence of a real excitation or relaxation process means that the response times of organic materials can be significantly faster than those of semi-conductor compounds. The only limitation is in principle the electronic response times themselves, and subpicosecond response times are predicted.
3. Fabrication of semiconductor MQW devices is complicated technically and requires significant investment in terms of equipment and expertise. In contrast, the processing of many polymeric compounds is relatively simple, making use of many standard processes of thin film fabrication, and lithographic modification of the surface topography.

Thus there may be room for both semiconductor and organic–polymeric compounds in the rapidly expanding technology of optical signal processing, with each finding its own niche.

2.2. MOLECULAR PROPERTIES

2.2.1. The origin of molecular nonlinearities β and γ

The organic materials found to exhibit the highest second-order nonlinear coefficients **β** are those which contain conjugated electron pathways, consisting of a series of alternating double and single covalent bonds. These may take the form, for example, of benzene or heterocyclic rings (i.e. rings containing other atoms, such as nitrogen or oxygen, as well as carbon), unsaturated chains or a combination of these. The conjugated systems contain π-electron orbitals that are delocalized, meaning that the electrons are not tightly bound to one particular atom but are shared within the conjugated region. The electron clouds of these loosely bound delocalized π electron systems are readily influenced by the electric field of the incident light, and this is the origin of the high polarizabilities. Additionally, in order to produce large second-order effects a high degree of asymmetry is required in the molecule. This can be achieved by the addition of chemical groups which tend to push electrons into the conjugated system (donor groups) and/or groups that withdraw electrons from the system (acceptor groups). The highest nonlinearities are found when donors and acceptors are situated at either end of an extended conjugated pathway, as can be seen in Figure 2.3, where —OMe, —NH_2, —NMe_2 groups are donors and —CN and — NO_2 groups are acceptors. Such a molecule can be visualized as an anharmonic well for the motion of electrons, as described for inorganic materials in Chapter 3.

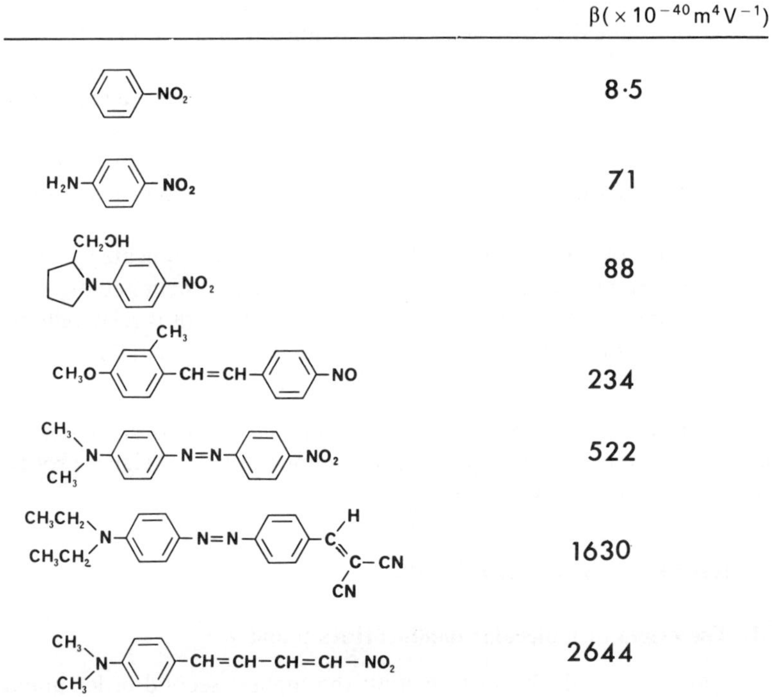

Figure 2.3 Molecular hyperpolarizabilities β for a range of conjugated moieties, as determined by EFISH experiments.

Most of the work that has taken place over the last ten years or so has concentrated on achieving high molecular nonlinear optical coefficients by trying to optimize the form of the conjugated pathway (in terms of both its length and its atomic structure), and maximizing the effect of the donor and acceptor groups, through their number, structure and position with respect to the conjugation path. The scope of organic chemistry for the synthesis of variations on this basic theme is enormous.

The situation is further complicated by the need to build into the molecules certain other properties. Particularly important is the transparency range of the materials in question. As will be discussed in more detail later, there is a general correlation between the magnitude of the nonlinearity of a molecule and its colour, or the position of the maximum optical absorption of that molecule. The same features (long conjugation path coupled with donor and/or acceptor groups) that imbue the high nonlinearities result in absorption bands extending far into the visible portion of the spectrum. Thus for example, in the design of molecules for use in second harmonic generation of laser diode sources, from wavelengths around 820 nm to 410 nm, the requirement of optical transparency at both these wavelengths severely restricts the chemist's scope.

The molecular requirements for high third-order nonlinearities $\boldsymbol{\gamma}$ have been less thoroughly studied. Many studies of third-order effects have been on polymeric materials rather than on individual molecules, and often it is then just the bulk nonlinearity $\boldsymbol{\chi}^{(3)}$ that is quoted. Since the length of a single polymer chain is not usually known (and there will be a large range of chain sizes present in any sample) it is more sensible to define $\boldsymbol{\gamma}$ as the nonlinearity associated with a single repeat unit of the polymer chain.

Because $\boldsymbol{\gamma}$ is a fourth-rank tensor, there are fewer symmetry constraints on the coefficients than was the case for $\boldsymbol{\beta}$. In particular, the coefficients may be nonzero in a symmetric molecule. It appears, experimentally, that as for $\boldsymbol{\beta}$ the highest nonlinearities $\boldsymbol{\gamma}$ are associated with molecules having large conjugated π-electron systems, and that again, in general, the greater the length of the conjugation pathway the greater the nonlinearity. Different studies have suggested power law dependences of γ on the number of atoms N in the conjugated path of the form (Pierce, 1988):

$$\gamma = N^3 \tag{2.7}$$

However, there is also evidence that the nonlinearity will eventually saturate as N increases to large values.

Because of the above-mentioned symmetry considerations, it is not necessary to have donor and acceptor groups grafted onto the conjugated pathway to generate nonzero γ values. However, it appears that the nonlinearity is again enhanced if such groups are present. They will, as for β, presumably increase the anharmonicity of the potential well experienced by the mobile π-electrons. For example, the γ value of p-nitroaniline (a benzene ring with an NO_2

acceptor group and NH_2 donor group) is of the order of 200 times that of the unsubstituted benzene ring alone.

2.2.2. Measurement of molecular nonlinearities

It is not in general possible to make direct measurements of molecular nonlinearities—meaning that we cannot take a single molecule and measure all the different components of its nonlinear optical coefficient $\boldsymbol{\beta}$ or $\boldsymbol{\gamma}$. It is instead necessary to take some, preferably well-defined, aggregation of molecules, to measure the bulk nonlinearity $\chi^{(2)}$ or $\chi^{(3)}$ and to infer the molecular coefficients from a knowledge (or model) of the molecules within the sample. This essentially involves working backwards through equations 2.3 or 2.4. It is also therefore necessary to know, or to be able to estimate, the values of the local field factors $f^{n\omega}$ contained within these equations.

One method that has been extensively used for the measurement of both $\boldsymbol{\beta}$ and $\boldsymbol{\gamma}$, is the so-called 'dcSHG' or electric field induced second harmonic (EFISH) experiment (Levine and Bethea, 1975; Singer and Garito, 1981; Oudar, 1977). In this case the aggregation used is a solution of a known concentration of the molecules of interest in a well-characterized solvent. A pulsed electric field is applied across the solution simultaneously with an incident laser pulse. A second harmonic signal is generated in the solution via a third-order process as outlined below. If the molecules are centrosymmetric (and therefore have $\beta = 0$) the value of γ can be derived directly from the second harmonic intensity. However, if the molecules are polar, having a non-zero β coefficient, then the molecular electric dipoles are aligned to a small extent by the applied d.c. electric field. The degree of order induced is sufficient to produce an additional contribution to the second harmonic generation. Analysis is in this case somewhat more complicated, as shown below.

The induced polarization $p^{2\omega}$ at the harmonic frequency, for the case when all electric fields (i.e. the applied d.c. electric field and the optical fields associated with the polarized laser beam) are parallel and along the Z-axis is given by (Levine and Bethea, 1975)

$$p^{2\omega} = N f^0 (f^{\omega})^2 f^{2\omega} \bar{\gamma} E^0 (E^{\omega})^2 \tag{2.8}$$

where N is the number of active molecules per unit volume, f are the local field factors, E^0 is the amplitude of the d.c. pulse, E^{ω} is the amplitude of the optical field of frequency ω, and $\bar{\gamma}$ is an average third-order coefficient defined by

$$\bar{\gamma} = \gamma + \beta_z \mu_z / 5kT \tag{2.9}$$

Here k is the Boltzmann constant, T the absolute temperature,

$$\gamma = 1/5\, (\gamma_{xxxx} + \gamma_{yyyy} + \gamma_{zzzz} + 2\gamma_{xxyy} + 2\gamma_{xxzz} + 2\gamma_{yyzz}) \tag{2.10}$$

is a scalar combination of the individual third-order tensor coefficients and

$$\beta_z = \beta_{zzz} + (\beta_{zxx} + \beta_{zyy} + 2\beta_{xzx} + 2\beta_{yzy})/3 \tag{2.11}$$

is a vector component of the β tensor, along the direction of μ_z, the ground state dipole moment of the molecule. The molecular axes x, y, z are chosen so that z lies along the dipole moment $\boldsymbol{\mu}$.

It can be seen from the above that if $\beta_z = 0$ then γ can be determined directly. In molecules for which β is nonzero, it is usually the case that the second term in equation 2.9 is significantly greater than the first, and so it is possible to ignore the effect of γ. It is possible to take both terms into account, for example by repeating the experiment at a number of different concentrations. Alternatively, the γ coefficent may be determined independently, for example using third harmonic generation. Care must be taken when using this procedure, as dispersion effects will mean that the γ values obtained by the two different techniques will not necessarily be the same. The design of the cell for this experiment is particularly important, and has been described in detail in the literature (Oudar, 1977). In principle the experiment is straight forward but in practice it can be difficult to perform and also to evaluate the results. Third-order nonlinearities in both the cell and the solvent must be accounted for (which complicates the equations). If there is absorption at either the fundamental or the harmonic wavelength the signal contrast is much reduced and the analysis further complicated. For full analysis the dielectric constants and refractive indices of both solvent and solution, at ω and 2ω, and the molecular dipole moments must be known for calculation of the local field factors.

Finally, a significant limitation of the method is that it is only the 'vector component' β_z or the scalar γ that is determined. Nonetheless, a large number of compounds have been characterized using this technique, particularly in terms of their β_z coefficients. A comprehensive listing of these results has been published by Nicoud and Twieg (1986).

Molecular coefficents can also be deduced from bulk measurements on other solid state samples such as single crystals (where the molecular orientations can be accurately determined) or Langmuir–Blodgett films. In both cases estimation of the local field is the major source of uncertainty.

2.2.3. Calculation of molecular nonlinearities

An alternative to this experimental approach is to develop theoretical techniques for the calculation of the molecular nonlinearities. Methods based on a perturbation theory description of the nonlinearities developed in the early days of nonlinear optics by Ward (1965) have more recently been applied to molecular systems (Lalama and Garito, 1979; Docherty *et al.*, 1985).

Within this description the second-order molecular nonlinearity $\boldsymbol{\beta}$ applicable to second harmonic generation is given by the rather complicated formula

$$\beta^{2\omega}_{ijk} + \beta^{2\omega}_{ikj} = \frac{-e^3}{4\hbar^2} \times \sum_{n'}\sum_{n}\Bigg\{(r^j_{gn'}r^i_{n'n}r^k_{ng} + r^k_{gn'}r^i_{n'n}r^j_{ng})$$

$$\left[\frac{1}{(\omega_{n'g}-\omega)(\omega_{ng}+\omega)} + \frac{1}{(\omega_{n'g}+\omega)(\omega_{ng}-\omega)}\right]$$

$$+(r^i_{gn'}r^j_{n'n}r^k_{ng} + r^i_{gn'}r^k_{n'n}r^j_{ng})$$

$$\left[\frac{1}{(\omega_{n'g}+2\omega)(\omega_{ng}+\omega)} + \frac{1}{(\omega_{n'g}-2\omega)(\omega_{ng}-\omega)}\right]+$$

$$+(r^j_{gn'}r^k_{n'n}r^i_{ng} + r^k_{gn'}r^j_{n'n}r^i_{ng})\left[\frac{1}{(\omega_{n'g}-\omega)(\omega_{ng}-2\omega)} + \frac{1}{(\omega_{n'g}+\omega)(\omega_{ng}+2\omega)}\right]\Bigg\} \tag{2.12}$$

where the terms $r^j_{gn'}$ are transition moments between the ground state and various excited states of the molecule (related to the strength of the optical absorption associated with transitions between these states), and the $\omega_{n,g}$ are the frequencies of the transitions, which are directly related to the wavelengths at which absorptions take place. Fortunately all these terms can be readily calculated using standard techniques of quantum chemistry, and the compilation of the components of the $\boldsymbol{\beta}$ tensor from these terms is then a straightforward task. Lalama and Garito (1979) have applied CNDO semiempirical methods, using limited configuration interaction, to the calculation of the β values of a number of simple molecules, and this work has later been extended to cover a much greater range of molecules (Docherty *et al.*, 1985; Morley *et al.*, 1987a–c). This latter program has been parametrized by comparison of experimental and calculated values of the transition energies and dipole moments (related to the transition moments r_{gn}) for a series of molecules having a large variation in size and complexity. After adjustment of the parameters, the correlation obtained between calculated and experimental values of the $\boldsymbol{\beta}$ coefficients is excellent, as shown in Figure 2.4. In contrast to the EFISH experiment described above, all components of the $\boldsymbol{\beta}$ tensor are obtained by this method, and it is simple to calculate, for example, the wavelength dependence of the coefficients, or by simple extension the coefficients for related processes such as the electro-optic effect.

It can be seen that equation 2.12 contains terms in the denominator of the form $\omega_{ng} - \omega$ and $\omega_{ng} - 2\omega$. The presence of these terms means that if either the energy of the fundamental beam incident on the molecule, or that of the second harmonic generated by the molecule, approaches the energy of one of the molecular transitions there will be a large increase in the β value ('resonant enhancement'). In practice this may only be advantageous in the case of very thin films (as described in section 2.5), as in general the increase in the β value

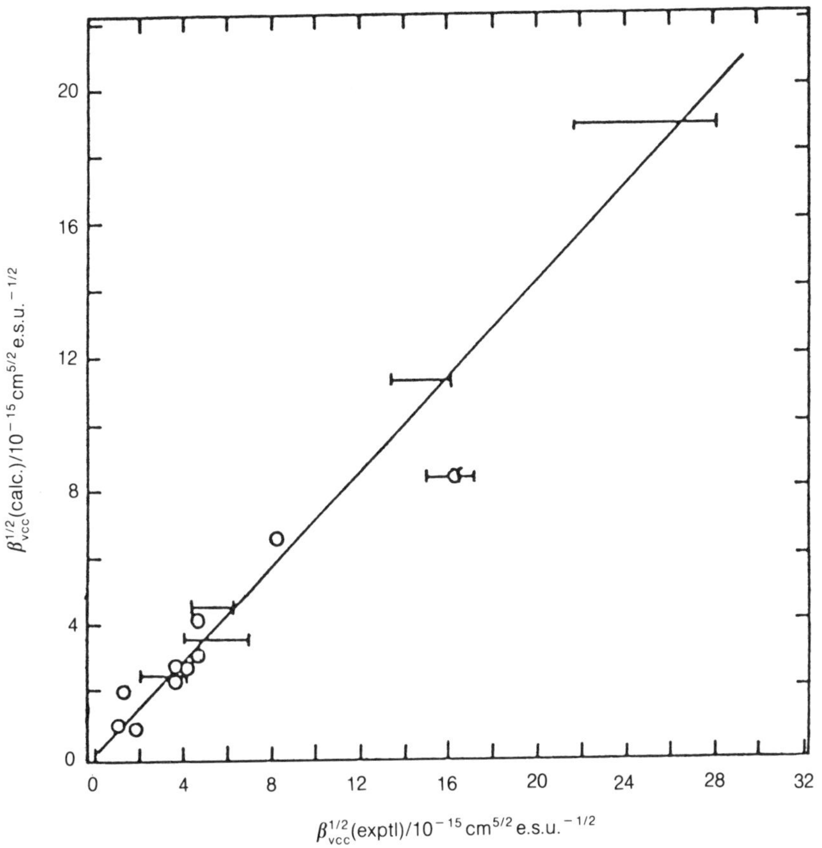

Figure 2.4 Correlation of experimental values for the vector component β_z of the molecular hyperpolarizability with those calculated using a semiempirical technique (Docherty *et al.*, 1985) for a range of molecules.

will be outweighed by the increased absorption of the fundamental or harmonic in the material, which can also give rise to harmful heating of the sample. In general it will be advantageous to work at wavelengths as close to the absorption peak as is possible, within the constraints of optical transparency set by the particular application targeted.

Because of the different energies (wavelengths) of the interlevel transitions for different molecules, the best comparison of the intrinsic molecular nonlinearities can be obtained using a zero-energy coefficient $\beta(0)$, given by equation 2.12 in the limit that $\omega \rightarrow 0$. The program described above is a very useful tool for the comparison of families of molecules, to study for example the effect of chemical changes on the nonlinearity, or the prediction of trends within a series of similar molecules. A great deal of time can be saved by avoiding the chemical systhesis of unoptimized molecules.

An example of the use of the program is given by a comparison of the effects of molecular length on the nonlinearity of two systems, namely the polyphenyls and the polyenes (Figure 2.5) (Morley *et al.*, 1987a). In both cases a single acceptor and donor group are situated at either end of the molecule. As shown in Figure 2.5, the calculations show (as expected) that as the chain length is increased in both cases there is an increase in the β value, and that this increase in β flattens off as the chain becomes more extended. However, the parameter that is really of importance in the determination of the potential macroscopic nonlinearities is not β but the hyperpolarizability density $\rho = \beta/V$ (where V is the molecular volume), as it is this that relates directly to the bulk nonlinearity $\chi^{(2)}$ via equation 2.3. Figure 2.5 also shows ρ as a function of chain length n for the two systems and in this case there is a significant difference. For the polyphenyls, ρ increases only as far as $n = 2$ or 3, and then decreases rapidly, as the increase in β is outweighed by the increase in molecular volume V. For the polyenes there is a monotonic rise in the ρ value up to very long chain lengths, as the increase in β dominates the increase in V. The ρ values predicted for these structures are at least an order of magnitude greater than those of previously reported systems.

Finally, a better feeling for the origin of the molecular nonlinearities may be obtained by the simplification of equation 2.12 using the approximation of a two-level system (Oudar and Chemla, 1977). Although this may at first seem unreasonable for such a complex entity as a molecule containing upward of 30 atoms, in practice it is often the case that one of the molecular excited states has a dominant contribution to the nonlinearity. This state usually corresponds to a major π–π^* transition, involving a redistribution of the delocalized electrons of the conjugated-electron chain. Under this two-level approximation equation 2.12 reduces to

$$\beta \approx Ef\delta\mu/[E^2 - (2\hbar\omega)^2][E^2 - (\hbar\omega)^2] \qquad (2.13)$$

where $\hbar\omega$ is the fundamental laser energy, E and f the energy and oscillator strength of the dominant excited state and $\delta\mu$ is the difference between the excited and ground state dipole moments. The dispersion of the β value (i.e. variation with wavelength) can be easily understood using this model, and has been studied in some detail for simple molecules (Teng and Garito, 1983).

A similar expression to equation 2.12 can be written for the tensor coefficients γ (Ward, 1965). However, in this case there are many more terms to take into account, and the terms themselves are somewhat more complicated. Attempts have been made to use the same semiempirical methods for the calculation of γ, but have in general met with little success. It appears that for accurate predictions of γ it is necessary to include many more of the possible excited state configurations in the calculation than it

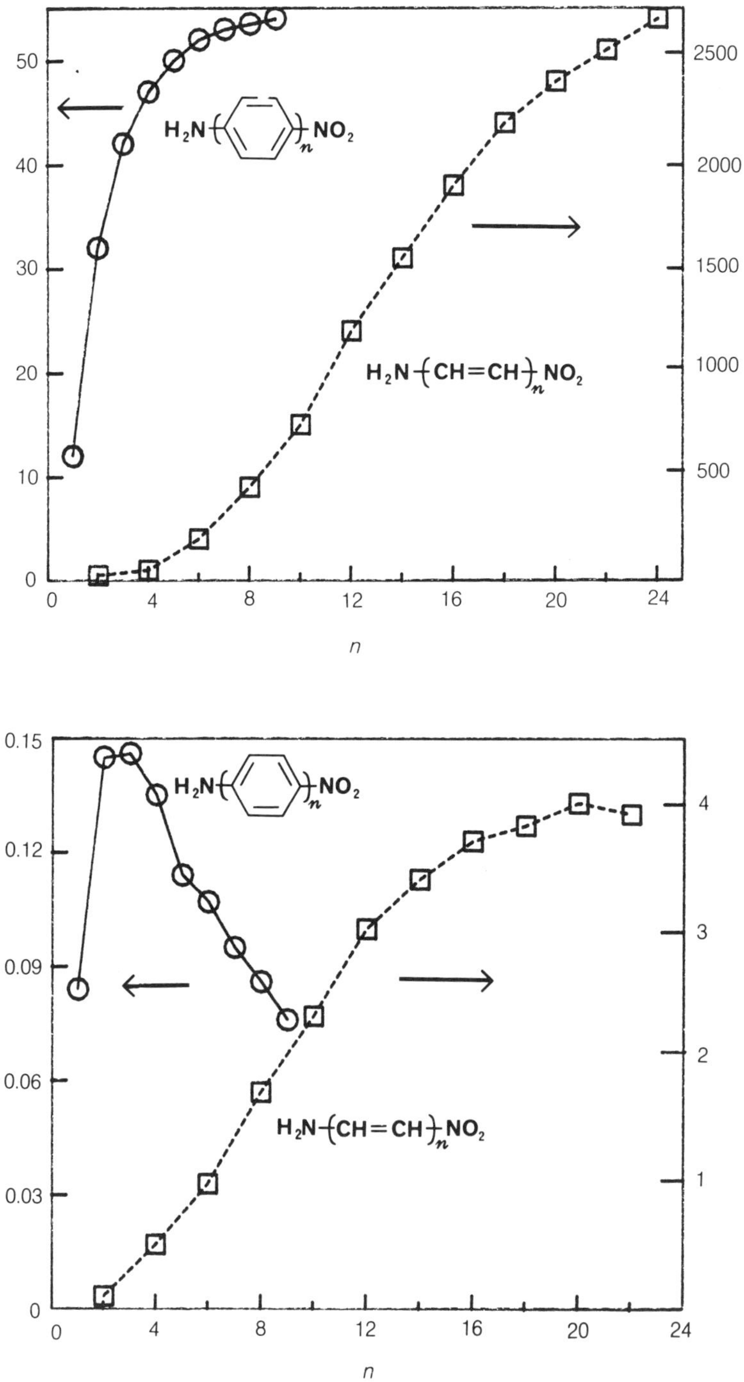

Figure 2.5 (a) Molecular hyperpolarizability β and (b) hyperpolarizability density ρ for donor–acceptor substituted polyphenyls and polyenes as a function of the length of the conjugated pathway (Morley *et al.*, 1987a).

is for β, where the summations can usually be limited to the lowest energy π–π^* transitions (Garito *et al.*, 1988).

The techniques described in this section thus enable us to design, synthesize and characterize a large range of highly nonlinear molecules, and to optimize the molecular structures in preparation for incorporation in some form of solid state aggregation.

2.3. SINGLE CRYSTALS

There have been many studies of nonlinear optical organic single crystals reported over the past few years, and these have been concerned almost exclusively with second-order processes. The only crystalline systems that have been studied in any detail for their third-order properties are the

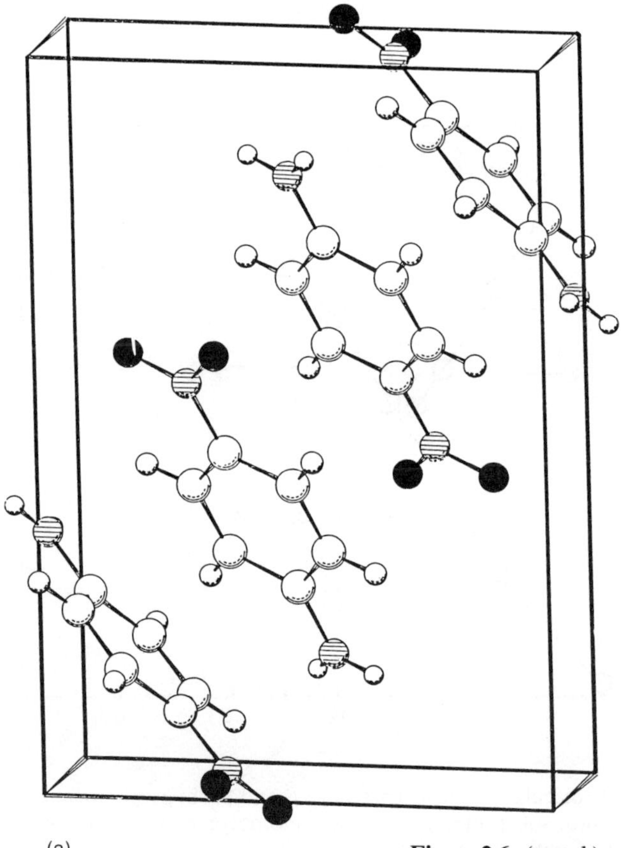

(a) **Figure 2.6.** (*contd.*)

polydiacetylenes, and these will be described in a following section on polymeric materials.

Single crystals have a number of advantages as a method of aggregating the nonlinear molecules. Firstly, a high density of active molecules can be achieved, in contrast to the polymeric or Langmuir–Blodgett systems described below. Secondly, in principle a very high degree of order can be achieved, and thirdly the crystals can be of very high optical quality (in terms for example of the degree of light scattering induced within the material).

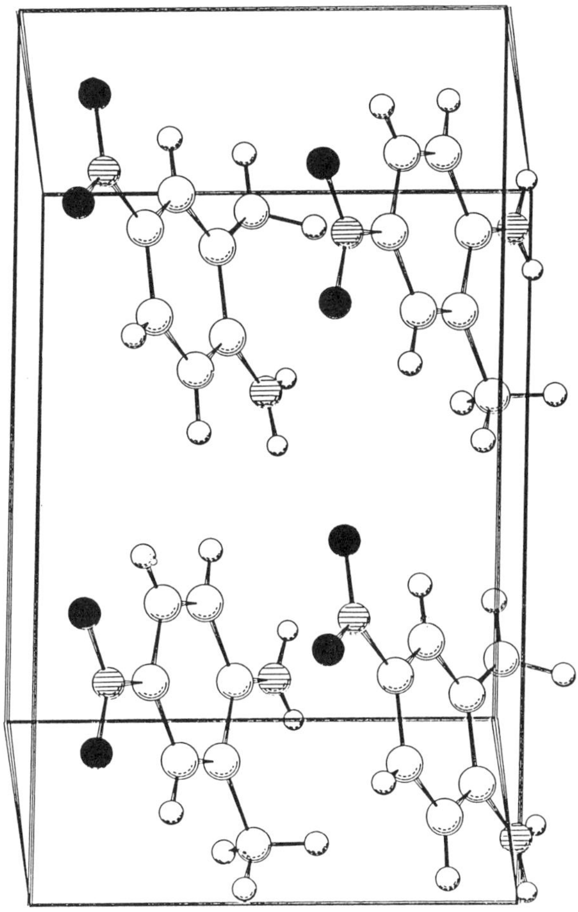

(b)

Figure 2.6 (a) The crystal structure of pNA, showing the centrosymmetric arrangement of molecules which results in all components of $\chi^{(2)}$ being zero. (b) The crystal structure of MNA, in which the molecules are aligned in a polar, parallel manner, leading to high $\chi^{(2)}$ values.

The main disadvantages lie in the difficulty of fabrication of single-crystal devices—the crystals themselves may take a long time to grow to the size required for device construction, and the cutting and polishing processes required to prepare the input and output faces may be technically complicated. However, perhaps the biggest problem lies in the fact that it is not possible to predict in advance what the crystal structure for any particular molecule will be. In fact the majority of active molecules will crystallize in centrosymmetric structures resulting in $\boldsymbol{\chi}^{(2)}$ coefficients of zero. This tendency is presumably enhanced because of the highly polar nature of strongly nonlinear molecules—the electrical interaction energies will be minimized when the molecules are arranged in a head-to-tail manner.

The factors governing the molecular arrangements in crystals are very complex, and small changes in molecular structure can have large effects on the crystal symmetry. For example, the simple molecule nitroaniline (*p*-NA) forms a centrosymmetric crystal structrue as shown in Figure 2.6, but the addition of a single CH_3 group to the ring, to form the molecule MNA, results in a crystal structure in which all the molecules are arranged nearly parallel to each other (Figure 2.6) and which has very high nonlinearities.

Another factor that greatly influences the observed crystal structure is the method of crystal growth used. As will be described in more detail below, the method used is usually that of solution growth. It is often found that different crystal structures may be obtained (polymorphism) when the growth is performed from different solvents or even from the same solvent but under different growth conditions.

The ideal arrangement of the molecules within the crystal structure is dependent on the application envisaged. The simplest situation is for the case of the transverse electro-optic effect. In this case a change in the refractive index associated with light polarized in one direction is produced by the application of an electric field along that same direction. Considering then a 'one-dimensional' molecule, which has the major component of its $\boldsymbol{\beta}$ tensor associated with a single direction, z, the maximum bulk transverse electro-optic effect will be obtained by aligning all these molecules with their z-axes parallel to each other and applying the electric field along this direction. This will result in a large χ_{zzz} coefficient (and a related high electro-optic coefficient r_{zzz}), and in much smaller values for all the other components of the $\boldsymbol{\chi}^{(2)}$ tensor.

The requirements are much more complicated for applications such as second harmonic generation, where phase matching is needed. Phase matching is achieved by having equal refractive indices at the fundamental and harmonic wavelengths, and since all materials are to a greater or lesser extent dispersive this generally means that the fundamental and harmonic beams cannot be polarized parallel to each other. This in turn means that it is off-diagonal terms in the $\boldsymbol{\chi}^{(2)}$ tensor (such as χ_{xyy}) that must in this case be maximized, rather than on-diagonal terms as in the transverse electro-optic

Table 2.1 Optimum alignment of vector-like second-order nonlinear optical molecules (Zyss and Oudar, 1982)

Crystal point group	*NLO coefficient*	θ	Φ
2	d_{21}	54.74	—
m	d_{12}	54.74	0
*mm*2	d_{31}	54.74	0
62*m*	d_{22}	90	60
6,3	d_{22}	90	—
3*m*	d_{22}	90	30
32	d_{11}	90	60
222	d_{31}	54.74	0

θ is the angle that the axis of the one-dimensional molecules (assumed to lie along the dipole moment μ of the molecule) makes to the polar axis (or in the case of point group *m* to the normal to the mirror plane). Φ is the azimuthal angle that the projection of the axis of the one-dimensional molecule onto the plane perpendicular to the polar axis makes with the crystal *Z*-axis. A — indicates that no direction is preferred.

effect described above. The implications of this in terms of molecular arrangement have been studied in some detail by Zyss and Oudar (1982). The requirements are dependent on the overall symmetry (space group) of the crystal structure. Their findings, again for the example of a one-dimensional molecule, are summarized in Table 2.1, where the optimal alignment of the molecules, with respect to a given crystal axis, is given for a number of common crystal space groups. In general these optimum structures can be described in terms of a herring-bone-like arrangement of the molecules, with all the dipoles having a net alignment in one direction (the polar axis). An example of such a structure (that of the crystal NPP) is shown in Figure 2.7.

Although as stated above it is not possible to predict in advance the crystal structure of a given molecule, there are a number of approaches that have been used to try to increase the probability of achieving an active crystalline structure. One technique has been to design molecules having particularly low dipole moments to try to reduce the unfavourable electrical interaction energies (Zyss *et al.*, 1981). In general, however, such molecules tend also to have low nonlinearities. The incorporation of groups into the molecules which favour the formation of hydrogen bonds has also been utilized. The hydrogen bonds may then tend to counteract the effect of the electrical interactions. The molecule NPP shown in Figure 2.7 is an example of this strategy (Zyss *et al.*, 1984). This molecule also has a chiral group (the prolinol heterocyclic terminal ring). The inclusion of such a group (where a single carbon atom has four different groups attached to it) means that the crystal structure must be noncentrosymmetric, but does not guarantee that there will be favourable alignment of the molecules.

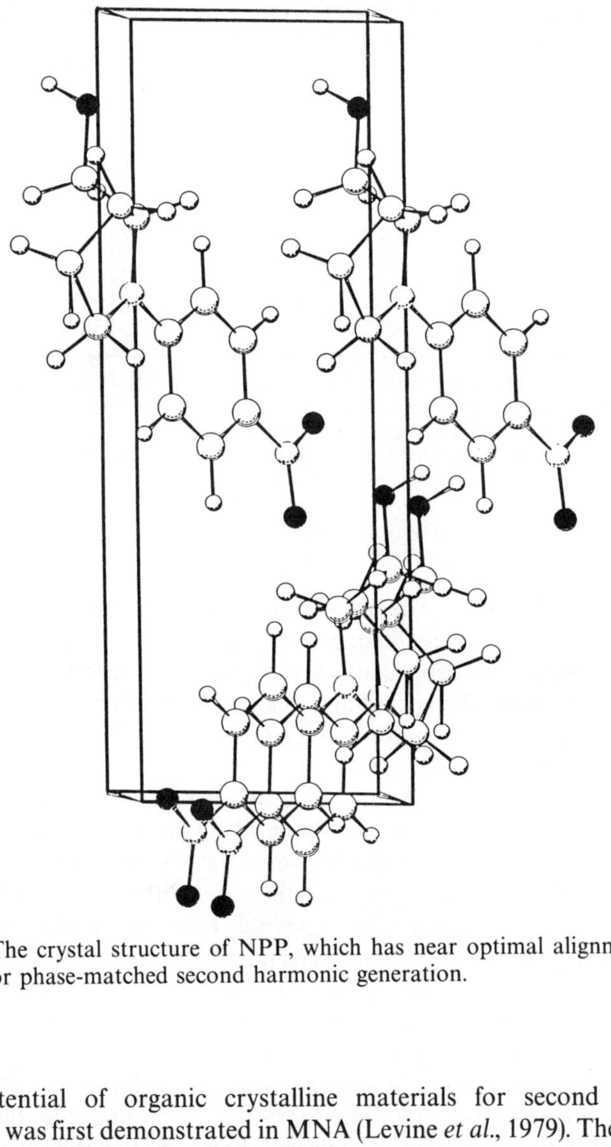

Figure 2.7 The crystal structure of NPP, which has near optimal alignment of the molecules for phase-matched second harmonic generation.

The potential of organic crystalline materials for second harmonic generation was first demonstrated in MNA (Levine *et al.*, 1979). The molecule MNA has a relatively modest β value but, as described earlier (Figure 2.6), there is a strong alignment of the molecules within the crystal structure. The molecules lie at an angle of about 10° to the polar axis, which is a long way from the ideal angle of about 54° predicted by Zyss and Oudar. Thus although the on-diagonal nonlinearity $\chi^{(2)}_{111}$ is 500 pm V^{-1}, (compared with 34 pm V^{-1} for lithium niobate), the highest phase-matchable coefficient is $\chi^{(2)}_{122}$, at 76 pm V^{-1}. This still results in a figure of merit for phase-matchable SHG 45 times that of lithium niobate.

More recently a number of other organic materials have been extensively characterized. Other nitroaniline derivatives (having similar molecular properties to MNA) have been described. Prominent amongst these are DAN (Kerkoc *et al.*, 1989), NPP and PNP (Sutter *et al.*, 1989). In all these cases the molecular arrangements are much closer to the optimum—for PNP and NPP the molecular axes are within 5° of the predicted optimum, and so these structures probably represent the best that can be achieved with this family of molecular structures. The highest phase-matchable coefficient for PNP is $\chi^{(2)}_{211}$, at about 100 pm V^{-1}, and the corresponding figure of merit for the direction in which phase matching occurs is 525 (pm $V^{-1})^2$.

Higher nonlinearities can be obtained by utilizing molecules having higher intrinsic nonlinearities. However, as already described, in the case of second harmonic generation the range of useful materials is limited by the increase in absorption in the visible portion of the spectrum that occurs as the nonlinearities are increased. Two larger molecules; that have been characterized are NMBA (Bailey *et al.*, 1989) and MMONS (Bierlein *et al.*, 1990). Both form yellow crystals, which are just sufficiently transparent at 532 nm to allow SHG from Nd:YAG sources at 1.064 μm. The highest coefficient in MMONS is $\chi^{(2)}_{333}$ (368 pm V^{-1}) with phase matching being possible using the coefficient $\chi^{(2)}_{223}$ (142 pm V^{-1}). The phase-matched figure of merit is in this case 850 (pm $V^{-1})^2$.

Fewer materials have been developed for the commercially important target of frequency-doubling laser diode sources from about 820 nm down to 410 nm. The problem is to produce materials having high nonlinearities but remaining transparent at both fundamental and harmonic wavelengths. The well-studied compound urea satisfies the transparency criteria, but has a rather low nonlinearity ($\chi^{(2)}_{122} = 2.8$ pm V^{-1}). It does, however, also have a high laser damage threshold. The reason for the low nonlinearity is that the urea molecule is small, and does not contain any extended conjugation pathway. Another organic molecule with no significant conjugation is L-arganine phosphate monohydrate (LAP). This too has a wide transparency range and high optical damage threshold (Yokotani *et al.*, 1989), but once again its nonlinearity is too low to make it a serious contender for laser diode applications. Preliminary results on a number of more conjugated molecules have recently appeared in the literature. Typical of these is the chalcone derivative EMC (Kitaoka *et al.*, 1990), which is transparent down to 430 nm, and has an effective nonlinear optical coefficient $\chi^{(2)}$ of about 11 pm V^{-1}. There remains much work to be done to produce fully characterized crystals suitable for use in this demanding application.

The first organic crystal to be studied in terms of its electro-optic properties was *meta*-nitroaniline (*m*-NA; Stevenson, 1973). The molecular β coefficient of this molecule will be smaller than that of MNA because of the less favourable relative positioning of the donor (NH_2) and acceptor (NO_2) groups. The electro-optic r_{33} coefficient was found to be about 16.7 pm V^{-1} at 633 nm, this being comparable with the coefficients in the inorganic KD*P. Two other quantities are often used to compare the efficiency of electro-optic materials, these being the **electro-optic figure of merit** $n^3r/2$, and the **half-wave voltage** V_π or the **reduced half-wave voltage** V_π^*. The half-wave voltage is the voltage that must be applied to the crystal, when positioned between polarizers, in order to switch from maximum to minimum transmission. It is the voltage required to induce a phase shift of π between two orthogonally polarized waves within the crystal. It is in general dependent on the crystal dimensions, and so a better comparison between different materials is obtained by using the reduced half-wave voltage, which is just the half-wave voltage for a unit cube of the compound.

The compound MNA, described above for second harmonic generation, also has a very large electro-optic coefficient, because of the near-parallel arrangement of the molecules. The coefficient r_{11} is about $67\,\mathrm{pm\,V^{-1}}$ (Lipscombe *et al.*, 1981). The reduced half-wave voltage measured for MNA was 1.3 kV, compared with a value of about 3 kV for lithium niobate. The electro-optic coefficients r and the second-order susceptibility $\chi^{(2)}(\omega;\ \omega, 0)$ associated with the electro-optic effect are related by the equation

$$\chi^{(2)}(\omega; \omega, 0) = r/2n^4 \tag{2.14}$$

The terms in parentheses after $\chi^{(2)}$ denote the frequencies of the electric fields involved in the process under consideration—the first being the output frequency, and the second and third being the two input frequencies. Thus the electro-optic effect can be considered as the interaction of an optical wave of frequency ω with a d.c. field (frequency 0) resulting in output also at frequency ω. In contrast, second harmonic generation is the interaction of two waves of frequency ω, resulting in an output of frequency 2ω, so the associated susceptibility is written $\chi^{(2)}(2\omega; \omega, \omega)$. In MNA, as in *m*-NA and most other organic crystals studied to date, it was found that the second-order susceptibility $\chi^{(2)}(\omega; \omega, 0)$ determined from the electro-optic coefficient and that determined by SHG are very similar in magnitude. This supports the view outlined in previous sections that the origin of the nonlinearities in organic materials is predominantly electronic. In inorganic materials such as lithium niobate or KDP, which are ferroelectric crystals, low frequency electric fields are able to produce motion in the relatively loosely bound ferroelectric atoms, as well as the electrons. This atomic motion will also contribute to the electro-optic effect. At high (i.e. optical) frequencies the relatively heavy atoms cannot follow the oscillations in the electric field, and so only the electronic component of the nonlinearity will be evident. Thus in these materials the susceptibilities associated with the electro-optic effect $\chi^{(2)}(\omega; \omega, 0)$ and second harmonic generation $\chi^{(2)}(2\omega; \omega, \omega)$ are found to be significantly different.

Two organic crystals with significantly higher electro-optic coefficients have been described more recently. Both are formed from molecules having a much longer conjugation length than the nitroaniline derivatives detailed

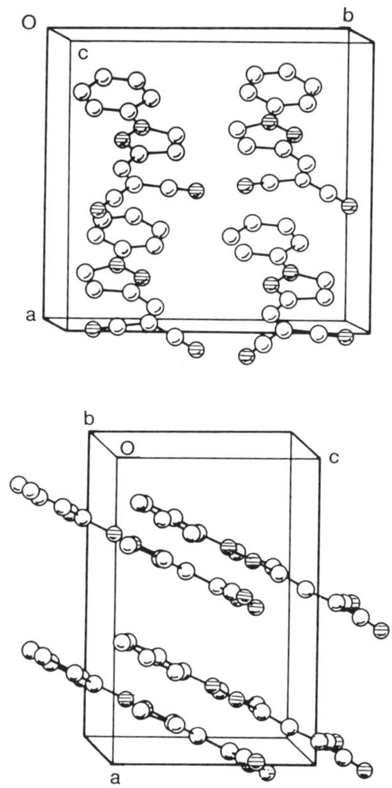

Figure 2.8 The crystal structure of DCNP, in which the molecules are ideally arranged for the transverse electro-optic effect.

above. As a result of this the crystals are red in colour, and therefore have limited applications in the visible portion of the spectrum. However, many applications of the electro-optic effect will be associated with optical communications, where the wavelengths of operation are in the near infrared. These materials are well suited to operation at these wavelengths. The compound SPCD (Yoshimura, 1987) is an ionic molecular crystal. A reduced half-wave voltage of about 500 V was measured for this crystal, at 633 nm.

The crystal structure of the second compound DCNP (Allen *et al.*, 1988a; Allen, 1989a) is interesting in that the molecules have in this case near-perfect alignment (Figure 2.8). Thus the electro-optic coefficient r_{33} associated with this alignment direction is maximized, and all other components will be small. This is probably the ideal crystal structure for the transverse electro-optic effect. The electro-optic properties of this material have been determined as a function of wavelength, and the reduced half-wave voltage and electro-optic coefficient r_{33} are shown in Figure 2.9. The dispersion of

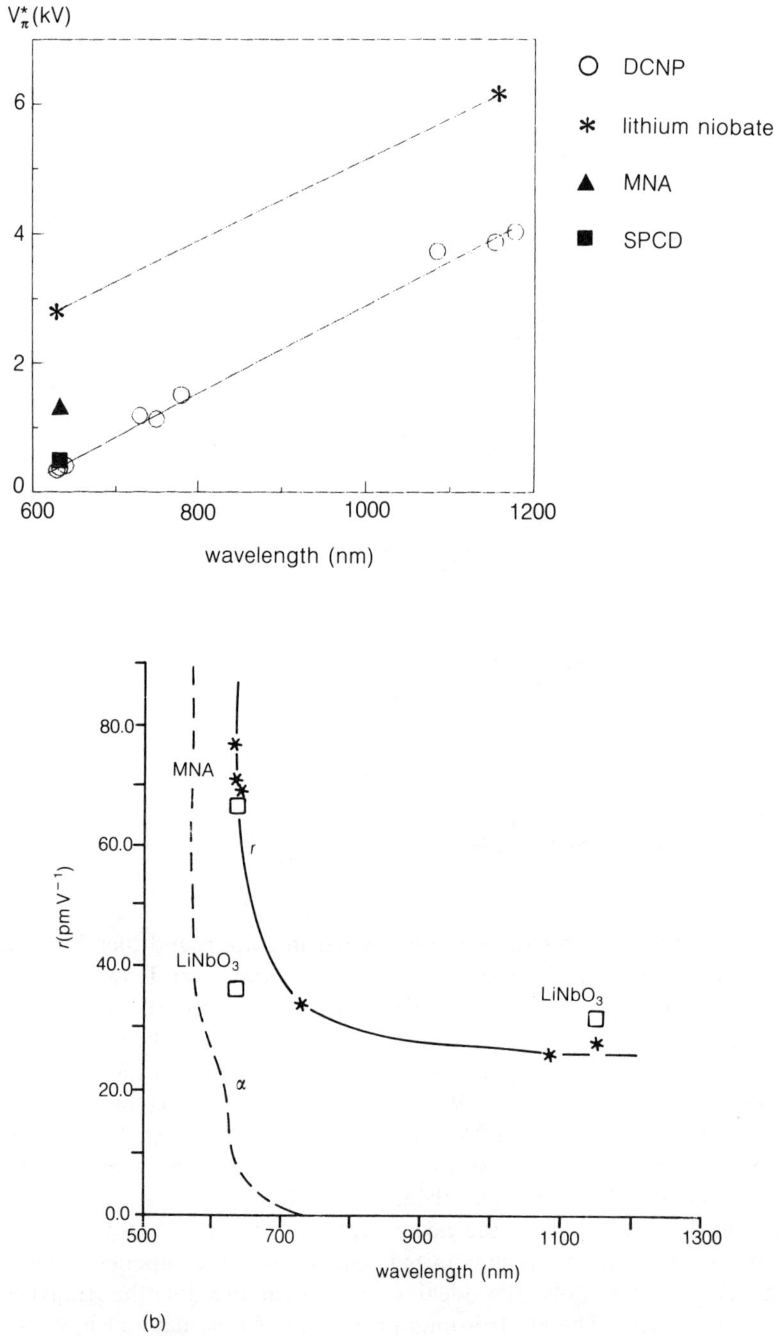

Figure 2.9 (a) Reduced half-wave voltage and (b) electro-optic coefficient r_{33} of DCNP as a function of wavelength.

the nonlinearity with wavelength is well illustrated in this figure. The absorption edge of the crystalline material is also shown. At 633 nm, which is very close to the absorption edge itself, the electro-optic coefficient is very high, and the reduced half-wave voltage (370 V) is about one-eighth that of lithium niobate. However, at longer wavelengths, remote from the absorption edge, there is no resonant enhancement and the nonlinearity is significantly reduced. In lithium niobate, however, there is no absorption in the visible part of the spectrum and so dispersion of the nonlinearity is much less pronounced. Nevertheless, even at 1.3 μm the half-wave voltage of DCNP is half that of lithium niobate.

Another attractive application in nonlinear optics is the generation of continuously tunable laser radiation by the process of optical parametric generation. This process is essentially the inverse of a frequency-mixing process. Each photon of the pump beam is split into two photons of lower energy, known as the signal and idler beams. The process must be phase matched, and the relative energies of the idler and signal photons are determined by these phase-matching conditions, and are therefore dependent on the precise crystal geometry used. Wavelength tuning can be achieved, for example, by rotating the crystal so as to change the phase-matching conditions in a controlled manner. Crystals of the compound urea (Rosker and Tang, 1985; Rosker *et al.*, 1985) have been used for generation of light over a wide wavelength range (498 nm–1.24 μm when pumped with the 355 nm third harmonic of an Nd:YAG laser). The wide transparency range and high optical damage threshold of urea make it well suited to these optical parametric applications.

Growth, and subsequent cutting and polishing, of cystals of the above compounds of sufficient size and high enough optical quality to permit accurate measurements to be made, and ultimately for the production of saleable devices, is a whole science in itself. Organic crystals are usually grown either from solution (by controlled evaporation of solvent or controlled lowering of temperature), from the melt (typically by variations on Bridgman growth) or by sublimation (Karl, 1989; Badan *et al.*, 1986; McArdle and Sherwood, 1987). Obviously melt growth is limited to those compounds which are stable at their melt temperature. Higher optical quality can usually be achieved by solution growth, as the thermal stresses involved in cooling crystals down from the melt temperature, in the Bridgman technique, can result in large stress-induced birefringences and the introduction of dislocations and other defects. A prerequisite for the formation of crystals of high optical quality is the use of high purity starting materials. A variety of methods have been used for this initial purification stage, including multiple recrystallization, sublimation, chromatography, evaporation and (for thermally stable materials) zone refining.

Solution growth methods have been used to produce large defect-free crystals (Bailey *et al.*, 1988). Usually methods involve the controlled lowering of the temperature of a supersaturated solution of the material of interest, in which a seed crystal is suspended. The temperature must be accurately controlled by some form of feedback mechanism. The reduction in temperature is programmed so as to maintain a constant level of supersaturation. Equipment of varying degrees of sophistication may be designed, incorporating for example complex stirring programmes, and computer-controlled temperature lowering. Alternatively, the level of supersaturation may be controlled by evaporation of the solvent. These methods are generally less easy to control precisely than the temperature lowering techniques.

Melt growth methods can also produce sizeable crystals—boules of 1–2 cm diameter are common. As mentioned above, the quality of the crystals obtained is generally poorer than those from solution growth. Often stresses are induced in the crystals during the cooling process, and these can even result in fracture of the crystals. Small areas of high quality crystal obtained by melt growth can be selected and used as seeds in subsequent solution growth experiments.

Crystals obtained by sublimation growth are usually small (less than 1 cm in dimension), although they can be of high optical quality. These methods have been used much less frequently than the two above.

In summary, much progress has been made in the development of crystalline organic nonlinear optical materials, but the complexities of crystal growth and device fabrication have meant that to date no devices have been commercialized in these materials. Nevertheless, many of the compounds described above have great potential for further development.

2.4. POLYMERS

2.4.1. Poled polymers for second-order nonlinearities

A potentially simple scheme for the fabrication of a processable nonlinear optical film in which the active molecules are at least partially aligned so as to give some second-order nonlinear optical effects consists of the formation of a solution of a highly nonlinear molecule in a polymer known to have good linear optical properties. A film of this material is then produced by conventional techniques such as spin coating, casting, or printing. The incorporated molecules will at this stage be randomly oriented, and will therefore show no second-order nonlinear optical effect. A noncentrosymmetric alignment of the molecules can in principle be obtained by then heating the polymer close to its glass transition temperature T_g, so that motion of the polymer chains and the dopant molecules is possible, followed by the application of a strong d.c. electric field across the sample. This field will orient the highly polar nonlinear optical molecules via the interaction with their permanent dipole moments. The temperature is then lowered back to well below T_g, with the field still applied, so that relaxation of the aligned molecules is restricted by the glassy nature of the host polymer. Alternatively, the film can be formed from a monomeric or partially polymerized host, again with the dopant nonlinear molecule incorporated, an electric field applied to orient the dopant molecules, and then full polymerization implemented (thermally or photochemically) while the field is still applied (Hubbard *et al.*, 1989).

Although simple in principle, there are a number of problems associated with this approach. Firstly, although the molecules themselves have high $\boldsymbol{\beta}$ coefficients, the nonlinearity of the film can never be high unless the dopant molecules constitute a significant fraction of the total material. In practice the highly polar molecules have low solubilities in most polymer hosts and it can prove difficult to incorporate more than 10–15% of dopant into the host. Use of higher concentrations can lead to problems of recrystallization within the film on standing.

The second problem relates to the degree of order that can be obtained in the poling process. The degree of orientational order present in any system can be described using a distribution function (Singer *et al.*, 1987), $G(\Omega)$. The electrically poled polymeric systems considered here will in general be symmetric about the direction of the applied electric field, and so the distribution function will depend only on the polar angle θ that any given molecule makes with the field. The distribution function can then be expanded as a power series in terms of Legendre polynomials $P_n(\cos\theta)$, and associated coefficients A_l:

$$G(\theta) = \Sigma A_l P_l(\cos\theta) \tag{2.15}$$

In guest–host systems the distribution function, under the application of an electric field of strength E, is

$$G(\theta, \mathbf{E}) = \exp[-(U - \mathbf{m}^*\cdot\mathbf{E})/kT]/\int \exp[-(U - \mathbf{m}^*\cdot\mathbf{E})/kT]\,\mathrm{d}(\cos\theta) \tag{2.16}$$

where U is the zero-field interaction energy and $\mathbf{m}^*$ the local-field-corrected dipole moment of the molecules involved. It can be shown (Singer *et al.*, 1987) that the second-order nonlinear optical susceptibility can then be written in terms of the order parameters $\langle P_l \rangle$ of the system before poling:

$$\chi^{(2)}_{ijk} = (NE/kT)\Sigma_{m=0}\sigma^{2m}ijk\langle P_{2m}\rangle \tag{2.17}$$

where the tensor coefficients σ include a contribution from both dipole moments and molecular susceptibilities. The equations may be simplified by making the assumption that the molecules are 'one dimensional', having only a single significant component of the molecular susceptibility β_{zzz}, and only one component m^*_z of the dipole moment. For many nonlinear molecules this is a reasonable approximation, with the z-axis being associated with the vector between the donor and acceptor substituents. In this case, and when the electric field is applied along the z-axis in the laboratory frame of reference, the explicit expressions for $\chi^{(2)}_{333}$ and $\chi^{(2)}_{113}$ are

$$\chi^{(2)}_{333} \sim N\beta^*_{zzz}(m^*_z E/kT)(1/5 + 4\langle P_2\rangle/7 + 8\langle P_4\rangle/35) \tag{2.18}$$

$$\chi^{(2)}_{113} \sim N\beta^*_{zzz}(m^*_z E/kT)(1/15 + \langle P_2\rangle/21 - 8\langle P_4\rangle/70) \tag{2.19}$$

So for an initially isotropic polymer, for which $\langle P_2\rangle$ and $\langle P_4\rangle$ are both zero, $\chi^{(2)}_{333}$ is three times larger than $\chi^{(2)}_{113}$. In, for example, liquid crystalline phases, the order parameters can become quite large. In the limit of $\langle P_2\rangle$ and $\langle P_4\rangle$ being equal to unity, χ_{333} is increased by a factor of five compared with that of an isotropic polymer, and the off-diagonal coefficients become zero.

We can use similar techniques to calculate the average degree of alignment achieved by the electric field. Thus

$$\langle \cos\theta\rangle = \int_0^{\pi/2} \cos\theta \exp(-\mathbf{m}^*\cdot\mathbf{E}/kT)\sin\theta\, d\theta \Bigg/ \int_0^{\pi/2} \exp(-\mathbf{m}^*\cdot\mathbf{E}/kT)\sin\theta\, d\theta \tag{2.20}$$

For the case when $\mathbf{m}^*\cdot\mathbf{E} \ll kT$ this reduces to (Meredith *et al.*, 1982)

$$\langle \cos\theta\rangle \approx m^*_z E_z/5kT \tag{2.21}$$

Assuming a molecular dipole moment of 8 debye (typical for a 'large' nonlinear optical molecule), and an applied field of $10^8\ \mathrm{V\,m^{-1}}$ (corresponding to the experimental situation of the application of 400 V across a 4 μm sample), this order parameter factor is about 0.23 at room temperature, corresponding to the molecules making an average angle of 77° to the direction of the applied field. The maximum nonlinearity of the film is in this case limited by the small degree of order that can be induced. Given that applied fields of this magnitude are the highest obtainable without producing electrical breakdown in the material, the magnitude of the Boltzmann factor can be increased either by increasing μ or by decreasing the temperature T. The latter course of action is not feasible when poling must occur around T_g, as any reduction in T_g would increase the relaxation of the nonlinear optical effect at room temperature. The main problem with increasing the dipole moment μ is that in general there is an associated decrease in solubility of the molecule in the host polymer, increasing the first problem mentioned above.

Finally, it is clearly desirable for there to be little relaxation of the nonlinear optical coefficient with time. This means that the host polymer must be chosen to have as high a T_g as possible, subject to processing constraints and also to the question of temperature at which poling will occur as addressed above.

Early reports in this area described the doping of a liquid crystalline polymer with nonlinear optical stilbene molecules (DANS) (Meredith *et al.*, 1982), in an attempt to utilize the enhancement of $\chi^{(2)}$ described above. A nonlinearity $\chi^{(2)} \approx 1.3\,\mathrm{pm\,V^{-1}}$ for a polymer containing 2% DANS and poled at high temperature with a field of $1.3\,\mathrm{V\,\mu m^{-1}}$ was reported. A strong temperature dependence of induced second harmonic generation was observed in this case, resulting in a decrease in efficiency as the temperature was increased. This was attributed to the formation of complexes between neighbouring DANS molecules, or between the DANS molecules and the liquid crystalline side groups resulting in local cancellation of the molecular nonlinearities.

Nonlinear optical molecules have also been incorporated into a polymethylmethacrylate (PMMA) polymer. The advantage of conventional polymers such as PMMA over liquid crystalline polymers lies in their optical clarity. This is of great importance in the fabrication of waveguides. Liquid crystalline materials in general have a characteristic texture due to boundaries between regions of local order, resulting in local variations of refractive index which scatter light. Although this effect is often less severe in liquid crystalline polymers than in monomeric materials, it nevertheless sets a limit to the quality of waveguide that can be achieved.

Electro-optic and second harmonic generation experiments have been carried out on samples consisting of 10% MNA in PMMA (Khanarian *et al.*, 1988), giving values of $\chi^{(2)}(\omega;\omega,0)$ and $\chi^{(2)}(2\omega;\omega,\omega)$ of $0.6\,\mathrm{pm\,V^{-1}}$ and $0.7\,\mathrm{pm\,V^{-1}}$ respectively (compared with $240\,\mathrm{pm\,V^{-1}}$ for single-crystal MNA and $34\,\mathrm{pm\,V^{-1}}$ for single-crystal $LiNbO_3$).

Complementary work on solutions of an azo dye (Disperse Red, DR1) concentrated on second harmonic measurements (Small *et al.*, 1987; Singer *et al.*, 1986). The molecular β value for this compound should be significantly higher than that of MNA. Using concentrations of about 15% dopant, a $\chi^{(2)}(2\omega;\omega,\omega)$ of $5\,\mathrm{pm\,V^{-1}}$ was obtained which compares favourably with the value of $9.2\,\mathrm{pm\,V^{-1}}$ for single crystals of the inorganic compound KDP, commonly used for second harmonic generation.

A big improvement over these 'doped polymer' systems can be achieved by chemically grafting the nonlinear optical units onto the polymer itself, in the form of 'side chains' as shown in Figure 2.10. Firstly, a much higher concentration of NLO units can be achieved, with in principle 100% of possible substitution sites being occupied by the active molecules. Because of the chemical bond between molecule and polymer backbone, the tendency of the molecules to aggregate or recrystallize at high concentrations is much reduced. The alignment process for these types of polymers is exactly as described above. Poling with the electric field has usually been found to be most effective at temperatures just below the glass transition temperature of the polymer. Because of the link between active molecule and polymer backbone the long-term stability of these side chain polymers is much better than that of the doped polymer systems. It is much harder for the active

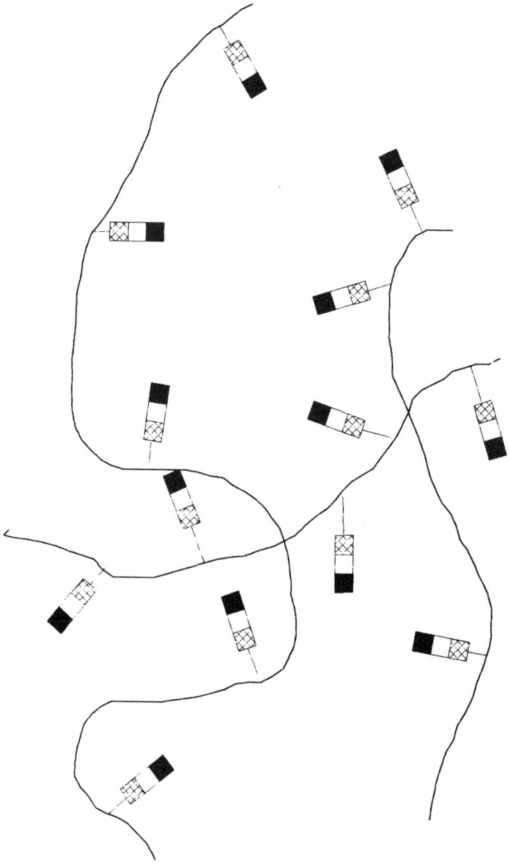

Figure 2.10 A schematic representation of a side chain polymer.

molecules to rotate back to their unaligned positions. Recently there has been rapid progress in the development of new side chain polymer systems, and in the technology of the poling process. Some of the most active polymers are shown in Figure 2.11.

The poling electric field can be applied either conventionally, using electrodes deposited on either side of the active film, or using a corona discharge (Eich *et al.*, 1989b; Hampsch *et al.*, 1990). In the corona method a high voltage (typically 10 kV) is applied to a needle, or array of needles, positioned a short distance (2–3 cm) above the surface of the film. An electrode placed below the film is connected to earth. The corona discharge from the needle point results in the build up of a large static charge on the surface of the film, which induces a strong electric field across the film and produces the desired orientation of the polar molecules within the film. There

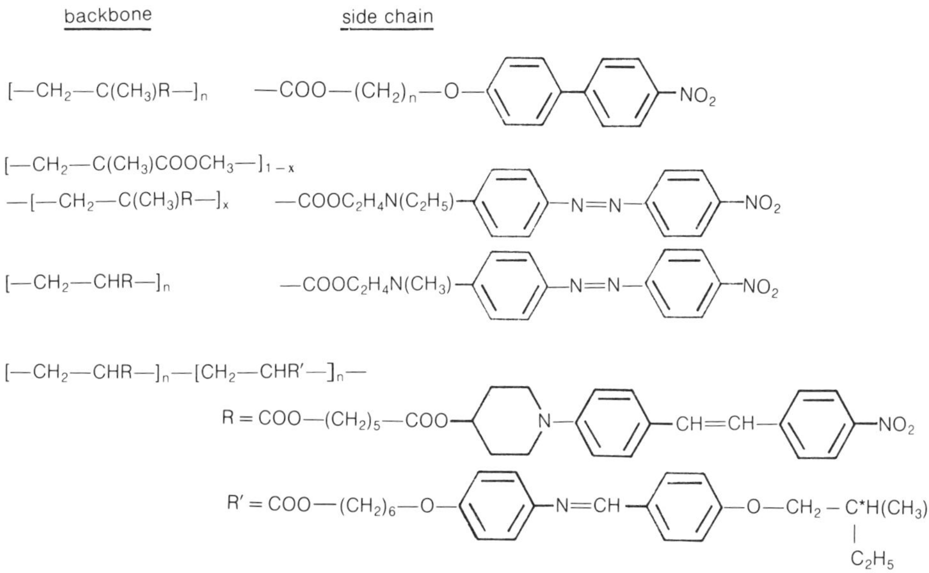

Figure 2.11 Structures of some of the nonlinear optical side chain polymers reported in the literature.

is some evidence that higher fields can be applied by this technique than by the use of electrodes. One reason for this is that a local electrical breakdown in the film will generally be less catastrophic in the corona technique.

Even better long-term stability of the induced orientation should be achieved in systems which allow the active molecules to be locked into place at more than one position. Eich *et al.* (1989a), Jungbauer *et al.* (1990) and Allen *et al.* (1990) have described cross-linkable nonlinear optical molecules, as shown in Figure 2.12. In this case the method of producing an active film is slightly different. The starting material is film (formed by spin coating) of a low molecular weight (i.e. monomeric) material. The molecules have attached to them substituents that contain groups that may be cross-linked thermally. Thus if an electric field is applied to the film, to orient the molecules, and the film is then heated up, the molecules should cross-link and form a rigid network, which will continue to hold the molecules in place once the electric field has been removed. The rigidity of the network, and hence the ultimate stability of the induced nonlinearities, will depend on the extent to which the cross-linking takes place. The work quoted above suggests that indeed it is the case that the cross-linked systems are more thermally stable than their side chain counterparts.

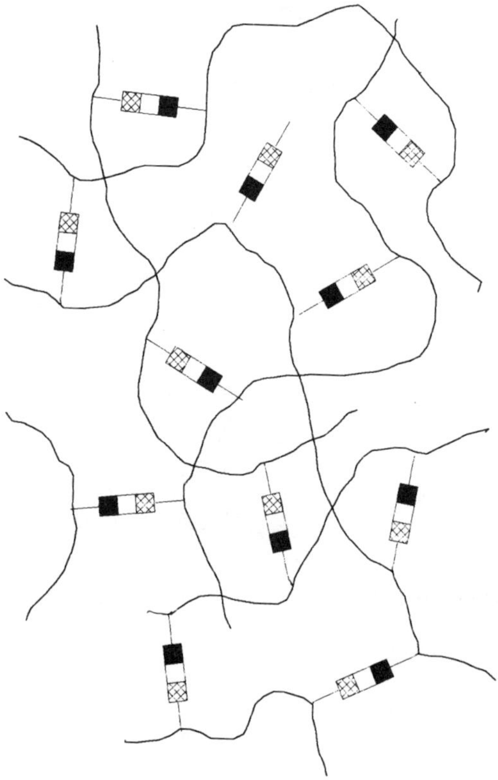

Figure 2.12 A schematic representation of a cross-linked nonlinear optical polymer.

Most work on these electrically poled films has been directed at the production of electro-optic waveguide devices. Examples of such devices, shown in Figure 2.13, include a Mach Zehnder interferometer and evanescent coupler. These devices have applications in the modulation and switching (routing) of optical signals in optical communications. In these applications the new polymeric materials are in direct competition with devices fabricated from inorganic crystals such as lithium niobate, or potassium titanyl phosphate (KTP), which are described elsewhere in this book. The principle of operation of all such devices depends on the modification of the optical path length of some part of the device, using a change in refractive index induced by an applied electric field via the electro-optic effect. As well as the potential for very high nonlinearities, already described in some detail in earlier sections, the polymeric compounds offer a number of other advantages for these applications.

1. The processing steps involved in the fabrication of channel waveguide

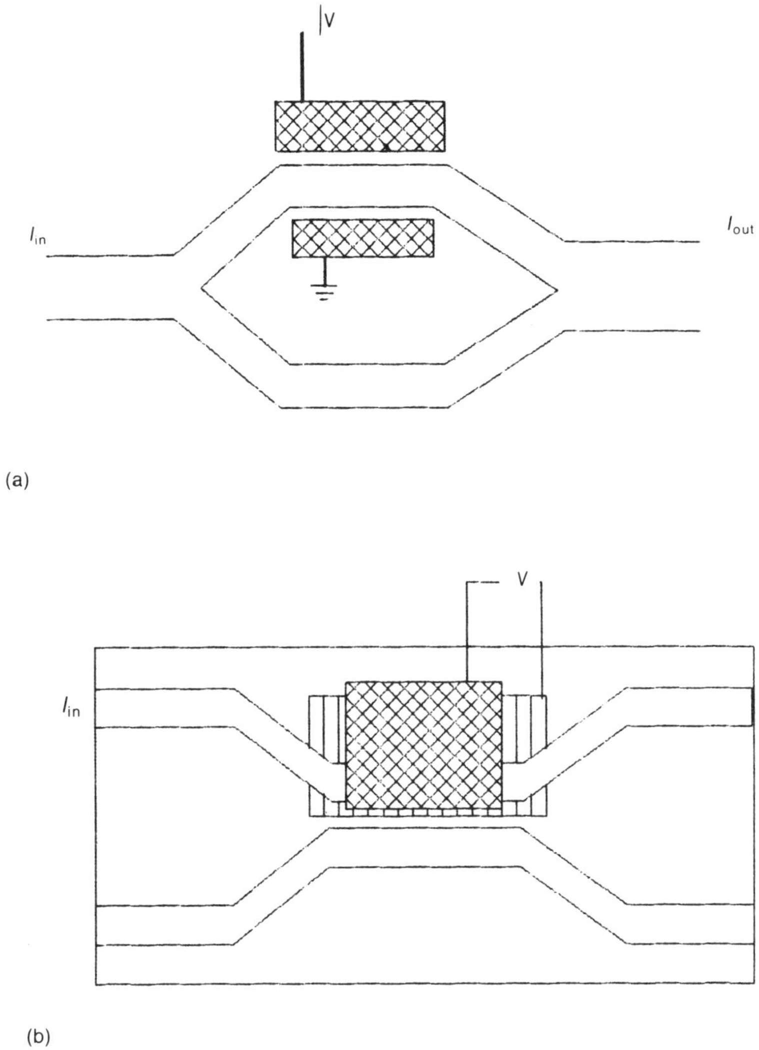

Figure 2.13 Typical integrated optics waveguide devices. Electrode regions are shown hatched. (a) Mach–Zehnder interferometer; (b) directional coupler.

devices out of the polymers are technologically much simpler than those for crystalline materials such as lithium niobate. Crystals of lithium niobate are grown from the melt at high temperatures (>1000 °C), and the waveguides are then defined by either indiffusion of titanium ions or ion exchange. In either case the depth of the waveguide is limited to about 5 μm. Films of the polymeric materials can be deposited by conventional techniques such as spin or dip coating. The poling process takes place at relatively low temperature, and channel waveguides can also be fabricated by a number of low temperature, low technology techniques as described below.

2. The polymeric materials have significantly lower dielectric permittivities than the (generally ferroelectric) inorganics. Lithium niobate, for example, has a relative permittivity of about 28, whereas that of a polymer is typically about 3–4. This results in shorter time constants for capacitive electrode configurations, resulting in faster switching times. It is also advantageous for modulator devices operating at high frequencies (in the gigahertz range). Here the electrical signal must usually be applied as a travelling wave, using a microwave strip-line configuration for the electrodes. If the dielectric permittivities of the waveguide at microwave and optical frequencies are very different (as is the case for lithium niobate) then the electrical and optical signals become out of phase with each other, again limiting the frequency range of operation.
3. Polymeric materials have greater potential for integration with other components. They could, for example, be fabricated directly on top of electronic circuitry.

Despite the relatively high level of dilution of the molecular nonlinearities by the polymer chains, a number of materials having electro-optic coefficients comparable with or greater than those of lithium niobate at the technologically important wavelength of 1.3 μm have been reported (Mohlmann *et al.*, 1989; Hass *et al.*, 1989). At shorter wavelengths the electro-optic coefficients are even higher, owing to the onset of resonant enhancement, but the optical losses of waveguides also increase, owing to increased molecular absorption.

A number of methods can be used to define channel waveguides in the polymeric films. The requirement for channel waveguiding is that a channel, or stripe, within the waveguide layer must have a higher effective refractive index than the surrounding material. The channel can be produced physically, by for example using lithographic techniques to remove all the waveguiding layer except the required stripe, producing a ‘rib waveguide’, or by etching a channel into the substrate before the deposition of the waveguide layer. It can also be produced by local modification of the material’s refractive index. For example, photobleaching (Diemeer *et al.*, 1990) has been used to reduce the refractive index of the film in all regions except at the desired position of the waveguide. A change in refractive index of about 0.03 can be achieved by this technique. Alternatively, the change in orientation that occurs during the poling has also (Thackara *et al.*, 1988) been used to define the waveguide stripe. In this case the refractive index for light polarized parallel to the direction of the poling field is locally increased, and so if the electric field is just applied over a track of the material, using lithographically defined electrodes, light of this polarization will be confined within the track. Light polarized perpendicular to the poling field direction will not be guided.

There have been substantial advances in the properties of electrically poled polymers over the past two or three years, and it should not be long before prototype devices utilizing these materials become generally available.

2.4.2. Third-order nonlinearities in polymer films

As stated in section 2.1.2, considerably less effort has been placed to date in the development of third-order materials than for those for second-order applications. In the field of polymeric films, work has concentrated on the measurement of the third-order coefficeints $\chi^{(3)}$ of a fairly limited range of polymeric materials, with a small amount of work reported on the fabrication of waveguide devices out of suitable third-order materials. A fairly thorough groundwork of theoretical studies has, however, been laid (Stegeman *et al.*, 1988, Bennion, 1989, Boardman and Twardowski, 1989), in which a number of potential all-optical waveguide devices, fabricated out of thin films with high third-order nonlinearities, have been identified. These theoretical studies have highlighted the properties required for device fabrication, and it is clear that high nonlinearities themselves are not sufficient. Any material under consideration must combine these with good optical transmission, good processability and other desirable features.

Two experimental techniques have been used, in the main, for the assessment of the third-order properties of polymeric films—third harmonic generation (THG) and degenerate four-wave mixing (DFWM). In principle both techniques can be used to measure the nonlinearities on very fast time-scales (subpicosecond) and over a wide range of wavelengths. However, direct comparison of the results from the two techniques can be difficult as the nonlinearities measured are different, being $\chi^{(3)}(3\omega; \omega, \omega, \omega)$ and $\chi^{(3)}(\omega; \omega, -\omega,\omega)$ respectively. As in the case of second-order susceptibilities described earlier, in this notation the first parameter in parentheses refers to the output frequency, while the others denote the input frequencies of the three photons involved in a third-order process. Thus for third harmonic generation three photons of frequency ω result in an output photon at 3ω, whereas in DFWM the output and input photons are all at the same frequency. They will only be comparable if there are no absorption bands present between, or in the region of, ω and 3ω for the material under consideration.

Two broad classes of materials have been studied. The first consists of a number of materials originally of interest for their electrical conduction properties. This group includes polyacetylene, polydiacetylenes and, for example, polythiophene derivatives. Polyacetylene and polydiacetylenes show the highest third-order nonlinearities recorded amongst organic materials. The second class of materials consists of a range of ladder and chain polymers. This group is characterized by lower nonlinearities, but better optical and mechanical properties. The experimental results for the two classes are summarized below.

Polydiacetylenes (PDAs) are the class of organic polymers that have been studied most intensely in terms of their third-order nonlinearities. This family of materials has been extensively studied over the past 10 years (Bloor and Chance, 1985). The polymer has an extended conjugated backbone, onto which a variety of pendent side groups may be attached (Figure 2.14(a)). The high nonlinearities observed in many polydiacetylenes are derived from this conjugated pathway. The nature of the attached groups will control the properties of the polymer—for example determining whether single-crystal polymers (PTS) may be obtained, or whether the polymer will be soluble in common solvents, or even whether Langmuir–Blodgett films (next section) can be formed from the polymer. Third-order nonlinearities in PDAs have been measured over a range of wavelengths, allowing the role of resonant enhancement to be explored. In pioneering experiments Sauterat *et al.* (1976) used third harmonic generation on single-crystal PDAs to show that $\chi^{(3)}$ is

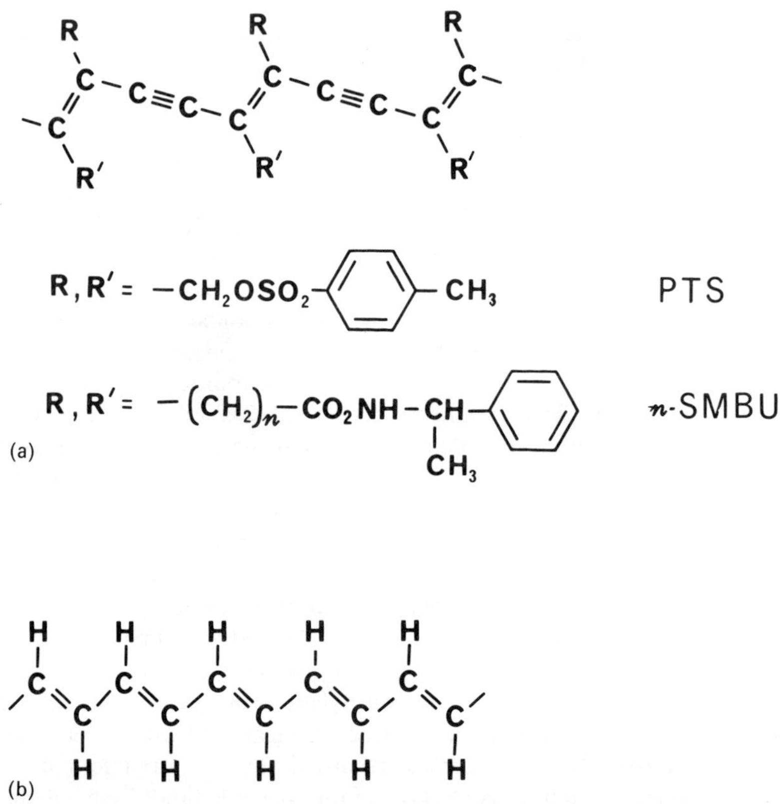

Figure 2.14 Structures of linear conjugated third-order nonlinear optical polymers: (a) polydiacetylene; (b) polyacetylene.

increased as the absorption edge is approached. Further studies by Carter *et al.* (1985), this time using DFWM, obtained a resonant $\chi^{(3)}$ of about 9×10^{-9} esu in the same material, compared with an off-resonant value of 5×10^{-10} esu. The response time was found to increase as resonance was approached, owing to the involvement of real excitation and relaxation processes. Off-resonance the response times were found to be faster than the 300 fs resolution of their laser, confirming the fact that the purely electronic nonlinearities present in these materials are capable of very fast operation. The nonlinearities in ordered polydiacetylenes (single crystals or Langmuir–Blodgett films) are found to be very anisotropic, with the component parallel to the direction of the polymer chains being as much as two orders of magnitude higher than the component perpendicular to the chains.

Soluble PDAs have also been used to form waveguides by spin coating (Jackel *et al.*, 1988; Mann *et al.*, 1988). In this case the films are disordered, and the nonlinearity is significantly lower than the highest values found in the ordered systems described above. Nevertheless, the much improved processability of these soluble polymers makes them attractive for further investigation. Channel waveguide fabrication has been demonstrated in the above reports. The low level of the nonlinearities, however, means that long interaction lengths will be required to make a useful device.

Polyacetylene (Figure 2.14(b)) has for many years been studied as a conducting polymer. More recently (Kajzar *et al.*, 1987; Sinclair *et al.*, 1987) its nonlinear optical properties have also been studied, using third harmonic generation. As for the polydiacetylenes described above, measurements as a function of wavelength show a large resonant enhancement in the region of the two-photon absorption peak. Here the magnitude of $\chi^{(3)}$ is about $(1–5) \times 10^{-9}$ esu—comparable with the best polydiacetylenes. High laser beam irradiances (up to $10\,\text{GW}\,\text{cm}^{-2}$) were used in the experiments without damaging the film. In both these experiments the polyacetylene was unoriented—the polymer chains having a random angular distribution within the film. Since the nonlinearity of a given chain will be highly anisotropic, with $\chi^{(3)}$ being much greater parallel to the chain than perpendicular to it, the nonlinearity of the unoriented film will be an average of the parallel and perpendicular components of $\chi^{(3)}$, and will be significantly reduced from the maximum possible value. Experiments with oriented films (Drury, 1988) have shown that a much higher value for a single component of $\chi^{(3)}$ can indeed be obtained in such samples. One disadvantage of polyacetylene films is their poor processability, stability and mechanical properties.

The second group of third-order polymers include a range of rigid rod and ladder polymers such as those shown in Figure 2.15. In these polymers the conjugated pathways are forced to remain in a single plane by chemical bonding, and this can lead to an increase in the overall conjugation length. In linear polymers such as those described above, distortions in the linear chain can

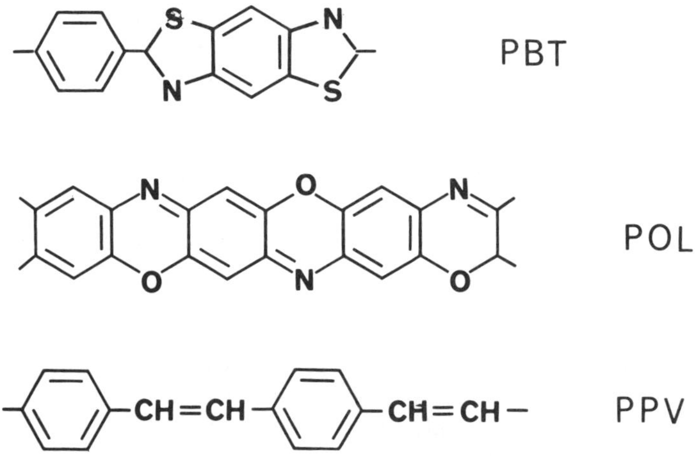

Figure 2.15 Structures of two-dimensional nonlinear optical polymers.

lead to a disruption of the conjugation path, and a reduction in the nonlinearity. $\chi^{(3)}$ values as high as 10^{-9} esu have been reported in ladder polymers (Dalton, 1989). PBT is reported to have very good mechanical properties and high stability. Films of PBT have good optical quality (Rao *et al.*, 1986a,b). The maximum off-resonant nonlinearity, measured by DFWM, is about 9×10^{-12} esu—considerably smaller than that of polyacetylene or the polydiacetylenes. However, its other properties mean that it is still an interesting material.

A number of other polymers, not really fitting into either of the above categories, have also been studied. An example of one such material, having very promising properties, is polyphenylenevinylene (Figure 2.15(c)). High quality films of this polymer have been formed by spin coating of a soluble precursor, followed by thermal treatment in vacuum. Measurements using both THG and DFWM result in a value of $\chi^{(3)}$ of about 10^{-10} esu (Bubeck *et al.*, 1989a), and subpicosecond response times were recorded.

Thus, although little progress has been made to date in the fabrication of nonlinear optical devices based on third-order effects in polymeric films, the basic characterization of such films is progressing steadily, and an understanding of the molecular and structural basis of high nonlinearities is being gradually formulated.

2.5. LANGMUIR–BLODGETT FILMS

The Langmuir–Blodgett deposition technique provides another route to the formation of molecularly ordered thin films, and has the added advantage that in principle the structure of the film can be designed on a monolayer scale to provide whatever properties, in terms of for example refractive index or nonlinear optical coefficients, are required. The technique has been known since the 1920s (Blodgett, 1935), but has been developed more fully only over the last 20 years. Applications have been considered in many branches of chemistry, biology and physics, but over recent years much interest has focused on the nonlinear optical properties of Langmuir–Blodgett films (Roberts, 1985).

Molecules suitable for Langmuir–Blodgett deposition (Allen, 1989b) have one hydrophilic end (typically having a —COOH group) and one hydrophobic end (usually a long fatty chain such as $—C_{12}H_{25}$). A drop of solution containing the molecule of interest is deposited on the surface of a liquid subphase (e.g. water which may or may not contain cations such as Ca^{2+}, Cd^{2+}) contained within a specially designed 'trough'. This trough has moveable barriers round its perimeter so that surface area of the subphase may be altered. The Langmuir–Blodgett molecule forms a monomolecular coverage on the subphase (Figure 2.16), but does not dissolve because of the hydrophobic end group. The surface area of the trough is then reduced, compressing the monolayer. As the molecules are pushed closer together, the hydrophobic ends are expelled from the water surface (Figure 2.16(b)), and the molecules become closely packed on the surface. Compression continues until the molecules are tightly packed on the subphase, forming what is essentially a two-dimensional crystalline layer, one molecule thick. Attempts at further compression will result in collapse of the monolayer. The progress of the monolayer formation can be followed by monitoring the surface pressure of the film. A typical surface pressure against area per

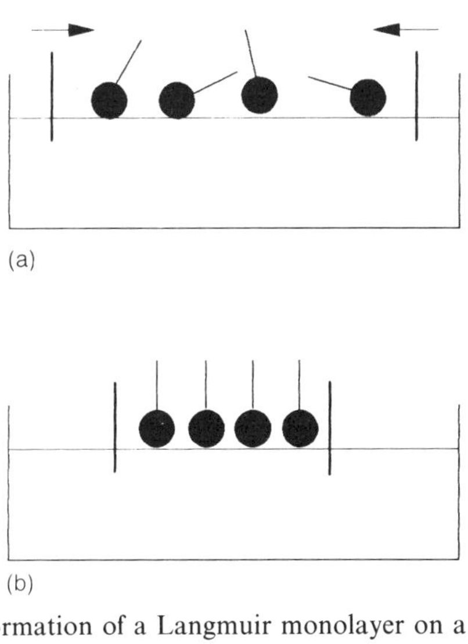

Figure 2.16 Formation of a Langmuir monolayer on a water surface.

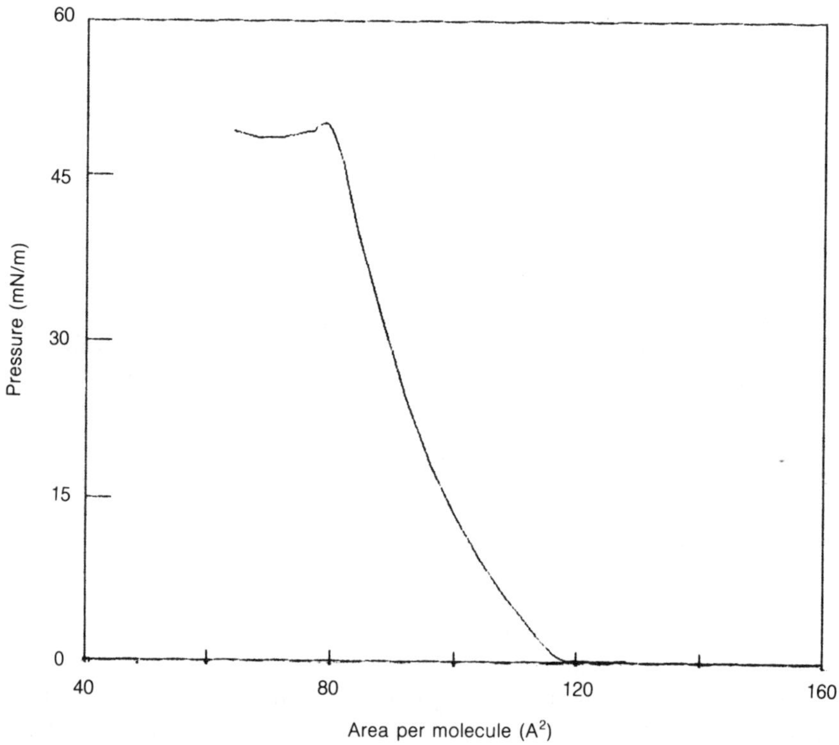

Figure 2.17 Typical surface pressure against area isotherm for the compression of a Langmuir monolayer.

molecule isotherm is shown in Figure 2.17. The shallow slope at the high area end corresponds to Figure 2.16(a). As the molecules pack together the increase in pressure becomes more rapid, with the steep part of the curve corresponding to the formation of the crystalline monolayer. The horizontal portion of the curve at the low area end corresponds to the collapse of the monolayer.

If a substrate, such as a glass microscope slide, is now inserted into or withdrawn from the subphase, whilst the surface pressure is maintained (by a feedback system) on the steep part of the pressure–area isotherm, the monolayer is transferred to the surface of the substrate. This process may be repeated until the desired number of monolayers has been deposited. A number of different geometries of deposited film are possible, depending on the nature of the molecule and of the substrate. The most common geometry occurs when the monolayer is deposited on both insertion and withdrawal of the substrate (Figure 2.18(c)). This is denoted Y-type deposition and in this case a centrosymmetric arrangement of molecules is built up as successive

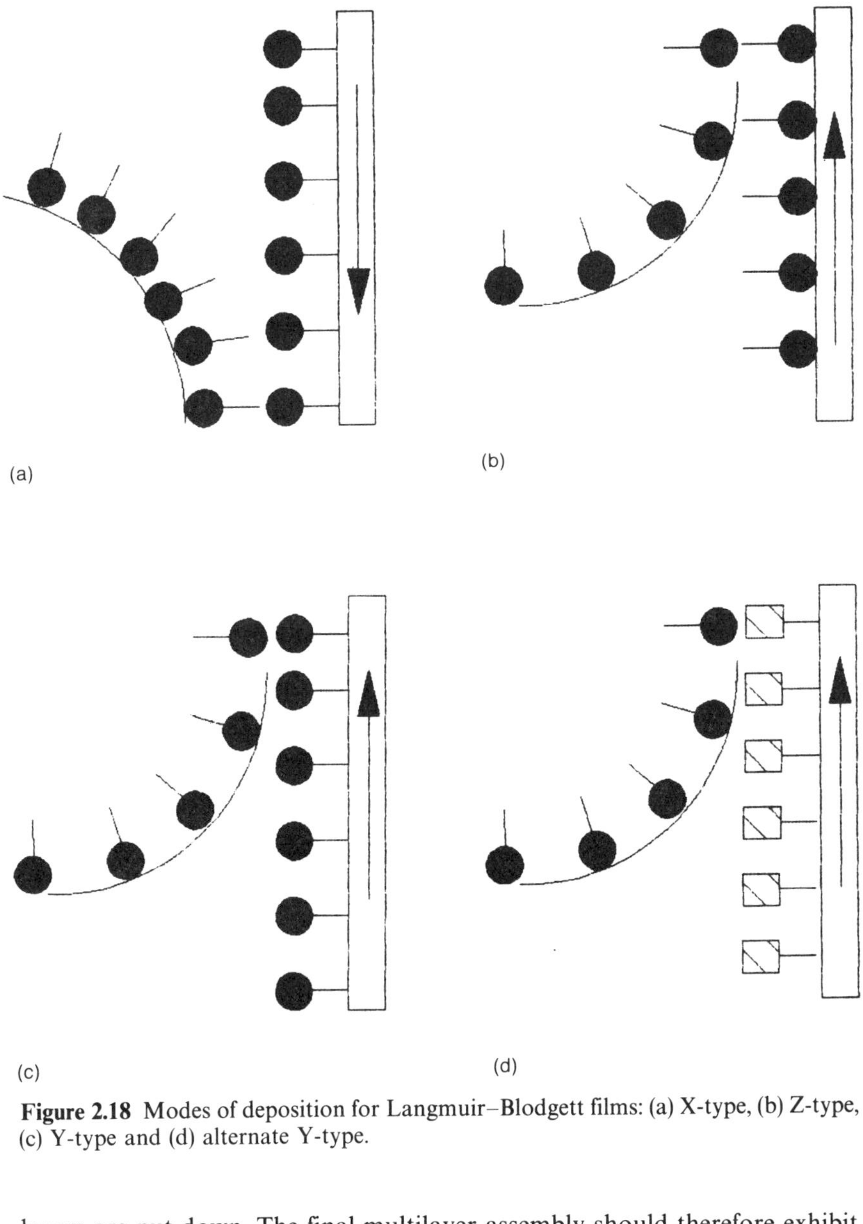

Figure 2.18 Modes of deposition for Langmuir–Blodgett films: (a) X-type, (b) Z-type, (c) Y-type and (d) alternate Y-type.

layers are put down. The final multilayer assembly should therefore exhibit no second-order nonlinear optical effects, but may have nonzero third-order coefficients. If molecules are deposited only on insertion of the substrate (X-type deposition, Figure 2.18(a)) or only on withdrawal (Z-type deposition, Figure 2.18(b)) a noncentrosymmetric multilayer is formed. However, these films are often unstable, having a tendency to rearrange themselves with time

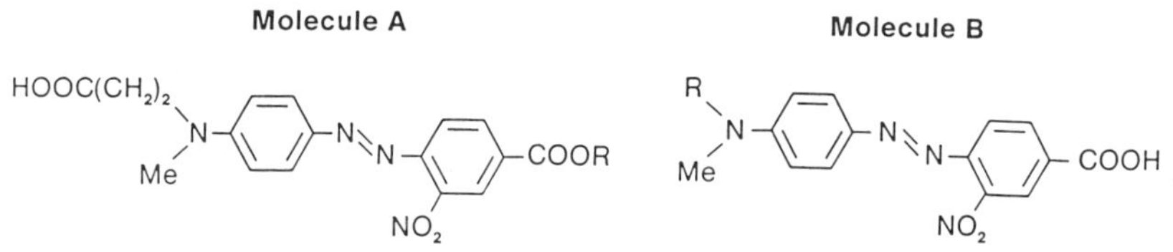

Figure 2.19 A complementary pair of molecules for deposition as an alternate Y-type multilayer. R represents a $—C_{12}H_{25}$ hydrophobic chain.

into a Y-type configuration. A more stable noncentrosymmetric film can be obtained if two different molecules are used. A specially adapted trough is needed to allow monolayers of the two molecules to be formed on different parts of the trough, and to allow for insertion of the substrate through one of these monolayers and withdrawal through the other. In this case a noncentrosymmetric film of the form shown in Figure 2.18(d) is produced (alternating Y-type, or AY-type). Structures have been described where the design of the molecules within the AY structure is such that all monolayers have a nonlinear optical element, and these elements are all aligned in the same direction so as to maximize the overall second-order nonlinearity (Neal *et al.*, 1986; Ledoux *et al.*, 1988b). An example of such a system is shown in Figure 2.19. These compounds show that it is possible to design molecules satisfying the requirements for both high optical nonlinearities and also for stable Langmuir–Blodgett deposition. The basic nonlinear optical unit is common to both molecules. However, the positions of the hydrophobic chain and the hydrophilic —COOH group are interchanged between the two molecules, so that on deposition the nonlinear units are additive.

The experimental arrangement for the assessment of these Langmuir–Blodgett films (deposited on a glass microscope slide) by second or third harmonic generation is shown in Figure 2.20(a). The slide is mounted on a rotation stage and irradiated with the fundamental beam from a pulsed Nd:YAG laser at 1.06 μm (or Raman shifted to 1.91 μm). The harmonic intensity generated is measured as a function of the angle between the incident beam and the normal to the substrate. A typical intensity profile for the case of second harmonic generation is shown in Figure 2.20(b) and third harmonic generation produces similar fringed output. In both cases there is an oscillating pattern contained within a broader envelope. The oscillations are due to interference between harmonic signals generated by the films on the front and back faces of the substrate. The signal from the front face passes through the slide as the fundamental. Because of the dispersion of the refractive index of the substrate, the two harmonic signals have a phase difference which depends on the path length through the substrate, and therefore on the angle of incidence.

The nonlinearity of the film $\chi^{(2)}$ or $\chi^{(3)}$ is obtained by comparing the envelope of the signal with that from a standard sample (usually a quartz

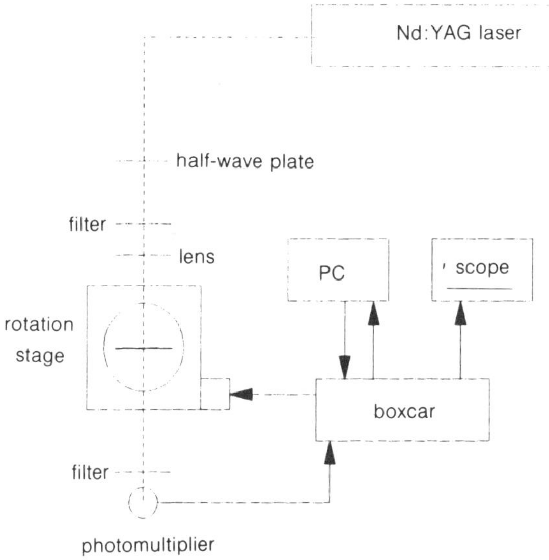

(a)

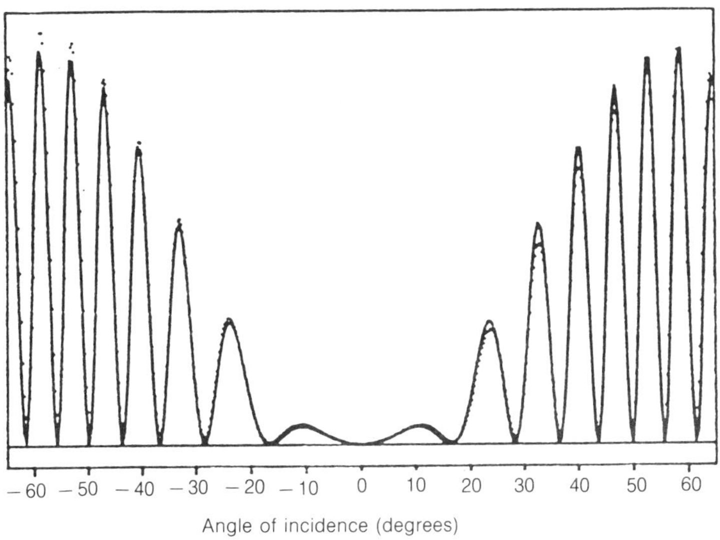

(b)

Figure 2.20 (a) Experimental arrangement for the measurement of second harmonic generation in Langmuir–Blodgett films. (b) Typical experimental output, along with the theoretically fitted curve (———).

plate or wedge for second harmonics and fused silica for third harmonics). By recourse to standard equations of electromagnetic wave propagation, a theoretical model of the envelope function as a function of the relative magnitudes of different coefficients in the $\boldsymbol{\chi}^{(2)}$ tensor can be developed, and fitted to the experimental results. Using different combinations of input and output polarizations allows most of the independent coefficients to be determined. If simple models for the orientation of the molecules within the film are assumed, it is then possible to estimate both the molecular $\boldsymbol{\beta}$ or γ coefficients and the orientation of the nonlinear optical sections of the molecule within the film (Allen *et al.*, 1988b).

For example, the system shown in Figure 2.19 has now been extensively characterized in terms of its second-order nonlinearity. AY-type films consisting of up to 34 monolayers (about 100 nm thick) have been deposited. These AY-type films show very high nonlinearities ($\chi^{(2)} = 336\,\mathrm{pm\,V^{-1}}$; Ledoux *et al.*, 1988), partly as a result of resonant enhancement (Section 2.1.2) occurring because of the proximity of the second harmonic wavelength of 530 nm to the absorption maximum of these molecules (about 480 nm). Nevertheless, the 100 nm thick film mentioned above generates approximately the same second harmonic intensity as does a 2 nm slab of quartz crystal.

Table 2.2 The molecular structures of a number of molecules which form Langmuir–Blodgett films, and their experimental molecular hyperpolarizabilities (β)

		β *(nm⁴ V)*
Azobenzene	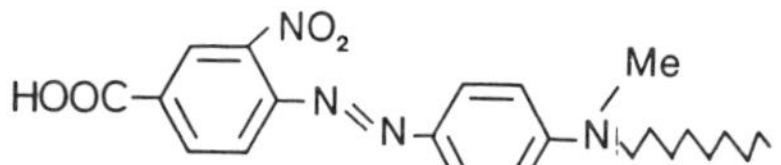	0.12
Stilbazene	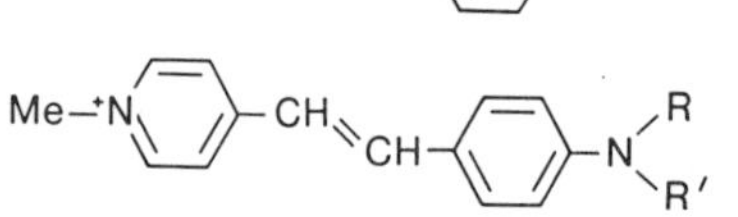	0.84
Merocyanine	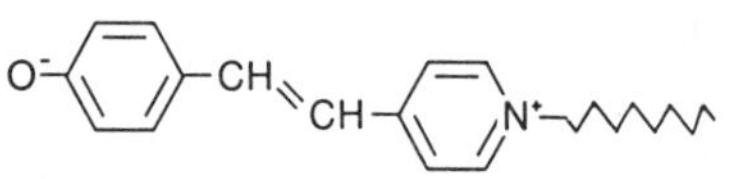	1.13
Hemicyanine	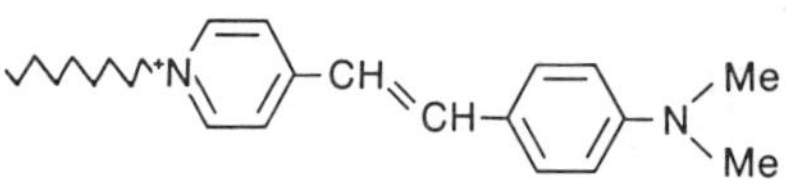	0.11
Phenylhydrazone	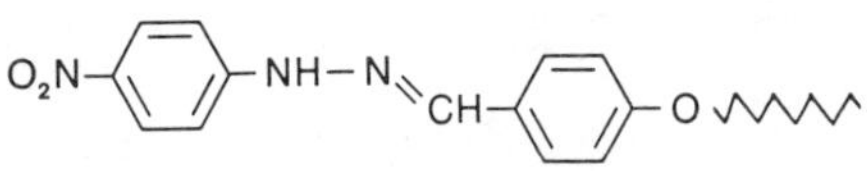	0.17
Polyene	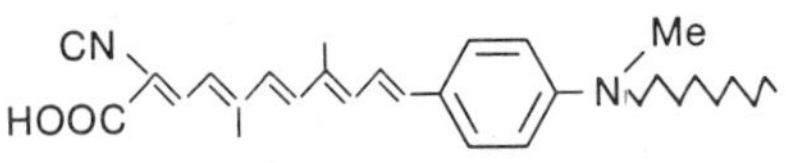	0.35

Many other chromophores have also been studied in terms of their second harmonic generation efficiencies, and the measured nonlinearities of some of these are given in Table 2.2.

Langmuir–Blodgett films have also proved a useful vehicle for the study of third-order properties of a number of compounds. A number of polydiacetylene films have been deposited in this way (Kajzar *et al.*, 1983; Kajzar and Messier, 1985a), as have a merocyanine dye (Kajzar *et al.*, 1986) and rigid rod phthalocyanine molecules (Bubeck *et al.*, 1989b). As in the case of polymeric films described in the previous section, the highly conjugated polydiacetylene molecules showed the highest nonlinearities, but in all these molecules the presence of alkyl chains needed in order to produce the film resulted in a decrease in the nonlinearity compared with 'unfunctionalized' forms of the molecules found in the crystalline or polymeric state.

Although the Langmuir–Blodgett technique is very attractive, giving close control over the thickness and composition of the film created, there are a number of problems to be overcome before it can be routinely used for the fabrication of nonlinear optical waveguide devices. The major one of these problems is the optical quality of the film. On a molecular scale the monolayer is a two-dimensional crystal, with tight packing of the molecules, but macroscopically the films become polycrystalline, consisting of many small crystallites, typically of dimensions 1–20 μm (Chollet and Messier, 1983). The boundaries between these crystallites scatter light, and produce very high optical losses when attempts are made to propagate light along the length of the film. Much work is now going on to understand the nature of the internal structure in Langmuir–Blodgett films and to reduce the internal scattering. Recently encouraging progress has been reported, using both polymeric Langmuir–Blodgett molecules (Tredgold *et al.*, 1988; Young *et al.*, 1989), and also more conventional monomeric molecules (Bossard *et al.*, 1989). In the latter case, the observation of waveguided light propagating distances of up to a few centimetres along a thick (>500 layer) Langmuir–Blodgett film has been reported. The film also has high nonlinear optical coefficients, as measured by second harmonic generation.

2.6. CONCLUSIONS

In this chapter the potential advantages of organic materials for second- and third-order nonlinear optical processes have been described and the origin of their high nonlinear optical coefficients explained in terms of molecular structure. An attempt has been made to review the current state of progress in the development of these organic materials from the status of an 'exciting new discovery' to that of well-characterized components of nonlinear optical devices, capable of being used in efficient production processes. This field of research can no longer be described as in its infancy. Indeed, almost every chemical and electronics company in Europe, the USA and Japan now has a research programme in this area. Many of these research programmes are now producing 'prototype devices' which, on a laboratory scale, have performances superior to the equivalent devices currently available, fabricated from inorganic materials such as lithium niobate. However, yet further increases in efficiency and performance will be required in order to persuade manufacturers that it is worthwhile abandoning technologies that they now know well (despite the limitations of these technologies), and investing in the new processes required for exploitation of the organic materials. In fact, there is still much research and development needed before techniques such as Langmuir–Blodgett deposition can even be considered on a production scale.

Nevertheless, great advances have been made over the last five years, and, given the amount of effort being expended on this topic world-wide, it can only be a matter of time before nonlinear optical devices utilizing organic materials are commonplace.

REFERENCES

Allen, S. (1989a) In *Organic Materials for Nonlinear Optics* (eds R. A. Hann and D. Bloor), RSC, London, pp. 137–50.

Allen, S. (1989b) In *Materials for Non-linear and Electro-optics* (ed. M. H. Lyons), Institute of Physics Conference Series No. 103, pp. 163–74.

Allen, S., McLean, T. D., Gordon, P. E., *et al.* (1988a) *J. Appl. Phys.*, **64**, 2583.

Allen, S., McLean, T., Gordon, P. E., Robin, P., and Ledoux, I. (1988b) *Proc. Soc. Photo-Opt. Instrum. Eng.*, **971**, 206.

Allen, S., Bone, D. J., Carter, N., *et al.* (1990) *RSC Special Publication No. 91*, in Organic material for Non-linear optics II (eds. R. A. Hann, D. Bloor), Royal Society of Chemistry, pp. 235–41.

Badan, J., Hierle, R., Perigaud, A., and Vidakovic, P., (1986) In *Nonlinear Optical Properties of Organic Molecules and Crystals*, Vol. 1 (eds D. S. Chemla and J. Zyss), Academic Press, London, pp. 297–356.

Bailey, R. T., Cruikshank, F. R., Guthrie, S. M. G., *et al.* (1988) *Opt. Commun.*, **65**, 229.

Bailey, R. T., Bourhill, G. H., Cruikshank, F. R., *et al.* (1989) In *Materials for Non-linear and Electro-optics* (ed. M. H. Lyons), Institute of Physics Conference Series No. 103, pp. 119–25.

Bennion, I. (1989) In *Materials for Non-linear and Electro-optics* (ed. M. H. Lyons), Institute of Physics Conference Series No. 103, pp. 1–12.

Bierlein, J. D., Cheng, L. K., Wang, Y., and Tam, W. (1990) *Appl. Phys. Lett.*, *56*, 423.

Blodgett, K. B. (1935) *J. Am. Chem. Soc.*, **57**, 1007.

Bloor, D. and Chance, R. R. (eds) (1985) *Polydiacetylenes: Synthesis, Structure and Electronic Properties*, Martinus Nijhoff, Dordrecht.

Boardman, A. D., and Twardowski, T. (1989) *Phys. Rev. A*, **39**, 2481.

Bossard, C., Drecher, G., Tieke, B. and Gunter, P. (1989) *Proc. Soc. Photo-Opt. Instrum. Eng.*, **1017**, 141.

Bubeck. C., Kaltbeitzel, A., Lenz, R. W., *et al.* (1989a) In *Nonlinear Optical Effects in*

Organic Polymers (eds. J. Messier *et al.*), Kluwer, Dordrecht, pp. 143–8.

Bubeck, C., Neher, D., Kaltbeitzel, A., *et al.* (1989b) In *Nonlinear Optical Effects in Organic Polymers* (eds. J. Messier *et al.*), Kluwer, Dordrecht, pp. 185–93.

Carter, G. M., Thakur, M. K., Chen, Y. J. and Hryniewicz, J. V. (1985) *Appl. Phys. Lett.*, **47**, 457.

Cassidy, C., Halbout, J.-M., Donaldson, W. and Tang, C. L. (1979) *Opt. Commun.*, **29**, 243.

Chemla, D. S. and Zyss, J. (eds) (1986) *Nonlinear Optical Properties of Organic Molecules and Crystals*, Vols. 1 and 2, Academic Press, Orlando, FL.

Chollet, P. A. and Messier, J. (1983) *Thin Solid Films*, **99**, 197.

Dalton, L. R. (1989) *Nonlinear Optical Effects in Organic Polymers* (eds J. Messier *et al.*), Kluwer, Dordrecht, pp. 123–42.

Diemeer, M. B. J., Suyten, E. M. M., Trommel, E. S., *et al.* (1990) *Electron. Lett.*, **26**, 379.

Docherty, V. L., Pugh, D. and Morley, J. O. (1985) *J. Chem. Soc., Faraday Trans, 2*, **81**, 1179.

Drury, M. R. (1988) *Solid State Commun.*, **68**, 417.

Eich, M., Reck, B., Yoon, D. Y., *et al.* (1989a) *J. Appl. Phys.*, **66**, 3241.

Eich, M., Sen, A., Looser, H., *et al.* (1989b) *J. Appl. Phys.*, **66**, 2559.

Garito, A. F., Helfin, J. R., Wong, K. Y. and Zamani-Khamiri, O. (1988) *Proc. Soc. Photo-Opt. Instrum. Eng.*, **971**, 2.

Haas, D., Yoon, H., Man, H.-T., *et al.* (1989) *Soc. Photo-Opt. Instrum. Eng.*, **1147**, 222.

Hampsch, H. L., Torkelson, J. M., Bethke, S. J., Grubb, S. G. (1990) *J. Appl. Phys.*, **67**, 1037.

Hubbard, M. A., Marks, T. J., Yang, J. and Wong, G. J. (1989) *Chem. Mater.*, **1**, 167.

Hurst, M. and Munn, R. W. (1986) *J. Mol. Electron.*, **2**, 35.

Jackel, J. L., Schlotter, N. E., Townsend, P. D., *et al.* (1988) *Proc. Soc. Photo-Opt. Instrum. Eng.*, **971**, 239.

Jungbauer, D., Reck, B., Twieg, R., *et al.* (1990) *Appl. Phys. Lett.*, **56**, 2610.

Kajzar, F. and Messier, J. (1985a) *Thin Solid Films*, **132**, 11.

Kajzar, F. and Messier, J. (1985b) *Phys. Rev. A*, **32**, 2352.

Kajzar, F., Messier, J., Zyss, J. and Ledoux, I. (1983) *Opt. Commun.*, **45**, 133.

Kajzar, F., Messier, J., Girling, I. R. and Peterson, I. R. (1986) *Electron. Lett.*, **22**, 1230.

Kajzar, F., Etemad, S., Baker, G. L. and Messier, J. (1987) *Solid State Commun.*, **63**, 1113.

Karl, N. (1989) In *Materials for Non-linear and Electro-optics* (ed. M. H. Lyons), Institute of Physics Conference Series No. 103, pp. 107–18.

Kerkoc, P., Zgonik, M., Sutter, K., *et al.* (1989) In *Materials for Non-linear and Electro-optics* (ed. M. H. Lyons), Institute of Physics Conference Series No. 103, pp. 133–8.

Khanarian, G., Che, T., Martino, R. N., *et al.* (1988) *Soc. Photo-Opt. Instrum. Eng.*, **824**, 72.

Kitaoka, Y., Sasaki, T., Nakai, S., *et al.*, (1990) *Appl. Phys. Lett.*, **56**, 2074.

Lalama, S. J. and Garito, A. E. (1979) *Phys. Rev. A*, **20**, 1179.

Ledoux, I., Josse, D., Fremaux, P., *et al.*, (1988) *Thin Solid Films*, **158**, (160), 217.

Levine, B. F. and Bethea, C. G. (1975), *J. Chem. Phys.*, **63**, 2666.

Levine, B. F., Bethea, C. G., Thurmond, C. D., *et al.* (1979) *J. Appl. Phys.*, **50**, 2523.

Linford, G. J. and Clendening, C. W. (1986) *Proc. Soc. Photo-Opt. Instrum. Eng.*, **642**, 110.

Lipscombe, G. F., Garito, A. E. and Narang, R. S. (1981) *J. Chem. Phys.*, **75**, 1509.
Mann, S., Oldroyd, A. R., Bloor, D., *et al.*, (1988) *Proc. Soc. Photo-Opt. Instrum. Eng*, **971**, 245.
McArdle, B. J. and Sherwood, J. N. (1987) In *Advanced Crystal Growth* (eds P. M. Dryburgh, B. Cockayne and K. C. Barraclough), Prentice-Hall, Englewood Cliffs, NJ.
Meredith, G. R., VanDusen, J. G. and Williams, D. J. (1982) *Macromolecules*, **15**, 1385.
Mohlmann, G. R., Horsthuis, W. H. G., van der Horst, C. P. J. M., *et al.* (1989) *Proc. Soc. Photo-Opt. Instrum. Eng.*, **1147**, 245.
Morley, J. O., Docherty, V. J. and Pugh, D. (1987a) *J. Chem. Soc., Perkin Trans.* **II**, 1351.
Morley, J. O., Docherty, V. J. and Pugh, D. (1987b) *J. Chem. Soc., Perkin Trans.* **II**, 1357.
Morley, J. O., Docherty, V. J. and Pugh, D. (1987c) *J. Chem. Soc., Perkin Trans.* **II**, 1361.
Neal, D. B., Petty, M. C., Roberts, G. G., *et al.* (1986) *Electron. Lett.*, **22**, 460.
Nicoud, J. F. and Twieg, R. J. (1986) In *Nonlinear Optical Properties of Organic Molecules and Crystals*, Vol. 2 (eds. D. S. Chemla and J. Zyss), Academic Press, Orlando, FL, p. 255.
Oudar, J. L. (1977) *J. Chem. Phys.*, **67**, 445.
Oudar, J. L. and Chemla, D. S. (1977) *J. Chem. Phys.*, **66**, 2664.
Pierce, B. M. (1988) *Proc. Soc. Photo-Opt. Instrum. Eng.*, **971**, 25.
Rao, D. N., Burzynski, R., Xin Mi and Prasad, P. N. (1986a) *Appl. Phys. Lett.*, **48**, 387.
Rao, D. N., Swiatkiewicz, J., Chopra, P., Ghoshal, K. and Prasad, P. N. (1986b) *Appl. Phys. Lett.*, **48**, 1187.
Roberts, G. G. (1985) *Adv. Phys.*, **34**, 475.
Rosker, M. J. and Tang, C. L. (1985) *J. Opt. Soc Am. B*, **2**, 691.
Rosker, M. J., Cheng, K. and Tang, C. L. (1985) *IEEE J. Quantum Electron.*, **21**, 10.
Sauterat, C., Hermann, J.-P., Frey, R., Pradere, F., Ducuing, J., Baughman, R. H. and Chance, R. R. (1976) *Phys. Rev. Lett.*, **36**, 956.
Sinclair, M., Moses, D., Heeger, A. J., Vilhelmsson, K., Valk, B. and Salour, M. (1987) *Solid State Commun.*, **61**, 221.
Singer, K. D. and Garito, A. F. (1981) *J. Chem. Phys.*, **75**, 3572.
Singer, K. D., Sohn, J. E. and Lalama, S. J. (1986) *Appl. Phys. Lett.*, **49**, 248.
Singer, K. D., Kuzyk, M. G. and Sohn, J. E. (1987) *J. Opt. Soc. Am. B*, **4**, 968.
Small, R. D., Singer, K. D., Sohn, J. E., Kuzyk, M. G. and Lalama, S. J. (1987) *Proc. Soc. Photo-Opt. Instrum. Eng*, **682**, 160.
Stegeman, G. T., Right, E. M., Finlayson, N. *et al.* (1988) *J. Lightwave Technol.*, **4**, 2069.
Stevenson, J. L. (1973) *J. Phys. D.*, **6**, L13.
Sutter, K., Bossard, C., Baraldi L. and Gunter, P. (1989) In *Materials for Non-linear and Electro-optics* (ed. M. H. Lyons), Institute of Physics Conference Series No. 103, pp. 127–32.
Teng, C. C. and Garito, A. F. (1983) *Phys. Rev. B*, **28**, 6766.
Thackara, J. I., Lipscombe, G. F., Stiller, M. A., Ticknor, A. J. and Lytel, R. (1988) *Appl. Phys. Lett.*, **52**, 1031.
Tredgold, R. H., Young, M. C. J., Jones, R., Hodge, P., Kolinsky, P. and Jones, R. J. (1988) *Electron. Lett.*, **24**, 308.
Ward, J. (1965) *Rev. Mod. Phys.*, **37**, 1.
Wherrett, B. S. and Smith, S. D. (1985) In *Optical Bistability, Dynamical Nonlinearity*

and Photonic Logic, Royal Society, London, 1985.

Williams, D. J. (1983) In *Nonlinear Optical Properties of Organic and Polymeric Materials*, ACS Symposium Series No. 233, American Chemical Society, Washington, DC.

Williams, D. J. (1984) *Angew. Chem., Int. Edn. Engl.*, **23**, 690.

Yokotani, A., Sasaki, T., Yoshida, K. and Nakai, S. (1989) *Appl. Phys. Lett.*, **55**, 2692.

Yoshimura, T. (1987) *J. Appl. Phys.*, **62**, 2028.

Young, M. C. J., Tredgold, R. H. and Hodge, P. (1989) In *Organic Materials for Non-linear Optics* (eds R. A. Hann and D. Bloor), RSC, London, pp. 354–60.

Zyss, J., and Oudar, J. L. (1982) *Phys. Rev. A*, **26**, 2028.

Zyss, J., Chemla, D. S. and Nicoud, J. F. (1981) *J. Chem. Phys.*, **74**, 4800.

Zyss, J., Nicoud, J. E. and Coquillay, M. (1984) *J. Chem. Phys.*, **81**, 4160.

3 Semiconductors

Alan Miller

3.1. INTRODUCTION

Since the invention of the transistor in 1947, semiconducting materials have brought about technological advances which would have been impossible to imagine at the outset, and which now influence our daily lives in everything from home entertainment to space exploration. The basis for the success of semiconductor electronic devices is the sensitivity of their electrical properties to the density of free carriers, which may be controlled by doping, electrical injection or optical excitation. However, the fundamental optical properties of semiconductors are also altered by changes in carrier density so that the optical excitation of electrons between energy bands and/or the spatial transfer of charge carriers within the semiconductor leads to nonlinear optical phenomena (A. Miller *et al.*, 1981; Jain, 1981; Klingshirn and Haug, 1981; Alfano, 1984; Gibbs, 1985; Haug, 1988).

Many types of optical nonlinearities have been studied in semiconductors. Here we are primarily interested in processes which cause changes of absorption coefficient and refractive index as a function of light intensity. We are therefore concerned with effects which arise from the third-order susceptibility, $\chi^{(3)}$ (or an effective susceptibility, $\chi^{(3)}{}_{\mathrm{eff}}$, in the case of dynamic optical nonlinearities). Real and imaginary parts of the susceptibility, χ, describe refraction and absorption linked by causality via Kramers–Kronig relations. These relations may be modified (D. A. B. Miller *et al.*, 1981; Hutchings *et al.*, 1992) to account for optical nonlinearites when laser excitation of a quasi-equilibrium of electrons and holes results in normalized material parameters; thus, refractive index changes, Δn, and absorption changes, $\Delta\alpha$, are connected by the integral expression,

$$\Delta n(\omega, \xi) = \frac{c}{\pi} \mathscr{P} \int_0^\infty \frac{\Delta\alpha(\omega', \xi)}{\omega'^2 - \omega^2} \mathrm{d}\omega' \tag{3.1}$$

where, ξ represents the 'cause' of the change in absorption and $\mathscr{P}$ denotes the principal part of the integral. Note that changes in refractive index at a given frequency are induced by changes in absorption at other frequencies.

Four categories of nonlinear absorption and refraction may be identified in semiconductors, which fall within different regimes of response time and sensitivity. The time constants and sensitivities determine the potential application. We distinguish (a) nonresonant bound electronic, (b) band gap resonant, (c) charge transfer, and (d) optothermal phenomena.

The fastest nonlinear optical processes are nonresonant interactions between electromagnetic radiation and bound electrons. The bound electrons which interact with light occupy the outer shell, valence states. These interactions are usually adequately described classically whereby an intense light field produces anharmonic components in vibrations of the bound electrons. The alternative quantum mechanical approach describes the interaction in terms of virtual electronic transitions between energy bands. The response is essentially instantaneous but relatively weak. Recent theoretical work has determined a fundamental link between the bound electron, optical Kerr effect (nonlinear refraction) and two-photon absorption, Raman, and a.c. Stark effects based on semiconductor band theory, but applicable to a wide range of solid state materials (Sheik-Bahae *et al.*, 1990a). This has helped to characterize these interactions and to determine the requirements for ultrafast nonlinear optical switching devices.

Resonant or dynamic optical nonlinearities occur when the photon energy coincides with the energy of an interband, intraband or excitonic transition (Wherrett and Higgins, 1982; Wherrett, 1983). The optically induced redistribution of carriers between bands can result in relatively sensitive nonlinear optical responses in many instances. The build-up (or switch-on) time is very rapid using short optical pulses. The recovery (or switch-off) time is set by the recombination rate, diffusion coefficient or sweep-out rate in the presence of an electric field. Time scales range from microseconds to femtoseconds depending on the precise conditions. Coherence between the light and the free carriers is rapidly lost because of scattering. Band filling and exciton saturation in semiconductors have proved valuable mechanisms for detailed investigations of all-optical bistability and associated switching schemes at moderate laser power levels.

Resonant optical nonlinearities can be enhanced by allowing the spatial separation of the two carrier types after optical generation, to produce changes in refractive index via electro-optic effects, which are described by $\chi^{(2)}$. Carrier motion in usually driven by the concentration gradient created by the interference of two laser beams. For instance, optical excitation of either electrons or holes from deep impurity or defect levels within the band gap of a semiconductor can generate free carriers which move from regions of high light intensity and retrap at regions of low intensity. The response times (rise and fall) are thus set by carrier transport properties. Although this photorefractive effect is better known in large gap insulators, semiconductors show good figures of merit with fast responses (Glass, 1990). Applications are in phase conjugation, real time holography and image processing (Fisher,

1983). Photorefraction and its applications are fully discussed elsewhere in this volume. We note here that the electro-optic effect in semiconductors can be resonantly enhanced by the Franz–Keldysh effect. Further, multiple quantum wells and superlattices lead to other types of charge transport devices such as self-electro-optic devices (SEEDS) based on the quantum confined Stark effect (D. A. B. Miller *et al.*, 1984; D. A. B. Miller, 1990).

The fourth type of optical nonlinearity of interest arises from the effect of heating induced by a laser beam (Smith *et al.*, 1984; Olbright *et al.*, 1984). In most semiconductors, thermally induced changes in the optical constants are larger than resonant electronically induced effects even when relatively short pulses are employed. Most optothermal effects in semiconductors can be understood in terms of a shift of the fundamental absorption edge to longer wavelengths (a few exceptions give the opposite shift) resulting in an increase in the refractive index below the edge. The index change for a given optical power is enhanced at photon energies nearer to the band energy because of higher disperison and absorption from band tail states in this region. Typical recovery times are in the millisecond regime although nanoseconds are possible for very small spot sizes and good heat sinking. The first demonstrations of all-optical regenerative loop circuits have employed arrays of bistable devices employing optothermal effects (Smith *et al.*, 1987; Mandel *et al.*, 1987).

In all categories described above, the initial change in optical constants can be very rapid when short excitation pulses are employed. For intrinsically absorptive nonlinearities, the recovery is set by a combination of material parameters and experimental conditions. The nonlinear optical susceptibility formalism developed for frequency mixing is carried over to describe resonant, dissipative interactions in terms of effective third-order susceptibilities $\chi^{(3)}_{\text{eff}}$ to allow comparison of the magnitudes of the different phenomena. The magnitude of $\chi^{(3)}$ or $\chi^{(3)}_{\text{eff}}$ in semiconductors range from 10^{-13} up to 10 esu.

The first concern when considering a given semiconductor is its transparency to a particular laser frequency whether resonant or nonresonant optical nonlinearities are being employed . Figure 3.1 gives band gap energies for most common semiconductors in the visible and near-infrared compared with the major laser output frequencies. Additional ranges may be covered by alloying different semiconductors. For instance, $Ga_xIn_{1-x}As_yP_{1-y}$ provides band gaps in the important 1.3 and 1.55 μm regions to provide compatibility with low loss optical fibres, and the band gap energy of $Cd_xHg_{1-x}Te$ spans the range from zero to 1.5 eV. Quantization of energy levels in low dimensional structures also provide another method of shifting band gap energies. Meeting the conditions for optical switching devices invariably requires a trade-off between size of optical nonlinearity, response speed and background absorption. Semiconductors offer a great deal of flexibility with these parameters. Careful consideration must therefore be given to the choice

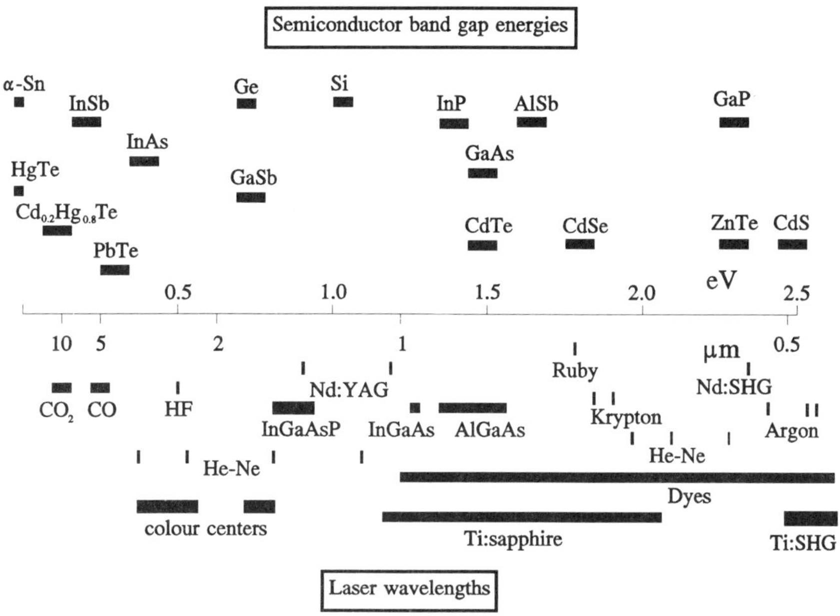

Figure 3.1 Semiconductor band gap energies (the 0 to 300 K range is shown) compared to the common laser output frequencies.

of semiconductor, wavelength of operation and the detailed design of the switching device.

3.2. EXPERIMENTAL TECHNIQUES

Our essential requirement for an all-optical device is a material which provides some nonlinear dependence of output level on the intensity of the incident light. The basic experiment is to monitor the optical power either transmitted by or reflected from the semiconductor as a function of input power. However, many techniques are available which can characterize the many subtleties of light–material interactions. Figure 3.2 illustrates the principal methods.

Two-beam, excite–probe measurements can be carried out in the time or

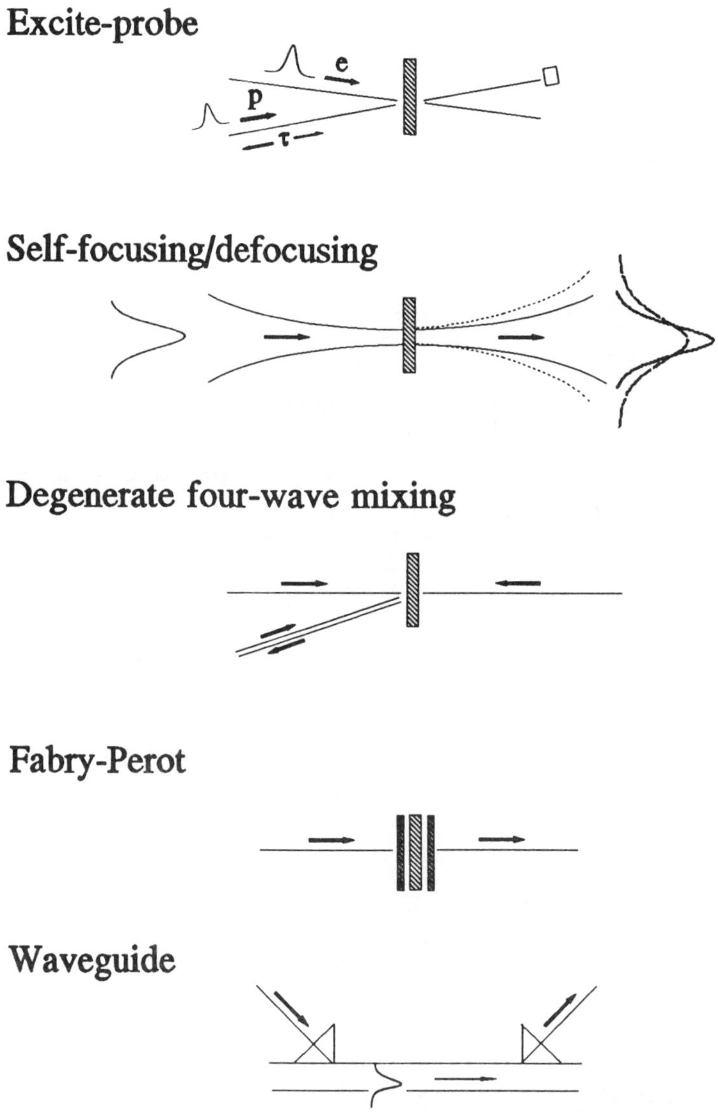

Figure 3.2 Techniques for studying dynamical optical nonlinearities in semiconductors.

frequency domain to determine time constants associated with the carrier dynamics in resonant nonlinear optical interactions. Time-resolved measurements are more popular because of advances in mode-locked lasers which now provide pulse durations down to a few femtoseconds. Time resolution is achieved by altering the arrival time of the probe pulse by incorporating a variable, spatial time delay. Recombination times, scattering rates, trapping times and carrier sweep-out rates can be measured.

Degenerate four-wave mixing has the advantage of being a sensitive interferometric technique for the determination of nonlinear optical constants and can also provide temporal information. Two beams interfere to produce a spatial modulation in optical properties and a probe beam diffracts from this temporary grating (Eichler *et al.*, 1986). The grating may be either absorptive or refractive. By employing ultrashort pulses and ensuring that the two excitation pulses coincide in time and space in the sample, arrival of the probe pulse can be time delayed to monitor the dynamics of the grating. The decay of the grating can proceed by either recombination and/or carrier diffusion thus yielding spatial information and a direct measure of carrier diffusion rates.

Spatial distortions of laser beams were used in the earliest observations of band gap resonant optical nonlinearities in narrow gap semiconductors (D. A. B. Miller *et al.*, 1978). Analysis of near or far field beam shapes provides values of phase changes and nonlinear refraction in the nonlinear medium. Recent variations of this technique have shown that the measurement sensitivity can approach interferometric methods (Sheik-Bahae *et al.*, 1990b).

By placing the nonlinear medium within a Fabry–Pérot interferometer, feedback is provided such that optical bistability can be achieved (Karpushko and Sinitsyn, 1978; D. A. B. Miller *et al.*, 1979; Gibbs *et al.*, 1979; Klingshirn and Haug, 1981). The critical switching conditions for optical bistability have proved testing for models of semiconductor optical nonlinearities and much detailed information has been gathered from these types of measurements.

An alternative configuration employs waveguides (LiKamWa *et al.*, 1985; Stegeman and Wright, 1990). Coupling into a waveguide is sensitive to refractive index and thus nonlinear refraction when the intensity dependence is measured. Directional coupler switching devices employing two closely spaced waveguides give a measure of refractive index changes. Four-wave mixing and optical bistability have also been studied in a waveguide format.

3.3. OPTICAL EXCITATION

3.3.1. Linear absorption

A survey of most semiconductors of interest reveals a common band structure consisting of a lowest conduction band having a predominantly atomic s-like zone centre wavefunction, and three p-like upper valence bands consisting of heavy hole, light hole and spin–orbit split bands. Figure 3.3 shows interband transitions in reciprocal (wavevector) space which lead to absorption of radiation in semiconductors. Vertical arrows represent absorption of a photon (vertical because the wavevector of the photon is small), while other lines

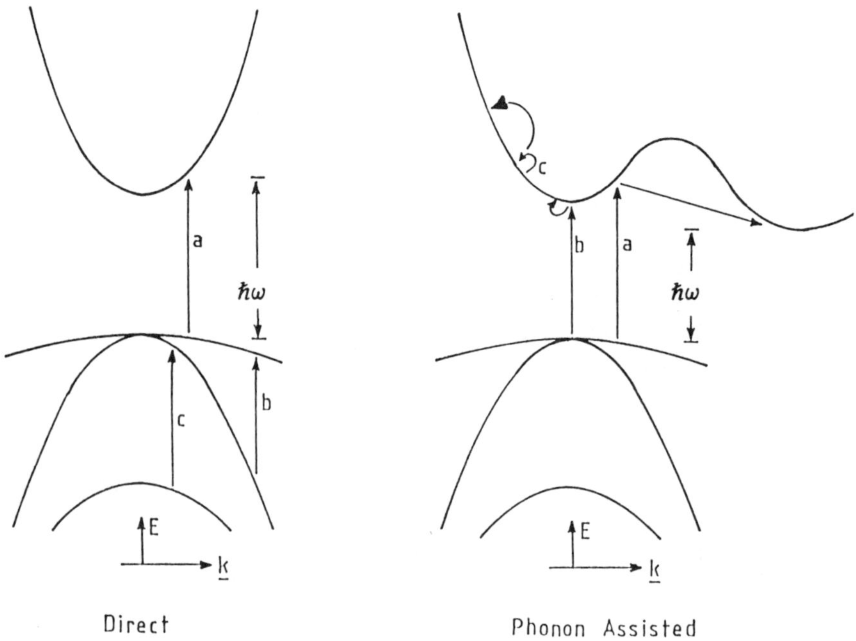

Figure 3.3 Direct and indirect optical transitions in semiconductors.

indicate phonon assisted (indirect) transitions. Direct, single-photon transitions occur between an initially full valence band and an empty conduction band for photon energies in excess of the band gap. Interband absorption in semiconductors is calculated from the first-order perturbation of the single-electron Hamiltonian by the radiation:

$$H(t) = \frac{e}{mc}\mathbf{A}\cdot\mathbf{p} \tag{3.2}$$

where **A** is the vector potential of the radiation and **p** is the electron momentum. e is the electron charge, m the free electron mass and c the velocity of light. The transition rate, $W(\omega)$, is calculated using Fermi's golden rule for optically coupled pairs of valence and conduction band states integrated over all possible vertical transitions in the first Brillouin zone taking account of all contributing bands at the light frequency, ω. Thus,

$$W(\omega) = \frac{2\pi}{\hbar}\left(\frac{2\pi e^2 I}{ncm^2\omega^2}\right)\sum_{v,c}\int\frac{\mathrm{d}^3\mathbf{k}}{(2\pi)^3}|\xi\cdot\mathbf{M}_{c,v}(\mathbf{k})|^2\,\delta[E_c(\mathbf{k}) - E_v(\mathbf{k}) - \hbar\omega] \tag{3.3}$$

where $\mathbf{M}_{c,v}(\mathbf{k})$ is a momentum matrix element for a given transition, often assumed slowly varying in **k** for valence to conduction band transitions, ζ is a unit vector representing the polarization of the light, and n is the refractive index.

The absorption coefficient, α, defined through the Beer's law dependence of intensity, I, on propagation distance, z, is given by

$$\alpha(\omega) = -I^{-1}\frac{dI}{dz} = \frac{\hbar\omega}{I}W(\omega) \tag{3.4}$$

The absorption coefficient for valence to conduction band transitions including contributions from both the heavy and light hole bands given by $\mathbf{k}\cdot\mathbf{p}$ theory is

$$\alpha(\omega) = \frac{4\sqrt{2}e^2}{3ncm\hbar^2}\left(\frac{mP^2}{\hbar^2}\right)\frac{(\hbar\omega - E_g)^{1/2}}{\hbar\omega}(m_{rh}^{3/2} + m_{rl}^{3/2}) \tag{3.5}$$

where P is the Kane momentum matrix element which is essentially constant for zinc blende structured semiconductors, and m_{rh} and m_{rl} are the reduced masses of heavy and light holes with the conduction electrons respectively.

In principle, Coulomb contributions should be taken into account for all interband transitions since electron–hole pairs are being created at the same location in real space. Excitonic effects can appear as absorption features just below the fundamental edge of semiconductors and as an enhancement of the absorption coefficient above the edge. Clearly resolved excitonic absorption features are seldom observed at room temperture in bulk III–V semiconductors since the excitons are effectively thermalized. For instance, in narrow gap semiconductors such as InSb, the exciton binding energy is small which implies large orbital radii such that Coulomb interactions are effectively screened at very moderate background carrier densities. This also applies to below-gap exciton absorption features in larger gap semiconductors; however, above-edge excitonic contributions produce major modifications to the absorption in materials such as GaAs. The theory developed by Elliot gives a description of the continuum Coulomb absorption enhancement of the form

$$\alpha_{ex}(\omega) = \frac{\pi C R^{1/2} e^{\pi\gamma}}{\sinh(\pi\gamma)} \tag{3.6}$$

where C is defined by equation 3.5 if we write $\alpha(\omega) = C(\hbar\omega - E_g)^{1/2}$,

$$\gamma = \left(\frac{R}{\hbar\omega - E_g}\right)^{1/2} \tag{3.7}$$

and R is the hydrogen atom Rydberg modified to account for the reduced effective mass, m_r, and dielectric constant, ε,

$$R = \frac{m_r e^4}{2\hbar^2\varepsilon^2} \tag{3.8}$$

Direct interband transitions are also possible between the three valence bands in the p-doped semiconductors and observed in the 10 μm wavelength region for light to heavy hole transitions. The like parity of the light and heavy hole wavefunctions makes these transitions weak except at larger

k-values where s and p intermixing of the wavefunctions increases, resulting in matrix elements approximately proportional to the electron wave-vector, **k**.

Free carriers, whether created by doping or optical excitation, have a significant influence on optical properties. Optically generated carriers are employed in a number of nonlinear optical devices but they also lead to detrimental losses in other devices. Absorption by conduction band electrons requires the participation of a phonon to satisfy momentum conversion within a single band and is therefore relatively weak. Typical absorption cross-sections for electrons are in the range $\sigma_e \sim 10^{-17}$–$10^{-18}\,\mathrm{cm}^2$. Absorption due to optically excited free carriers is normally dominated by holes since intervalence band transitions are possible without the need for a third particle. Absorption cross-sections for holes are typical in the range $\sigma_h \sim 10^{-15}$–$10^{-17}\,\mathrm{cm}^2$.

3.3.2. Multiphoton absorption

Two-photon absorption (TPA) provides a means of generating large densities of carriers in semiconductors at wavelengths in the transparency regime. Large, relatively uniform, carrier densities can be created at high laser intensities (typically in excess of $100\,\mathrm{kW\,cm^{-2}}$) resulting in changes in absorption and refractive index. TPA can be treated by second-order perturbation theory. Although in general the two photons can have different frequencies, electron–hole pair generation requires only one laser with a photon energy larger than half the band gap. In this case, the transition rate is given by,

$$W(\omega) = \frac{2\pi}{\hbar}\left(\frac{4\pi^2 e^4 I^2}{n^2 c^2 m^4 \omega^4}\right)\sum_c \int \frac{\mathrm{d}^3\mathbf{k}}{(2\pi)^3}\left|\sum_t \frac{\zeta\cdot\mathbf{M}_{ct}(\mathbf{k})\,\zeta\cdot\mathbf{M}_{ti}}{E_t - E_v - \hbar\omega}\right|^2 \delta[E_c(\mathbf{k}) - E_v(\mathbf{k}) - 2\hbar\omega] \tag{3.9}$$

where t is an intermediate (virtual) state in any valence or conduction band, such that momentum is conserved for each individual transition, but states with energies far from the lowest conduction band and upper valence band will only have a weak contribution because of the resonant denominator. For instance, a model including the lowest conduction band and three valence bands gives 12 possible intermediate states for any pair of valence and conduction band states (Lee and Fan, 1974; Pidgeon *et al.*, 1979; Weiler, 1981). Examples of possible two-photon transitions are shown in Figure 3.4.

The degenerate two-photon absorption coefficient, β, is given by the relation,

$$\beta(\omega) = \frac{2\hbar\omega}{I^2} W(\omega) \tag{3.10}$$

Since free carriers are invariably generated by the two-photon absorption

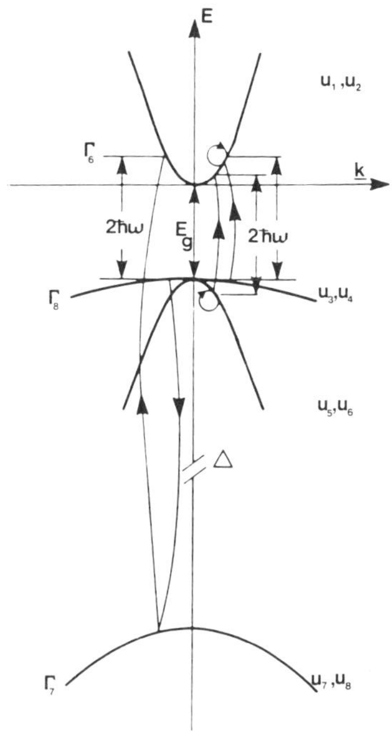

Figure 3.4 Transitions contributing to two-photon absorption in semiconductors (Pidgeon *et al.*, 1979).

process, it is important to include the effects of free carrier absorption, thus,

$$-\frac{\mathrm{d}I}{\mathrm{d}z} = \alpha I + \beta I^2 + (\sigma_e + \sigma_h)\Delta N \tag{3.11}$$

TPA is a nonlinear process in which interband transitions are completed by the absorption of pairs of photons and can be related to the imaginary part of $\chi^{(3)}$. Interband TPA is actually forbidden at the zone centre of most common semiconductors because the wavefunctions of the band extrema at zero electron wavevector have opposite parity; however, transitions to states above the minimum energy gap become allowed because of mixing of the s- and p-like wavefunctions at finite wavevectors.

TPA coefficients have proved difficult to measure accurately. Measurements are sensitive to spatial and temporal beam quality, generated free carriers give a significant absorption contribution and nonlinear refraction causes defocusing of the transmitted beam. However, measurements which take account of the above effects now give very good agreement with a theoretical model based on $\mathbf{k}\cdot\mathbf{p}$ theory over a large range of semiconductors. A two-band $\mathbf{k}\cdot\mathbf{p}$ model gives the frequency dependence of the degenerate TPA absorption coefficient as (Wherrett, 1984)

$$\beta = K \frac{P}{n_0^2 E_g^3} F_2\left(\frac{\hbar\omega}{E_g}\right) \tag{3.12}$$

where,

$$K = \frac{2^9 \sqrt{2} \pi e^4}{\hbar c^2} \tag{3.13}$$

and

$$F_2(x) = \frac{(2x-1)^{3/2}}{(2x)^5} \tag{3.14}$$

This relation has been generalized for higher-order multiphoton absorption (Wherrett, 1984). Measured TPA coefficients in a large number of semiconductors are compared with this model in Figure 3.5 (VanStryland *et al.*, 1985).

Biexcitons also offer potential for large optical nonlinearities via two-photon absorption. Excitonic molecules or biexcitons are created by two photon excitation, $2\hbar\omega = E_{Bi} = 2E_{ex} - B$ where E_{ex} and E_{Bi} are the energies of the

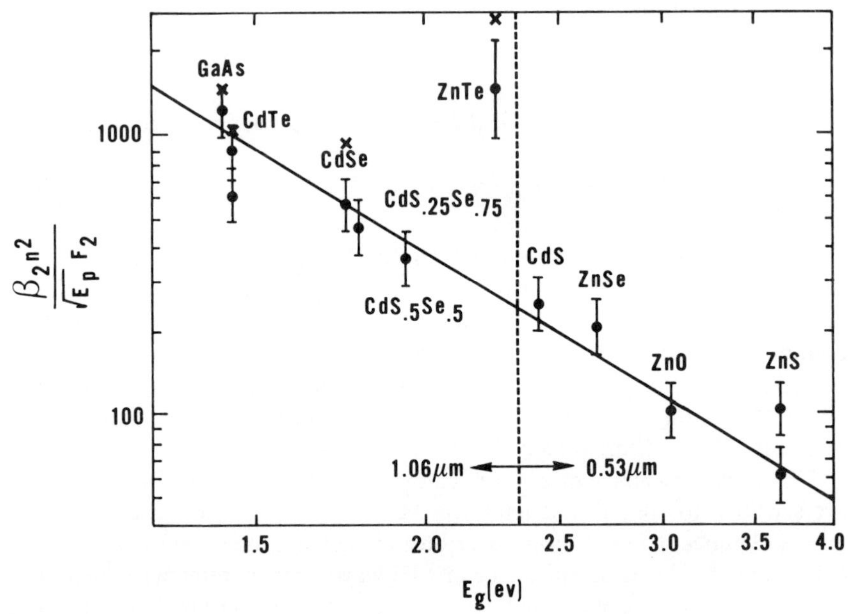

Figure 3.5 Measured two-photon absorption coefficients in semiconductors compared to the theoretical prediction given by eq. 3.12 (Van Stryland, 1988).

exciton and biexciton respectively and B is the binding energy of the two excitons in the molecule. Two-photon transitions to the biexciton state are always allowed if the exciton is dipole active, and thus the biexciton is observed below the single exciton energy. Since the biexciton transition has a large dipole moment resonantly enhanced by the single exciton providing an intermediate state, the biexciton has a large nonlinearity associated with it. Most studies of biexciton optical nonlinearities have been in CuCl and CdS (Klingshirn and Haug, 1981; Kalt *et al.*, 1986).

3.3.3. Carrier dynamics

Consider now a direct valence to conduction band transition which initially creates an electron in some state above the bottom of the conduction band leaving a hole in a state below the top of the valence band. The electron and hole are likely to be rapidly scattered by one of three interactions, (a) ionized impurities, (b) phonons, or (c) carrier–carrier collisions. These scattering processes ensure that the electron and hole will remain in their initial excited states for less than a picosecond. Even relatively pure semiconductors show ionized impurity scattering rates on the order of a picosecond. Optical phonons also exhibit scattering rates of around a picosecond at room temperature. Carrier–carrier scattering depends on concentration and therefore on the level of excitation. Scattering times can reach 100 fs for carrier densities of around $10^{17}\,cm^{-3}$, which is a typical level of excitation in an all-optical devices employing free carrier optical nonlinearities.

Ionized impurity scattering elastically randomizes the momentum of electrons and holes in reciprocal space. Carrier–carrier scattering produces a thermalized Fermi-like distribution with a characteristic temperature above that of the lattice for both carrier types. (Carriers may also be cooled below the lattice temperature by excitation to excess energies less than the lattice temperature.) Energy exchange with phonons cools (or heats) the distribution to thermal equilibrium with the lattice. Complete thermalization of the carriers to the lattice temperature will typically take on the order of a few picoseconds at room temperature. This time scale may be altered by screening of some of the interactions at high carrier densities or by cooling bottlenecks. For instance, energy transfer from hot carriers to optical phonons may change significantly at high excitation because optical to acoustic phonon relaxation is slow, and thus the temperature of the optical phonon distribution may rise to the carrier temperature thus reducing the energy transfer rate. In addition, screening of the electron–phonon interaction may be effective when zone centre (long wavelength) phonons are involved in the cooling process.

Carrier recombination takes place on longer time scales via both radiative and nonradiative routes. This is typically nanoseconds or longer but depends on many factors including the specific band structure details, crystal quality, doping, temperature, and excitation level. Time scales can thus vary not only from one semiconductor to another but also within one crystal depending on the specific conditions. During this time scale, the excess carriers can move significant distances from regions of high density towards low density at velocities determined by the mobility of the carriers. Since optical excitation creates equal numbers of electrons and holes, the carrier motion is characterized by the ambipolar diffusion coefficient, D_a, which is dominated by the slower hole mobility. The diffusion length defined by the distance the carriers can diffuse before they recombine is given by

$$L_D = (D_a \tau_R)^{1/2} \tag{3.15}$$

where τ_R is the carrier recombination time. This parameter may be compared with the relevant dimension within the sample or device to determine whether local or nonlocal conditions may be assumed. This parameter may be the laser spot in a single-beam experiment, the cross-section of the mode in a waveguide, or the distance between fringes in a transient grating or phase conjugation configuration. The diffusion length also determines how closely devices can be packed without interference.

3.4. BOUND ELECTRON NONLINEARITIES

Modulation of the polarization of a material at optical frequencies demands an essentially instantaneous response (bound electron interaction). These nonlinearities are characterized by susceptibilities defined by expanding the induced polarization in terms of the electric field.

The third-order susceptibility describes four-wave mixing phenomena such as third harmonic generation and also accounts for nonlinear refraction, two-photon absorption, stimulated Raman scattering and absorption saturation through real and imaginary parts.

$$\mathbf{P} = (\chi^{(1)} + \chi^{(3)}|\mathbf{E}|^2)\mathbf{E} \tag{3.16}$$

(Note here we use Gaussian units consistent with the literature on this topic.)

Under nonresonant conditions, the nonlinear refraction can be viewed in terms of virtual, non-energy-conserving transitions whereby an excited population exists only as long as the optical field is present. During their brief existence in the excited state, electrons remain coherent with the light field and induce changes in the optical constants of the material. Nonlinear refraction defined to third order is known as the optical Kerr effect and described by a parameter n_2,

$$n = n_0 + \Delta n = n_0 + n_2 I \tag{3.17}$$

where the intensity of the light, I, is related to the field amplitude, $\mathbf{E}_0$, by

$$I = \frac{c n_0 |\mathbf{E}_0|^2}{8\pi} \tag{3.18}$$

We can relate n_2 to the real part of a total degenerate, $\chi^{(3)}$, via the relation,

$$n_2 = \frac{8\pi^2 \operatorname{Re}[\chi^{(3)}]}{cn_0^2} \tag{3.19}$$

An instantaneous contribution to nonlinear refraction arises as a direct consequence of TPA (Sheik-Bahae *et al.*, 1990a). This may be understood as follows. The first photon in the two-photon interaction causes the transition of an electron to a virtual intermediate state which induces an absorption change given by $\Delta\alpha = \beta I$. The corresponding alteration in refractive index can then be calculated by carrying out a Kramers–Kronig integration, equation 3.1, if the frequency response of the two-photon absorption coefficient is known. When the nondegenerate version of equation 3.12 is employed (within a two-band TPA model), a Kerr coefficient, n_2, can be derived,

$$n_2 = K\frac{\hbar c P}{n_0^2 E_g^4} G_2\left(\frac{\hbar\omega}{E_g}\right) \tag{3.20}$$

where the dispersion function, G_2, is given by

$$G_2(x_2) = \frac{2}{\pi}\int_0^\infty \frac{F_2(x_1;x_2)\,\mathrm{d}x_1}{x_1^2 - x_2^2} \tag{3.21}$$

A full description requires inclusion of Raman terms and linear and quadratic Stark effects (Sheik-Bahae *et al.*, 1991). Different functional forms of G_2 are obtained for these processes which can be combined to give the dispersion of n_2 as shown in Figure 3.6. The quadratic Stark effect makes a significant contribution to the calculation at photon energies near resonance with the band edge. This model gives excellent agreement with measured values of bound electronic nonlinear refraction, not only in a wide range of semiconductors, but also for other large gap solid state materials as shown in Figure 3.6. It is important to realize that this contribution to the refractive index extends to photon energies of less than half the band gap energy even although there is no TPA, just as linear refraction below the gap may be calculated by including (virtual) transitions at frequencies above the edge. The band gap dependence of the nonlinear refraction is found to be consistent with $n_2 \propto E_g^{-4}$. There is a sign change from positive to negative predicted at around 0.7 of the band gap consistent with measurements (note in particular the two values for ZnSe at 1.06 and 0.53 μm), and there is a peak in the nonlinear refraction predicted at about half the band gap energy.

The significance of these results is that the bound electronic nonlinear refraction can be accurately predicted along with the TPA coefficient. Figures of merit for optical switching depend on the magnitude of the nonlinear refraction divided by the absorption which we might assume is TPA under nonresonant conditions. Thus, the optimum wavelength range for the operation of optical switching devices such as bistable elements and waveguide switches

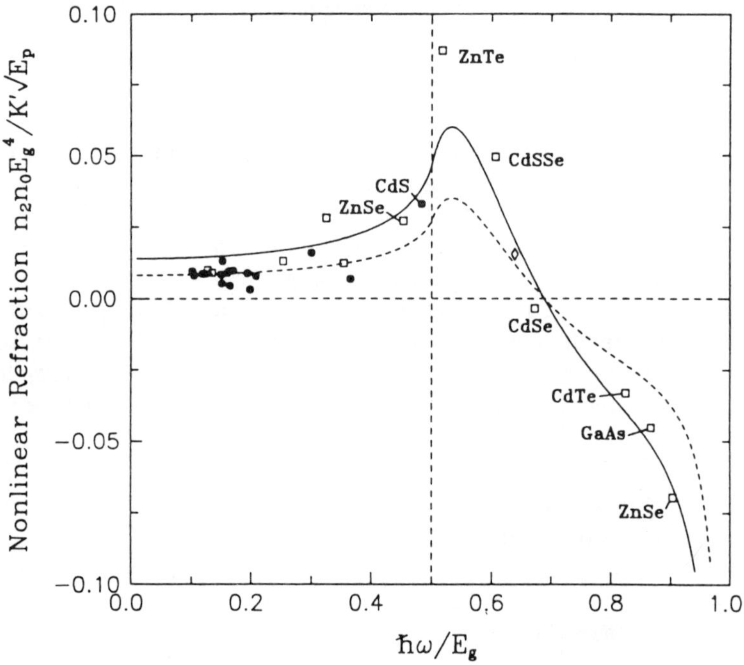

Figure 3.6 Dispersion of n_2 given by equation 3.20 as compared to measured values. (Sheik–Bahae, 1991).

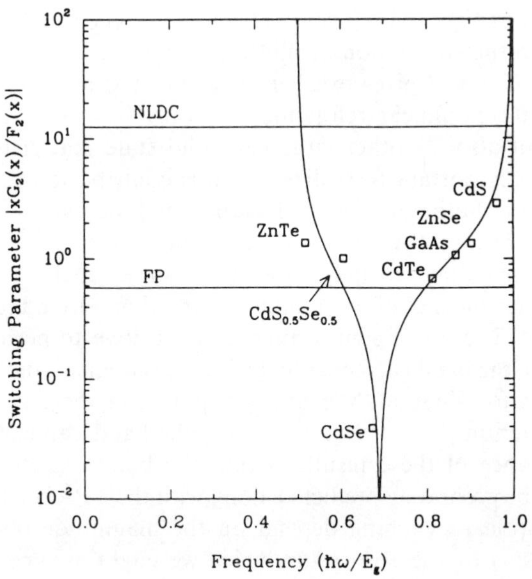

Figure 3.7 Predicted figure of merit for switching in Fabry–Perot and directional coupler based on *k.p* model for n_2 and two photon absorption (Sheik–Bahae, 1991).

(DeLong and Stegeman, 1990) can be determined. Figure 3.7 shows switching parameters for Fabry–Perot devices and directional couplers with results for several semiconductors (Sheik-Bahae *et al.*, 1991). This shows that only frequencies below half the band gap energy and frequencies close to resonance are attractive for optical switching. However, band tail absorption and resonant nonlinearities will dominate close to resonance and thus effectively mask the faster optical nonlinearities. Away from resonance, three-photon and multiphoton induced free carrier absorption may be significant.

3.5. FREE CARRIERS

Early research on free carrier optical nonlinearities in semiconductors concentrated on the effects of nonparabolicity of conduction band electrons (Wolff and Pearson, 1966). In n-doped, narrow gap semiconductors with small effective masses which exhibit a large degree of nonparabolicity, very large third-order nonlinearities were used to demonstrate four-wave mixing at wavelengths around 10 μm (Patel *et al.*, 1966). However, a major breakthrough for optical switching was the realization that optical excitation of semiconductors close to the band gap energy could produce very sensitive nonlinear optical effects due to the generation of free carriers (D. A. B. Miller *et al.*, 1978). Excess free carriers can be efficiently photogenerated by absorption into the band tail states with subsequent scattering into the band. Alternatively, multiphoton absorption may be employed to generate the carriers. We may characterize these optical nonlinearities as either absorptive or refractive; in this case, the relationship between changes in absorption and refractive index is given by the Kramers–Kronig relation, equation 3.1, with $\xi = \Delta N$, the generated free carrier density.

These 'real' carrier populations result in intensity-dependent changes in the first-order susceptibility, $\Delta\chi^{(1)}$. This situation is most conveniently described by transition probabilities combined with rate equations, describing the dynamics, but, since the intensity of the light is proportional to the square of the electric field, an effective third-order susceptibility, $\chi^{(3)}_{\text{eff}}$, is often quoted to enable comparison of the magnitude of different types of third-order optical nonlinearities, i.e.

$$\mathbf{P}(\omega) = (\chi^{(1)} + \chi^{(3)}_{\text{eff}}|\mathbf{E}(t)|^2)\mathbf{E}(\omega) = (\chi^{(1)} + \Delta\chi^{(1)})\mathbf{E}(\omega) \tag{3.22}$$

Nevertheless, care must be taken with the use of the effective susceptibility approach if the density of excess carriers is not proportional to the intensity. For instance, Kerr-type nonlinear refraction cannot be assumed if the carrier lifetime is density dependent or the absorption is nonlinear at the excitation wavelength.

Resonant optical nonlinearities may also be distinguished as being local or nonlocal. In the former case, the material excitation is limited to the location of the optical field causing the excitation, i.e. any spreading of the excitation is small compared with the laser spot size or fringe spacing for two or more beams. Under nonlocal conditions, the excitation (electronic or thermal) may diffuse to adjacent regions; therefore, the experimental conditions together with material properties determine whether local or nonlocal conditions apply. At high excitation levels, the ambipolar diffusion coefficient will change as the semiconductor moves from nondegenerate to degenerate conditions.

The local excess carrier density, ΔN, is given by the continuity equation for each carrier type. For completeness, we include single- and two-photon absorption, density-dependent recombination and diffusion

$$\frac{\mathrm{d}N}{\mathrm{d}t} = \frac{\alpha I}{\hbar\omega} + \frac{\beta I^2}{2\hbar\omega} - \frac{\Delta N}{\tau_\mathrm{R}} - r_1\Delta N^2 - r_2\Delta N^3 + \nabla\cdot(D_\mathrm{a}\nabla\Delta N) \tag{3.23}$$

where r_1 and r_2 give bimolecular and Auger recombination rates respectively.

3.5.1. Dynamic nonlinear absorption

Generated free carriers may alter the optical constants of a semiconductor via several contributions. Depending on the conditions, absorption can increase or decrease with increasing carrier density. Saturation of interband absorption can be attributed to either state-filling or band-filling mechanisms which may be distinguished by response time. Both effects result from Pauli exclusion, in other words, blocking of interband transitions because of

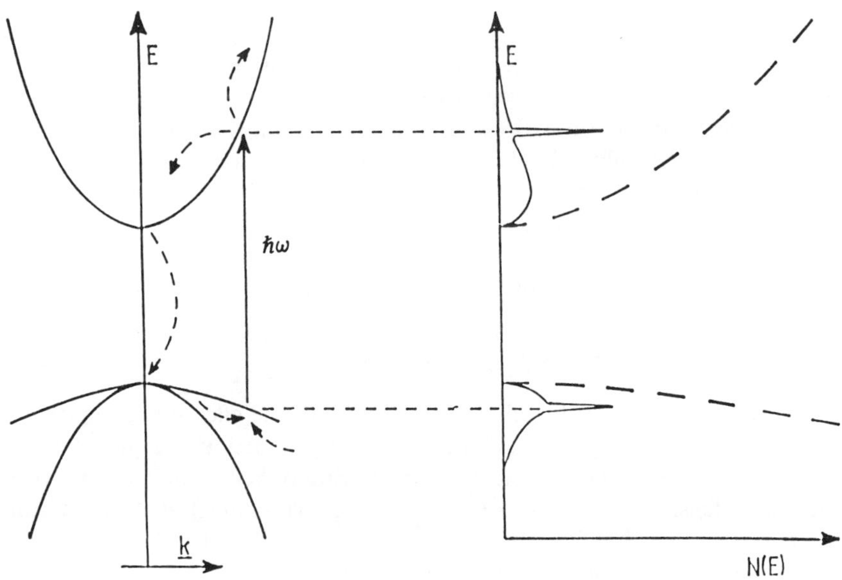

Figure 3.8 Illustration of state-filling absorption saturation.

electrons occupying the optically coupled states. State filling describes blocking of only the optically coupled energy levels thus producing a spike in the density of filled states, as illustrated in Figure 3.8. The frequency width of the saturation (or hole burning) is determined by the momentum or phase relaxation time, T_2. Because scattering rates in semiconductors are so fast, state-filling saturation relaxes very rapidly, typically on subpicosecond timescales, and therefore requires high power, ultrafast pulses for observation (Smirl *et al.*, 1982; Tang and Erskine, 1983; Oudar *et al.*, 1984; Lin *et al.*, 1988). Employing polarized light induces an anisotropic saturation since states in specific directions of **k** space will be accessed (Smirl *et al.*, 1982, 1983).

State-filling absorption saturation has also been observed in p-type semiconductors in which the transitions take place between the light and heavy hole valence bands (Gibson *et al.*, 1972; Keilman, 1976), (Figure 3.9). Relaxation is by emission of phonons in this case, and is of the order of 1 ps. This fast nonlinearity has been studied using pump–probe measurements in the frequency regime and employed to mode lock CO_2 lasers by placing a slice of p-type germanium within a laser cavity.

Absorption saturation is more easily achieved by band filling which results from the thermally relaxed population (Figure 3.10) (Nurmikko, 1976a,b). In this case, scattering redistributes optically excited carriers to the lowest energy levels causing a blue shift of the absorption edge (dynamic Moss–Burstein effect (Moss, 1980)). Thus, a quasi-equilibrium of electrons and holes results in normalized material parameters. This provides a much larger nonlinear

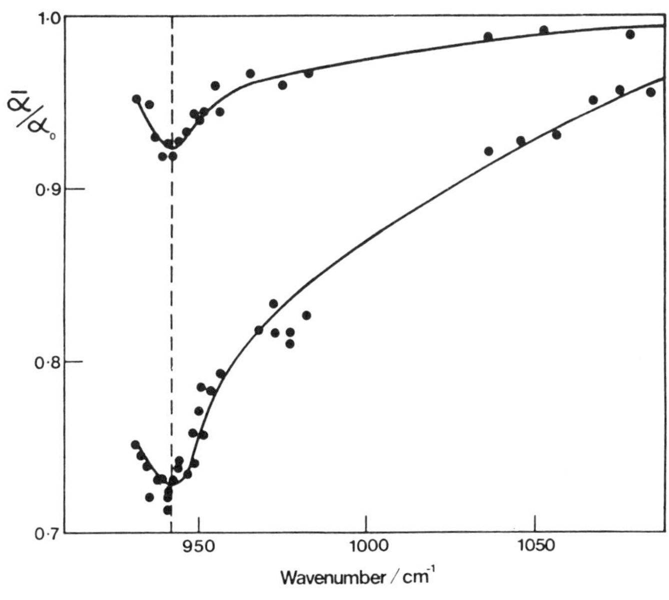

Figure 3.9 State-filling saturation in *p*-type germanium (Keilman, 1976).

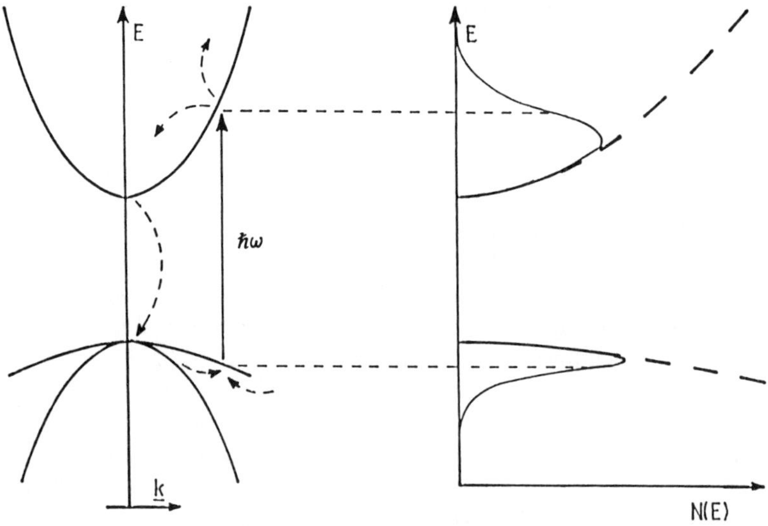

Figure 3.10 Illustration of band-filling absorption saturation.

optical response at the expense of a slower relaxation time determined by carrier recombination, τ_R, or, in some instances, diffusion or carrier sweep-out rates. Absorption changes associated with band filling can be estimated starting with the expression for linear absorption and taking account of the probability of a state being filled via the thermalized electron and hole Fermi population factors, $f_e(E_c)$ and $f_h(E_v)$ for the optically coupled conduction band state, E_c, and valence band state, E_v. Thus,

$$\alpha(\hbar\omega) = \alpha_0(\hbar\omega)[1 - f_e(E_c) - f_h(E_v)] \tag{3.24}$$

where α_0 is the absorption coefficient for filled (empty) valence (conduction) bands. The Fermi population factors are determined from the three parameters, carrier temperature, carrier density and quasi-fermi level for each carrier type. Thus absorption changes and other related free carrier optical nonlinearities depend parametrically on electron–hole density rather than the laser field strength *per se*. Assuming that the hole effective mass is much greater than the electron mass, the absorption change as a function of carrier density, N, is given by

$$\frac{d\alpha(\hbar\omega)}{dN} = -\alpha_0 \frac{d f_e(E_c)}{dN} \tag{3.25}$$

Within the Boltzman approximation,

$$\frac{d f_e(E_c)}{dN} = \frac{1}{N_c} \exp\left(\frac{E_g - E_c}{kT}\right) \tag{3.26}$$

where

$$N_c = 2\left(\frac{m_c^* k T}{2\pi\hbar^2}\right)^{3/2} \tag{3.27}$$

Band filling optical nonlinearities can be observed under CW laser excitation and are well understood in narrow gap semiconductors which have small effective masses, m_c^*. It becomes necessary to consider the role of Coulomb effects in larger band gap semiconductors. The primary interest in excitons follows from large changes in absorption and refractive index associated with saturation of bound exciton absorption features. These features are better resolved in the more ionic II–VI semiconductors and in III–V low dimensional structures. For example, saturation of bound excitons in II–VI semiconductors can provide very low saturation intensities. Figure 3.11 shows saturation of CdS at a temperature of 2 K where a saturation power of only 3.6 μW (58 W cm^{-2}) was deduced for this inhomogeneously broadened exciton absorption (Dagenais, 1983). This system decays radiatively in 500 ps,

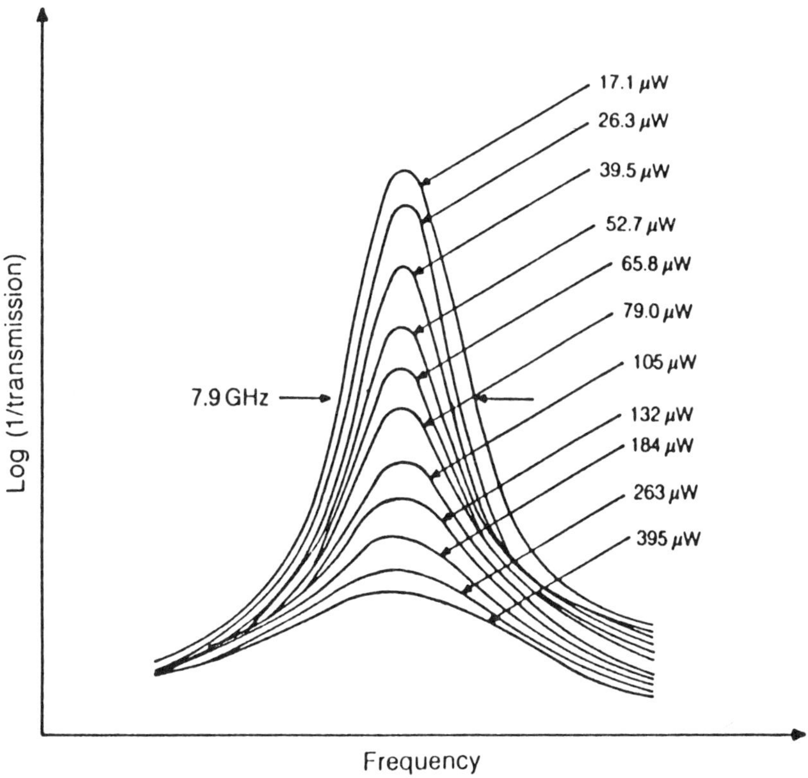

Figure 3.11 Saturation of the absorption of the bound exciton in CdS (Dagenais, 1983).

providing a sensitive and fast optical nonlinearity which allows optical bistability.

The above-edge Coulomb absorption enhancement is also screened at high carrier densities and needs to be incorporated in any model of free carrier optical nonlinearities in large gap semiconductors. Banyai and Koch (1986) have developed a successful, partly phenomenological theory which includes plasma screening of the Coulomb interaction. Their starting point is the choice of Hulthen potential to replace the screened Coulomb potential resulting in an absorption coefficient based on the Elliot theory of the form,

$$\alpha(\omega) = A \tanh[(B/2)(\hbar\omega - \mu)][b(\omega) + c(\omega)] \quad (3.28)$$

where μ is the chemical potential for the electron–hole system and $b(\omega)$ and $c(\omega)$ are the contributions from the bound states and the continuum states respectively. A further consequence of generating large densities of free carriers is a normalization of the band gap to lower energies (red shift). Thus photons of energy just less than the gap can experience an increase or decrease in absorption depending on whether band-filling saturation or band gap renormalization is dominant. These factors have been evaluated for several semiconductors and the results for GaAs (Lee *et al.*, 1986) are shown in Figure 3.12. Here, experimental absorption curves in Figure 3.12(a) measured by Lee *et al.* at different carrier excitation levels are compared with theory, Figure 3.12(b). The model takes into account band filling, screening of the Coulomb enhancement and band gap renormalization.

The opposite to saturation is induced absorption which may again result from a variety of different mechanisms in semiconductors. Multiphoton absorption, as already discussed, leads to increased absorption at high intensities. Induced free carriers from one- or multiphoton absorption will cause a dynamic increase in absorption. The nonlinear absorption due to multiphoton processes and generated free carriers can be distinguished by analysis of the pulse shapes (A. Miller *et al.*, 1979) or by using different mode-locked laser pulse widths (Baggess *et al.*, 1986). Under certain circumstances, the heating associated with free carrier absorption can lead to an avalanche generation of carriers (Figueira *et al.*, 1976; Jamison and Nurmikko *et al.*, 1979). Here a rapid cutting-off in transmission occurs when laser-heated free carriers are energetic enough to excite additional carriers across the band gap by carrier–carrier collisions (Figure 3.13).

3.5.2. Dynamic nonlinear refraction

The Kramers–Kronig relation, equation 3.1, shows that changes in absorption within one wavelength band will always be accompanied by some change of refractive index at other wavelengths. For example, a laser beam with a

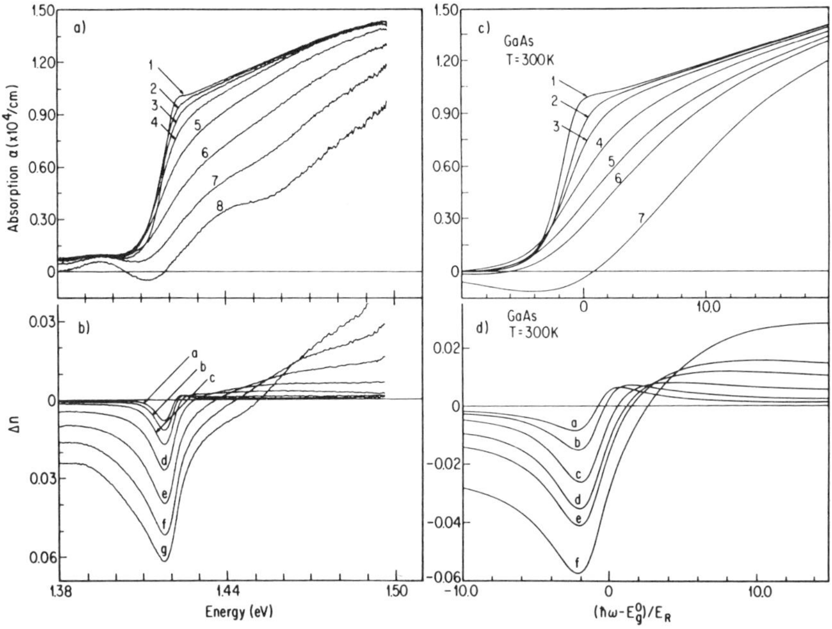

Figure 3.12 Experimental and theoretical absorption and refractive index spectra for ($T = 330$ K) GaAs at room temperature (Lee *et al.*, 1986). Figure (a) shows measured absorption spectra for different excitation powers, 1 to 8, up to 5 mW focused to a 15 μm diameter. The dispersion changes, Δn, shown in (b) are calculated from (a) using the Kramers–Kronig relations. Calculated absorption spectra are shown in (c) for carrier densities corresponding to the excitation conditions, 1 to 8 in (a), (d) shows the corresponding Kramers–Kronig calculated dispersion curves for (c).

frequency just less than the band gap energy of a semiconductor will experience a change in refractive index if a second intense beam with photon energy larger than the band gap were to cause blocking of the interband transitions. In fact the second beam is not necessary because any band tail absorption which the first beam encounters will generate free carriers from scattering into the band. Band tail excitation may thus be expected to produce absorption changes due to free carrier absorption, band filling, band gap renormalization and Coulomb screening.

The simplest contribution comes from long wavelength free carrier absorption which is successfully described by the Drude–Lorentz classical model

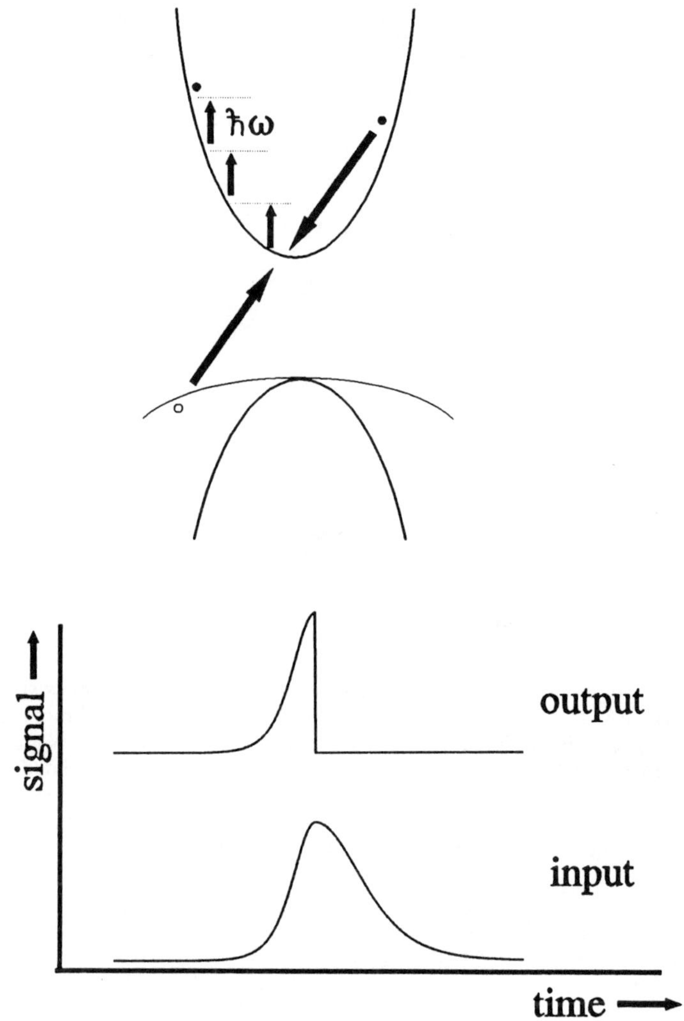

Figure 3.13 Illustration of avalanche carrier generation mechanism for n-doped semiconductors.

for a free carrier plasma modified by the effective mass of the electrons, m_c^*, of the semiconductor (Auston *et al.*, 1978). This predicts a negative refractive index change since it results from an increase in absorption at longer wavelengths, equation 3.1. Because the optical nonlinearity is strictly carrier dependent rather than light dependent, it is useful to define a refractive index change per generated free carrier per unit volume, n_p, which is independent of carrier lifetime and absorption coefficient, thus,

$$n_p = -\frac{2\pi e^2}{n_0 m_e^* \omega^2} \simeq -\frac{8\pi e^2}{3n_0}\frac{P^2}{E_g^3}\left(\frac{E_g}{\hbar\omega}\right)^2 \tag{3.29}$$

This Drude approximation is found to suffice in indirect gap semiconductors (Woerdman and Bolger, 1969; Eichler and Massman, 1982; Jain and Steel, 1980), e.g. Si and Ge, and in most instances when the photon is not too close to resonance with the band edge (e.g. two-photon generation of free carriers (Baggess *et al.*, 1986; Craig *et al.*, 1985)), such that band filling becomes dominant, and thus,

$$\Delta n = n_p \Delta N \tag{3.30}$$

The effect of blocking of interband transitions in the spectral region above the band gap energy also results in a negative refractive contribution below the edge, which can be calculated from the absorption change above the gap, equations 3.5 and 3.25, using the Kramers–Kronig integration, equation 3.1. A widely used expression is obtained for this 'band-filling' nonlinear refraction cross-section per generated carrier, n_b, assuming photon energies close to the band gap energy, and that the Boltzman approximation holds (D. A. B. Miller *et al.*, 1981),

$$n_b = -\frac{4\sqrt{\pi}e^2}{3n_0}\frac{P^2}{kT(\hbar\omega)^2}J\left(\frac{m_r}{m_c}\frac{\hbar\omega - E_g}{kT}\right) \tag{3.31}$$

where

$$J(a) = \int_0^\infty \frac{\sqrt{x}e^{-x}}{x-a}\,dx \tag{3.32}$$

$J(a)$ is a thermodynamic integral giving the resonance with the band gap and m_r is the reduced mass. The $x^{1/2}$ factor reflects the contribution of the density of band states. The $J(a)$ function is shown as the upper curve in Figure 3.14. Below the gap, the negative plasma and band-filling contributions

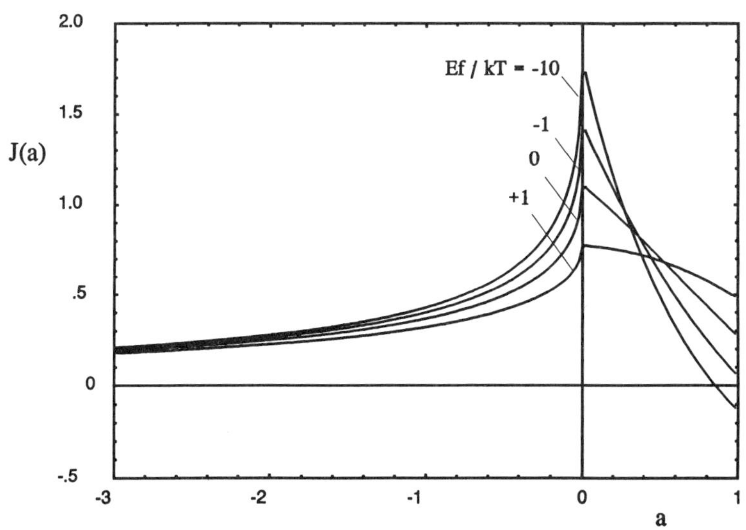

Figure 3.14 The J(a) function of equations 3.32 and 3.35.

add with the latter dominating at photon energies close to resonance. Combining both contributions with $\hbar\omega \sim E_g$ gives

$$n_2(h\omega) \simeq -\frac{8\pi e^2 P^2}{3n_0\ E_g^4}\left[1+\frac{1}{2\sqrt{\pi}}\frac{E_g}{kT}J(a)\right]\alpha(h\omega)\tau_R \quad (3.33)$$

We see that a strong band gap dependence favours narrow gap materials. The Drude term predicts $n_2 \propto E_g^{-4}$ while band filling gives $n_2 \propto E_g^{-3}$; low temperature also enhances the band-filling contribution since the electrons will be concentrated at the bottom of the band. Band-filling optical nonlinearities have been well studied in the narrow gap semiconductors InSb (D. A. B. Miller *et al.*, 1981) InAs (Poole and Garmire, 1984) and CdHgTe (Hill *et al.*, 1982). Figure 3.15 shows far field beam profiles of an initially Gaussian beam from a CW CO laser after being self-defocused in a crystal of InSb (D. A. B. Miller *et al.*, 1978). Beam distortion measurements have been used to determine band gap nonlinear refraction in InSb and successfully compared with the band-filling model (D. A. B. Miller *et al.*, 1981) (Figure 3.16). Figure 3.17 shows the effect of moving a thin slice of CdHgTe through a focus on the far field beam width of a CW CO_2 laser (Hill *et al.*, 1982). This

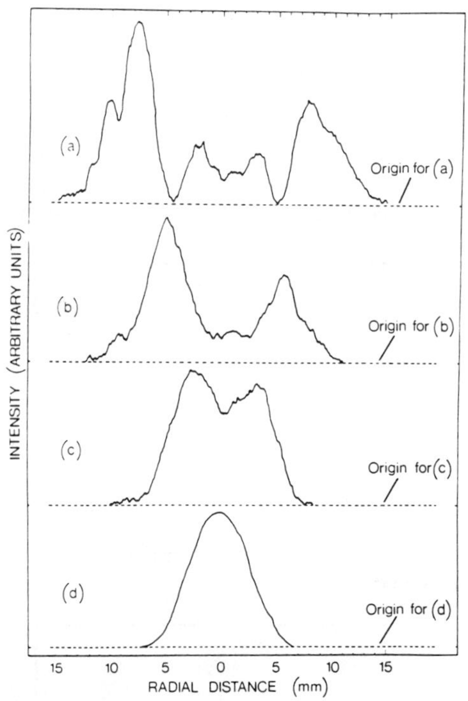

Figure 3.15 Beam profiles of a cw CO laser after transmission through InSb as a function of power due to self-defocusing (Miller *et al.*, 1978).

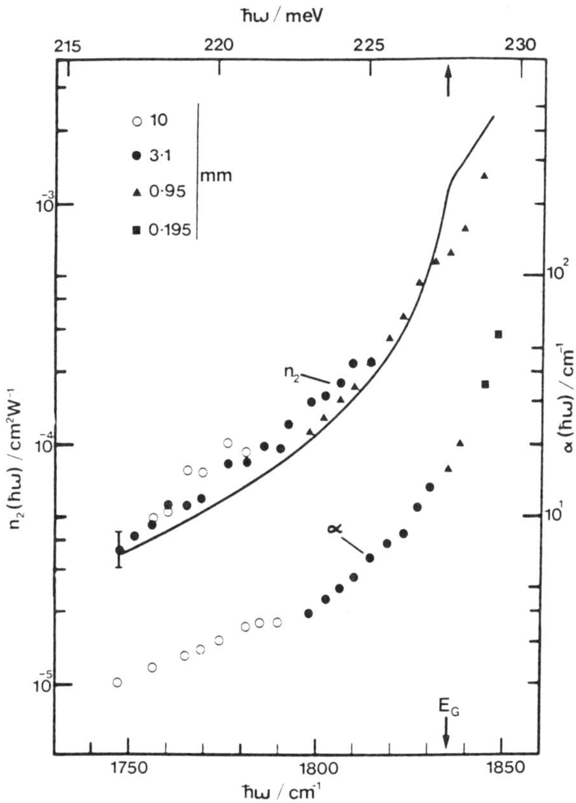

Figure 3.16 Experimental and predicted band gap resonant nonlinear refraction in InSb at 77 K (Miller *et al.* 1981).

'z-scan' technique has been refined (Sheik-Bahae *et al.*, 1990b) as a simple but very sensitive technique for determining the magnitude of nonlinear refraction in optical materials.

The light hole band can also have a significant contribution to band-filling nonlinear refraction. Including this band, $J(a)$ can be replaced by

$$J(a) \rightarrow J(a_{\mathrm{hh}}) + \frac{m_{\mathrm{rl}}}{m_{\mathrm{rk}}} J(a_{\mathrm{lh}}) \tag{3.34}$$

in equation 3.31, where a_{hh} and a_{lh} are defined by the optically coupled levels in the heavy and light hole bands respectively. Retaining the full Fermi function instead of taking the Boltzman approximation gives the result

$$J(a) = \frac{\mathrm{d}\eta_{\mathrm{fe}}}{\mathrm{d}N} \int_0^\infty \frac{x^{1/2} \exp(x - \eta_{\mathrm{fe}})\, \mathrm{d}x}{(x-a)[1 - \exp(x - \eta_{\mathrm{fe}})]^2} \tag{3.35}$$

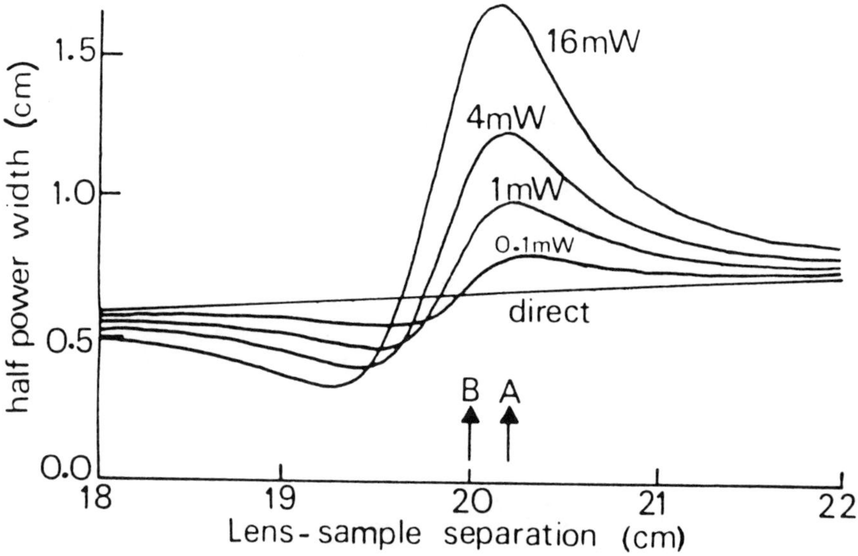

Figure 3.17 '*z*-scan' of CdHgTe at 175 K under band gap resonant conditions (Hill *et al.*, 1982).

where

$$\eta_{fe} = \frac{E_{fe} - E_g}{kT} \tag{3.36}$$

This modified function is plotted in Figure 3.14 for several values of n_{fe} within the carrier degeneracy regime, where higher values correspond to larger carrier densities.

In larger gap semiconductors, the effects of Coulomb screening and band gap renormalization need to be reconsidered. A numerical integration of the absorption changes, as discussed in section 3.5.1, using the Kramers–Kronig relation, equation 3.1, allows a prediction of the nonlinear refraction. For example, Figures 3.12(c) and 3.12(d) show the results of Kramers–Kronig integrations for the experimental and theoretical absorption curves of GaAs (Lee *et al.*, 1986). Other III–V and II–VI semiconductors have been successfully modelled in the same way.

For photon energies in excess of half the band gap energy, TPA will generate free carriers which contribute to nonlinear refraction in the same way as for single-photon transitions, i.e. Drude, band-filling, band gap renormalization and other many-body mechanisms. Ignoring dynamical effects and saturation of the nonlinearity, we may expect the two-photon-

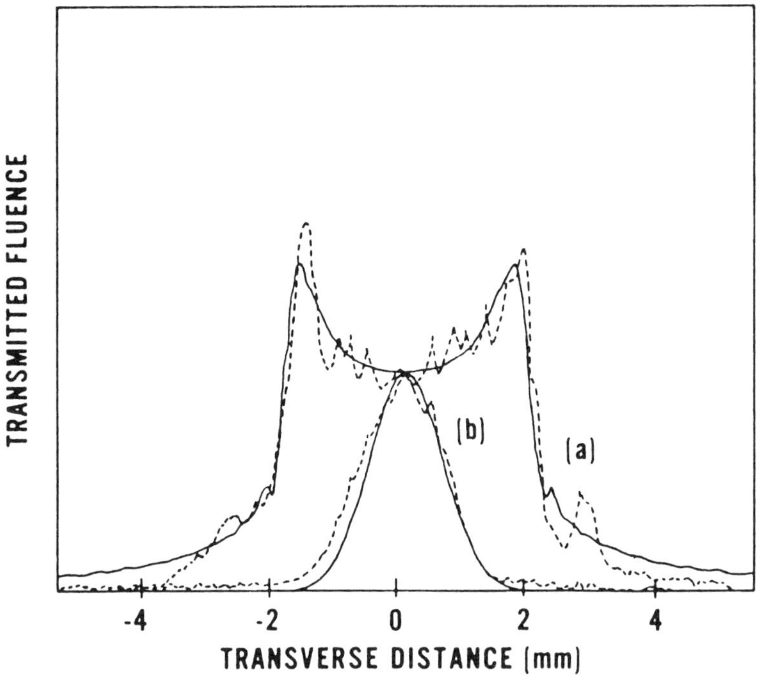

Figure 3.18 Beam distortion due to self-defocusing produced by two-photon generated carriers in CdS (Guha *et al.*, 1985).

induced free carrier nonlinear refraction to have an intensity dependence equivalent to a $\chi^{(5)}$ optical nonlinearity. This contribution has been measured in large gap II–VI semiconductors by self-focusing and degenerate four-wave mixing techniques. An example of the distortion in the spatial profile of 1.06 μm laser pulses after transmission through CdSe due to self-defocusing (Guha *et al.*, 1985) is shown in Figure 3.18. Relatively high intensities are needed to induce this effect in large gap II–VIs since the band gap dependence of both TPA and free carrier nonlinear refraction does not favour larger band gaps.

In general, if the recombination time of the carriers is independent of density, the refractive index change resulting from the free carriers generated from single- or multiphoton absorption will be given by

$$\Delta n \propto I^q \tag{3.37}$$

where q denotes the number of photons involved in the generation process and I is the laser intensity. However, this power dependence will be modified by density-dependent recombination processes. For example, Auger carrier recombination, equation 3.23, is a three-particle interaction which gives a one-third power dependence of the recombination time on carrier density when above the intrinsic density, N_i (A. Miller and Parry, 1984). Thus, at high carrier densities, equation 3.37 becomes (A. Miller and Craig, 1987)

$$\Delta n \propto I^{q/3} \quad \Delta N \gg N_i \tag{3.38}$$

Nonlinear refraction in narrow gap CdHgTe shows power dependences, q, of 1/3, 2/3 and 1 under conditions of single- and two-photon absorption, carrier density independent and Auger recombination (Craig and Miller, 1986).

3.6. OPTOTHERMAL EFFECTS

Any absorption of light invariably leads to some heating of the lattice. The resulting temperature rise can produce significant absorption and hence refractive index changes at very moderate optical power levels. Only in narrow gap semiconductors such as InSb and InAs do the electronically induced optical nonlinearities dominate over thermal effects under CW illumination. Thermal effects may be viewed as a nuisance if they compete with faster electronic effects; however, some of the most successful optically bistable devices developed for prototype all-optical circuits have exploited the sensitive thermal nonlinear optical effects in nonlinear optical interference filters based on II–VI semiconductors.

Lattice heating is produced by the dissipation of the excess energy of optically generated electrons into acoustic lattice vibrations and typically occurs on picosecond time scales. The change of optical thickness, nL, due to heating can have two components,

$$\Delta(nL) = nL\left(\frac{1}{n}\frac{dn}{dT} + \frac{1}{L}\frac{dL}{dT}\right)\Delta T \tag{3.39}$$

i.e. a contribution from the thermo-optic coefficient and a contribution from the thermal expansion. The thermo-optic effect normally dominates the thermal expansion. The thermo-optic coefficient may be considered to consist of a contribution which arises from a shift of the absorption edge plus a background component. For frequencies close to the band gap of a semiconductor, the former dominates, thus,

$$\frac{dn}{dT} = \frac{dn}{dE_g}\frac{dE_g}{dT} \tag{3.40}$$

The temperature dependence of semiconductor band gap energies results from two competing processes, the electron–phonon interaction and the effect on the band structure of changing the average spacing between the atoms.

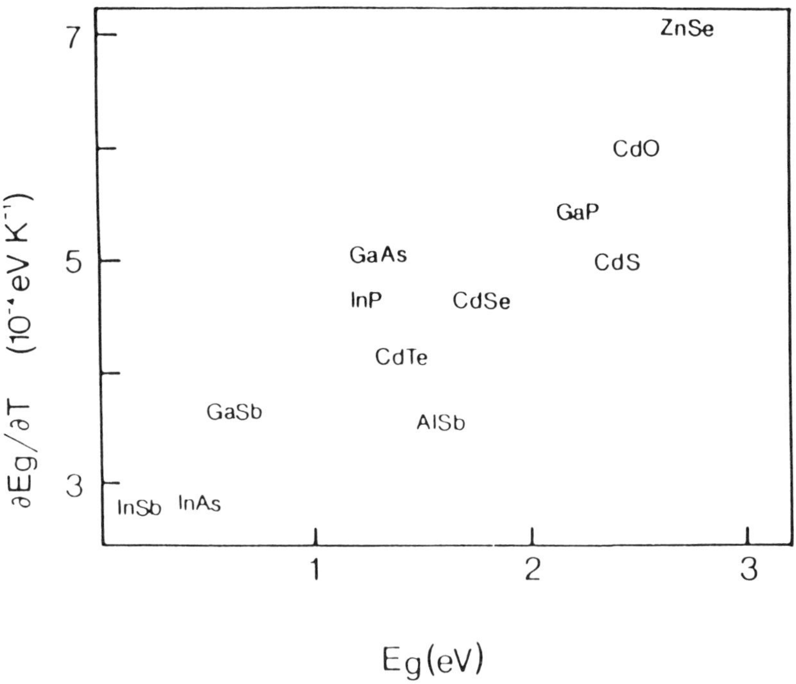

Figure 3.19 Temperature dependence of band gaps for several semiconductors (Wherrett *et al.*, 1988).

Figure 3.19 shows energy gap shifts with temperature for some III–V and II–VI crystalline semiconductors at room temperature. For most common semiconductors, the band gap shift with temperature is negative and in the range from -3×10^{-4} to -7×10^{-4} eV K^{-1}. A few semiconductors, notably narrow gap CdHgTe and lead salts such as PbSnTe, exhibit an increase in band gap with temperature.

An estimate of the refractive index change with temperature rise, ΔT, for photon energies below the band gap can be obtained from a Kramers–Kronig transformation, equation 3.1, of an absorption edge shifted to longer wavelengths by the temperature increase, i.e. $\xi = \Delta T$ in equation 3.1, which gives (Wherrett *et al.*, 1988).

$$\frac{\mathrm{d}n}{\mathrm{d}E_g} = C' \frac{e^2}{n_0 P E_g} g\left(\frac{\hbar\omega}{E_g}\right) \tag{3.41}$$

where

$$g(x) = R_e \frac{\left[4\left(2 - \sqrt{1+x} - \sqrt{1-x} + \left(\frac{1}{\sqrt{1+x}} - \frac{1}{\sqrt{1-x}}\right)x\right)\right]}{2x^2} \tag{3.42}$$

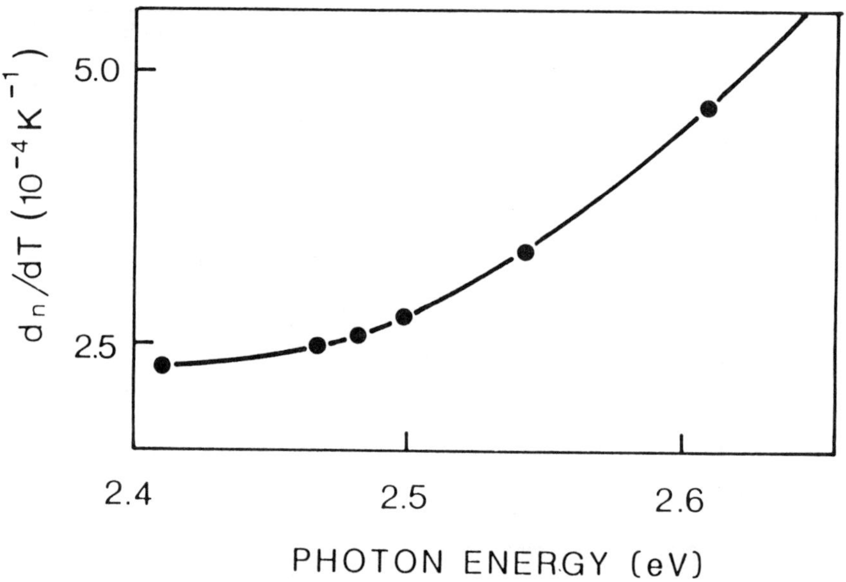

Figure 3.20 Thermo-optic coefficient as a function of wavelength for ZnSe (Wherrett *et al.*, 1988).

Values of refractive index change with temperature, dn/dT, are typically on the order of $10^{-4}\,K^{-1}$ with only a weak band gap resonance (Figure 3.20). The thermo-optic coefficient is normally positive (negative if dE_g/dT is positive) and scales with band gap close to the edge according to

$$\frac{dn}{dT} = E_g^{-1/2}\left(\frac{E_g}{E_g - \hbar\omega}\right)^{1/2} \tag{3.43}$$

3.7. CONCLUSIONS

Semiconductors offer a rich variety of passive and dynamically nonlinear optical phenomena. For signal processing applications, values of the effective third order susceptibility, $\chi^{(3)}$, span the range from 10^{-13} esu to ~ 10 esu with response times from microseconds down into the femtosecond regime. The more sensitive optical nonlinearities in semiconductors have allowed the first demonstrations of all-optical circuits. They provide a means of exploring the fundamental physics of carrier dynamics and transport on femtosecond time scales in pump–probe measurements using mode-locked lasers. Many optical nonlinearities have been investigated recently for possible applications in all-optical devices such as bistable switches, waveguide directional couplers and phase conjugate reflectors discussed in other chapters.

REFERENCES

Alfano, R. R. (ed.) (1984) *Semiconductors Probed by Ultrafast Laser Spectroscopy*, Academic Press, Orlando, FL.

Auston, D. H., McAfee, S., Shank, C. V., Ippen, E. P. and Teschke, O. (1978) *Solid State Electron.*, **21**, 147.

Banyai, L. and Koch, S. W. (1986) *Z. Phys. B*, **63**, 283.

Baggess, T. F., Bohnert, K. M., Mansour, K. *et al.* (1986) *IEEE J. Quantum Electron*, **22**, 360.

Craig, D. and Miller, A. (1986) *Opt. Acta*, **33**, 397.

Craig, D., Kar, A. J., Mathew, J. G. and Miller, A. (1985) *IEEE J. Quantum Electron.*, **21**, 1363.

Dagenais, M. (1983) *Appl. Phys. Lett.*, **43**, 742.

DeLong, K. W. and Stegeman, G. I. (1990) *Appl. Phys. Lett.*, **57**, 2063.

Eichler, H. J. and Massman, F. (1982) *J. Appl. Phys.*, **53**, 3237.

Eichler, H. J., Gunter, P. and Pohl, D. W. (1986) *Laser Induced Dynamic Gratings*, Springer Series in Optical Sciences, Vol. 50, Springer, Berlin.

Figueira, J. F., Cantrell, C. D., Forman, P. R. and Riak, J. P. (1976) *Appl. Phys. Lett.*, **28**, 398.

Fisher, R. A. (ed.) (1983) *Optical Phase Conjugation*, Academic Press, New York.

Gibbs, H. M. (1985) *Optical Bistability: Controlling Light with Light*, Academic Press, Orlando, FL.

Gibbs, H. M., McCall, S. L., Venkatesan, T. N. C., Gossard, A. C., Passner, A. and Wiegman, W. (1979) *Appl. Phys. Lett.*, **35**, 451.

Gibson, A. F., Rosito, C. A., Raffo, C. A. and Kimmett, M. F. (1972) *Appl. Phys. Lett.*, **21**, 356.

Glass, A. M. (ed.) (1990) *Charge Transport Optical Nonlinearities*, in *Opt. Quantum Electron.* (Spec. Issue). **12**.

Guha, S., VanStryland, E. W. and Soileau, M. J. (1985) *Opt. Lett.*, **10**, 285.

Haug, H. (ed.) (1988) *Optical Nonlinearities and Instabilities in Semiconductors*, Academic Press, San Diego, CA.

Hill, J., Parry, G. and Miller, A. (1982) *Opt. Commun.*, **43**, 151.

Hutchings, D., Sheik-Bahae, M., Hagan, D. J. and VanStryland, E. W. (1992) *Opt. Quantum Electron.*, **24**, 1.

Jain R. K. (1981) *Opt. Eng.*, **21**, 199.

Jain, R. K. and Steel, D. G. (1980) *Appl. Phys. Lett.*, **37**, 1.

Jamison, S. A. and Nurmikko, A. V. (1979) *Phys. Rev. B*, **19**, 5185.

Kalt, H., Renner, R. and Klingshirn, C. (1986) *IEEE J. Quantum Electron.*, **22**, 1312.

Karpushko, F. K. and Sinitsyn, G. V. (1978) *J. Appl. Spectrosc.*, **30**, 1323.

Keilman, F. (1976) *Appl. Phys.*, **14**, 29.

Klingshirn, C. and Haug, H. (1981) *Phys. Rep.*, **70**, 316.

Lee, C. C. and Fan, H. Y. (1974) *Phys. Rev. B.*, **9**, 3502.

Lee, Y. H., Chavez-Pirson, A., Koch, S. W., Gibbs, H. M., Park, S. H., Morhange, J., Jeffrey, A., Peyghambarian, N., Banyai, L., Gossard, A. C. and Weigmann, W. (1986) *Phys, Rev. Lett.*, **57**, 2446.

LiKamWa, P., Sitch, J. E., Mason, N. J., Roberts, J. S. and Robson, P. N. (1985) *Electron. Lett.*, **21**, 26.

Lin, W. Z., Schoenlein, R. W., Fujimoto, J. G., and Ippen, E. P. (1988) *IEEE J. Quantum Electron.*, **24**, 267.

Mandel, P., Smith, S. D. and Wherrett, B. S. (eds) (1987) *From Optical Bistability Towards Optical Computing: the EJOB Project*, North-Holland, Amsterdam.

Miller, A. and Craig, D. (1987) In *Optical Properties of Narrowgap Low-dimensional Structures* (eds C. M. Sotomayor Torres, J. C. Portal, J. C. Mann and R. A. Stradling), Plenum, New York, p. 149.

Miller, A. and Parry, G. (1984) *Opt. Quantum Electron.*, **16**, 339.

Miller, A., Johnston, A., Dempsey, J., Smith, J., Pidgeon, C. R. and Holah, G. D. (1979) *J. Phys. C.*, **12**, 4839.

Miller, A., Miller, D. A. B. and Smith, S. D. (1981) *Adv. Phys.*, **30**, 697.

Miller, D. A. B. (1990) *Opt. Quantum Electron.*, **22**, S61.

Miller, D. A. B., Mozolowski, M., Miller, A. and Smith, S. D. (1978) *Opt. Commun.*, **27**, 133.

Miller, D. A. B., Smith, S. D. and Johnston, A. (1979) *Appl. Phys. Lett.*, **35**, 658.

Miller, D. A. B., Seaton, C. T., Prize, M. E. and Smith, S. D. (1981) *Phys. Rev. Lett.*, **47**, 197.

Miller, D. A. B., Chemla, D. S., Damen, T. C., Gossard, A. C., Wiegman, W., Wood, T. H. and Burrus, C. A. (1984) *Appl. Phys. Lett.*, **45**, 13.

Moss, T. S. (1980) *Phys. Status Solidi B*, **101**, 555.

Nurmikko, A. V. (1976a) *Opt. Commun.*, **16**, 365.

Nurmikko, A. V. (1976b) *Opt. Commun.*, **18**, 522.

Olbright, G. R., Peyghambarian, N., Gibbs, H. M., Macleod, H. A. and VanMilligan, F. (1984) *Appl. Phys. Lett.*, **45**, 1031.

Oudar, J. L., Migus, A., Hulin, D., Gillon, G., Etchepare, J. and Antonetti, A. (1984) *Phys. Rev. Lett.*, **53**, 384.

Patel, C. K. N., Slusher, R. E. and Fleury, P. A. (1966) *Phys. Rev. Lett.*, **17**, 1011.

Pidgeon, C. R., Wherrett, B. S., Johnston, A. M., Dempsey, J. and Miller, A. (1979) *Phys. Rev. Lett.*, **42**, 1785.

Poole, C. D. and Garmire, E. (1984) *Opt. Lett.*, **9**, 356.

Sheik-Bahae, M., Hagan, D. J. and VanStryland, E. W. (1990a) *Phys. Rev. Lett.*, **65**, 96.

Sheik-Bahae, M., Said, A. A., Wei, T. H., Hagan, D. J. and VanStryland, E. W. (1990b) *IEEE J. Quantum Electron.*, **26**, 760.

Sheik-Bahae, M., Hutchings, D. C., Hagan, D. J. and VanStryland, E. W. (1991) *IEEE J. Quantum Electron.*, **26**, 1296.

Smirl, A. L., Boggess, T. F., Wherrett, B. S., Perryman, G. P. and Miller, A. (1982) *Phys. Rev. Lett.*, **49**, 933.

Smirl, A. L., Boggess, T. F., Wherrett, B. S., Perryman, G. P. and Miller, A. (1983) *IEEE J. Quantum Electron.*, **19**, 690.

Smith, S. D., Mathew, J. G. H., Taghizadeh, M. R., Walker, A. C., Wherrett, B. S. and Hendry, A. (1984) *Opt. Commun.*, **51**, 357.

Smith, S. D., Walker, A. C., Tooley, F. A. P. and Wherrett, B. S. (1987) *Nature (London)*, **325**, 27.

Stegeman, G. I. and Wright, E. W. (1990) *Opt. Quantum Electron.*, **22**, 95.
Tang, C. L. and Erskine, D. J. (1983) *Phys. Rev. Lett.*, **51**, 840.
Van Stryland, E. W., Vanherzeele, H., Woodall, M. A. *et al.* (1985) *Opt. Eng.*, **24**, 613.
Van Stryland, E. W., Wu, Y. Y., Hagan, D. J. (1988) *J. Opt. Soc. Amer BS*, p. 1980.
Weiler, M. H. (1981) *Solid State Commun.*, **39**, 937.
Wherrett, B. S. (1983) *Proc. R. Soc. London, Ser. A.*, **390**, 373.
Wherrett, B. S. (1984) *J. Opt. Soc. Am. B*, **1**, 67.
Wherrett, B. S. and Higgins, N. A. (1982) *Proc. R. Soc. London, Ser. A*, **379**, 67.
Wherrett, B. S., Walker, A. C. and Tooley, F. A. P. (1988) In *Optical Nonlinearities and Instabilities in Semiconductors* (ed. H. Haug), Academic Press, San Diego, CA.
Woerdman, J. P. and Bolger, B. (1969) *Phys. Lett A 30*, p. 164.
Wolff, P. A. (1969) and Pearson, G. E. (1966) *Phys. Rev. Lett.*, **17**, 1015.

4 The Photorefractive Effect

Andrew K. Powell, Trevor J. Hall, Hussain Imam
and David A. Fish

4.1. INTRODUCTION

It is remarkable that photorefractive crystals, such as barium titanate, display such enormous optical nonlinearities with the minimum of optical intensity levels, typically no more than that output from a helium–neon laser (Feinberg, 1988). The reason for this is that the magnitude of the photorefractive effect is virtually independent of the overall illuminating intensity, giving rise to a sensitive, large and saturated nonlinear effect. It is not altogether surprising, then, that these materials have the reputation of being slow in response, considering that they often only ever experience the low photon rates produced by the lowest of power continuous wave lasers. Nevertheless, one would be hard pressed to demonstrate any optically based system, incorporating optically nonlinear devices, that was not largely photorefractive in nature. Optical systems that have been demonstrated with photorefractive materials include implementations of associative memories, reconfigurable optical interconnections and even the optical implementations of neural-type processing architectures. Apart from the ferroelectric photorefractive materials such as barium titanate ($BaTiO_3$) (Neurgaonkar, *et al.*, 1990) and strontium barium niobate (SBN) (Salamo, 1986), the ever-increasing family of photorefractive crystals includes the sillenite crystals of bismuth silicon oxide (BSO) (Stepanov *et al.*, 1982; Feinburg, 1988), bismuth germanium oxide (BGO), and bismuth titanium oxide (BTO) (Stepanov and Petrov, 1984), along with many of the common semiconductors, such as gallium arsenide (GaAs) (Walsh, 1990), cadmium telluride (CdTe) (Bylsma *et al.*, 1987) and indium phosphide (InP).

The photorefractive effect is a nonlinear optical effect that describes an intensity-dependent refractive index in a material that is a photoconductor. Under nonuniform illumination conditions, photogenerated charge carriers in a photoconductor migrate from the bright regions and subsequently become trapped in darker regions of the material (Figure 4.1). This charge displacement results in the build up of a space charge field that can induce refractive index changes in a photoconductor that also has an electro-optic

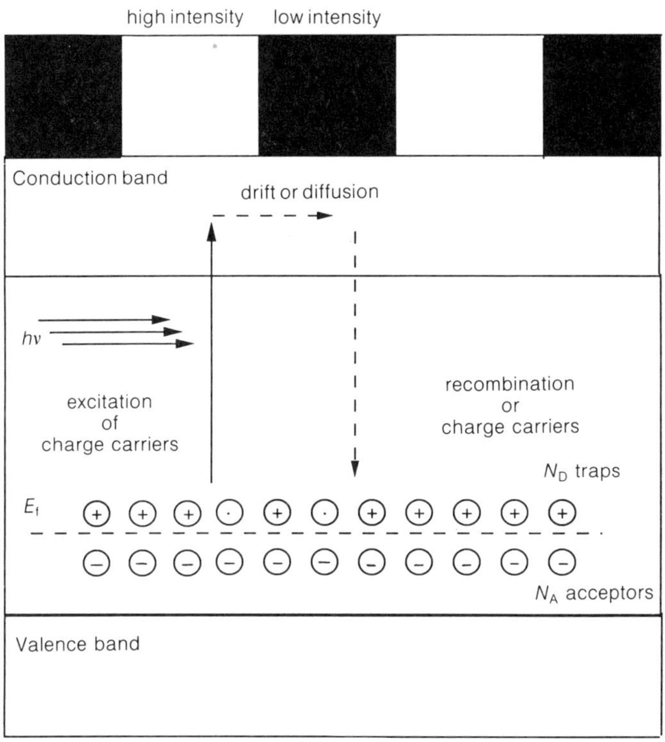

Figure 4.1 The process of redistribution of charge carriers resulting from nonuniform illumination.

effect. A nonlinear intensity-dependent refractive index is thus achieved (Feinburg *et al.*, 1980) (Figure 4.2). This description can be extended to include not just the aforementioned crystals but also the range of materials that exhibit carrier transport assisted nonlinearities (CATAN) (Garmire *et al.*, 1989). These include many types of liquid crystal light valves (LCLVs) and even multiple quantum well self-electro-optic devices (SEEDs). In the case of the LCLVs, photoconductive materials such as BSO are often used in conjunction with electrically birefringent materials such as liquid crystals, to produce an optically addressed spatial light modulator. In the case of multiple quantum well SEEDs, a large electro-optic effect is engineered by the fabrication of the well structure, which gives rise to sharp exciton absorption features at room temperature. The application of an electric field can shift the exciton absorption peaks in wavelength, and can thus be used to modulate the light transmission through the device. The bistable switching SEED devices are formed by arranging that the photoconductivity of the device alters the applied field and hence the light transmission (Miller, 1986). Both

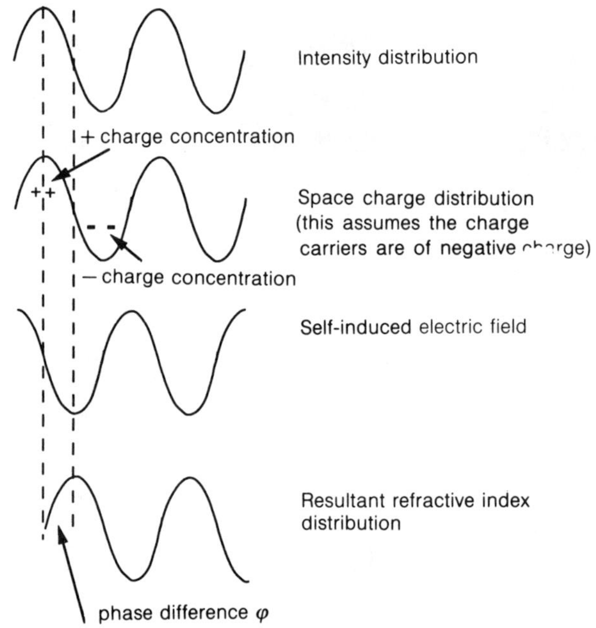

Figure 4.2 The relative phases of the intensity distribution, space charge distribution, the space charge field and the refractive grating.

the LCLV and SEEDs are used in conjunction with an electric field and external electric circuit to modify their refractive properties. In photorefractive crystals, the electric circuit is intrinsic, utilizing the properties of both charge diffusion and drift.

In this chapter, it is intended to review the basic physics of the photorefractive effect, considering both the photoconductive and the electro-optic properties. Methods of enhancing and optimizing the photoconductive properties and the electro-optic properties will be considered in turn. The advantages and disadvantages of various electric field enhancement methods will then be considered, including the effect that they have upon the time response of the photorefractive effect.

4.2. THE BASIC PHYSICS OF THE PHOTOREFRACTIVE EFFECT

The photorefractive effect was first observed in 1967, in the electro-optic material lithium niobate (Chen, 1967). What at first appeared to be optical damage of the material was found to be a reversible effect and photorefractive behaviour was identified. The slow response of material was due to metastable charge trapping in the material with correspondingly poor charge transport properties. Since that time, photorefractive materials with much larger effects than that of lithium niobate have been discovered and perhaps the most interesting of these is the ferroelectric crystal barium titanate. This material exhibits such large optical nonlinearities that strong asymmetric beam fanning (Giuliano, 1981; Feinberg, 1982b) and even self-pumped phase conjugation can be demonstrated with a simple helium–neon laser (Giuliano, 1981; Feinberg, 1982a). Self-pumped phase conjugation is the process where the barium titanate crystal acts as a mirror to a sole input beam, with the remarkable property of reversing the wavefront of the incident light field (Hall and Powell, 1990). The low light levels that can be used demonstrate the fact that the magnitude of the photorefractive effect is independent of the incident optical intensity, making the effect equivalent to a saturated third-order nonlinear process. With most other third-order nonlinear intensity dependent processes, saturation does not occur until large optical intensities are achieved, often well above the point of damage of the material. This is particularly the case with the third-order effects in organic molecules. The speed of response of photorefractive materials is, however, proportional to the illuminating intensity. It is necessary to consider the photoconductive interaction in some detail, particularly in order to understand the properties of materials such as barium titanate.

4.2.1. The photoconductive interaction

The principal behaviour of photoconductive materials subject to nonuniform illumination may be understood by considering only the illumination due to a cosinusoidal intensity pattern I, as brought about by the interference of two coherent light beams, of intensities I_1 and I_2:

$$I = I_0\{1 + m\cos[\mathbf{K}\cdot\mathbf{x} + \phi(t)]\} \tag{4.1}$$

where m is the modulation ratio of the two beams and is equal to $2(I_1 I_2)^{1/2}/I_0$ and I_0 is the total intensity $(I_1 + I_2)$. $\mathbf{K}$ is the wavevector of the interference pattern with $|\mathbf{K}| = 2\pi/\Lambda$ and Λ is the spacing of the fringes of the interference pattern. The interference pattern need not be stationary and $\phi(t)$ describes the time-dependent phase development of the fringe pattern as it evolves through the photoconductive material.

To interpret the effect that nonuniform illumination has upon the charge distribution in the photoconductor, it is useful to proceed by considering the band transport model of Kukhtarev *et al.* (1979). The equations that describe the band transport model arise from the consideration of the charge generation rate and recombination rate in a photoconductive medium. Considering the simple case of one type of photocarrier species, the equations are expressions for the charge continuity for the immobile ionized donors, density N_D^+, the mobile charge carriers, density n, and Poisson's equation.

They take the forms

$$\frac{\partial N_D^+}{\partial t} = (N_D - N_D^+)\sigma I - \gamma_R N_D^+ n \tag{4.2}$$

$$\frac{\partial n}{\partial t} = \frac{\partial N_D^+}{\partial t} + \nabla\cdot(D\,\nabla n + \mu n\mathbf{E}) \tag{4.3}$$

$$\nabla\cdot(\varepsilon\mathbf{E}) = q(N_D^+ - N_A - n) \tag{4.4}$$

where N_D is the number density of the charge donors with N_D^+ the number density of ionized donors. σ is the photo-excitation cross-section of an occupied donor and γ_R is a recombination rate. D is the diffusivity of the free carriers and μ is their mobility. N_A is the number density of acceptor sites with q being equal to the charge of the carrier, $\mathbf{E}$ the space charge field and ε the material permittivity. The equations form a set of nonlinear differential equations that descirbe the charge distribution within a photoconductor under illumination. In general, there are no analytic solutions to the equations but, with suitable approximations, predictions can be made about the behaviour of the photoconductive response of photorefractive materials.

To proceed further with these equations, it is necessary to make some simplifications. It is useful to express the equations in a dimensionless form and to collect all constants into normalized parameters (Hall *et al.*, 1985). One way to proceed is to consider the probabilities of donor occupation due to the illumination. We define the probability of donor occupation W as

$$W = \frac{N_D - N_D^+}{N_D} \tag{4.5}$$

This takes the value W_c at the normalizing uniform intensity I_c. The corresponding value of the free carrier concentration n_c is used to normalize the actual free carrier concentration n.

It is also useful to define a normalization electric field $\mathbf{E}_c$ and this then allows us to define a normalized spatial frequency, v. To do this we need to consider the contributions to the space charge field from the diffusion process of the charge carriers and the saturation of the charge donor sites. These two effects together determine the response of the space charge field to the spatial frequency content of the illuminating pattern. The saturation field corresponds to the largest field that can be supported by a finite number of charge donor sites. A convenient point to normalize the band transport equations is where the diffusion and saturation fields, both functions of the grating spacing, have the same value (Figure 4.3). This happens for a unique value of grating wavevector $\mathbf{K}$ of magnitude K_c and has a corresponding space charge field of value E_c. The diffusion field E_D will increase with

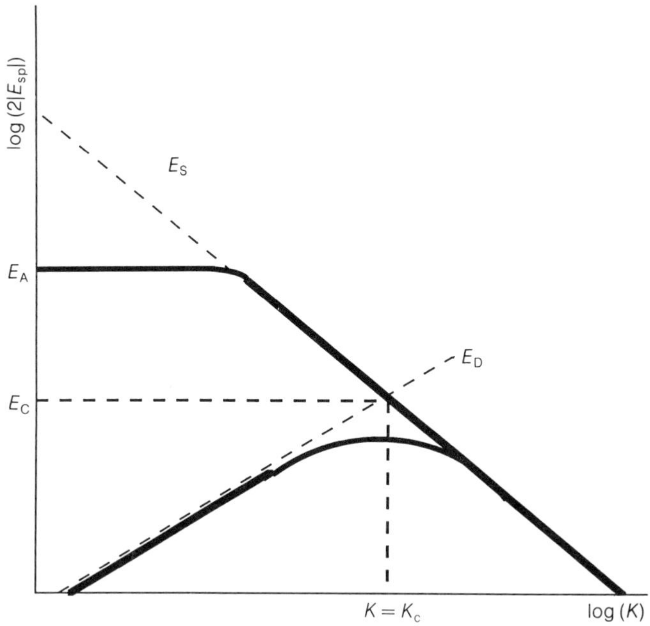

Figure 4.3 The form of the space charge field.

increased spatial frequency and with a one-dimensional analysis

$$E_D = \frac{D}{\mu} K \tag{4.6}$$

The saturation field E_S will increase with increased number of donors or decreased spatial frequency. It will also depend upon the probability of donor occupation, hence

$$E_S = \frac{W_c(1 - W_c)qN_D}{\varepsilon K} \tag{4.7}$$

Thus

$$K_c = \frac{\mu}{D} \frac{W_c(1 - W_c)qN_D}{\varepsilon} \tag{4.8}$$

giving

$$E_c = E_D(K_c) = E_S(K_c) \tag{4.9}$$

A normalized spatial frequency parameter v may now be defined as

$$v = \frac{|\mathbf{K}|}{K_c}, \quad v_c = \frac{2\pi}{K_c} \tag{4.10}$$

The band transport equations may now be expressed for one dimension and with the normalized parameters:

$$\frac{W}{W_c}\{1 + m\cos[\xi + \phi(t)]\} - \frac{(1-W)n}{1-W_c} - \theta v^2 \frac{\partial^2 E}{\partial\xi\,\partial t} = 0 \tag{4.11}$$

$$\frac{\partial}{\partial\xi}\left(v\frac{\partial n}{\partial\xi} + nE + \frac{\partial E}{\partial t}\right) = 0 \tag{4.12}$$

$$v\frac{\partial E}{\partial\xi} + \frac{W - W_c}{W_c(1-W_c)} = 0 \tag{4.13}$$

where $\xi = \mathbf{K}\cdot\mathbf{x}$ and $\theta^2 = D\tau_R K_C^2$ with $\tau_R = [(1-W_c)N_D\gamma_R]^{-1}$. E and n are in units of E_c and n_c and I_0 is chosen to be equal to I_c.

Solutions for the space charge field may now be obtained by expanding W, E and n as a power series in the modulation ratio m. For a constant

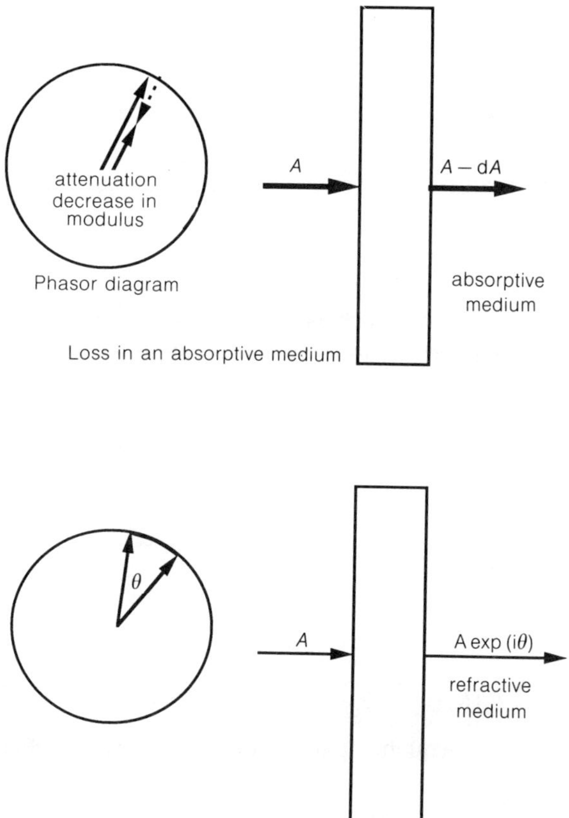

(a)

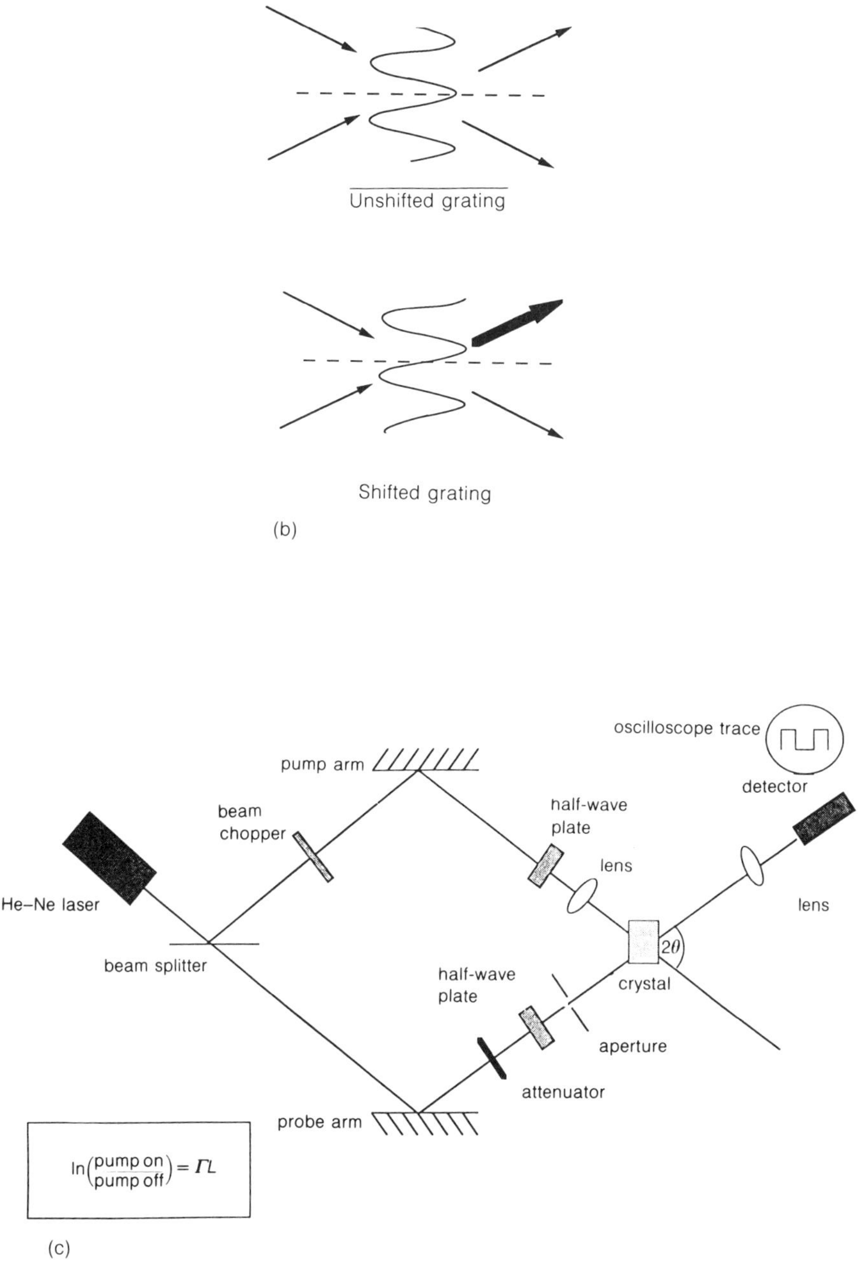

Figure 4.4 (a) Phase change associated with refractive media. (b) Introduction of a $\pi/2$ phase shift between the illuminating intensity and the refractive index grating provides the necessary phase relationship to establish energy exchange. (c) Two-wave mixing experiment.

applied field E_A, we can obtain expressions for the zeroth- and the first-order terms in the expansion of m:

$$E^{(0)} = E_A, \quad E^{(1)} = E_1 \exp(i\xi) + E_1^* \exp(-i\xi) \tag{4.14}$$

Truncating the expansions after the first order corresponds to the linear in modulation approximation which neglects terms of higher spatial frequency but is found to be accurate in most experimental situations. A differential equation for the first component of the space charge field, E_1, may be derived from the normalized band transport equations, and hence from equation 4.14 provides information on the size, phase and dynamics of the fundamental component of the space charge field in the linear in modulation approximation:

$$\left(\frac{\partial}{\partial t} + s\right) E_1 = sF \exp[i\phi(t)] \tag{4.15}$$

F is the steady state value of E_1 and determines the magnitude and phase of the steady state space charge field,

$$F = -\frac{1}{2} \frac{i\nu + E_A}{1 - i\nu(E_A + i\nu)} \tag{4.16}$$

and s is a complex frequency which determines the dynamics of the space charge field:

$$s = \frac{1}{\tau_g} + i\omega_g \tag{4.17}$$

In equation 4.15 s is in units of $\tau_c = \varepsilon/q\mu n_c$, the dielectric relaxation time of the grating at intensity I_c. $1/\tau_g$ and ω_g are the real and imaginary components of s. In the simple case of a stationary grating and no applied field ($E_A = 0$), the solution to equation 4.15 is purely imaginary with

$$E_1 = F = -\frac{i}{2} \frac{\nu}{1 + \nu^2} \tag{4.18}$$

The space charge field is thus phase shifted by $\pi/2$ radians from the interference pattern and this is the optimum situation to allow energy exchange between interfering beams, in a two-wave mixing interaction (Figure 4.4).

Depending upon the sign of the relevant electro-optic coefficient of the photorefractive material, a refractive grating is induced that is either in phase or in antiphase with the space charge field and hence $+\pi/2$ or $-\pi/2$ out of phase with the interference pattern. Although the application of an external field E_A increases the magnitude of the space charge field, it does not necessarily preserve the optimum $\pi/2$ phase shift. In the case of an applied d.c. field, this phase shift is indeed lost but can be restored by using a moving illuminating fringe pattern (Imbert *et al.*, 1988). Alternatively, alternating electric fields may be applied (Walsh *et al.*, 1990) and the advantages of doing this will be discussed later.

4.2.2. The electro-optic interaction

The effect of generating large space charge fields in electro-optic crystals is that large refractive index gratings may be induced. In the case of barium titanate, refractive gratings with Δn greater than 10^{-4} can be induced. As will be described later, there are limits to the maximum enhancement of the space charge field that can be achieved in photorefractive materials, which makes optimizing the size of electro-optic coefficients of photorefractive materials an interesting topic. In many cases, particularly the semiconductor materials, the resonant tuning of the illuminating laser light to the band gap excitation energy can greatly enhance the electro-optic properties of the material (Partovi *et al.*, 1990). This then generates large photorefractive nonlinearities, that in many cases can be used in conjunction with picosecond pulsed lasers to give fast response times also, making bulk photorefractive materials attractive candidates for optically nonlinear parallel processing and interconnection applications. It is therefore useful to understand the electro-optic properties of photorefractive crystals (Shih and Yariv, 1982) and this is done here with particular reference to the ferroelectric crystal, barium titanate, and to the cubic semiconducting crystal, gallium arsenide.

Barium titanate is a tetragonal crystal of symmetry point group $4mm$ at room temperature (Klein, 1989). It is a uniaxial crystal with a large component to its electro-optic tensor, reported to be as high as $1600\,\mathrm{pm\,V^{-1}}$. The large anisotropic optical properties give rise to pronounced asymmetric beam fanning and other nonlinear optical phenomena such as self-pumped phase conjugation.

The ferroelectric nature of barium titanate arises from the spontaneous polarizability of the unit cell of the crystal, when in its tetragonal phase. This is responsible for the generation of the ferroelectric domains, which are regions in the crystal where the spontaneous polarizations of many unit cells are aligned. In its tetragonal phase, the crystal is typical of many ferroelectric crystals, such as orthorhombic crystal potassium niobate ($KNbO_3$), where the ferroelectric properties arise from a heavy metal ion, such as the titanium, sitting at the centre of a unit cell. The symmetric electrostatic potential trap formed from the negatively ionized oxygen atoms, one at each of the six faces of the cube surrounding the heavy metal ion, allows the heavy ion to move through large displacements under the influence of applied electric fields or forces. The high state of oxidation of the heavy metal ion also gives rise to a large dipole moment and hence the large spontaneous polarization (Klein, 1989).

The tetragonal phase of barium titanate only exists between approximate temperatures of 9 and 120 °C, at which point the crystal transforms phase to either an orthorhombic phase or a cubic phase, respectively. The fact that the tetragonal phase only exists over a small temperature range, around that of room temperature, gives rise to a lattice that is under a certain degree of internal stress, which is borne out by the way the crystal will shatter when accidently dropped, or subjected to temperatures below 9 °C.

The spontaneous polarizations of the unit cells of barium titanate may be directed along any of the six directions directed towards a neighbouring oxygen atom. The second-order nonlinear response or hyperpolarizability is dependent upon the polarizability of the crystal and is responsible for a macroscopic electro-optic effect. This effect is reversed when the direction of the polarizability of the unit cells of the crystal is reversed (Figure 4.5). Interruptions to the domains can be caused by crystallographic defects and impurities, normally confining the domain size to microscopic extents. For the crystal to have a net overall electro-optic effect, it is necessary for most of the ferroelectric domains to become aligned to form, ideally, one single domain. This may be achieved by the application of mechanical pressure, or the application of electric fields at temperatures near to the transition temperature for the cubic phase, or the Curie temperature.

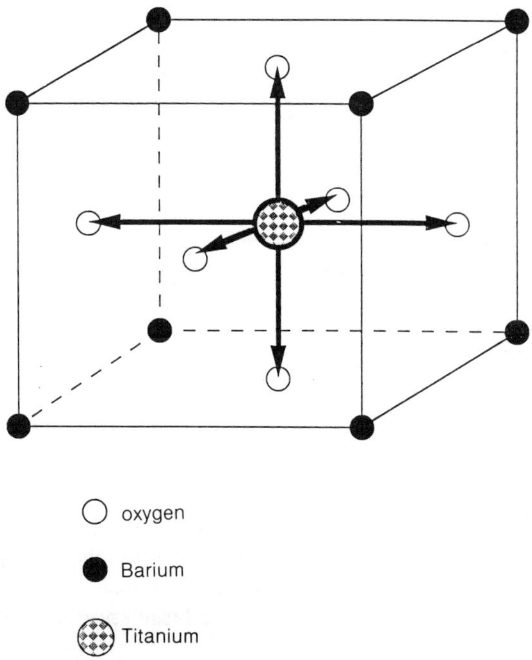

Figure 4.5 Barium titanate unit cell.

We shall now consider light propagation through an anisotropic medium such as barium titanate. To describe anisotropic scattering, holographic readout and non-paraxial effects a full vector analysis is required as described by Foote (1987). With a uniaxial crystal, there are two orthogonal polarization modes that can propagate through the crystal, for a given wave normal. An index ellipsoid may be constructed that can be used to find the refractive indices for the two orthogonal polarization modes and the ellipsoid may be described by the equation (Foote, 1987)

$$\left(\frac{x}{n_x}\right)^2 + \left(\frac{y}{n_y}\right)^2 + \left(\frac{z}{n_z}\right)^2 = 1 \tag{4.19}$$

where x, y and z denote the principal dielectric axes and n_x, n_y, n_z are the refractive indices associated with each of these directions. In the absence of an applied field, the index ellipsoid for uniaxial crystals is of the simpler form

$$\left(\frac{x}{n_o}\right)^2 + \left(\frac{y}{n_o}\right)^2 + \left(\frac{z}{n_e}\right)^2 = 1 \tag{4.20}$$

where the refractive indices for two directions are the same and represent the refractive index experienced by a wave of ordinary polarization. The third direction has a different value of refractive index, as is experienced by a wave of extraordinary polarization that travels perpendicular to the optic axis. An electric field induced perturbation to this index ellipsoid determines the form of the electro-optic tensor for a uniaxial material and it is this that is of importance for the photorefractive effect. Thus the symmetries that apply to the index ellipsoid of a uniaxial material also affect the symmetries of the electro-optic tensor denoted by R_{ijk}. In a lossless medium where reciprocity must apply, the following symmetry exists:

$$R_{ijk} = R_{jik} \tag{4.21}$$

and a contracted notation is adopted, where the 27 elements of R_{ijk} may be expressed by 18 components r_{ij} where the relationship between the two tensors is given by

$$\begin{aligned}
r_{1k} &= R_{11k} \\
r_{2k} &= R_{22k} \\
r_{3k} &= R_{33k} \\
r_{4k} &= R_{23k} = R_{32k} \\
r_{5k} &= R_{13k} = R_{31k} \\
r_{6k} &= R_{12k} = R_{21k}
\end{aligned} \tag{4.22}$$

Thus for barium titanate we have

$$r_{ij} = \left\{\begin{matrix} \bullet & \bullet & r_{13} \\ \bullet & \bullet & r_{23} \\ \bullet & \bullet & r_{33} \\ \bullet & r_{42} & \bullet \\ r_{51} & \bullet & \bullet \\ \bullet & \bullet & \bullet \end{matrix}\right\} \tag{4.23}$$

where ● denotes a zero element and with barium titanate's crystal class $r_{13} = r_{23}$ and $r_{51} = r_{42}$.

The perturbation to the inverse permittivity tensor is given by $\Delta\varepsilon_{ij}^{-1} = R_{ijk}E_k^0$. The rotational symmetry of this class of materials requires that only two components of an electric field E_k^0 be specified for a field in a general direction. If the three components of E_k^0 are denoted by E_x^0, E_y^0, and E_z^0 then E_x^0 can be set equal to zero. The perturbation to the inverse permittivity tensor induced from such a field can be expressed as

$$\Delta\varepsilon^{-1} = \begin{pmatrix} r_{13}E_z^0 & 0 & 0 \\ 0 & r_{13}E_z^0 & r_{42}E_y^0 \\ 0 & r_{42}E_y^0 & r_{33}E_z^0 \end{pmatrix} \tag{4.24}$$

The modified equation for the index ellipsoid then takes the form

$$(\varepsilon_{xx}^{-1}\,\Delta\varepsilon_{xx}^{-1})x^2 + (\varepsilon_{yy}^{-1}\,\Delta\varepsilon_{yy}^{-1})y^2 + (\varepsilon_{zz}^{-1}\,\Delta\varepsilon_{zz}^{-1})z^2 + 2\,\Delta\varepsilon_{xz}^{-1}\,xz + 2\,\Delta\varepsilon_{yz}^{-1}\,yz + 2\,\Delta\varepsilon_{xy}^{-1}\,xy = \varepsilon_0^{-1} \tag{4.25}$$

where the indices i and j now run over x, y and z to correspond to the coordinate system. The numbered indices are left on the electro-optic coefficients for the sake of convention. Thus the perturbed index ellipsoid for barium titanate is of the form

$$\left(\frac{1}{n_0^2} + r_{13}E_z^0\right)(x^2 + y^2) + \left(\frac{1}{n_e^2} + r_{33}E_z^0\right)z^2 + 2r_{42}E_y^0 yz = 1 \tag{4.26}$$

New axes may be found for the index ellipsoid and this may be done in the following way (Foote, 1987). Defining the matrix $\boldsymbol{M}$ as

$$\boldsymbol{M} = \begin{pmatrix} m_1 & 0 & 0 \\ 0 & m_2 & m_4 \\ 0 & m_4 & m_3 \end{pmatrix} \tag{4.27}$$

with

$$m_1 = \frac{1}{n_0^2} + r_{13}E_z^0 = m_2 \tag{4.28}$$

$$m_3 = \frac{1}{n_e^2} + r_{33}E_z^0 \tag{4.29}$$

$$m_4 = r_{42}E_y^0 \tag{4.30}$$

the eigenvalues of $\boldsymbol{M}$ will satisfy the set of homogenous equations

$$\boldsymbol{M} = \begin{vmatrix} m_1 - \lambda & 0 & 0 \\ 0 & m_2 - \lambda & m_4 \\ 0 & m_4 & m_3 - \lambda \end{vmatrix} = 0 \tag{4.31}$$

giving eigenvalues

$$\lambda^{(1)} = \frac{1}{n_0^2} + r_{13}E_z^0 \tag{4.32}$$

$$\lambda^{(2)} = \frac{1}{n_0^2} + r_{13}E_z^0 - \frac{r_{42}^2(E_y^0)^2}{1/n_0^2 - 1/n_e^2 + (r_{13} - r_{33})E_z^0} \tag{4.33}$$

$$\lambda^{(3)} = \frac{1}{n_e^2} + r_{33}E_z^0 - \frac{r_{42}^2(E_y^0)^2}{1/n_0^2 - 1/n_e^2 + (r_{13} - r_{33})E_z^0} \tag{4.34}$$

allowing a new index ellipsoid to be expressed along new axes as

$$\lambda^{(1)}x'^2 + \lambda^{(2)}y'^2 + \lambda^{(3)}z'^2 = 1 \tag{4.35}$$

from which the refractive indices for the new orthogonal components are

$$n_e(\theta)' \approx n_e(\theta) - \frac{[n_e(\theta)]^3}{2}(r_{13}E_z^0\cos^2\theta + r_{33}E_z^0\sin^2\theta - r_{42}E_y^0\sin\theta\cos\theta) \tag{4.36}$$

$$n_0 \approx n_0 - n_0 r_{13}E_z^0 \tag{4.37}$$

With the electric field directed perpendicular to the direction of propagation then

$$n_e(\theta)' = n_e(\theta) - \frac{n_e(\theta)}{2}|\mathbf{E}^0|\sin\theta(r_{13}\cos^2\theta + r_{33}\sin^2\theta + 2r_{42}\cos^2\theta) \tag{4.38}$$

and

$$n_0(\theta)' = n_0(\theta) - \frac{n_0^3}{2}r_{13}|\mathbf{E}^0|\sin\theta \tag{4.39}$$

where E_y^0 and E_z^0 are components of $\mathbf{E}^0$. With the electric field parallel to the propagation direction then

$$n_e(\theta)' = n_e(\theta) - \frac{n_0^3}{2}|\mathbf{E}^0|\cos\theta(r_{13}\cos^2\theta + r_{33}\sin^2\theta - 2r_{42}\sin^2\theta) \tag{4.40}$$

and

$$n_0(\theta)' = n_0(\theta) - \frac{n_0^3}{2} r_{13} |\mathbf{E}^0| \cos\theta \tag{4.41}$$

The perturbation to refractive indices of the polarization eigenmodes brought about by the electric field may be expressed more generally as

$$\Delta n = \frac{n^3}{2} r_{\mathrm{eff}} |\mathbf{E}^0| \tag{4.42}$$

where r_{eff} may be computed from the components of the electro-optic tensor, which depends upon orientation and the direction of the incident wave, and with barium titanate the dominant component to the electro-optic coefficient is r_{42}.

We now turn our attention to those photorefractive crystals that have a cubic crystal class. These include the sillenite crystals such as BSO and BTO whose $\bar{2}3m$ symmetry gives rise to the property of optical activity. Optical activity causes the undesirable effect of rotating the polarization state of light on passage through the crystal and not preserving an optimum interaction geometry. The cubic semiconductors such as GaAs have a $\bar{4}3m$ symmetry that disallows optical activity and therefore enables an optimum state of polarization, of incident light waves, to be maintained throughout the crystal's volume, therefore preserving the optimum interaction geometry throughout the crystal with respect to the electro-optic tensor.

Using GaAs as an example, we will show how the application of an electric field induces optical birefringence (Walsh, 1990; Fabre *et al.*, 1988). The field may, of course, be that of the space charge field brought about by non uniform illumination. Without an applied field, GaAs is isotropic (Nye, 1979). The application of the electric field linearly perturbs the dielectric tensor. As before this perturbation can be expressed as

$$\Delta\varepsilon_{ij}^{-1} = \varepsilon_{ij}^{-1}(E) - \varepsilon_{ij}^{-1}(0) = R_{ijk} E_k^0 \tag{4.43}$$

The index ellipsoid for the perturbed system can thus be expressed as:

$$[\varepsilon_{ij}^{-1}(0) + R_{ijk} E_k] x_i x_j = \frac{1}{\varepsilon_0} \tag{4.44}$$

Using the symmetry property of equation 4.21, the reduced form for the electro-optic tensor for a crystal of $\bar{4}3m$ symmetry thus becomes

$$\begin{pmatrix} 0 & 0 & 0 \\ 0 & 0 & 0 \\ 0 & 0 & 0 \\ r_{41} & 0 & 0 \\ 0 & r_{52} & 0 \\ 0 & 0 & r_{63} \end{pmatrix} \tag{4.45}$$

where $r_{41} = r_{52} = r_{63}$ and the form for the index ellipsoid in the principal coordinate system becomes

$$\frac{1}{n_x^2} + \frac{1}{n_y^2} + \frac{1}{n_z^2} + 2r_{41}E_x^0 yz + 2r_{41}E_y^0 xz + 2r_{41}E_z^0 xy = 1 \tag{4.46}$$

There are two important cases of interest and these correspond to

1. when an electric field is applied along the direction of the z principal axis, and
2. when the applied field is directed along one of the induced principal axes.

In the first of these cases, when the electric field **E** is applied in the [001] direction, the field may be described in terms of its components

$$\mathbf{E} = \begin{pmatrix} 0 \\ 0 \\ E_z^0 \end{pmatrix} \tag{4.47}$$

Letting E_z^0 become E^0 then gives the following equation for the perturbed index ellipsoid:

$$\frac{x^2 + y^2 + z^2}{n_0^2} + 2r_{41}E^0 xy = 1 \tag{4.48}$$

for which a new set of axes can be found (x', y', z') such that the perturbed ellipsoid can be expressed as

$$\frac{x'^2}{n_{x'}^2} + \frac{y'^2}{n_{y'}^2} + \frac{z'^2}{n_{z'}^2} = 1 \tag{4.49}$$

As the z-axis can be seen to be left unchanged, the new axes x', y' are confined to a plane so

$$x = x' \cos\theta - y' \sin\theta \tag{4.50}$$

$$y = x' \sin\theta + y' \cos\theta \tag{4.51}$$

Substituting these equations into the equation for the index ellipsoid and setting $\theta = 45^\circ$ to eliminate the cross-term we have

$$x'^2\left(\frac{1}{n_0^2} + r_{41}E^0\right) + y'^2\left(\frac{1}{n_0^2} - r_{41}E^0\right) + \frac{z^2}{n_0^2} = 1 \tag{4.52}$$

New crystallographic axes at 45° to the original axes have been induced by the field and are equivalent to a rotation of the principal axes through this angle. The refractive indices corresponding to the new axes are as follows:

$$n_{x'} = n_0 - \frac{n_0^3 r_{41} E^0}{2} \tag{4.53}$$

$$n_{y'} = n_0 + \frac{n_0^3 r_{41} E^0}{2} \tag{4.54}$$

It is important to note that, for the above orientation, Δn may be positive or negative and hence the direction of energy exchange brought about by photorefractive beam coupling may by reversed by the changing of polarization states of the beams or the orientation of the crystal.

A second important case exists, that affects the form of photorefractive beam coupling in materials such as GaAs. Instead of arranging for an electric field to be directed in the direction of the z-axis, it can be directed at 45° between the x and y axes, which is an equivalent direction to the induced axes of the first example. With such an arrangement, the components of the electric field, which is directed in the [110] direction, may be given as

$$\mathbf{E} = \begin{pmatrix} (1/\sqrt{2})E^0 \\ (1/\sqrt{2})E^0 \\ 0 \end{pmatrix} \tag{4.55}$$

The resultant index ellipsoid can be seen to be

$$\frac{x^2 + y^2 + z^2}{n_0^2} + \sqrt{2}E^0 r_{41} yz + \sqrt{2}E^0 r_{41} xz = 1 \tag{4.56}$$

in a similar way as before.

A suitable coordinate transformation is required to regain the simplest expression of the index ellipsoid corresponding to a new z-axis, z^0, perpendicular to the electric field and thus directed in the $[\bar{1}10]$ direction. The transformation is described by

$$x = \frac{1}{\sqrt{2}}(x' \cos\phi - y' \sin\phi + z') \tag{4.57}$$

$$y = \frac{1}{\sqrt{2}}(x' \cos\phi - y' \sin\phi - z') \tag{4.58}$$

$$z = z' \sin\phi + y' \cos\phi \tag{4.59}$$

where ϕ is again set to 45° and corresponds to the rotation of the induced axes x' and y' from the principal axes. This then gives the new index ellipsoid as

$$x'^2\left(\frac{1}{n_0^3} + r_{41}E^0\right) + y'^2\left(\frac{1}{n_0^3} - r_{41}E^0\right) + \frac{z'^2}{n_0^2} = 1 \tag{4.60}$$

The refractive indices for the induced principal axes are thus

$$n_{x'} = n_0 - \frac{n_0^3}{2} r_{41} E^0 \tag{4.61}$$

$$n_{y'} = n_0 + \frac{n_0^3}{2} r_{41} E^0 \tag{4.62}$$

$$n_{z'} = n_0 \tag{4.63}$$

The effect of the electro-optic nature of materials upon the polarization state of light will now be considered. The polarization state of light has a critical effect upon the magnitude of the electro-optic effect that it experiences. This electro-optic effect, described by the effective electro-optic element r_{eff}, depends not only upon the crystal orientation and the direction of the applied electric field, but also on the direction of the electric field vector of the light field with respect to the crystal axes.

The perturbation to the refractive index, brought about by the application of an electric field and its influence on a particular polarization state of incident wave, can be seen from the previous two examples to be of the form

$$\Delta n = \frac{n_0^3}{2} r_{\text{eff}} E^0 \tag{4.64}$$

where the sign and the magnitude of the r_{eff} term will depend on the above-mentioned parameters.

Considering the result of two interacting waves in a GaAs crystal leads us to a vector wave-mixing problem that has six components to consider, for the determination of the effective electro-optic effect experienced by waves of arbitrary linear polarization states. To induce a maximum fringe modulation for a maximum space charge field in a photorefractive material, it is necessary for the interacting waves to be of the same polarization state, hence reducing the six possible orientations of the beam polarizations to the crystal axes to a set of three duplicated terms of the form

$$2(\mathbf{a}.\hat{\mathbf{x}}_3)(\mathbf{b}.\hat{\mathbf{x}}_1)(\mathbf{c}.\hat{\mathbf{x}}_1) + 2(\mathbf{a}.\hat{\mathbf{x}}_1)(\mathbf{b}.\hat{\mathbf{x}}_3)(\mathbf{c}.\hat{\mathbf{x}}_1) + 2(\mathbf{a}.\hat{\mathbf{x}}_1)(\mathbf{b}.\hat{\mathbf{x}}_1)(\mathbf{c}.\hat{\mathbf{x}}_3) \tag{4.65}$$

where **a**, **b**, **c** are the unit vectors in the crystal axes and $\hat{\mathbf{x}}_1$, $\hat{\mathbf{x}}_2$, $\hat{\mathbf{x}}_3$ correspond to the orthogonal directions of the electric field, which in this case can be considered to be the induced space charge field.

Thus we now have r_{eff} in terms of r_{41}, the dominant term in the electro-optic tensor for $\bar{4}3m$ cubic crystals. This may be described by some function F of the polarization states and the relative directions of the principal axes of the crystals.

$$r_{\text{eff}} = r_{41} F(\hat{\mathbf{x}}_1, \mathbf{a}, \hat{\mathbf{x}}_2, \mathbf{b}, \hat{\mathbf{x}}_3, \mathbf{c}) \tag{4.66}$$

For the examples described previously, one with the electric field in the z-direction and one with the **E** field directed along the direction of one of the induced axes of the crystals, we may now specifically find values of F. In the first case we have

$$F = \tfrac{1}{2}(1 + \cos 2\phi) \tag{4.67}$$

which does not allow a change of sign for the refractive index perturbation experienced. For the second orientation, however, we have

$$F = \sin 2\phi \tag{4.68}$$

where Δn is reversed periodically with ϕ. In the first case, ϕ is the angle the electric field vector makes with the nominally vertical direction z as it sweeps through the $(\bar{1}10)$ plane and, in the second example, as the electric field vector sweeps through the (110) plane.

From our knowledge of the electro-optic effect and the vector dependences of waves interacting in photorefractive crystals, it is possible to find what orientations give rise to an optimum effect. The form for the effective electro-optic coefficient r_{eff} is

$$r_{\text{eff}} = 2\mathbf{f}.\hat{\mathbf{x}}_3 r_{41} \tag{4.69}$$

in a $\bar{4}3m$ material, where $\mathbf{f}$ is

$$\mathbf{f} = (\mathbf{b}.\hat{\mathbf{x}}_1)(\mathbf{c}.\hat{\mathbf{x}}_1)\mathbf{a} + (\mathbf{a}.\hat{\mathbf{x}}_1)(\mathbf{c}.\hat{\mathbf{x}}_1)\mathbf{b} + (\mathbf{a}.\hat{\mathbf{x}}_1)(\mathbf{b}.\hat{\mathbf{x}}_1)\mathbf{c} \tag{4.70}$$

To maximize the expression for r_{eff}, $\hat{\mathbf{x}}_3$ and $\mathbf{f}$ must be parallel, given $\hat{\mathbf{x}}_3$ is a unit vector in the direction of $\mathbf{f}$, i.e.

$$\hat{\mathbf{x}}_3 = \frac{\mathbf{f}}{(\mathbf{f}.\mathbf{f})^{1/2}} \tag{4.71}$$

giving

$$r_{\text{eff}} = 2[(\mathbf{b}.\hat{\mathbf{x}}_1)^2(\mathbf{c}.\hat{\mathbf{x}}_1)^2 + (\mathbf{a}.\hat{\mathbf{x}}_1)^2(\mathbf{c}.\hat{\mathbf{x}}_1)^2 + (\mathbf{a}.\hat{\mathbf{x}}_1)^2(\mathbf{b}.\hat{\mathbf{x}}_1)^2]^{1/2} r_{41} \tag{4.72}$$

This expression corresponds to an r_{eff} that cannot change sign with orientation. The expression is maximized for a polarization unit vector $\hat{\mathbf{x}}_1$ of the form

$$\hat{\mathbf{x}}_1 = \frac{1}{\sqrt{3}}(\mathbf{a} + \mathbf{b} + \mathbf{c}) \tag{4.73}$$

giving the corresponding direction for the electric field unit vector $\hat{\mathbf{x}}_3$ of

$$\hat{\mathbf{x}}_3 = \frac{1}{\sqrt{3}}(\mathbf{a} + \mathbf{b} + \mathbf{c}) \tag{4.74}$$

i.e. this is in the same direction, and a resultant effective electro-optic component of

$$r_{\text{eff}} = \frac{2}{\sqrt{3}} r_{41} = 1.15 r_{41} \tag{4.75}$$

which is 15% higher than the r_{41} component alone and may be utilized by the appropriate interaction geometry.

When the orientation is chosen that is consistent with an electro-optic effect sign change with a rotation of polarization state, it is convenient to express the effective electro-optic component as

$$r_{\text{eff}} = \hat{\mathbf{x}}_1 . \boldsymbol{R} . \hat{\mathbf{x}}_1 \tag{4.76}$$

where the components of $\boldsymbol{R}$ are

$$\boldsymbol{R} = \begin{pmatrix} 0 & \mathbf{c}.\hat{\mathbf{x}}_3 & \mathbf{b}.\hat{\mathbf{x}}_3 \\ \mathbf{c}.\hat{\mathbf{x}}_3 & 0 & \mathbf{a}.\hat{\mathbf{x}}_3 \\ \mathbf{b}.\hat{\mathbf{x}}_3 & \mathbf{a}.\hat{\mathbf{x}}_3 & 0 \end{pmatrix} \tag{4.77}$$

It is useful to consider the eigenvalues in $\boldsymbol{R}$ to find maxima in r_{eff}. The sum of the three eigenvalues is zero, as $\boldsymbol{R}$ is traceless. The symmetry of the orientation demands that r_{eff} has two equal but opposite maxima, requiring that the third eigenvalue be zero. Hence

$$\text{Det}(\boldsymbol{R} - \lambda \mathbf{I}) = 0 \tag{4.78}$$

where $\boldsymbol{I}$ is the identity matrix and to ensure that $\lambda = 0$ is a solution

$$(\mathbf{a}.\hat{\mathbf{x}}_3)(\mathbf{b}.\hat{\mathbf{x}}_3)(\mathbf{c}.\hat{\mathbf{x}}_3) = 0 \tag{4.79}$$

which can occur when $\hat{\mathbf{x}}_3$ is perpendicular to any of the principal axes. Choosing $\hat{\mathbf{x}}_3$ to be in the plane defined by the axes **a** and **b** then $\boldsymbol{R}$ reduces to

$$\boldsymbol{R} = \begin{pmatrix} 0 & 0 & \mathbf{b}.\hat{\mathbf{x}}_3 \\ 0 & 0 & \mathbf{a}.\hat{\mathbf{x}}_3 \\ \mathbf{b}.\hat{\mathbf{x}}_3 & \mathbf{a}.\hat{\mathbf{x}}_3 & 0 \end{pmatrix} \tag{4.80}$$

simplifying the eigenvalue problem to

$$-\lambda[\lambda^2 - (\mathbf{a}.\hat{\mathbf{x}}_3)^2] + \lambda(\mathbf{b}.\hat{\mathbf{x}}_3)^2 = 0 \tag{4.81}$$

giving the nonzero eigenvalues

$$\lambda^2 = (\mathbf{a}.\hat{\mathbf{x}}_3)^2 + (\mathbf{b}.\hat{\mathbf{x}}_3)^2 \tag{4.82}$$

maximizing r_{eff} for

$$\hat{\mathbf{x}}_3 = \frac{1}{\sqrt{2}}(\mathbf{a} + \mathbf{b}) \tag{4.83}$$

and giving for r_{eff}

$$r_{\text{eff}} = \sqrt{2}(\mathbf{c}.\hat{\mathbf{x}}_1)[(\mathbf{a}.\hat{\mathbf{x}}_1) + (\mathbf{b}.\hat{\mathbf{x}}_1)]r_{41} \tag{4.84}$$

A maximum value of r_{eff} occurs when x_1 is constrained in the $(\bar{1}10)$ plane and when it is at 45° to the c-axis. In this case

$$\mathbf{a}.\hat{\mathbf{x}}_1 + \mathbf{b}.\hat{\mathbf{x}}_1 = \tfrac{1}{2} \tag{4.85}$$

and

$$\mathbf{c}.\hat{\mathbf{x}}_1 = \frac{1}{\sqrt{2}} \tag{4.86}$$

giving

$$r_{\text{eff}} = r_{41} \tag{4.87}$$

This section has shown how knowledge of the electro-optic nature of photorefractive crystals can be used to optimize the magnitude of the nonlinear effect, particularly for the experimental geometries of two- and four-wave mixing.

4.3. TECHNIQUES FOR SPACE CHARGE FIELD ENHANCEMENT

Enhancement of the photorefractive effect by the application of large d.c. fields is now a well-established technique (Walsh, 1990; Kumar, 1987a; Refregier *et al.*, 1985). When d.c. applied fields are used in conjunction with moving fringe pattern techniques, large nonlinear optical beam coupling strengths can be achieved. Electric field enhancement techniques, however, are not just confined to d.c. static fields but include time-varying fields; this has important consequences for the phase relationship of the space charge with respect to the illuminating interference pattern.

4.3.1. Static fields and moving fringes

When d.c. fields are applied to photorefractive crystals, there is a net drift of charge carriers in the direction of the electric field that allows a larger space charge field to be built up than would occur with the diffusion process alone. The finite lifetime of the photorefractive grating introduces a change in phase relationship between the illumination intensity pattern and the space charge field (Hall *et al.*, 1985). This can be effectively described in terms of a complex space charge field, the components of which depend upon the magnitudes of the diffusion-only field, any externally applied field and the grating relaxation time in the material (and hence the illuminating intensity).

As the $\pi/2$ phase shift is lost with an increased applied d.c. field, so is the mechanism of energy exchange in the two-wave mixing process. The technique of a moving fringe pattern is used to restore the optimum $\pi/2$ phase relationship between the illuminating pattern and the space charge field. The technique requires that one of the two beams in a two-wave mixing interaction is detuned in frequency, so that a moving interference pattern is achieved. When the speed of the moving fringe pattern is made to match the speed of response of the material, the space charge field and hence the refractive grating are once again in quadrature with the illuminating pattern.

The moving grating situation may again be analysed by equation 4.15, but this time with a phase development $\phi(t) = -\omega t$, where ω is the detuning frequency of one of the interacting beams. The space charge field E_1 then has the dependence $E_1 \exp(-\mathrm{i}\omega t)$ with

$$E_1 = \frac{s}{s - \mathrm{i}\omega} F \tag{4.88}$$

where s is again the complex frequency of the form (Hall *et al.*, 1985)

$$s = \frac{1}{\tau_g} + \mathrm{i}\omega_g \tag{4.89}$$

In the resonance condition when $\omega = \omega_g$, the magnitude of the space charge field is at a maximum. The optimum frequency detuning for a two-wave mixing interaction is given when the imaginary component of the space charge field ($\mathrm{Im}(E_1)$) is at a maximum, corresponding to a frequency ω_m, and this is given by the relationship

$$\omega_m \tau_g = [1 + (\omega_g \tau_g)^2]^{1/2} \tag{4.90}$$

With large enhancements ($\omega_g \tau_g \gg 1$) this is just the resonance condition that maximizes the space charge field and in this case

$$s - \mathrm{i}\omega = s - \mathrm{i}\omega_g = \frac{1}{\tau_g} \tag{4.91}$$

and

$$E_1 = (1 + \mathrm{i}\omega_g \tau_g) F \approx \mathrm{i}\omega_g \tau_g F \tag{4.92}$$

Enhancement of the space charge field can only be achieved up to the point of saturation, the point where charge donor sites become exhausted. The saturation field increases with lower spatial frequencies, v, as more donors become available in the physical space within one period of the grating. Efficient space charge field enhancement is thus confined to low spatial frequencies. In these circumstances the following approximations will hold true:

$$E_A \gg 1, \quad vE_A \ll 1, \quad E_A \gg v \tag{4.93}$$

giving

$$F \approx -\tfrac{1}{2} E_A \tag{4.94}$$

which is a real quantity, illustrating that the optimum $\pi/2$ phase relationship has been preserved and also an enhancement factor $\xi = \omega_g \tau_g$ has been achieved. In terms of physical properties, ξ may be expressed as (Walsh *et al.*,

1990)

$$\xi = \omega_g \tau_g = P \triangleq \frac{L_{E_A} K}{1 + (L_D K)^2} \tag{4.95}$$

which we shall define here as the enhancement factor and in this example adopts the value P. $L_{E_A} = \mu \tau_R E_A$ is the drift length of the charge carriers in the material, $L_D = (D\tau_R)^{1/2}$ is the diffusion length, K is the magnitude of the grating wavevector and E_A is the value of the applied field.

Despite being able to achieve large enhancements of the space charge field while maintaining an optimum phase shift, experimental difficulties still arise. The two main problems are firstly the problem of introducing large uniform d.c. electric fields into the photorefractive crystal, which is usually made difficult by electrode contact problems and varied intensity levels across the photoconductor. The second problem arises from the effect of absorption in the material.

The first of the aforementioned problems is overcome when a.c. enhancement voltages are applied across the crystal, as the electric field is introduced through the capacitive coupling across the electrodes, overcoming the problems of resistive contacts, Schottky barrier effects and electrode creep into the material. For this reason too, uniform illumination of the photorefractive material is no longer crucial for a uniform field distribution throughout. The second main problem of absorption is an even larger drawback to the d.c. field and moving fringe enhancement technique in practical photorefractive materials and the reasons for this are now discussed.

The characteristic complex frequency $s = 1/\tau_g + i\omega_g$ of a photorefractive material is intensity dependent, as discussed earlier. This means that the optimum fringe detuning frequency cannot be maintained through an absorbing material. This also has consequences when considering two-wave mixing based image amplification applications (Hall *et al.*, 1985; Tschudi *et al.*, 1986). Differing intensity regions of the image will require differing amounts of frequency detuning for optimum amplification and such a problem could not be easily solved. In an absorbing material (thickness $0 < z < L$, and absorption coefficient α) the overall intensity will fall off as $I_0(z) = I_0(0) \exp(-\alpha z)$, assuming a strong undepleted pump beam. As ω_g and τ_g depend on τ_c, the dielectric relaxation time, which in turn is just inversely dependent on the inensity I_0, ω_g and $1/\tau_g$ depend on I_0: hence

$$s(z) = s(0) \exp(-\alpha z) \tag{4.96}$$

This is a change in the characteristic frequency through the crystal due to absorption (Walsh, 1990; Walsh *et al.*, 1990). Considering the simplified undepleted pump approximation for the two-wave mixing interaction, coupled wave analysis tells us that for the signal beam

$$I_1(L) = I_1(0) \exp(\bar{\Gamma} L) \tag{4.97}$$

where

$$\bar{\Gamma}(z) = \frac{1}{L}\int_0^L \Gamma(z)\,\mathrm{d}z \tag{4.98}$$

and Γ is the two-wave mixing gain, depending on the imaginary part of the space charge field E_1. The enhancement factor in this case becomes:

$$\xi = \mathrm{Im}\left[\frac{1}{L}\int_0^L \frac{S(0)\exp(-\alpha z)}{S(0)\exp(-\alpha z) - \mathrm{i}\omega}\cdot \mathrm{d}z\right] \tag{4.99}$$

Evaluating the integral and taking the imaginary part, we have for the enhancement factor

$$\xi = \frac{1}{\alpha L}\tan^{-1}[(\omega - \omega_g)\tau_g] + \tan^{-1}\{[\omega - \omega_g \exp(-\alpha L)]\tau_g \exp(\alpha L)\} \tag{4.100}$$

For the case with absorption, this expression has a maximum when

$$\omega_m \tau_g = [1 + (\omega_g \tau_g)^2]^{1/2} \exp\left(-\alpha\frac{L}{2}\right) \tag{4.101}$$

and this is plotted in Figure 4.6, to show the critical effect the absorption

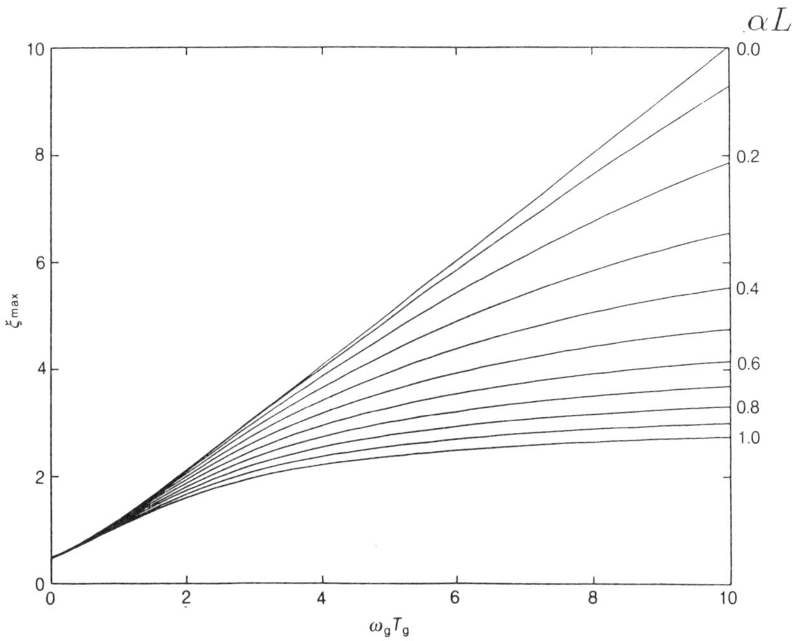

Figure 4.6 The effect of absorption on the enhancement factor.

has upon the maximum enhancement. With the approximation $\omega_g\tau_g \gg 1$,

$$\xi_{max} = \frac{1}{\alpha L}\tan^{-1} A \tan^{-1} B \tag{4.102}$$

where

$$A = \omega_g\tau_g\left[1 - \exp\left(-\frac{\alpha L}{2}\right)\right] \tag{4.103}$$

and

$$B = \omega_g\tau_g \exp\left(\frac{\alpha L}{2} - 1\right) \tag{4.104}$$

With low values of absorption (or short interaction length) $\alpha L \ll 1$ then

$$\xi_{max} \approx \frac{2}{\alpha L}\tan^{-1}\left[\omega_g\tau_g\left(\frac{\alpha L}{2}\right)\right] \tag{4.105}$$

and the enhancement of the space charge field becomes critically dependent upon the product $\omega_g\tau_g\alpha L/2$. When this is small ($\ll 1$) the previous expression can be seen to revert to $\xi_{max} \approx \omega_g\tau_g$. When $\omega_g\tau_g\alpha L/2 \gg 1$ then

$$\xi_{max} \approx \frac{\pi}{\alpha L} \tag{4.106}$$

Thus, in absorptive media, maximum enhancement may be achieved only in a thin region of the material. In extreme cases where long grating spacings are being used, it is possible that the volume diffractive nature of the enhanced grating is lost and multiple diffracted orders are seen as a self-diffracted process. Effects that can be described in this way have been reported in the literature.

In summary, d.c. applied fields can be applied to photorefractive materials to achieve large enhancement factors of the space charge field. In two-wave mixing beam amplification applications, care must be taken to preserve the optimum phase shift of the space charge field and this usually involves the use of moving fringe techniques. The effect of linear absorption can confine the large enhancements of the space charge field to a narrow interaction region in the photorefractive material and can, in extreme cases, give rise to thin grating scattering phenomena (Solymar and Cooke, 1981).

4.3.2. Alternating electric field enhancement techniques

Alternating electric field (a.c.) enhancement techniques can be used in place of the more established d.c. techniques, overcoming the requirement of detuning one of the beams in a two-wave mixing interaction, for optimum gain, while also alleviating the problems associated with the electrodes and nonuniform field distributions in the crystal (Kumar *et al.*, 1987b; Gan *et al.*, 1988; Kumar, 1987). The frequency of the applied a.c. field is not critical, provided that it lies between an upper and a lower limit. These limits arise because the period of the a.c. field must not be so short as to allow the field to be reversed within the recombination time of the photoexcited carriers. Individual charge carriers can then drift under the influence of this field, allowing a space charge field to be built up that is larger than that brought about by the diffusion-only process. On the long time scale, the period of the applied a.c. field must be suitably short so that there is no net motion of the charge carriers in a particular direction, as is the case with an applied d.c. field. This condition preserves the optimum $\pi/2$ phase shift between the illuminating pattern and the space charge field and this will be guaranteed if the period of the applied a.c. field is shorter than the grating relaxation time of the material, τ_g.

Although the grating relaxation time is intensity dependent, in most of the commonly used photorefractive materials it varies between values of a few milliseconds, for BSO, to a few microseconds for semiconductor GaAs, at typical power levels of 1 W/cm^{-2}. The free carrier lifetime is not so easily determined for the photorefractive materials but it is known to be of the order of nanoseconds. The frequency constraints thus placed upon the applied a.c. fields are such that they must be below several hundred megahertz, but above a few hundred kilohertz for the faster semiconductor photorefractive materials. For BSO two-wave mixing experiments at helium–neon laser optical power levels, 50 Hz mains frequency applied fields are suitable to achieve a.c. field enhancement of the two-beam coupling process (Stace, *et al.*, 1989). Alternating voltages applied to electrodes on the crystals, at the frequencies described, are capacitively coupled into the crystal to produce the fields, without ohmic contacts at the electrodes being a problem. The frequent reversal of the applied voltages eliminates electrode 'creep' into the crystal and hence electrode and crystal degradation.

In the range of applied field frequencies described above, enhancement of the photorefractive effect can be investigated by solving the differential equation 4.15 that describes the development of the space charge field. It is appropriate to average the differential equation over the relatively rapidly varying applied field, as compared with the grating lifetime, and this then gives

$$\frac{\partial E_1}{\partial t} + \langle s \rangle E_1 = \langle sF \rangle \tag{4.107}$$

with

$$E_1 = \frac{\langle sF \rangle}{\langle s \rangle} \tag{4.108}$$

Substituting the time-varying form for the electric field into the expressions for s and F such that E_A becomes $\hat{E}_A q(t)$, where $\hat{E}_A$ is the peak of the applied field and $q(t)$ describes the time variation of the field, and averaging over one period, we have that

$$\langle s\rangle = \frac{1+v^2}{1+\theta^2 v^2}\frac{1}{P}I_0(P) + \frac{1}{\theta^2}I_1(P) \tag{4.109}$$

and

$$\langle sF\rangle = -\frac{1}{2}\left[\frac{\mathrm{i}v}{1+\theta^2 v^2}\frac{1}{P}I_0(P) + \frac{1}{\theta^2 v}I_1(P)\right] \tag{4.110}$$

where P can be expressed as

$$P = \frac{\theta^2 v E_A}{1+\theta^2 v^2} = \frac{L_{\hat{E}_A}K}{1+L_D^2 K} \tag{4.111}$$

where $I_0(P)$ and $I_1(P)$ are the integrals

$$I_0 = \frac{1}{T}\int_{-T/2}^{T/2} \frac{P}{1-\mathrm{i}Pq(t)}\,\mathrm{d}t \tag{4.112}$$

and

$$I_1 = \frac{1}{T}\int_{-T/2}^{T/2} \frac{-\mathrm{i}Pq(t)}{1-\mathrm{i}Pq(t)}\,\mathrm{d}t \tag{4.113}$$

The values of these integrals are dependent on the nature of the applied waveform. The ratio $\langle sF\rangle/\langle s\rangle$, which gives the enhanced space charge field, then simply becomes of the form

$$E_1 = -\frac{1}{2}\,\frac{\mathrm{i}v + E'_A}{1-\mathrm{i}v(\mathrm{i}v + E'_A)} \tag{4.114}$$

where E'_A is an effective applied field given by

$$E'_A = \mathrm{i}\frac{I_1(P)}{I_0(P)}\hat{E}_A = \mathrm{i}\eta E_A \tag{4.115}$$

It should be noted that the form for the enhanced space charge field, E_1, is of the same form as for the diffusion-only case with the imaginary notation indicating that the $\pi/2$ phase shift between the illuminating interference pattern and enhanced space charge field is preserved.

It is convenient to restrict our attention to time-symmetric applied waveforms. We can then consider the difference in effect between varying applied waveforms by evaluating the integrals I_0 and I_1 with the dependence of the waveform expressed as the symmetric function $q(t)$. This is now done for the cases of applied square waveforms and cosinusoidal waveforms.

Square waveform applied a.c. fields give the maximum amount of enhancement of the space charge field. The square waveform may be defined over one period T as

$$q(t) = \begin{cases} -1 & \text{if} \quad -T/2 \leqslant t < 0 \\ 1 & \text{if} \quad 0 < t \leqslant T/2 \end{cases} \tag{4.116}$$

The integrals $I_0(P)$ and $I_1(P)$ can then be evaluated, giving the enhancement factor

$$\eta = \frac{I_1(P)}{I_0(P)} = P \tag{4.117}$$

and hence an enhancement to the space charge field of

$$E'_{\mathrm{A}} = \mathrm{i}P\hat{E}_{\mathrm{A}} \tag{4.118}$$

This is an equivalent enhancement to that achieved with the d.c. fields and moving fringe techniques but with the optimum $\pi/2$ phase shift of the space charge field ensured if the frequency of the applied field meets the requirements described previously. The effect of an applied a.c. field may be considered as an enhancement to the intrinsic diffusion process for the charge carriers as no net drift direction is established. It can also be shown that no applied field can exceed an enhancement factor of P and this factor may adopt large values, particularly in semiconductor photorefractive materials such as undoped GaAs. With a grating spacing where saturation of the photorefractive can be avoided (say 40 μm), values for P can be in excess of 30.

The integrals $I_0(P)$ and $I_1(P)$ can be evaluated for a range of applied a.c. waveforms. Practically, it may not be possible to generate 'ideal' square waves with an inifinite slew rate, so it is interesting to evaluate the effect of a cosinusoidally varying electric field, from which high voltages may be generated with a tuned resonant circuit. In the case of a cosinusoidal applied field, the enhancement factor of the space charge field becomes

$$E'_{\mathrm{A}} = \mathrm{i}\eta\hat{E}_{\mathrm{A}} = \mathrm{i}\frac{1}{P}[(P^2+1)^{1/2} - 1]\hat{E}_{\mathrm{A}} \tag{4.119}$$

This function is plotted in Figure 4.7, to compare the reduced amount of enhancement with that of the square wave field case. In the limit of large P the expression shows that the space charge field can only reach the maximum value of that of the applied field, i.e. $P_{\max} = 1$. The simplicity of generating cosinusoidal fields and some consequences arising from large depths of modulation in the space charge field, as discussed next, make cosinusoidal applied fields an attractive method for enhancing the photorefractive effect. The important $\pi/2$ phase shift between the space charge field and the illuminating intensity pattern is preserved.

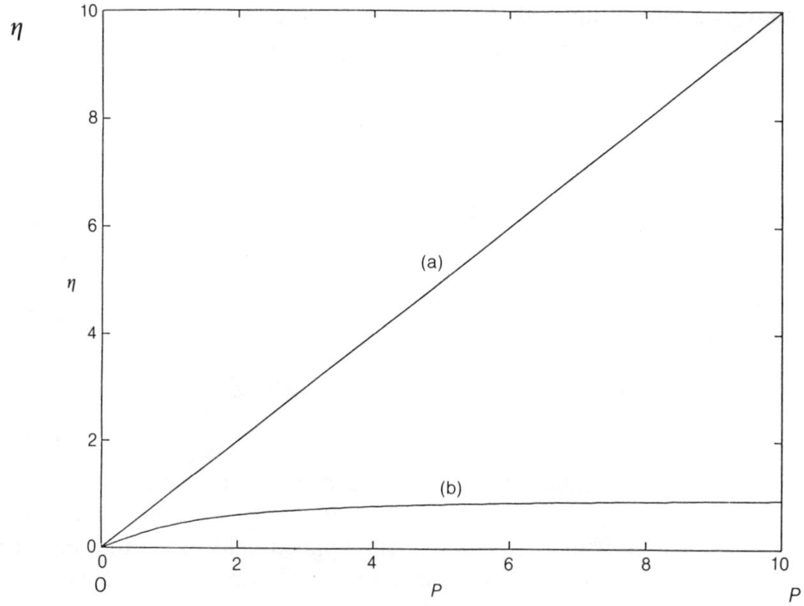

Figure 4.7 Enhancement factor for square wave fields (curve a) and cosinusoidal fields (curve b).

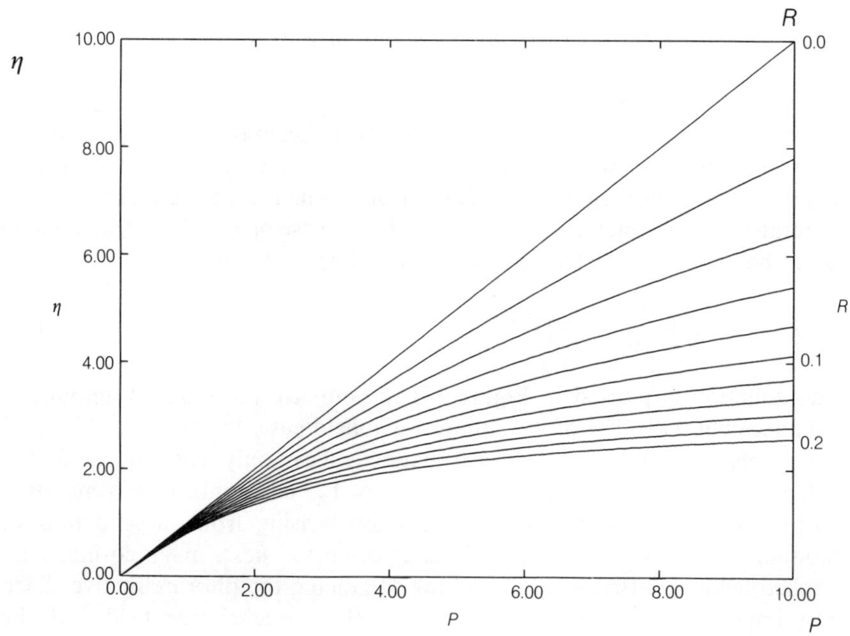

Figure 4.8 Fall-off in enhancement with a slew rate limited applied square wave field.

Another reason why cosinusoidal applied fields are advantageous to those of a square waveform is in the form of the spatial frequency response. Although the square wave applied fields offer the larger enhancement, this enhancement falls off rapidly over a range of spatial frequencies. The fall-off in enhancement is not so pronounced when using cosinusoidal fields and this makes their use more suitable for amplification of image bearing beams, containing spatial information at a range of frequencies. This effect is illustrated in Figure 4.14. Figure 4.8 shows the fall-off in the enhancement factor when a slew rate limited square waveform is applied as an enhancement field where R is the fraction of the period taken up with the rise and fall time. The enhancement becomes substantially reduced when the rise and fall times become greater than approximately 10% of the period of the applied waveform for the value of P chosen.

4.3.3. Effects at large modulation

Experimentally it has been shown that the large photorefractive nonlinear beam coupling gains, as measured in a two-wave mixing interaction, with applied d.c. fields, fall off as the two beams become comparable in intensity. This effect of large depths of modulation of the space charge field suggests that large optical gains can only be achieved in photorefractive materials in the regime of small signal beams (Swinburne *et al.*, 1989).

The saturation effect at large modulation depths, associated with large enhancements of the space charge field, arises because a point is reached where, to support such a space charge field, the free carrier density would have to acquire negative values. This, however, clearly does not happen and a limit is reached, for the maximum value of the space charge field at a particular depth of modulation. To understand the effect further it is necessary to look at numerical solutions of the band transport equations of Kukhtarev *et al.* Hall *et al.* (1985) have done this to investigate a whole range of nonlinear phenomena in photorefractive media, with applied d.c. fields, by expanding the band transport equations as a power series in the modulation ratio m. When this is done for the case of resonant moving fringes, care must be taken to limit the growth of the terms in the perturbative series and it is useful to consider the fundamental component of the space charge field as the expansion parameter to calculate the coefficients for the modulation depth in the series. Values for the applied field, saturation spatial frequency, dark donor occupation, drift length and ω, the detuning frequency, have to be supplied to evaluate the space charge field at various depths of modulation.

Numerically, the Fourier coefficients of the probability of donor occupation, free charge carrier density and space charge field at corresponding modulation depths can be computed. Figure 4.9 shows the computed values for the important first spatial component of the space charge field, for differing depths of modulation with large applied d.c. fields and no fringe detuning,

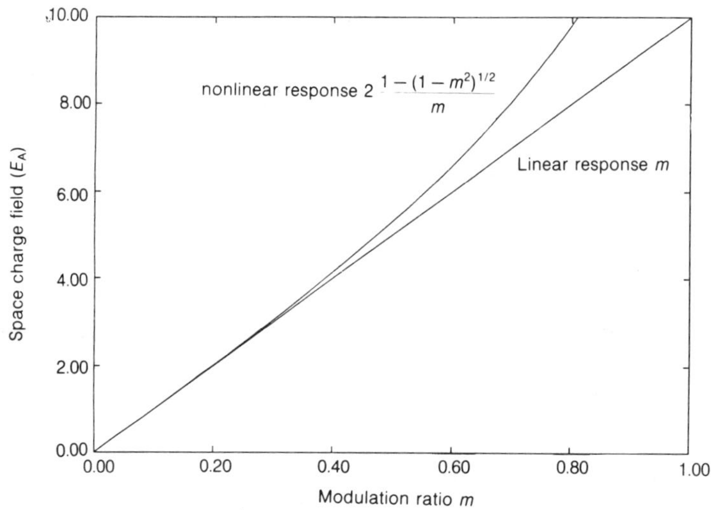

Figure 4.9 The large signal effect (d.c. field).

i.e. nonresonant enhancement. The graph shows the relationship

$$|E_1| \approx 2\left[\frac{1-(1-m^2)^{1/2}}{m}\right]E_A \tag{4.120}$$

from which the linear in modulation approximation is derived for small values of m and this is used to simplify the analysis of the band transport equations. When the value of m approaches 0.75, the departure from the linear in modulation case becomes significant.

Figure 4.10 shows the situation where one of the beams is detuned to give an enhancement factor of approximately ten times, i.e. $P = 10$, for the particular values chosen for E_A, v, and θ. At low depths of modulation, the first component of the space charge field rises rapidly, because of the large enhancement factor. When a modulation depth of only 0.1 is reached, the large signal effect can be seen to prevent the further build up of the space charge field. These results confirm the experimental investigations of Refregier *et al.* (1984).

The situation is similar with the case of applied a.c. fields. In general, the analysis of the large signal effects for arbitrary applied waveforms is a numerically demanding problem that does not lend itself to the Fourier series methods used in the case of the d.c. applied fields. In the case of an ideal square wave, a simplified analysis is possible giving similar results to those obtained numerically with the d.c. fields and moving fringes. This is done by first considering the rate equations for the generation and recombination

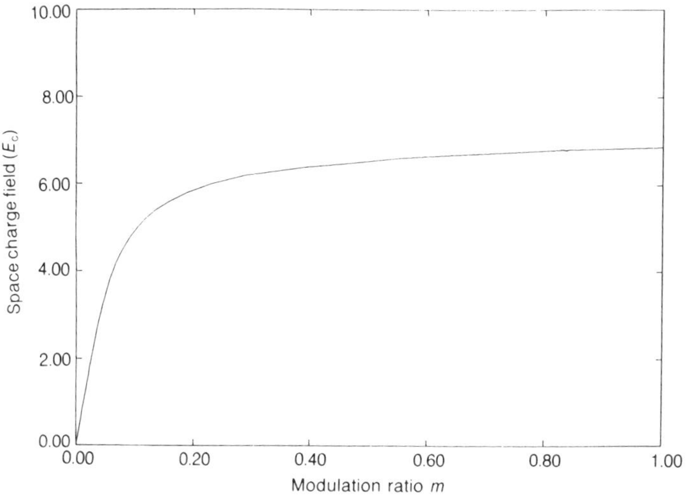

Figure 4.10 Large signal effects—d.c. applied field, and moving grating at the resonant frequency $\omega = \omega_a$.

in a photorefractive material where donor saturation and diffusion have been neglected:

$$g(I) - \frac{n}{\tau_R} + \mu\frac{\partial}{\partial x}(nE) = 0 \tag{4.121}$$

and

$$\frac{\partial N_D^+}{\partial t} = g(I) - \frac{n}{\tau_R} \tag{4.122}$$

where $g(I)$ is the intensity-dependent generation rate, n is the free carrier charge density and τ_R is the recombination time. N_D^+ is the number density of ionized donors, μ the mobility of the free charge carriers and E the space charge field. If we assume a cosinusoidal illumination pattern $I = I_0[1 + m \cos(\xi)]$, where $\xi = \mathbf{K}\cdot\mathbf{x}$, and if we normalize with respect to $g(I_0)\tau_R$ we obtain the differential equation

$$1 + m\cos\xi - n + P\frac{\partial}{\partial\xi}(nE) = 0 \tag{4.123}$$

where $P = L_{\hat{E}_A}K$ with $L_{\hat{E}_A} = \mu\tau_R\hat{E}_A$ for large enhancements.

Considering the space charge field and the free carrier density to be composed of a slowly varying component and a more rapidly varying component on the time scales of a period of the applied square waveform, then we have

$$E = \langle E \rangle + q(t) \tag{4.124}$$

$$n = \langle n \rangle + n(t) \tag{4.125}$$

Using these equations in the differential equation 4.122 and averaging, an expression for the time-averaged space charge field is obtained. Remembering that in the specific case of the applied square waveform $q^2 = 1$, we obtain

$$\langle E \rangle = -\frac{Pm\sin\xi}{1 + m\cos\xi}(1 - \langle E \rangle^2) - \frac{\partial}{\partial\xi}\langle E \rangle^2 \tag{4.126}$$

and this equation characterizes the space charge field inside a photorefractive material with an applied square waveform. This method has no knowledge of the free charge carrier density but the effect of the large modulation depths on the magnitude of the first component of the space charge field can be investigated. We may estimate the magnitude of the space charge field by substituting $\langle E \rangle = a\sin(\xi)$ into the above differential equation to find the following relationship:

$$a \approx Pm(1 - \tfrac{1}{2}a^2) \tag{4.127}$$

so that

$$a = \frac{[1 + 2(Pm)^2]^{1/2} - 1}{Pm} \tag{4.128}$$

and from this it may be seen that the magnitude of the first component of the space charge field has the following properties:

$$a \approx \begin{cases} Pm & \text{if} \quad Pm \ll 1 \\ \sqrt{2} & \text{if} \quad Pm \gg 1 \end{cases} \tag{4.129}$$

The large signal saturation of the space charge field is described by the $1 - \langle E \rangle^2$ term in the differential equation 4.126. The predicted amplitude of the first spatial Fourier component of the space charge field is plotted in Figure 4.11 and shows that saturation occurs as the product of the enhancement factor P and the modulation depth m becomes large. This was also found numerically to be the case with d.c. fields and moving fringes. The maximum value of the space charge field at large depths of modulation is found to be approximately $\sqrt{2}\hat{E}_A$.

For practical applications, applied a.c. square waveforms may not be as desirable as cosinusoidal fields because of the large signal saturation phenomena. The large enhancement factors possible with applied square waveforms may only be achieved when the signal beam is substantially lower in intensity than that of the pump beam in a typical two-wave mixing

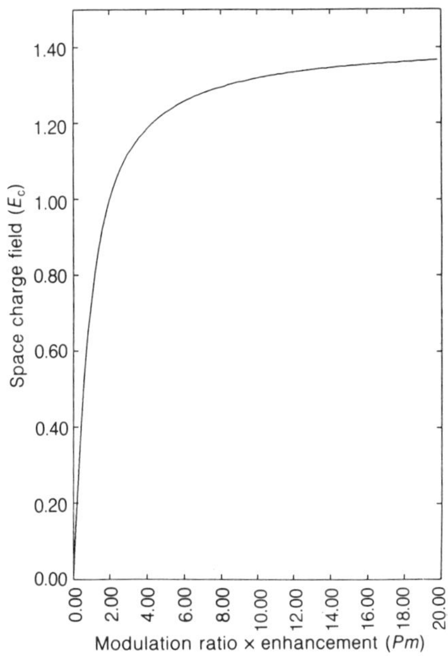

Figure 4.11 Large signal effects—a.c. field.

experiment. The maximum value that the space charge field can reach is approximately the value of the applied field. Large cosinusoidal fields can be generated resonantly, with little expenditure in energy in a tuned circuit incorporating the photorefractive crystal. The quality factor of the circuit may be as high as 100, so radio frequency fields of many tens of kilovolts can be generated and are suitable for use as enhancement fields for most photorefractive materials. The enhancement factor with cosinusoidal fields cannot exceed unity, which is of no disadvantage if large signal amplification is required, with magnitude of the space charge field reaching the value of the applied field.

Radio frequency electric fields have been applied to samples of the photorefractive material GaAs. With a field of $6\,\mathrm{kV\,cm^{-1}}$, at a frequency of approximately 780 kHz, two-beam coupling gain of $2.2\,\mathrm{cm^{-1}}$ was achieved (Walsh, *et al.*, 1989); this result is shown in Figure 4.12 and the point where the gain of the signal beam exceeds absorptive losses is indicated.

4.4. THE PHOTOREFRACTIVE TIME RESPONSE

The feeling that photorefractive materials are slow in response is not necessarily justified considering that such materials respond to continuous

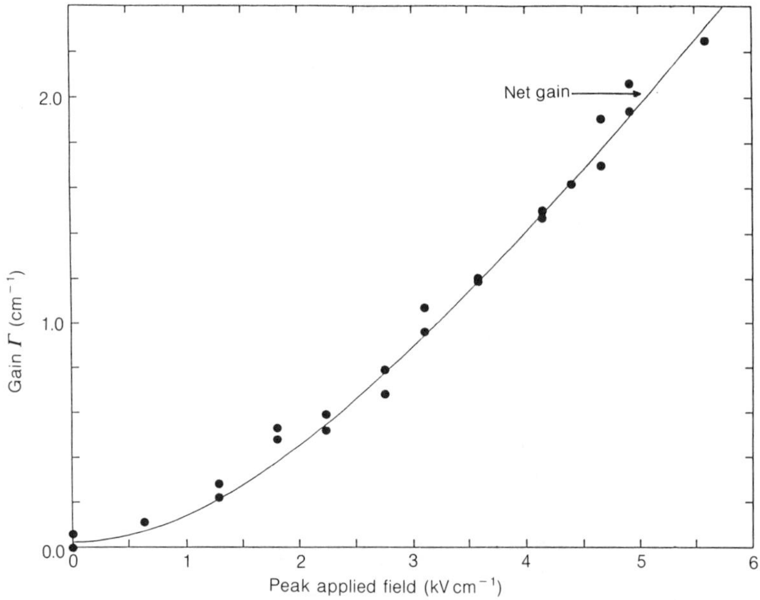

Figure 4.12 Gain exceeding absorption in GaAs with applied cosinusoidal fields.

wave lasers. It can be shown that photorefractive GaAs with a space charge field of $16\,kV\,cm^{-1}$ and responding on 25 ms time scales is limited in speed only by the arrival rate of the photons at an intensity of approximately $1\,W\,cm^{-2}$. Each arriving photon can only produce one excited charge carrier to contribute to the build up of the space charge field. When investigating the photorefractive effect in GaAs with the optical powers associated with a pulsed YAG laser system, significant photorefractive effects on picosecond time scales are observed (Valley *et al.*, 1989; Glass, 1990). The speed of response is limited by the same physics and is therefore comparable with the speed of response of any photoconductive electro-optic device, e.g. MQW SEEDs and LCLVs.

4.4.1. The photoconductive response

The effect of applying an electric field to photorefractive materials to enhance the space charge field slows down the photoconductive response of the material. This can be seen to arise from the extra photogenerated charge carriers and hence time that is required to build up the enhanced space charge field. An exception to this is the case of photorefractive materials that have poor carrier mobilities, such as barium titanate. In these materials the band transport model predicts a speed-up in photorefractive response with the application of electric fields (Hall *et al.*, 1985) and this effect is commonly used to improve the response speed for certain applications (Sayano *et al.*, 1990). It should be noted, however, that there is no enhancement of the space charge field for materials in this case. In the dimensionless band transport equations of Hall *et al.* (equations 4.11–4.13), it is the diffusion length parameter $\theta \approx \mu\tau_R E_C K_C$ that determines whether a speed up, or slow down and space charge field enhancement, is achieved when fields are applied. The former situation is the case when $\theta < 1$.

The application of either a.c. or d.c. electric fields modifies the time evolution of the space charge field. This can be described by the complex characteristic frequency s of the space charge field, which in the case of no applied field has the following dependence on the spatial frequency v and diffusion length parameter θ^2:

$$s = \frac{1+v^2}{1+\theta^2 v^2} \tag{4.130}$$

This gives rise to a real grating time constant of

$$\tau_g = \frac{1+\theta^2 v^2}{1+v^2}\tau_c \tag{4.131}$$

where τ_c is the dielectric relaxation time. Applied fields give a modified complex characteristic frequency s that may no longer be purely real and can be found by substituting the functions $I_0(P)$ and $I_1(P)$, evaluated for an appropriate waveform, into the expression for s. For a d.c. field this gives

$$\langle s \rangle = \frac{1}{\tau_g} + \mathrm{i}\omega_g = \frac{1}{1+P^2}\left(\frac{1+v^2}{1+\theta^2 v^2} + \frac{P^2}{\theta^2}\right) + \frac{\mathrm{i}P}{1+P^2}\left(\frac{1+v^2}{1+\theta^2 v^2} + \frac{1}{\theta^2}\right) \tag{4.132}$$

in units of τ_c. The complex time response gives rise to an oscillatory growth of the space charge field which is seen experimentally. When resonant moving fringes are used, the expression becomes real and gives a time constant of

$$\tau_g = (1+P^2)\left(\frac{1+v^2}{1+\theta^2 v^2} + \frac{P^2}{\theta^2}\right)^{-1} \tag{4.133}$$

in units of the dielectric relaxation time τ_c. This real time constant is also the case with applied square wave fields, which is not surprising, as the same amount of enhancement is achieved as with the d.c. field and moving fringe technique.

It is also interesting to examine the case with cosinusoidal applied fields. Again, evaluating the integrals $I_0(P)$ and $I_1(P)$ and finding the expression for the characteristic frequency s, which again is purely real, the time constant τ_g can be found:

$$\tau_g = (1+P^2)^{1/2}\left[\frac{1+v^2}{1+\theta^2 v^2} + \frac{(1+P^2)^{1/2}-1}{\theta^2}\right]^{-1} \tag{4.134}$$

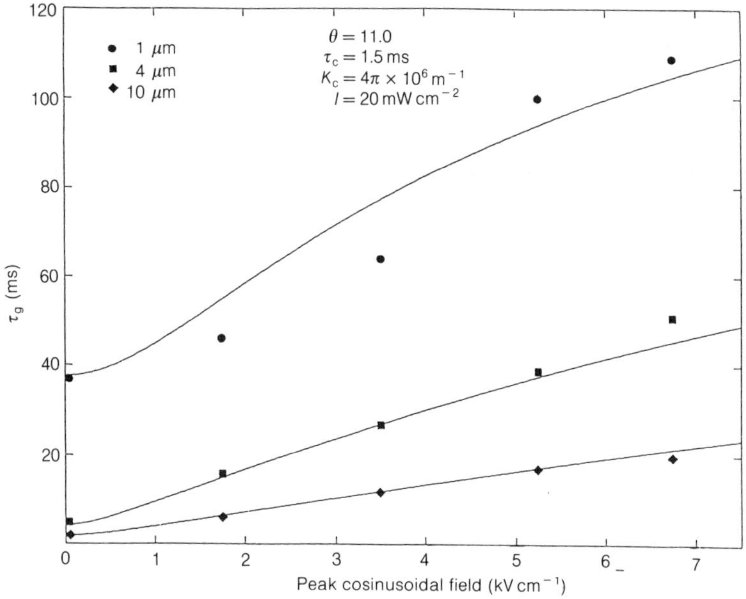

Figure 4.13 Slowing of the photoconductive effect with applied field.

in units of τ_c. Comparing this expression with that of the square waveform, it becomes apparent that the grating time constant slows down less rapidly with increasing P for the cosinusoidal applied fields. There is a trade-off between the amount of enhancement and the speed of response, again making the cosinusoidal field enhancement method more attractive.

In the limit of large electric enhancement, $\tau_g \approx \theta^2 \tau_c \approx (1 - W_c)/\sigma I_0$, where W_c is the probability of donor occupation at intensity I_0 and σI_0 is the rate of excitation of an occupied donor. Hence, in the limit of a large enhancement factor, the grating formation time constant is limited by the rate of photo-generation of carriers. Figure 4.13 shows experimental points and the theoretical curves for the grating formation time constant in photorefractive BSO with cosinusoidal applied fields.

4.4.2. The electro-optic response: critical slowing down

Apart from the slowing down effect that arises from the photoconductive effect in photorefractive materials, there is an electro-optical slowing effect analogous to the critical slowing down found with optical bistable switching elements. The effect of critical slowing down is not specific to a particular system but is a generic physical phenomenon.

The critical slowing down effect can be elegantly demonstrated in a particular four-wave mixing scheme, based on a photorefractive nonlinearity.

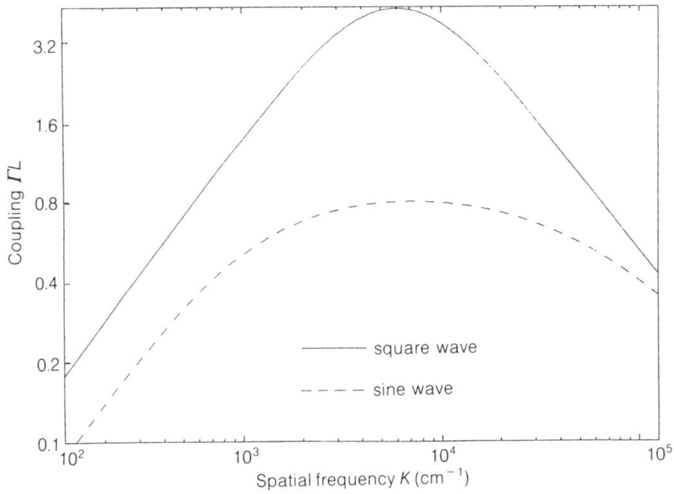

Figure 4.14 Spatial frequency dependence of two-beam coupling strength with sinusoidal and square wave applied fields (Note flatter response with sinusoidal applied field).

In this geometry the counterpropagating pump beams have orthogonal polarization states. A forward travelling third probe beam intersects the pump beams in the photorefractive crystal, and it shares the same polarization state as the forward travelling pump beam. The four-wave mixing interaction generates the fourth wave which returns back along the path of the probe wave as a phase conjugate beam. The photorefractive grating induced can be assumed to be transverse to the pump beams, with higher spatial frequency gratings being unimportant owing to saturation. In the case of a perfectly shifted refractive grating, the interaction can be considered as a forward and a backward two-wave mixing interaction. With a photorefractive material of cubic crystal class, the reversed polarization states of the forward and backward two-wave mixing interactions allow both the probe and the conjugate to experience gain. This gives rise to a resonance condition that gives the overall four-wave mixing interaction a theoretical infinite reflectivity for the probe beam at a particular strength of nonlinear interaction (ΓL), within the approximation of an undepleted pump beam analysis.

An analytical solution to the four-wave mixing coupled wave equations, assuming equal intensity undepleted pumps, predicts a reflectivity ρ for the interaction of

$$\rho = \frac{\Gamma L}{4 - \Gamma L} \tag{4.135}$$

which predicts an infinite reflectivity at $\Gamma L = 4$, a value that can be easily achieved in crystals of BTO with an applied cosinusoidal field of approximately $15\,\mathrm{kV\,cm^{-1}}$.

The time response of the four-wave mixing interaction at the resonance can be determined from the time development of the coupling strength γ. The solution for the reflectivity with a $\pi/2$ shifted grating, i.e. $\gamma = -\mathrm{i}\Gamma/2$, is given by equation 4.135. Since the problem is linear in the undepleted pumps approximation, the frequency response of the reflectivity (i.e. the response to a time-varying probe envelope) can be derived by considering the effect on γ of a frequency detuning of ω:

$$\gamma(\omega) = \frac{s\gamma}{s - \mathrm{i}\omega}, \quad s = \frac{1}{\tau_g} \tag{4.136}$$

hence

$$\begin{aligned}\rho(\omega) &= \frac{\Gamma L/(1 - \mathrm{i}\omega\tau_g)}{4 - \Gamma L/(1 - \mathrm{i}\omega\tau_g)} \\ &= \frac{\Gamma L}{4 - \Gamma L}\frac{1}{1 - \mathrm{i}\omega\tau_g'}\end{aligned} \tag{4.137}$$

where

$$\tau_g' = \tau_g \frac{4}{4 - \Gamma L} \tag{4.138}$$

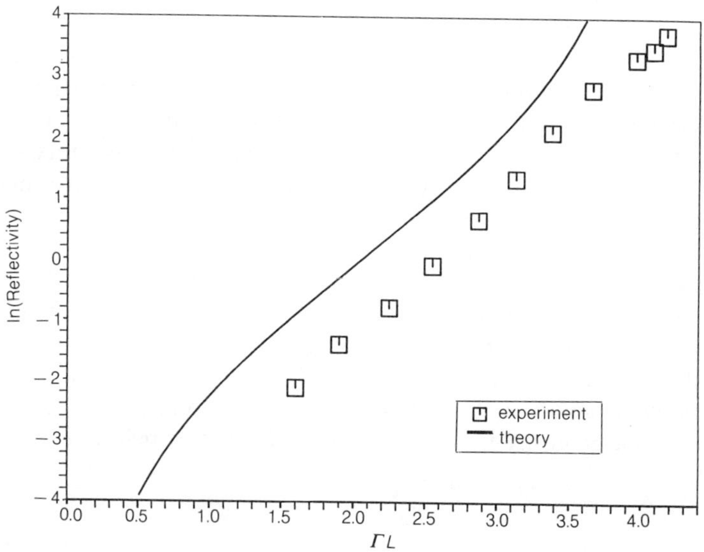

Figure 4.15 Large phase conjugate reflectivities—a comparison between experiment and theory.

This expression tends to infinity when the gain is adjusted to give the infinite reflectivity, i.e. $\Gamma L = 4$, and hence an infinite response time is experienced. This is the critical slowing down of the four-wave mixing interaction for which a constant gain–bandwidth product exists:

$$GB = \frac{\Gamma L}{4\tau_g} \tag{4.139}$$

The large reflectivities associated with this four-wave mixing interaction have been measured in our laboratory in a crystal of BTO; Figure 4.15. The time response of the interaction scales as $(1 + a\hat{E}_A)/(1 - b\hat{E}_A)$ with the size of the applied cosinusoidal field in a crystal of BTO which is as expected. There is an increase in the time constant of the photorefractive grating due to the photoconductive properties (the numerator of equation 4.138), as described earlier and there is another increase due to the critical slowing down phenomenon (the denominator of equation 4.138), explaining the observed results; Figure 4.16.

4.4.3. Relative merits of some nonlinear materials

Having now identified the underlying physics that governs the nonlinear behaviour of many types of nonlinear optical materials, it is now useful to

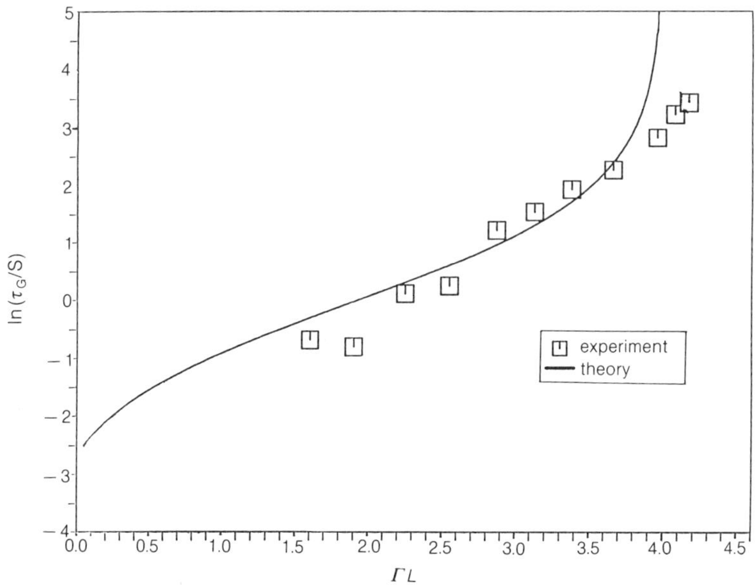

Figure 4.16 Critical slowing down—the natural logarithm of the effective grating time constant vs. coupling strength.

compare qualitatively the magnitude of nonlinear effects observed. A nonlinear optical material displaying an intensity-dependent refractive index needs ideally to induce a π phase shift on an interacting optical beam. Thus

$$\phi = \frac{\Delta n\, K}{\alpha} > \pi \tag{4.140}$$

This figure is not, however, a complete assessment of a material's performance as one should consider the number of photons required for the induced refractive change. In pixellated devices, where the limit in the area of the pixel must be no smaller than λ^2, because of diffraction effects, the number of photons arriving within the response time is given by

$$N = \frac{\alpha I \tau}{h\omega} \lambda^2 \tag{4.141}$$

where I is the intensity and τ is the response speed of the nonlinear device. The above figure should be as small as possible, for a particular size of nonlinearity, and it is possible to define a figure of merit M such that

$$M = \frac{\Delta n K/\alpha}{N} = \frac{2\pi hc\, \Delta n}{I \tau \alpha \lambda^4} \tag{4.142}$$

The strong dependence on the wavelength shows that short wavelengths are always desirable, but this is true for any material and, if a practical value of 1 μm is chosen, terms cancel in the expression leaving a figure of merit

$$M = \frac{\Delta n}{\alpha I \tau}, \quad \text{taking } \frac{2\pi hc}{(10^{-6})^4} \approx 1 \tag{4.143}$$

This figure can now be evaluated for a range of materials for comparison (Table 4.1).

Although the numbers chosen for the comparison are subjective, the figures of merit appear to be similar across the range of materials chosen. There appears to be some advantage with the SEED devices which is not altogether

Table 4.1 Estimated figures of merit

Parameter values	*PR $BaTiO_3$*	*PR GaAs*	*MQW SEED*	*Molecular polymer*
Δn	10^{-3}	10^{-5}	0.1	10^{-3}
α	0.2 cm^{-1}	1 cm^{-1}	10^4 cm^{-1}	10^2 cm^{-1}
I	1 W cm^{-2}	1 W cm^{-2}	1 KW cm^{-2}	1 GW cm^{-2}
τ	1 s	1 ms	10 ns	0.1 ps
λ	0.5 μm	1 μm	1 μm	1 μm
Merit (M)	0.8×10^{-7}	0.1×10^{-7}	10×10^{-7}	1×10^{-7}

correct, but they have been used for some of the fastest nonlinear optical switching currently achieved. With more precisely determined figures the apparent advantages of the MQW SEED devices may not be present. Interestingly, large gains can be achieved in the figures of merit if the nonlinear materials can be used at shorter wavelengths.

4.5. CONCLUSIONS

The band transport model of Kukhtarev *et al.* has been developed to describe the interaction of light with a photoconductor. The model has allowed the effect of a cosinusoidal intensity pattern on the charge distribution in the photoconductor along with its time evolution to be evaluated. When combined with the electro-optic properties of a photorefractive material, the photoinduced redistribution of charge can build up large internal electric fields of the order of several kV cm^{-1}, inducing substantial modulations in the refractive index, which can be as large as $\Delta n \sim 10^{-3}$ in photorefractive barium titanate. The redistribution of light resulting from a refractive grating index within a photorefractive material gives rise to a diverse range of nonlinear optical effects such as two- and four-wave mixing. The magnitudes of the nonlinear effects are essentially independent of the illuminating intensity, if dark thermal excitation of carriers can be neglected, and this makes the photorefractive effect a very sensitive nonlinearity. Examples of applications are given in Chapter 10.

Electric field techniques have been shown to offer large enhancements to the intrinsic photorefractive properties of a material, with the technique of d.c. fields and moving gratings preserving the $\pi/2$ phase relationship between the illuminating intensity pattern and the induced refractive index grating. Although cosinusoidal fields give less enhancement than square wave fields, they do have several advantages over the application of a square waveform field. These advantages are in the energy-efficient way that high voltages at radio frequencies can be generated with the use of a resonant circuit. With the reduced enhancement offered by a.c. cosinusoidal fields there is less critical slowing down of the photorefractive effect than with the square waveform field and the spatial frequency response is flatter for a wider range of spatial frequencies, making cosinusoidal enhancement fields possibly better suited for image amplification or optical information processing applications. Allied with these advantages is that large enhancements of the photorefractive effect space charge field can only be achieved at low signal beam levels, otherwise the space charge field can only be built up to approximately the same value as the applied field. To utilize pump beam energy efficiently in amplifying a signal beam in a two-wave mixing experiment, there is no disadvantage in using a cosinusoidal enhancement field, as the large fields that can be generated can outweigh any useful enhancement achieved with a square wave applied field. The main disadvantage of using a d.c. field and the moving fringe enhancement technique is that, in absorptive media, the optimum detuning frequency, which is intensity dependent, varies through the crystal, confining the maximum enhancement to a thin region of the crystal.

Apart from the slowing down of the photoconductive time response, with the application of enhancement fields, feedback from the electro-optic interaction gives rise to a critical slowing down phenomenon in a photorefractive medium. This effect is seen when high reflectivities are achieved in a resonant four-wave mixing experiment and is analogous to the critical slowing down observed in bistable switching at threshold levels. The significant result in the specific four-wave mixing example is that the gain multiplied by the speed of response, or the reciprocal of the time constant is a constant. This property appears to be generic through a range of nonlinear systems but may not be as elegantly shown.

Merit figures for a range of nonlinear materials and effects show that there is little to choose from in material performance, providing that response times are matched to the pulse duration of the chosen laser system. Photorefractive materials are often used with continuous wave lasers because of their sensitivity but many of the semiconductor photorefractive materials can exhibit large effects on picosecond time scales.

ACKNOWLEDGEMENTS

The authors are grateful to C. Stace and P. D. Foote, currently at the Sowerby Research Centre, British Aerospace, and to K. Walsh of the Sharp Laboratories of Europe Ltd., Abingdon (UK), for permission to reproduce some of their work.

REFERENCES

Bylsma, R. B., Bridenbaugh, P. M., Olson, D. H. and Glass A. M. (1987) Photorefractive properties of doped cadmium telluride, *Appl. Phys. Lett.*, **51** (12), 889–91.

Chen, F. S. (1967) *J. Appl. Phys.*, **38**, 3418.

Fabre, J. C., Jonathon, J. N. C. and Roosen G. (1988) $\bar{4}3m$ photorefractive materials in energy transfer experiments, *Opt. Commun.*, **65** (4), 257–60.

Feinberg, J. (1982a) Self-pumped, continuous wave phase conjugator using internal reflections, *Opt. Lett.*, **7** (10), 486–8.

Feinberg, J. (1982b) Asymmetric self-defocusing in an optical beam from the photorefractive effect, *J. Opt. Soc. Am.*, **72** (1), 46–51.

Feinberg, J. (1988) Photorefractive nonlinear optics, *Phys. Today*, **41** (10), 46–52.

Feinberg, J., Heiman, D., Tanguay, A. R. and Hellwarth, R. W. (1980) Photorefractive effects and light induced charge migration in barium titanate, *J. Appl. Phys.*, **51**, 1297–301.

Foote, P. D. (1987) Optically induced anisotropic light diffraction in photorefractive crystal, *PhD Thesis*, University of London.

Gan, X., Ye, S. and Sun, Y. (1988) Alternating electric field enhancement of two-wave mixing gain in photorefractive BSO, *Opt. Commun.*, **66** (2, 3), 155.

Garmire, E., Jokerst, N. M., Kost, A. *et al.* (1989) Optical nonlinearities due to carrier transport in semiconductors, *J. Opt. Soc. Am. B*, **6** (4), 579–87.

Giuliano, C. R. (1981) Applications of optical phase conjugation, *Phys. Today*, **34** (4), (April) 27–35.

Glass, A. M. (1990) Future trends in photorefractive and nonlinear optics, *Topical Meet. on Photorefractive Materials, Effects and Devices II, SFO, Aussois, January 17–19*, 1990.

Hall, T. J. and Powell, A. K. (1990) Phase conjugation. In *Nonlinear Optics in Solids* (ed. O. Keller), Springer Series on Wave Phenomena, Vol. 9, Springer, Berlin, pp. 235–64.

Hall, T. J., Jaura, R., Connors, L. M. and Foote, P. D. (1985) The photorefractive effect—a review, *Prog. Quantum Electron.*, **10**, 77.

Imbert, B., Rajbenbach, H., Mallick, S., Herriau, J. P. and Huignard, J. P. (1988) High photorefractive gain in two-beam coupling with moving fringes in GaAs:Cr crystals, *Opt. Lett.*, **13** (4), 327–9.

Klein, M. B. (1989) Photorefractive properties of $BaTiO_3$. In *Photorefractive Materials and Applications I* (eds P. Gunter and J. P. Huignard), Topics in applied Physics, Vol. 61 Springer, Berlin.

Kukhtarev, N. V., Markov, M. V., Odulov, S. G., Soskin, M. S. and Vinetskii, V. L. (1979) Holographic storage in electro-optic crystals. I: steady state, *Ferroelectrics*, **22**, 949–60.

Kumar, J. (1987) Photorefractive two-beam coupling with applied radiofrequency fields: theory and experiment, *J. Opt. Soc. Am. B*, **4** (7), 1079–82.

Kumar, J., Albanese, G. and Steier W. H. (1987a) Measurements of two-wave mixing gain in GaAs with a moving grating, *Opt. Commun.*, **63** (3), 191–3.

Kumar, J., Albanese, G., Steier, W. H. and Ziari M. (1987b) Enhanced two-beam mixing gain in photorefractive GaAs using alternating electric fields, *Opt. Lett.*, **12** (2), 120–2.

Miller, D. A. B. (1986) Electric field dependence of optical properties of quantum well structures. In *Electro-optic and Photorefractive Materials, Part I* (ed. P. Gunter), Springer Proceedings in Physics, Vol. 18, Springer, Berlin, pp. 35–49.

Neurgaonkar, R. R., Cory, W. K. and Oliver, J. R. (1990) Tungsten bronze crystals for photorefractive applications, *Topical Meet. on Photorefractive Materials, Effects and Devices II, SFO, Aussois, January 17–19, 1990.*

Nye, J. F. (1979) *Physical Properties of Crystals: Their Representation by Tensors and Matrices*, Clarendon, Oxford.

Partovi, A., Kost, A., Garmire, E. M., Valley, G. C. and Klein, M. B. (1990) Characterisation of band-edge photorefractive effect in compound semiconductors, *Topical Meet on Photorefractive Materials, Effects and Devices II, SFO, Aussois, January 17–19, 1990.*

Refregier, P., Solymar, L., Rajbenbach, H. and Huignard, J. P. (1984) Large-signal effects in an optical BSO amplifier, *Electron. Lett.*, **20** (16), 656–7.

Refregier, P., Solymar, L., Rajbenbach, H. and Huignard, J. P. (1985) Two beam coupling in photorefractive BSO crystals with moving grating: theory and experiments, *J. Appl. Phys.*, **58**, 45–7.

Salamo, G. (1986) Strontium barium niobate as a self-pumped phase conjugator, *Opt. Commun.*, **59** (5, 6), 417–22.

Sayano, K., Yariv, A. and Neurgaonkar, R. R. (1990) Order-of-magnitude reduction on the photorefractive response time in rhodium-doped $Sr_{0.6}Ba_{0.4}Nb_2O_6$ with a dc electric field, *Opt. Lett.*, **15** (1), 9–11.

Shih, C. C. and Yariv, A. (1982) A theoretical model of the linear electro-optic effect, *J. Phys. C*, **15**, 825–46.

Solymar, L. and Cooke D. J. (1981) *Volume Holography and Volume Gratings*, Academic Press, London.

Stace, C., Powell, A. K., Walsh, K. and Hall, T. J. (1989) Coupling modulation in photorefractive materials by applying a.c. electric fields, *Opt. Commun.*, **70**, 6.

Stepanov, S. I. and Petrov, M. P. (1984) Efficient phase conjugation in the photorefractive crystal $Bi_{12}TiO_{20}$, *Sov. Tech. Phys. Lett.*, **10**, 11.

Stepanov, S. I., Kulikov, V. V. and Petrov, M. P. (1982) Running holograms in photorefractive BSO crystals, *Opt. Commun.*, **44** (1), 19–23.

Swinburne, G. A., Hall, T. J. and Powell, A. K. (1989) Large modulation effects in photorefractive crystals, *IERE Int. Conf. on Holographic Systems, Components and Applications, Bath, IEE September 1989*, pp. 116–23.

Tschudi, T., Herden, A., Goltz, J., Klumb, H., Laeri, F. and Albers, J. (1986) Image amplification by two- and four-wave mixing in $BaTiO_3$ photorefractive crystals, *IEEE J. Quantum Electron.*, **22**, 8.

Valley, G. C., Dubard, J., Smirl, A. L. and Glass, A. M. (1989) Picosecond photorefractive response of GaAs:EL2, InP:Fe and Cd:Te:V, *Opt. Lett.*, **14**, 17.

Walsh, K. (1990) Photorefractive nonlinear optics in gallium arsenide, *PhD Thesis*, University of London.

Walsh, K. and Hall, T. J. (1989) Gain exceeding absorptive losses in photorefractive GaAs, *Appl. Opt.* **28**, 16–17.

Walsh, K., Powell, A. K., Stace, C. and Hall, T. J. (1990) Techniques for the enhancement of space charge fields in photorefractive media, *Opt. Soc. Am. B*, **7**, 288–303.

Optical bistability

5

Brian S. Wherrett and David C. Hutchings

5.1. INTRODUCTION

A bistable device is defined as one which, under steady state conditions and given a unique set of inputs, is capable of existing in either of two internal states and of producing either of two different stable outputs. At any given time the internal state and corresponding output depend on the history of operation of the device; by temporary adjustment of the inputs it should be possible to set or reset the device to either state. Thus the latest temporary set/reset signal determines the device output long after that signal has been removed—the system in effect remembers the nature of the temporary signal. Clearly the output-vs.-input response of such a device must be highly nonlinear. It is therefore not surprising that the concept of optical bistability was not discovered until the early 1960s when, following the advent of lasers, a vast range of nonlinear optical phenomena began to be investigated both experimentally and theoretically. Since that time optically bistable schemes have been studied for their intrinsic interest, for the role that bistability plays as a bridge between linear responses and the phenomena of instabilities and optical chaos, and for their device applications.

The use of bistables as optical memories or switches was suggested soon after 1960, as a basis for digital optical computing schemes. The recognition that similar, highly nonlinear but monostable responses, could be used for optical binary logic was implied by the earlier work but not discussed explicitly until the 1970s and has led to interest in digital optic numeric and image processing. The same two-dimensional (2D) parallel devices could equivalently be used as digital or analogue spatial light modulators, or as 2D image displays. Other applications in fast switching for optical communications networks, signal amplifiers in communications, and conversely for noise reduction and power limiting have also been mooted.

In this article we will concentrate on one form of optically bistable system—the nonlinearly dispersive Fabry–Pérot etalon; brief descriptions of alternative schemes are given in section 5.9. A Fabry–Pérot etalon consists typically of a plane-parallel-sided sample surrounded by partially reflecting mirrors. Following this introduction the fundamentals of such a cavity are presented in section 5.2 and the physics of nonlinearities of (i) electronic origin and (ii) thermal origin are covered in section 5.3. This structure allows us to consider a nonlinear system in section 5.4, having in mind precisely what is going on inside the active material of the Fabry–Pérot and therefore enabling both the steady state and the dynamics of the bistable switching to be considered. Device optmization is covered in sections 5.5 and 5.6. A brief description of experimental results for dispersive Fabry–Pérots is given in section 5.8 and for alternative bistability schemes in section 5.9, culminating in a discussion of achieved and predicted operating parameters for the sort of bistable plates that will form much of the basis for digital optical computing.

Bistability occurs in many branches of physics: mechanics, magnetism, electronics, optics, etc. A simple mechanical scheme serves to describe some of the basic features of bistability. The easily constructed device depicted in Figure 5.1(a) consists of a pointer mounted on a spindle at position A. Two pieces of elastic are connected to the pointer, at position B. One is attached to a fixed point C the other is free to move; it is held initially at some position on the x–y plane indicated. We are interested in the angle that the pointer makes to the vertical as the free elastic is moved. Thus for example if the tension is weak, corresponding to position y_1, and the elastic is moved from side to side (varying x) the pointer will follow. One produces the essentially linear response indicated by the straight dashed line in Figure 5.1(b). However, if the tension in the elastic is high, position y_2, then as the hold position is moved from the right to the left the pointer angle decreases slightly until at some critical point, with x to the left of centre (x_l), the pointer suddenly flips across from positive to negative angle. Conversely as the x position is retraced the pointer remains at negative angle until at an equivalent critical point, with x to the right of centre (x_r), the pointer jumps back to positive angle. The full line on Figure 5.1(b) shows the complete pointer response; a stable condition can be maintained at any place on this line, there being two possible stable states over the region from x_l to x_r; this is the region of bistability. In the bistable region the actual state at a given time depends on whether one approaches from the left or the right. There is also, at the midpoint, a steady state solution for which the pointer is vertical ($\theta = 0$). However, this is unstable to a slight movement of x. In fact over the full bistable region there is a third unstable steady state indicated by the chain-linked line in the figure. There are two 'control' parameters for this device—the positions x and y. Figure 5.1(b) gives the example of the family of response functions that is achieved for different y, as x is varied continuously. Note that if bistability is to be achieved then y (the tension) must exceed some critical value, and x must lie in a particular range of values. These same features will appear in the nonlinear Fabry–Pérot case, where x is replaced by the optical irradiance or power level incident on the etalon and y represents the initial detuning of the etalon with respect to a Fabry–Pérot transmission

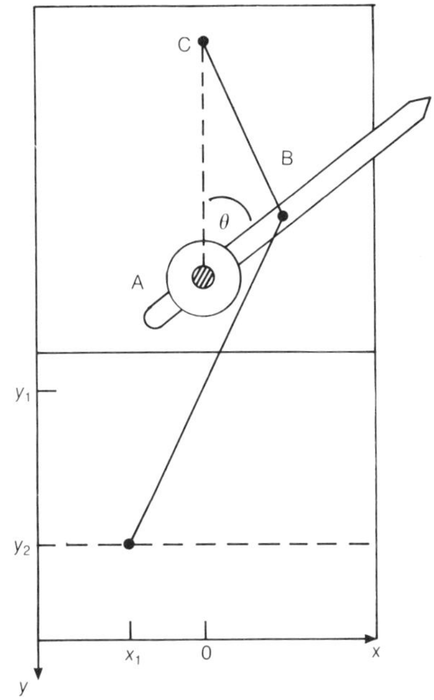

(a)

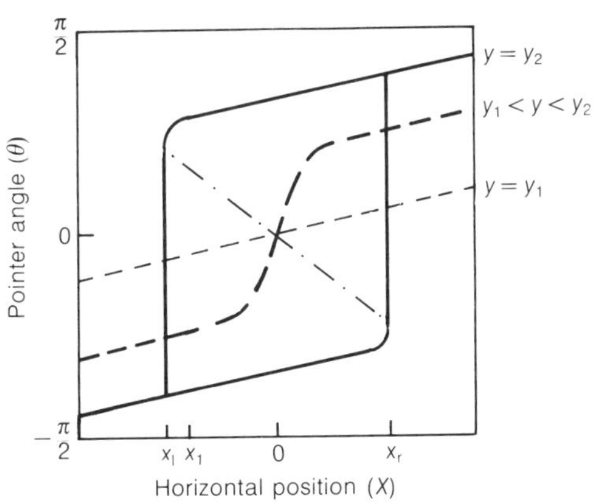

(b)

Figure 5.1 (a) Schematic of a mechanical device that demonstrates bistability; see text. (b) Example response functions for the device in (a): ---, linear response; **---**, nonlinear monotonic response; ———, hysteretic response ——·——, unstable steady state).

peak. There will be three steady state solutions for a range of irradiance, given a large enough initial detuning, and the central solution will be shown to be unstable. In the optical case it is also possible to achieve bistability at fixed irradiance (x) as the detuning parameter (y) is varied.

The first mention of optical versions of bistability in the scientific literature occurred in 1964, when schemes involving lasers with intracavity saturable absorbers were described conceptually (Kosomosky, 1964). If we consider such a scheme under lasing conditions the internal irradiance level is high and the absorber will be saturated; this is consistent with the achievement of the lasing threshold and therefore with the high internal irradiance. It is then supposed that a temporary external source can be used to quench the laser medium; this causes the system to drop below threshold, the internal irradiance falls and the absorber becomes operable. Hence when the external source is removed the presence of the absorption prevents laser action being reachieved. Hence for the same pumping conditions one either has lasing with the absorber saturated or no lasing with a large absorption—two, self-consistent, steady states. To switch back from below to above threshold one would introduce a second external source this time onto the absorber, in order to re-initiate the saturation. The first theoretical description of optical bistability also involved the use of nonlinear absorption. Szöke *et al.* (1969) formulated the response of a Fabry–Pérot containing a saturable absorber to a single laser input of variable irradiance. Again, two self-consistent CW solutions can be expected qualitatively for the same input level; one with internal irradiance high and the absorber inactive, and vice versa. To switch from one level to the other the input is either reduced or increased temporarily. Once again the system should 'remember' the last temporary signal. It was in an attempt to perform such an experiment, using a cell containing sodium vapour, that Gibbs *et al.* (1976) first observed CW dispersive optical bistability. Their results are reproduced in Figure 5.2; a family of nonlinear responses were observed, for different cavity lengths (equivalent to varying y above) as a function of the incident irradiance level (x). Following this experiment it was recognized that the high sensitivity of a Fabry–Pérot to small changes of the optical path length should be more effective in producing bistability than the weaker sensitivity to absorption changes. This has led to submilliwatt bistability in cavities containing thin (1–10 μm) layers of semiconductors or liquid crystals, as described in later sections of this article.

5.2. THE LINEAR FABRY–PÉROT ETALON

The simplest Fabry–Pérot cavity is a plane-parallel-sided, linear medium, in air, with no absorption; Figure 5.3. The electric fields (E_1' to E_2''') are evaluated in terms of the incident field E_1 by solving Maxwell's wave equation for the

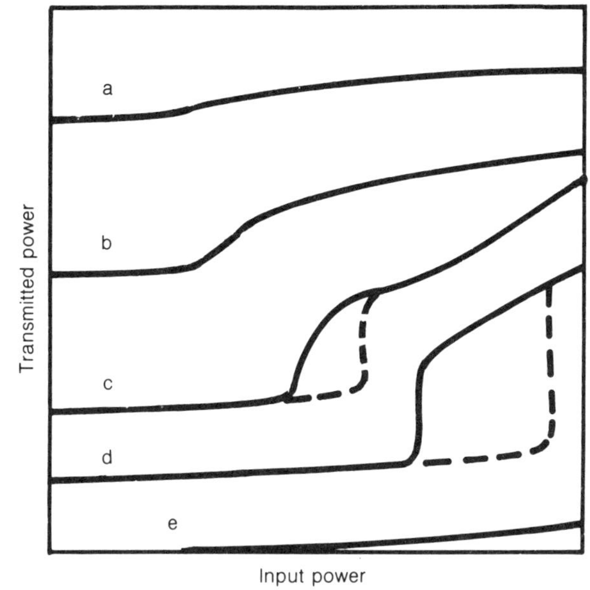

Figure 5.2 Dispersive optical bistability in Na vapour. After Gibbs *et al.* (1976). Curves a–e signify different cavity detunings.

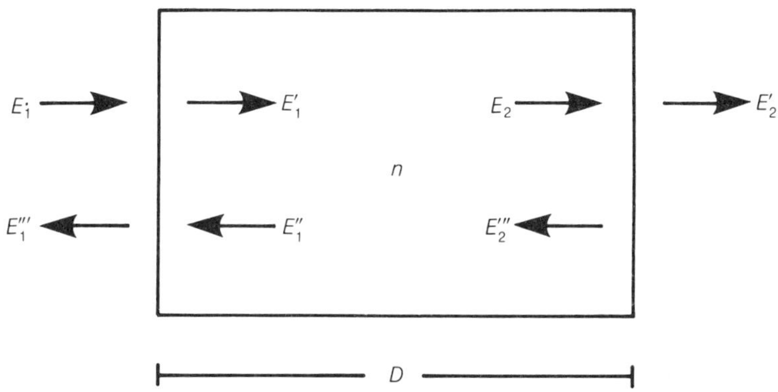

Figure 5.3 Boundary fields defined in a Fabry–Pérot cavity.

linear propagation inside the medium, with continuity of the tangential components of the E and H fields at the two medium–air interfaces. Thus

$$E_2 = E_1 \exp(\mathrm{i}k_0 D) \qquad E''_1 = E'''_2 \exp(\mathrm{i}k_0 D) \tag{5.1}$$

$$E_1 + E'''_1 = E'_1 + E''_1 \qquad E_2 + E'''_2 = E'_2 \tag{5.2}$$

$$E_1 - E'''_1 = nE'_1 - nE''_1 \qquad nE_2 - nE'''_2 = E'_2 \tag{5.3}$$

With the field definition $E(t) = E(\omega)\exp(-i\omega t) + \text{c.c}$, the irradiance in a medium of index n is

$$I = 2nc\varepsilon_0 |E(\omega)|^2 \tag{5.4}$$

and the solutions for the transmitted and average internal irradiance are

$$I_t = \frac{I_0}{1 + F\sin^2(k_0 D)}$$

$$I = \frac{1+R}{1-R} I_t \tag{5.5}$$

with surface reflectivity $R = [(n-1)/(n+1)]^2$ and $F = 4R/(1-R)^2$. Thus for example in InSb, with an index $n = 4$, the natural surface reflectivity is 36% and the internal irradiance 'on resonance' is twice the incident level. Figure 5.4 shows, for InSb, the classic Fabry–Pérot transmission as a function of the single-pass cavity phase, $\phi = k_0 D$. Resonance occurs, as usual in bounded systems, when the cavity length is a half integer number of optical wavelengths, $D = (M/2)\,\lambda_v/n$. Here we make use of the relation $k_0 = 2\pi n/\lambda_v$.

For more complicated cavities, described by front- and back-face reflectivities R_f, R_b, and including absorption, one obtains

$$I_t = \frac{(1-R_f)(1-R_b)a}{(1-R_\alpha)^2} \frac{I_0}{1 + F_\alpha \sin^2\phi}$$

$$I = \frac{(1+R_b a)(1-a)}{(1-R_b)\,\alpha D a} I_t \tag{5.6}$$

with $R_\alpha = (R_b R_f)^{1/2} a$, $F_\alpha = 4R_\alpha/(1-R_\alpha)^2$, $a = \exp(-\alpha D)$, and where ϕ includes phase changes at the surfaces (ϕ_f, ϕ_b),

$$2\phi = 2k_0 D + \phi_f + \phi_b = 4\pi n D/\lambda_v + \phi_f + \phi_b \tag{5.7}$$

A significant point to note is that the internal irradiance may be far higher

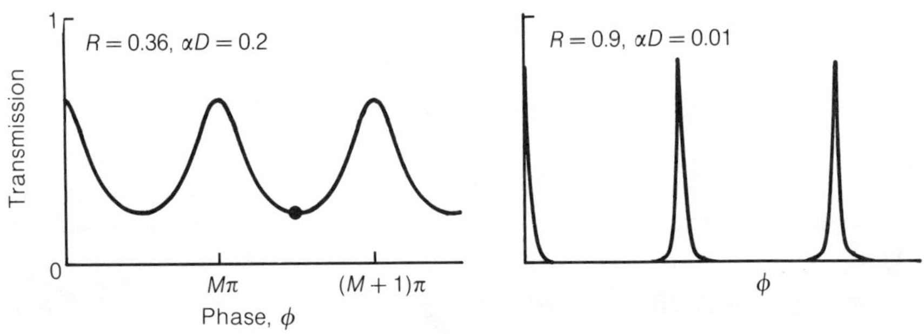

(a) (b)

Figure 5.4 Fabry–Pérot transmission for (a) InSb and (b) an $R = 0.9$ cavity.

than the incident for suitable cavity parameters. Thus with $\alpha D = 0.01$, $R_b \approx 1$, $R_f \approx R_b a$, the internal level is 50 times the incident and any nonlinearities should be considerably enhanced by the cavity feedback.

The cavity finesse, defined as the resonance full width at half-height divided by the peak separation, is $\mathscr{F} = \pi F^{1/2}/2$. From Figure 5.4(a), for example, it is immediately obvious why highly nonlinear responses might be expected from a Fabry–Pérot etalon. Suppose that the cavity phase, i.e. the optical path length, depends on the internal irradiance and that at low irradiance one is at an antiresonance position—marked by the full circle on the figure. If the incident irradiance, and hence the internal irradiance, is increased then the phase will alter and the cavity transmission will increase. However, the internal irradiance is directly proportional to that transmitted (equations 5.5 and 5.6). Hence the internal irradiance increases further. This process can avalanche so that from an initial low incident irradiance and low transmission one can in principle switch to a very much higher transmission (close to the resonance peak) for only a small additional signal. The high finesse response, Figure 5.4(b), should give similar avalanching for very small phase changes, given an initial phase detuning only slightly away from a peak.

The varying phase required here is achieved if either the physical thickness of the cavity, D, or the refractive index, n, is altered by the internal irradiance. In practice both occur; however, in almost all practical cases the index change dominates. We shall be concerned primarily with two physical origins for such a change: (i) the index change associated with carriers generated by the radiation in a semiconductor, and (ii) the thermo-optic index change due to absorption-induced temperature changes. Thus in general

$$\frac{\Delta(nD)}{nD} = \frac{1}{n}\frac{\partial n}{\partial N}\Delta N + \frac{1}{n}\frac{\partial n}{\partial T}\Delta T + \frac{1}{D}\frac{\partial D}{\partial T}\Delta T \tag{5.8}$$

where $\partial n/\partial N$ is the refractive cross-section (the index change per photo-generated carrier pair: σ_n), $\partial n/\partial T$ is the thermo-optic coefficient and $(1/D)\partial D/\partial T$ is the thermal expansion coefficient. The first two terms are addressed in the following section.

5.3. NONLINEAR REFRACTION

In nonlinear optics it has been conventional to describe nonlinearities by appropriate susceptibilities, through the polarization in the medium that is established by the incident field(s). Thus at the frequency ω we are interested in a polarization $P(t) = P(\omega)\exp(-i\omega t) + \text{c.c.}$ where

$$P(\omega) = \varepsilon_0[\chi^{(1)}E(\omega) + \chi^{(3)}|E(\omega)|^2E(\omega) + \cdots] \tag{5.9}$$

This contains those components of all powers of $E(t)$ that have $\exp(-i\omega t)$ time dependence. $\chi^{(n)}$ is known as an nth-order nonlinear susceptibility; it depends on the material and the radiation frequency, but not on the field strength. One can define a generalized dielectric constant, which has been used to compare $\chi^{(3)}$ with a nonlinear refractive index coefficient n_2:

$$\varepsilon(\omega, E) = 1 + P(\omega)/\varepsilon_0 E(\omega) \tag{5.10}$$

By expressing ε in terms of irradiance-dependent refraction and absorption coefficients,

$$\varepsilon^{1/2} = n_0 + n_2 I + \cdots + \mathrm{i}(\alpha_0 + \alpha_2 I + \cdots)c/2\omega \tag{5.11}$$

one obtains relations between the susceptibilities and these coefficients. In particular

$$n_2 = \frac{1}{4n_0^2 c\varepsilon_0} \operatorname{Re} \chi^{(3)} \tag{5.12}$$

Irradiance-, or intensity-, dependent refraction has been recognized since 1963, soon after the first applications of lasers to nonlinear optical phenomena (Franken and Ward, 1963). Early measured nonlinear refractive susceptibilities were of order 10^{-16} esu (H_2O) up to 10^{-12} esu (CS_2), corresponding to n_2 values between 10^{-19} $(\mathrm{W\,cm^{-2}})^{-1}$ and 10^{-15} $(\mathrm{W\,cm^{-2}})^{-1}$ (Mayer and Gires, 1964; Maker *et al.*, 1964). Closely associated $\chi^{(3)}$ values measured in semiconductors until the mid-1970s were all less than 10^{-9} esu. In the late 1970s, however, it was realized that by operating in frequency regions close to absorption resonances (e.g. in the band-tail of the semiconductor fundamental absorption edge) very large coefficients were achievable, large enough to lead to optical bistability at milliwatt power levels rather than at the kilowatt power levels typically employed in nonlinear optics at that time. The reason for these contrasting measurements lies not in the material but in the choice of frequency. Working in the transparency regime, nonlinear processes involve virtual transitions to intermediate states in which the material excitation can exist only on time scales determined by the uncertainty principle; $\Delta t \approx 1/\Delta\omega$, where $\hbar\Delta\omega$ is the energy mismatch between the transition frequency and the radiation frequency. Typically $\Delta t \approx 10$ fs. For resonant conditions, however, the excitation remains over the material recombination time scale, which may be as long as microseconds for electronic excitations and effectively milliseconds for thermal excitation. As a result, a given excitation level can be built up over a period of time, using far lower irradiances than in the virtual transition limit. It is the achieved excitation level that determines the magnitude of the change in the material optical properties, e.g. $\Delta n = \sigma_n \Delta N$.

5.3.1. Electronic nonlinearities

If the excitation recombination time is a constant, τ, then the excited state population is obtained from the rate equation:

$$\frac{\partial}{\partial t}\Delta N = \frac{\alpha_I I}{\hbar\omega} - \frac{\Delta N}{\tau} \tag{5.13}$$

In steady state

$$\Delta N = \alpha_I I\tau/\hbar\omega \tag{5.14}$$

where α_I is the interband absorption coefficient. Hence there is an effective nonlinear refractive index:

$$n_2 = \Delta n/I = \sigma_n \alpha_I \tau/\hbar\omega \tag{5.15}$$

For a two-level transition, time-dependent pertubation theory can be used to give the refractive cross-section:

$$\sigma_n = -\frac{4\pi}{n\hbar}|er_{10}|^2 \frac{1}{\omega_{10} - \omega} \tag{5.16}$$

where ω_{10} is the transition frequency and er_{10} is the transition electric dipole moment. In the semiconductor case the dipole moment is conventionally replaced by the Kane momentum matrix element, $P = |r_{cv}|E_g$, and one obtains (Wherrett *et al.*, 1988)

$$\sigma_n \approx -K\frac{e^2P^2}{nE_g^3}\frac{E_g}{E_g - \hbar\omega} \tag{5.17}$$

The constant K is determined by summing over bands and averaging over directions in k space. The significant factor in σ_n is the inverse cubic dependence on the band gap E_g. The coefficient P and the linear index n vary little from one material to another by comparison with E_g. As with all nonlinear phenomena, small gap semiconductors give larger effects (Wherrett, 1984a). The measurement of n_2 in InSb, close to the 5 μm band gap, with values as large as $-10^{-3}\,(\mathrm{W\,cm^{-2}})^{-1}$ (Miller *et al.*, 1981) is a consequence of this gap dependence and of the long carrier lifetime (200 ns–1 μs) of InSb; the corresponding $\chi^{(3)}$ value is 1 esu, 15 orders of magnitude greater than the original H_2O nonlinearity measurements.

In general the carrier lifetime depends on the carrier concentration itself, particularly at high densities, and equation 5.14 must be modified. Hence the cross-section result is a more general formulation than one using n_2. Alternatively one can express the index change in terms of the change in absorption coefficient, appealing to the Kramers–Krönig (KK) relation, which expresses causality:

$$\Delta n(\omega) = \frac{c}{\pi}\int_0^\infty \frac{\Delta\alpha(\omega')}{\omega'^2 - \omega^2}\,d\omega' \tag{5.18}$$

Physically the internal irradiance, at frequency ω, generates carriers which redistribute in the valence and conduction bands (e.g. forming a Boltzmann

Table 5.1 Semiconductor nonlinear refraction coefficients associated with electronic excitations

Material	*Temperature* (*K*)	*Wavelength* (μm)	$-\sigma_n$ (cm^3)	$-\partial n/\partial I$ ($cm^2\,kW^{-1}$)	*Reference*
CdHgTe	80	10.6	—	1	Miller *et al.* (1984)
CdHGTe	300	10.6	—	5×10^{-4}	Craig and Miller (1986)
InSb	300	10.6	—	1.1×10^{-4}	Mathew *et al.* (1985)
PbSnTe	50	10.6	—	$\sim 10^{-4}$	Kar and Wherrett (1986)
Te	300	10.6	$\sim 10^{-21}$	—	Staupendahl and Schindler (1982)
Ge	300	10.6	—	1.6×10^{-9}	Watkins *et al.* (1981)
InSb	80	5.5	2×10^{-18}	1	Miller *et al.* (1981)
InAs	80	3.1	—	2×10^{-2}	Poole and Garmire (1984)
InGaAsP	300	1.5	$\sim 10^{-20}$	—	Adams *et al.* (1985)
InGaAsP	300	1.3	$\sim 10^{-20}$	—	Sharfin and Dagenais (1985)
Si	300	1.06	$\sim 10^{-21}$	—	Eichler (1983)
Si	300	1.06	—	1.2×10^{-7}	Jain and Klein (1983)
CdSe	300	1.06	—	$\sim 10^{-9}$	
CdSe	300	1.06	—	$> 8 \times 10^{-10}$	Guha *et al.* (1985)
CdTe	300	1.06	—	$\sim 10^{-9}$	Kremenitskii *et al.* (1979)
GaAs	300	0.86	2×10^{-19}	0.2	Miller *et al.* (1983)
GaAs	80	0.82	—	0.1	Ovadia *et al.* (1985)
GaAs	300	0.82	—	9×10^{-4}	Lee *et al.* (1986)
GaAs	80	0.812	—	8.6×10^{-7}	Oudar *et al.* (1984)
CdSSe	300	0.56	—	2×10^{-8}	Jain and Lind (1983)
CdS	300	0.53	—	3.4×10^{-9}	
CdS	300	0.53	$\sim 10^{-19}$	—	Henneberger *et al.* (1987)
GaSe	80	0.514	—	1.7×10^{-4}	Bakiev *et al.* (1983)
CdS	2	0.487	—	0.1	Dagenais and Sharfin (1986)
CuCl	17	0.39	—	10^{-6}	Peyghambarian *et al.* (1983)

distribution in the nondegenerate limit). As a result the linear absorption is altered at all frequencies. The KK relation describes the index change at ω in terms of this change in the entire spectrum; it allows one to determine Δn purely from experimental absorption measurements. Both equation 5.18 and more especially equation 5.16 show that for frequencies below a saturating transition the refractive index decreases. Above the transition frequency an increasing refraction is observed. Nonlinear refraction coefficients measured for a number of materials are summarized in Table 1.

5.3.2. Thermal nonlinearities

The temperature rise (ΔT) of a sample of density ρ, specific heat C_p and thermal conductivity κ, obeys the dynamic equation

$$\frac{\partial}{\partial t}\Delta T = \frac{Q}{\rho C_p} + \frac{\kappa}{\rho C_p}\nabla^2(\Delta T) \tag{5.19}$$

For plane-wave irradiation the rate of heat input per unit volume is $Q = \alpha I$. To a first approximation in the thermal diffusion term one may replace $\nabla^2(\Delta T)$ by $-\Delta T/A$, where the effective area A depends on the sample heat sinking geometry. This results in a recombination time, $\tau^T = \rho C_p A/\kappa$. Under steady state conditions the resulting temperature change is

$$\Delta T = \alpha I \tau^T/\rho C_p \tag{5.20}$$

Table 5.2 Thermo-optic coefficients of materials used for optical bistability

Material	*Wavelength* (*μm*)	*n*	$\partial n/\partial T(K^{-1})$	*Reference*
PbSnSe	10.6	4	-8×10^{-3}	Kar *et al.* (1986)
CdHgTe	10.6	4	-1×10^{-3}	Craig *et al.* (1985)
Ge	10.6	4	4×10^{-4}	Golik *et al.* (1983)
InSb	10.6	4	4×10^{-4}	Walker and Hunter (1985)
InSb	5.5	4	4×10^{-3}	Tooley *et al.* (1985)
Si	1.06	3.5	2.4×10^{-4}	Eichler (1983)
GaAs	0.85	3.5	7×10^{-4}	Miller *et al.* (1986)
CdSSe	0.647	~2.5	3×10^{-6}	Gibbs *et al.* (1979)
$GeSe_2$	0.633	2.5	2.5×10^{-4}	Hajto and Janossy (1983)
GaSe	0.514	3	2.5×10^{-4}	Antonioli *et al.* (1979)
ZnSe	0.514	2.5	2.5×10^{-4}	Kar and Wherrett (1984)
ZnS	0.514	2.4	1.4×10^{-4}	Harris *et al.* (1977)
HgS	0.514	~3	8×10^{-5}	Ayrault *et al.* (1973)
ZnSe (thin film)	0.5–0.7	2.7–2.4	$(2–1) \times 10^{-4}$	Wherrett *et al.* (1988)

The equivalence of equations 5.19 and 5.20 to equations 5.13 and 5.14 is immediate, and one has an effective thermal index coefficient:

$$n_2^{\mathrm{T}} = \frac{\partial n}{\partial T}\frac{\alpha\tau^{\mathrm{T}}}{\rho C_{\mathrm{p}}} = \frac{\partial n}{\partial T}\frac{\alpha A}{\kappa} \tag{5.21}$$

In the vicinity of the band gap of semiconductors the thermo-optic coefficient is dominated by the thermal movement of the band edge. As shown in Table 5.2, however, the coefficient does not vary much from material to material. It can also be shown that there is little enhancement of $\partial n/\partial T$ as the edge is approached.

5.4. DYNAMICS AND STEADY STATE SOLUTIONS OF THE NONLINEAR ETALON

There are a number of timescales that are relevant to bistability. In practical devices the cavity round trip and build-up times are usually very short compared with the material response times. As a consequence it is assumed that the internal irradiation level adjusts instantaneously to the relatively slow excitation dynamics (i.e. for a given index change the irradiance is always taken to be that for a cavity of that index). This is known as adiabatic elimination of the field time dependence. Either of equations 5.13 or 5.19 then forms an ideal starting point to describe the two stable steady states that are achievable for a nonlinearly dispersive etalon. We will select the thermal case for illustration, substituting the Fabry–Pérot expression for the internal irradiance I, using equation 5.6:

$$\frac{\partial}{\partial t}\Delta T = \frac{BI_0}{1 + F_\alpha \sin^2(\phi_0 + b\,\Delta T)} - \frac{\Delta T}{\tau^{\mathrm{T}}} \tag{5.22}$$

The coefficient B contains cavity reflectivities and αD (Wherrett *et al.*, 1986), ϕ_0 is the initial detuning from the nearest cavity resonance at the ambient temperature, and b expresses the nonlinear refraction,

$$b = \frac{2\pi D}{\lambda_{\mathrm{v}}}\frac{\partial n}{\partial T} \tag{5.23}$$

In the dynamic equation the first term on the right-hand-side can be thought of as the driving force for temperature changes; the second term is the thermal sink. We are concerned with the optical response of the etalon, in particular the transmitted irradiance (equation 5.6):

$$I_{\mathrm{t}} = \frac{CI_0}{1 + F_\alpha \sin^2(\phi_0 + b\,\Delta T)} \tag{5.24}$$

It is important to note that I_{t} is directly proportional to the thermal driving forces. Equations 5.22 and 5.24 are key to understanding the steady state, stability and dynamic switching of nonlinear Fabry–Pérot etalons. They apply equally to the electronic case—simply substituting ΔN for ΔT—or to other excitation schemes. The equations are also expressable very simply in terms of the cavity phase change $\Delta\phi = b\,\Delta T$.

Figure 5.5(a) demonstrates the method of solution. The two contributions to $\partial(\Delta T)/\partial t$ are plotted against ΔT itself. The driving force is the Fabry–Pérot Airy function, of peak height equal to BI_0 and of value $BI_0/(1 + F_\alpha \sin^2 \phi_0)$

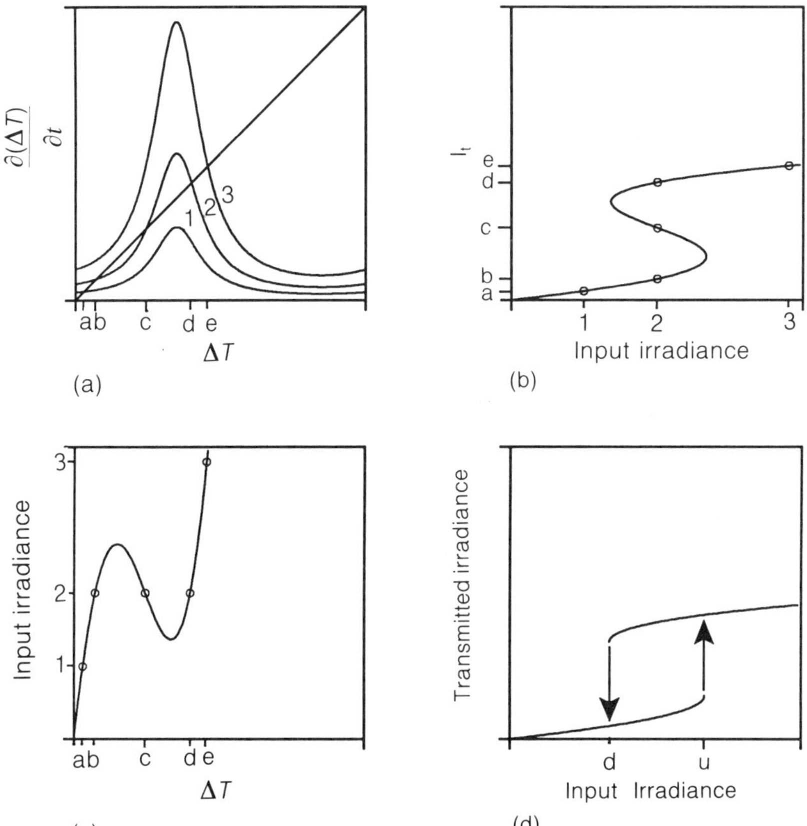

Figure 5.5 Illustration of graphical solutions for optical bistability, from the steady states of the dynamical equation 5.2. (a) Thermal driving term in $\partial(\Delta T)/\partial t$ for three irradiance levels, and the heat sink; plotted vs. the temperature change. (b) Intercepts obtained from (a), giving the transmitted vs. incident (scaled) irradiances by comparison of equations 5.24 and 5.22; showing non-monotonic solutions. (c) Input irradiance required to support a given temperature change obtained from the intercepts of (a). (d) Stable steady states in the transmitted irradiance response, showing switching and hysteresis (bistability).

at $\Delta T = 0$. The heat sink is simply a straight line of gradient $1/\tau^{\mathrm{T}}$. At low incident irradiance (curve 1) there is one intersection of the curve and the line. This is the steady state solution. The horizontal position of the intersect is the steady state temperature rise; the vertical position is, through equation 5.24, simply B/C times the transmitted irradiance.

At high incident irradiance (curve 3) there is again a single steady state, at higher ΔT and I_t. The temperature rises in this case are sufficient that the Fabry–Pérot resonance has been swept through the operating wavelength $(\phi_0 + b\,\Delta T) > 0$, whereas initially $\phi_0 < 0$.

Curve 2 shows the intermediate condition. For a range of I_0 values there are not one but three intersections of the Airy curve with the load line. There are in principle three self-consistent solutions for which the internal irradiance achieved for the given cavity phase is consistent with the phase that such an irradiance produces. One achieves bistability, however, because one of the steady state solutions is unstable.

A qualitative test of stability is obtained by assuming one particular steady state condition and perturbing the system slightly. Thus starting at position b, consider a small decrease in ΔT at fixed I_0. The driving force curve is then above the heat sink line and hence $\partial(\Delta T)/\partial t$ is positive, so that the temperature is restored towards that at position b. For a small increase in ΔT the curve falls below the line, $\partial(\Delta T)/\partial t$ is negative and again the system stabilizes onto b. The same argument applies to postion d and to the monostable solutions a and e that occur for the different I_0 values.

For position c, however, a small decrease in ΔT takes the curve below the line and $\partial(\Delta T)/\partial t$ is negative; the temperature will continue to fall until position b is reached. A small increase in ΔT from c leads to heating until position d is reached. Hence those intermediate steady state solutions, where the gradient of the line is less than that of the driving curve, are unstable; all other steady state solutions are stable.

In Figures 5.5(b) and 5.5(c) the steady state temperature and transmission are plotted as functions of I_0, showing the particular solutions discussed in Figure 5.5(a) and the regions of stability (positive gradient) and instability (negative gradient).

The above technique, based on the dynamic equations, and the graphical method introduced by Marburger and Felber (1978), that is based on the steady state equations alone, are particularly useful for an understanding of the forms of the nonlinear responses. In practice the simplest technique for plotting the precise response is to treat $b\,\Delta T$ as a dummy variable in equations 5.22 and 5.24. Thus for the steady states only,

$$I_0 = bB\tau^{\mathrm{T}}[1 + F_\alpha \sin^2(\phi_0 + x)] \tag{5.25a}$$

$$I_t = \frac{cx}{bB\tau^{\mathrm{T}}} \tag{5.25b}$$

It is a trivial computer exercise to step the x variable and to plot the resulting values of I_t vs. those of I_0.

Figure 5.5(d) shows what would be observed in practice as I_0 is gradually increased and then reduced. The transmission increases smoothly until at the switch-up irradiance I_u (for which the line is at the tangent to the Airy function) there is a sudden jump in the transmission, followed by a further smooth increase. On reducing I_0 the sudden jump down occurs when the top of the Airy curve is tangential to the line, at a lower, switch-down irradiance I_d. There is thus a hysteresis loop around the region of bistability, $I_d < I_0 < I_u$.

5.5. CHARACTERISTIC SWITCHING CONDITIONS

5.5.1. The critical switching irradiance

Figure 5.6 shows the range of possible transmission responses achievable for different initial detunings ϕ_0. If ϕ_0 is small one obtains nonlinear but monostable solutions, unless the irradiance is so high that the next order of Fabry–Pérot fringe is swept through the operating wavelength. Monostability

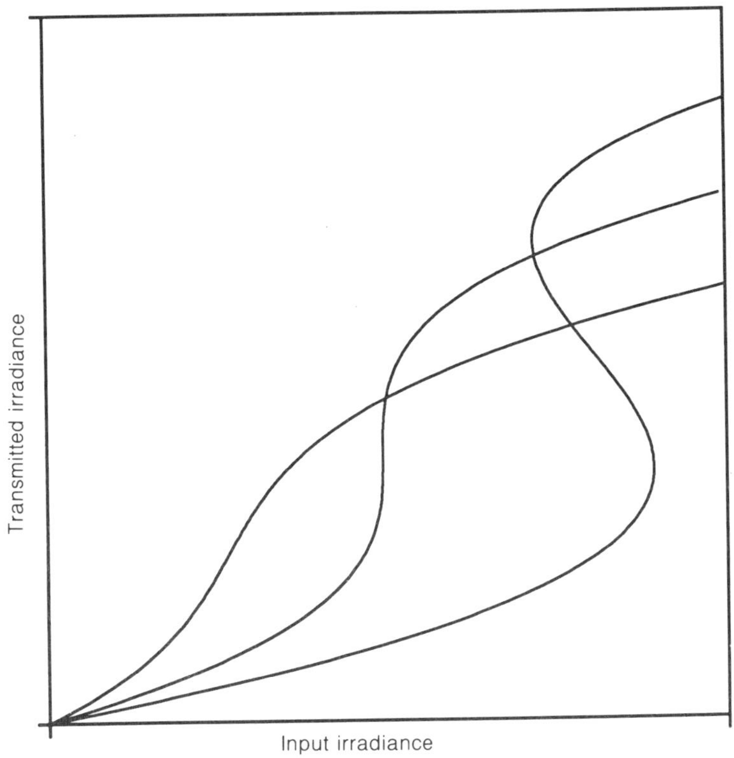

Figure 5.6 Family of steady state solutions to equation 5.22, obtained for three different initial detuning (ϕ_0) values.

will occur if the gradient of the thermal driving term does not exceed $1/\tau^{T}$ at any point of intersection. There is thus a critical condition: an initial detuning ϕ_c and a specific incident irradiance I_c, for which the intersection occurs at the point of inflection of the Airy function. Because the gradient of the curve is proportional to I_0, there is the possibility of bistability for $I_0 > I_c$, provided also that ϕ_0 is greater than ϕ_c, as shown in Figure 5.7. It is possible to obtain analytic solutions for the critical conditions, in the n_2 limit. These have the form (Miller, 1981)

$$\phi_c = f_\phi(F_\alpha) \tag{5.26}$$

$$I_c = \frac{\lambda_v \alpha}{2\pi |n_2|} f(R_f, R_b, \alpha D) \tag{5.27}$$

Note that the parameters ϕ_0 and I_0 correspond to the control parameters described in the introduction for mechanical bistability and that similar sets of responses are achieved. The unstable intermediate solutions were also

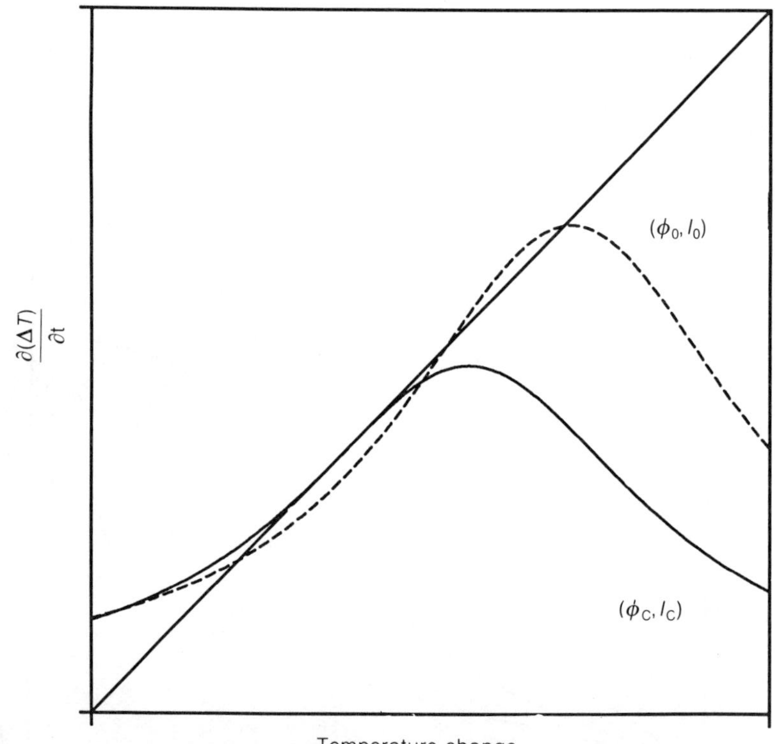

Figure 5.7 Critical conditions on I_0 and ϕ_0. The line passes through the point of inflection of the full curve, at a tangent. – – –, bistable conditions; $\phi_0 > \phi_c$, $I_0 > I_c$.

described for the mechanical example. The general form of the solutions as functions of ϕ_0 and I_0 forms a 'cusp catastrophe' with the cusp at the point ϕ_c, I_c.

For practical purposes one is usually interested in obtaining a low value of I_c, this being the incident irradiance that has to be exceeded in order to achieve bistability. Before considering the optimization of I_c it is, however, worth looking at the cavity internal irradiance value (I) that must be achieved for switching. In general the n_2 limit is achievable only if this internal irradiance is maintained below a certain level—either because of saturation of the fundamental nonlinearity (σ_n) or because the carrier recombination time decreases at higher densities effectively reducing n_2. Equally, higher-order thermo-optic terms could prevail in the thermal example. At the critical condition, I is given by the functional form

$$I_{ic} = \frac{\lambda_v \alpha}{2\pi|n_2|} \frac{g(F_\alpha)}{\alpha D} \tag{5.28}$$

The critical excitation levels are

$$\Delta N_c = \frac{\alpha \lambda_v}{2\pi|\sigma_n|} \frac{g(F_\alpha)}{\alpha D}$$

$$\Delta T_c = \frac{\alpha \lambda_v}{2\pi|\partial n/\partial T|} \frac{g(F_\alpha)}{\alpha D} \tag{5.29}$$

5.5.2. Optimization

5.5.2.1. Incident irradiances

To reduce the switching irradiances for a given material one must first design an optimized cavity. The optimum conditions will depend on whether or not n_2 is itself a function of the cavity length; this results in different conditions for the electronic and thermal cases.

Cavity optimization relies on achieving sufficient absorption to generate the relevant excitation but not so much that the finesse is destroyed. Figure 5.8(a) shows the cavity factor f of equation 5.27, plotted for fixed reflectivities and for varying αD. Pronounced minima exist at the conditions,

$$\alpha D = (2 - R_f - R_b)/4 \tag{5.30}$$

and the value of f at these minima decreases with increasing finesse. The implication is that, starting with an electronic nonlinearity with D-independent n_2, one would design a high finesse, small αD cavity:

$$I_c = \frac{\hbar c}{\tau} \frac{1}{|\sigma_n|} f \rightarrow \frac{\hbar c 3\sqrt{3}}{\tau|\sigma_n|} \alpha D \tag{5.31}$$

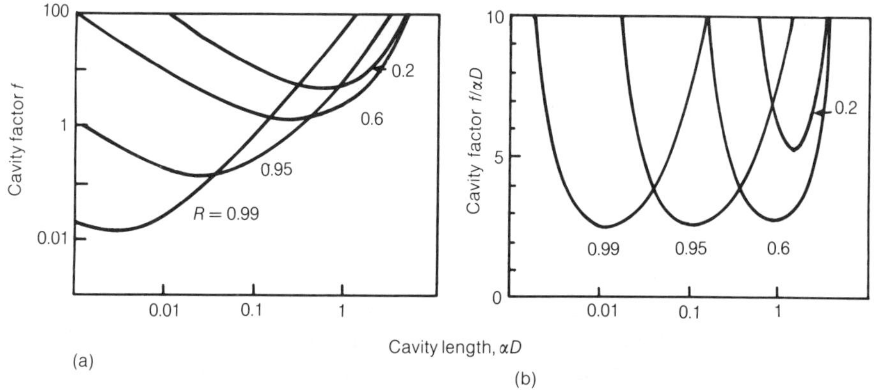

Figure 5.8 (a) The cavity factor f that determines the relative switching irradiances for different cavity reflectivities ($R_f = R_b = R$), plotted against the cavity length (αD); see equation 5.27 for optoelectronic nonlinearities. (b) The corresponding factor $f/\alpha D$ that pertains for optothermal nonlinearities.

The material factor peaks in the vicinity of the semiconductor band edge (Al-Attar *et al.*, 1986), thereby determining α at the optimum operating wavelength. One therefore knows the required D for chosen reflectivity conditions. The limit to the reduction of I_c is set by (i) a minimum value of D of approximately $\lambda_v/2$ below which there are no cavity resonances, (ii) fabrication restrictions either for low D or for high reflectivity coatings and (iii) the fact that unless carrier confinement is achieved (for example in quantum well material) the effective carrier lifetime will be reduced by surface recombination for small D (Frank and Wherrett, 1987).

If τ is dominated by surface carrier recombination it becomes effectively proportional to D. This same dependence is achieved for thermal heat sinking longitudinally through the interface into a substrate, as is appropriate in the interference filters described in section 5.8. In this case

$$I_c \propto \frac{f}{\alpha D} \tag{5.32}$$

The αD dependence is similar to that of f for given reflectivities (Figure 5.8) but now the minimum values do not decrease at high finesse; instead, $f/\alpha D$ tends to a fixed value of order 2.6, at the optimum conditions (Wherrett *et al.*, 1986):

$$\alpha D = 2 - R_f - R_b \tag{5.33}$$

5.5.2.2. Internal irradiances and excitations

At the minima of f (conditions 5.30 and 5.33) the function $g/\alpha D$ in equations 5.28 and 5.29 is essentially constant; in particular for high finesse cavities

$g/\alpha D = \sqrt{3}$ under condition 5.30 and $g/\alpha D = \sqrt{3}/2$ under condition 5.33. This result manifests the fact that in the electronic case the reduction in switching conditions achievable in high reflectivity cavities is almost entirely due to the enhancement of the internal fields. That is, to achieve the required internal irradiance I_{ic} one needs far lower incident irradiances if the cavity finesse is high. The required refractive index change is thus also independent of the cavity—under the optimum condition 5.30,

$$\Delta n_c = \frac{\sqrt{3}}{2\pi}\lambda_v \alpha \tag{5.34}$$

Typically we are interested in α values of 20–2000 cm^{-1} so that for visible radiation refractive index changes must be in the range 10^{-3}–10^{-7}. This explains why materials such as InSb with n_2 values as high as $10^{-3}\,W^{-1}\,cm^2$ are necessary if CW low power bistability is to be observed.

Result 5.34 also demonstrates clearly that if a given excitation (band or excitonic) is to be used to achieve bistability then the total index changes associated with the complete saturation of the near-resonance absorption must certainly be larger than $\alpha\lambda_v$. As α here refers to the total absorption coefficient (that associated with the excitation plus parasitic effects) the importance of removing background absorption is emphasized.

Under condition 5.33 the effective coefficient $|n_2^T|$ decreases at high finesse. Thus in the thermal case, in contrast with the above considerations, it is beneficial to use a relatively low finesse cavity if the internal irradiance is to be minimized.

5.6. SWITCHING POWERS AND POWER–TIME PRODUCTS

5.6.1. Powers and energies

In practice one must deal with finite beam sizes (e.g. Gaussian beams) rather than the plane waves discussed above. At first sight one would not expect bistability for such beams because, whilst the beam centre transmission will increase suddenly when the peak irradiance reaches the switch-up value, this will only effect an infinitesimally small spatial region. As the incident power is increased a gradually larger portion of the spatial beam profile will exceed the switching irradiance and a smoothly increasing power transmission would be expected. Similarly, on decreasing the incident power, as each part of the beam reduces below the switch-down irradiance value one expects the portion of the beam that is in the upper bistable branch to diminish smoothly. This is not the case, either in practice or if the full theoretical treatment is made. The reason is that transverse spatial effects, diffraction, carrier diffusion or thermal diffusion, couple different parts of the beam cross-section. Hence if the beam centre switches up then transverse coupling leads to higher excitation levels in the vicinity of the centre, and can cause a spread of the switching. This will occur essentially out to irradiance values at the switch-down level. The phenomenon is known as whole beam switching (Firth *et al.*, 1985) and leads to hysteretic responses very similar to those predicted from the plane wave treatment. In addition the trends (optimization) given by the plane wave model are obeyed qualitatively, though not in quantitative detail, for finite beams.

The time taken to switch is discussed in more detail in section 5.7. For practical devices the time is controlled by the material time constant, τ or τ^{T}. The product of the switch power (P) and switch time has been called the switching energy ($\mathscr{E}$). Whilst this does not necessarily represent the energy that must be dissipated at each switch (because only a fraction of the power is absorbed), it is a useful measure for the device. Perhaps more important, in view of the demand for high parallelism, is the switching power itself and hence the number of devices that could be operated simultaneously for a given available power.

The corresponding critical values, for Gaussian beams of $1/e^2$ irradiance radii equal to r_0, are

$$P_c = \frac{\pi}{2} r_0^2 I_c$$

$$\mathscr{E}_c = \frac{\pi}{2} r_0^2 \tau I_c \tag{5.35}$$

It is interesting to look at the limits to these parameters for (i) an idealized two-level system, (ii) electronic nonlinearities associated with carrier generation in semiconductors and (iii) thermal nonlinearities.

5.6.2. The ideal two-level system

The optimum value of f tends to $\alpha D 3\sqrt{3}$ at high finesse, with $D \geqslant \lambda_v/2n$. For a concentration of N completely isolated two-level systems in a cavity, from equation 5.31

$$I_{c(\min)} \rightarrow \frac{\hbar c}{\tau} \frac{1}{|\sigma_n|} \frac{\alpha \lambda_v}{n} = \frac{\hbar c 4N}{\tau} \frac{\gamma}{|\omega_{10} - \omega|} \tag{5.36}$$

where γ is the absorption linewidth.

The point is that there is no realistic theoretical limit; as one works further from the line centre and for very dilute systems then I_c, P_c and $\mathscr{E}_c$ can all decrease toward single-photon values. In practice parasitic absorption and the necessity for high reflectivity, very low loss mirrors reduce the device realization of such a system.

5.6.3. Semiconductor, electronic nonlinearities

The enhancement of the carrier refractive cross-section described by equation 5.17 is modified in the immediate vicinity of resonance and at finite temperatures. Thus for example in the Boltzmann limit (Wherrett, 1984b)

$$I_c \propto \frac{\hbar c}{e^2} \frac{n}{P^2} J^{-1}\left(\frac{E_g - \hbar\omega}{kT}\right) kT E_g^2 \frac{f}{\tau} \tag{5.37}$$

The factor J^{-1} has a minimum of order unity at $\hbar\omega = E_g$. Hence the minimum power for bistability, which will always occur for small spot sizes, is of order

$$P_c \approx \lambda_v^2 I_c = (\hbar c/E_g)^2 I_c \tag{5.38}$$

Because smaller spot sizes are achievable for shorter wavelengths the advantage of large σ_n or n_2 in small gap materials can in principle be lost; the minimum critical power level is essentially independent of material gap but is proportional to τ^{-1}, favouring long recombination times. Even this factor drops out in the characteristic switching energy:

$$\mathscr{E}_{c(\min)} \approx \frac{\hbar c}{e^2}\left(\frac{\hbar c}{P}\right)^2 kT f(R_f, R_b, \alpha D) \tag{5.39}$$

The importance of cavity optimization is emphasized here again; $\mathscr{E}_c/f$ has a value of order picojoules at room temperature. As the temperature is reduced the factor kT falls but must eventually be replaced by the dephasing lifetime of the excitation; nevertheless femtojoule limits are conceivable.

5.6.4. Thermal nonlinearities

In the thermal case consider a sample that is 'pixellated' so that individual elements are isolated from each other and have transverse dimension of the order of the radiation spot size (r_0). In this configuration the effective thermal sink area A is of order $r_0 D\kappa_s/\kappa$, where κ_s is the substrate conductivity. In the diffraction limit therefore

$$P_c \approx \frac{\alpha\lambda_v^2 \kappa_s}{|\partial n/\partial T|} \frac{1}{\beta} \frac{f}{\alpha D} \tag{5.40}$$

and the thermal conductivity time is of the form

$$\tau^T \approx \frac{\rho C_p}{\kappa_s} \lambda_v^2 \beta' \tag{5.41}$$

β and β' are thermal geometry factors of similar magnitudes. There is therefore considerable advantage, in this example, in going to short wavelength operation and to small active cavities,

$$\mathscr{E}_{\mathrm{c}} \approx \frac{\alpha\lambda_{\mathrm{v}}^4 \rho C_{\mathrm{p}}}{|\partial n/\partial T|} \tag{5.42}$$

To a first approximation $\partial n/\partial T$ is inversely proportional to the semiconductor band gap and the interband absorption coefficient is directly proportional. Hence the switching energy scales roughly as E_{g}^{-2}. A typical value for this energy, given that $|\partial n/\partial T|$ is of order $10^{-4}\,\mathrm{K}^{-1}$ for most semiconductors, particularly those with visible band gaps, and ρC_{p} is of order $10^{-3}\,\mathrm{K\,kg^{-1}\,J^{-1}}$, is $\mathscr{E}_{\mathrm{c}}/\alpha \approx$ picojoules, with α in cm^{-1}. It is important to operate with the minimum absorption values for which the cavity conditions on αD can be satisfied. Again in principle subpicojoule operation is conceivable, although typical observations to date (in unoptimized devices) have been nanojoule or above.

5.7. SWITCHING DYNAMICS

5.7.1. Switching times

The graphical method of section 5.4 can be used to give a qualitative description of the time dependence of either the transmission or the internal excitation level of a nonlinear Fabry–Pérot, and in particular a description of the phenomenon called critical slowing down.

A typical experimental situation is the so-called hold-and-switch configuration in which a CW (hold) beam is used to bias a device into the bistable region, as shown in Figure 5.9 by the input level I_{i}. An additional switching signal is such that is takes the total input irradiance beyond I_{u} (the switch-up level). After sufficient time the transmission will reach a level such that on removal of the switching signal the transmission will settle in the upper bistable branch. For theoretical analysis it is simpler to consider a step increase in the incident irradiance from I_{i} to I_{f} and to discuss the rate of excitation change. Again the thermal case will be used for illustration, assuming a constant thermal recovery time. Figure 5.10(a) shows the dynamics. The initial steady state is position (i). If the incident level is instantaneously raised to I_{f} then the system must eventually reach the steady state point (f), at a new temperature ΔT_{f} above ambient. The height of the I_{f} Airy curve above the sink line indicates the rate at which ΔT is changing at any given ΔT value. In the example shown $\partial(\Delta T)/\partial t$ is therefore moderately fast initially, then slows down in the region of proximity of the curve and the line, then speeds up dramatically as ΔT passes through the resonance condition at ΔT_{p} and finally slows down so that ΔT asymptotically approaches ΔT_{f}. This behaviour is shown in Figure 5.10(b). The corresponding transmission time dependence, taken from the height of the I_{f} Airy function, is shown in

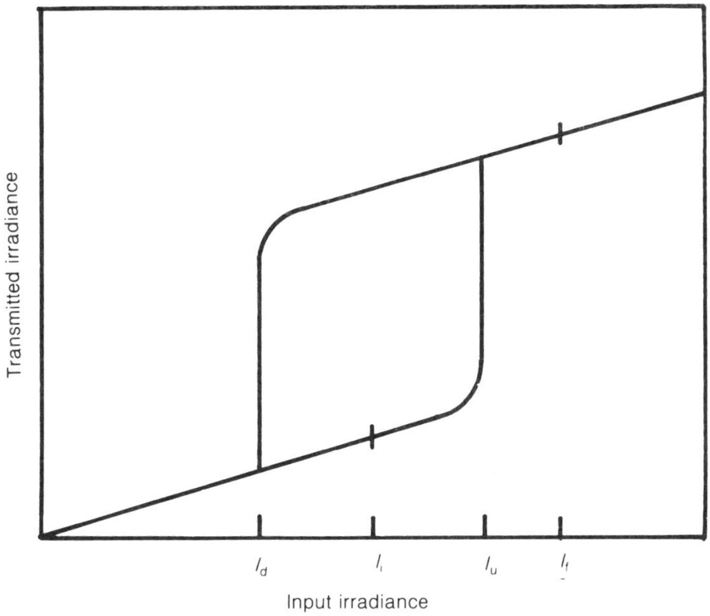

Figure 5.9 Schematic of the hold-and-switch conditions employed in bistable devices.

Figure 5.10(c). The transmission peak occurs at the point of inflection of Figure 5.10(b), where $\Delta T = \Delta T_p$.

There is therefore an observable characteristic switching time t_p at which the transmission peaks. It is guaranteed that the device will switch if the signal is removed after this time, and the system will settle to position d on Figure 5.10(a). In principle the switch-on will occur provided that the temperature change exceeds ΔT_c, so that if the input is reduced back to I_i the thermal driving term will take the system from c to d. Thus in general

$$t_{\text{switch-on}} = \int_{\Delta T_i}^{\Delta T_c} \frac{d(\Delta T)}{[BI_f/1 + F_\alpha \sin^2(\phi_0 + b\,\Delta T)] - \Delta T/\tau} = t(\Delta T_i, \Delta T_c) \quad (5.43)$$

A characteristic time for switching from $I_i = 0$ is

$$t_{on} = t(0, \Delta T_p) \quad (5.44)$$

where ΔT_p is such that $(\phi_0 + b\,\Delta T_p) = 0$.

The switching time is dramatically dependent on the final irradiance, I_f, and on the detuning, ϕ_0. Thus for the detuning used in Figure 5.10, if I_f is such that the line is almost tangential to the curve, as shown in Figure 5.11(a), then $\partial(\Delta T)/\partial t$ is very small, and it takes a long time for the temperature change to move through the region indicated. The switching time diverges to infinity as I_f is reduced to the value I_u of Figure 5.9, at which the switch

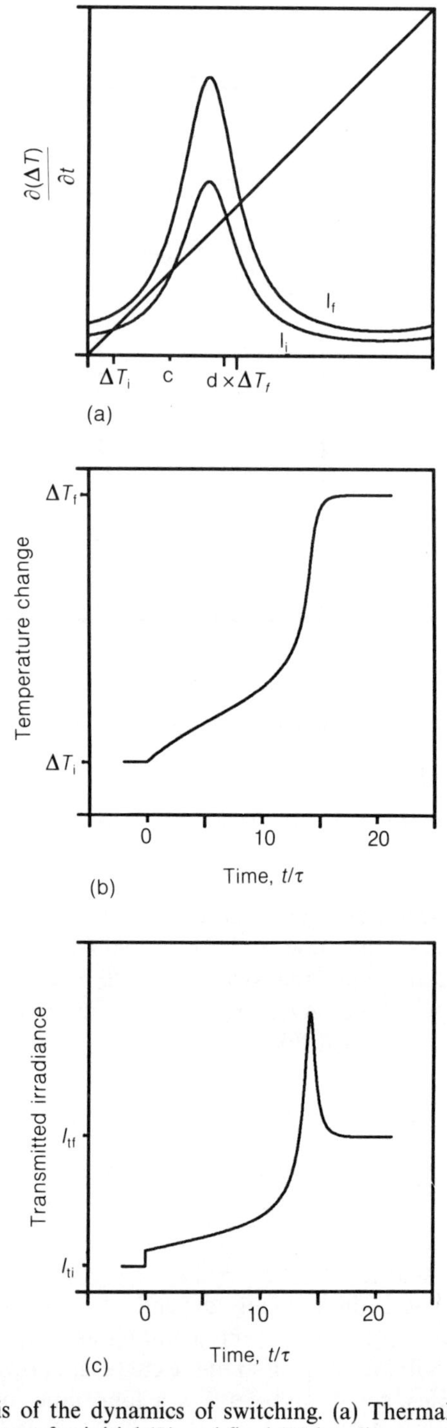

Figure 5.10 Analysis of the dynamics of switching. (a) Thermal driving force as a function of temperature for initial (I_i) and final (I_f) irradiances, and thermal sink. (b) Resulting time dependence of the temperature. (c) Corresponding time dependence of the transmitted irradiance.

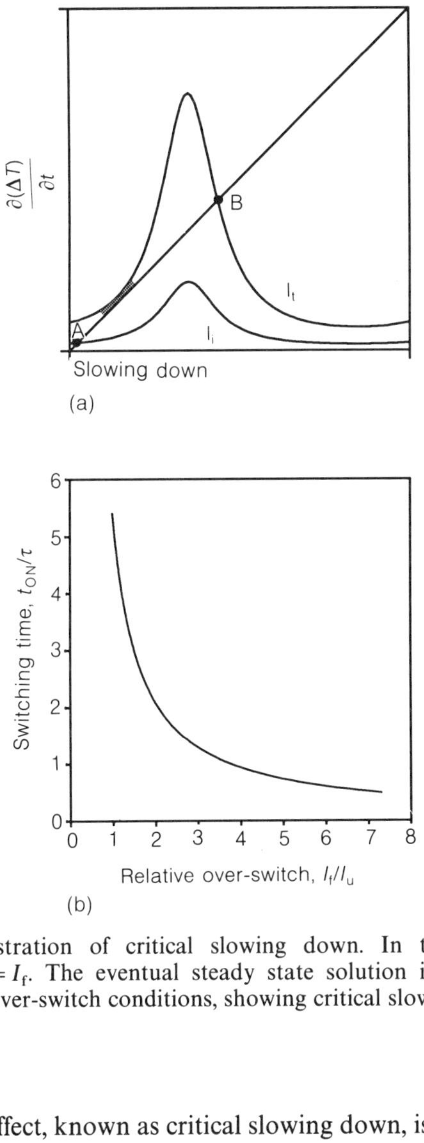

Figure 5.11 (a) Illustration of critical slowing down. In the region indicated $\partial(\Delta T)/\partial t \approx 0$ for $I = I_f$. The eventual steady state solution is at B, however. (b) Switch-on time vs. over-switch conditions, showing critical slowing down divergence for $I_f/I_u \rightarrow 1$.

is expected. This effect, known as critical slowing down, is particularly significant experimentally as it means that whilst a small incremental signal can in principle be used to cause switching it will be at a cost of device speed. Further, if one works close to the critical condition the curve and line are almost tangential over a longer ΔT range and the effect is accentuated. Note, however, that the initial proximity to the switch-up point for input I_i does not significantly alter the switch speed. Figure 5.11(b) shows the switching time as a function of the relative over-switch (I_f/I_u). For heavily overdrive cases, $I_f \gg I_u$, the limit to the switch time is set only by the time taken to achieve the desired excitation in the absence of recovery. Thus an extremely

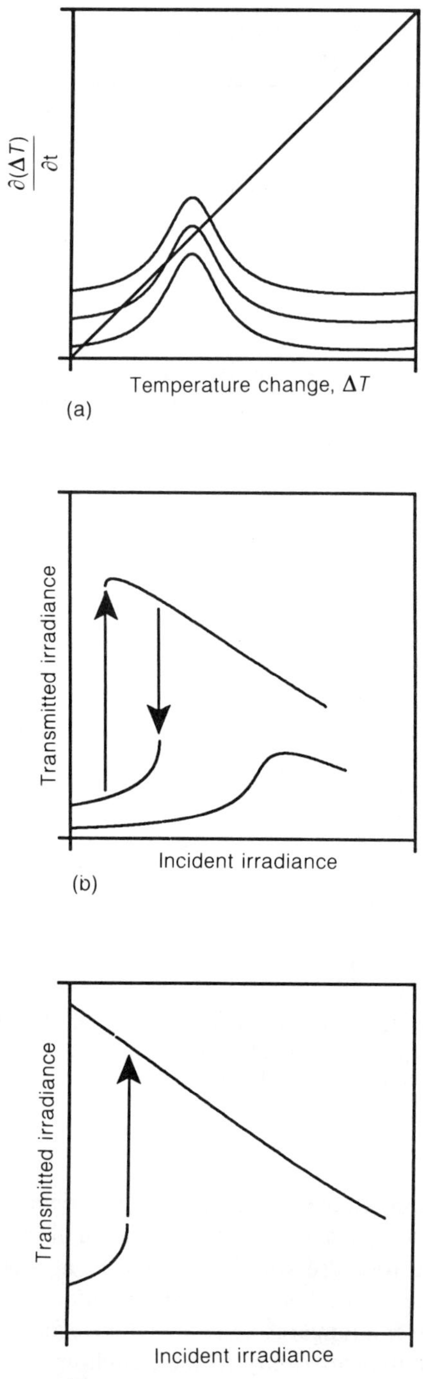

Figure 5.12 (a) Graphical solution for steady state of off-axis addressed nonlinear Fabry–Pérots. (b), (c) Resultant transmission responses vs. incident off-axis irradiance.

brief, high energy, pulse may be used to switch the transmission of a CW beam of level I_i, from low to high transmission, on extremely short time scales.

Note that similar effects to the above occur for switch-down from the upper bistable branch if the input level is reduced below I_d.

5.7.2. Off-axis address

Up to now all of the optical beams have been assumed to be collinear on the sample, of the same wavelength and polarization, and therefore experiencing the same feedback properties. Nonlinearity and bistability can also be achieved for a fixed irradiance beam experiencing feedback in the presence of a second, signal beam that may have different cavity resonances or indeed experience no feedback at all (e.g. one that is always entirely absorbed). The signal may even have a different physical form—for example electron beam or direct thermal heating of a thermo-optic cavity.

This configuration is known generically as 'off-axis address'—the axis referring to the propagation direction of the holding beam. In the extreme case that the additional signal experiences no feedback, in the thermal mode

$$\frac{\partial(\Delta T)}{\partial t} = \frac{BI_0}{1 + F_\alpha \sin^2(\phi_0 + b\,\Delta T)} + \sigma_E I_E - \frac{\Delta T}{\tau} \tag{5.45}$$

where I_E is the rate at which energy is supplied to the sample by the signal, per unit area, and σ_E is the temperature change per unit fluence, for the thermo-optic material in the cavity.

Similar graphical analysis is possible as for the on-axis case, but now I_0 is fixed and the Airy curve is vertically displaced by the additional input, rather than scaling with I_0 (Figure 5.12(a)). The transmitted irradiance varies between CI_0 and $CI_0/(1 + F_\alpha)$ as the address is varied; example nonlinear responses are shown in Figure 5.12(b). As the I_0, ϕ_0 combination could be set in a region of bistability it is also possible to achieve the response shown in Figure 5.12(c).

5.7.3. Transphasor operation

For detunings less than ϕ_c, the response is nonlinear but nonhysteretic. In this region the device may be used for signal gain or limiting, as indicated in Figure 5.13. If the system is biased to level I_0, then a small additional modulated signal can result in a large modulation of the total transmission. The dynamics of this operation obeys classic gain–bandwidth relations. An off-axis modulated address suffices to indicate the form of this relation:

$$\frac{d(\Delta T)}{dt} = \frac{BI_0}{1 + F_\alpha \sin^2(\phi_0 + b\,\Delta T)} + \sigma_E I_E(1 + \cos\Omega t) - \frac{\Delta T}{\tau} \tag{5.46}$$

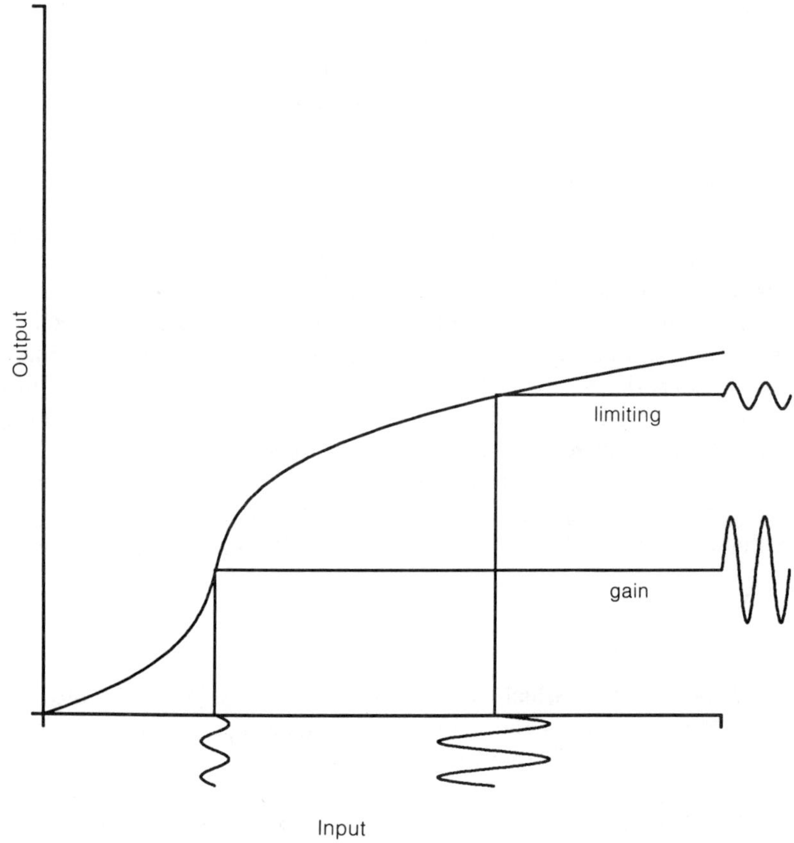

Figure 5.13 Non-hysteretic response characteristic with schematic of analogue signal gain and signal limiting.

If the modulated signal is weak then the solution will be of form

$$\Delta T(t) = \Delta T_0 + \sigma_E I_E[\beta + g\cos(\Omega t + \theta)] \tag{5.47}$$

where ΔT_0 is the temperature rise in the absence of the modulated signal. A perturbation treatment gives (Tooley *et al.*, 1985)

$$\beta = \left[1 + \frac{BI_0F_\alpha \sin^2(\phi_0 + b\,\Delta T_0)}{1 + F_\alpha \sin^2(\phi_0 + b\,\Delta T_0)}\right]^{-1} \tag{5.48}$$

$$g = \frac{B}{(1 + \Omega^2\tau^2/\beta^2)^{1/2}} \tag{5.49}$$

$$\theta = -\tan^{-1}(\Omega\tau/\beta) \tag{5.50}$$

The device gain at frequency Ω, defined as the amplitude of modulation of the transmitted hold beam divided by the amplitude of the modulated signal, I_E, is directly proportional to g. The point is that the gain depends on Ω only through the denominator in equation 5.49. In particular this means that the product of the zero Ω gain and the frequency at which the gain drops to one-half of this value—the gain–bandwidth product—is independent of the initial cavity detuning; it is a direct measure of the recombination time. This is another manifestation of the slowing-down phenomenon. For ϕ_0 close to the critical value the CW gain is at a maximum but the response rate is at its slowest.

By contrast to the gain experienced for a small modulated signal added to the bias at I_0, a limiting or noise reduction analogue mode is achieved for a bias I_l. Again noise or signal modulation in excess of a characteristic frequency will not experience the device nonlinearity. This results in the possibility of using bistable devices for high bandwidth (rapid modulation) transparent transmission, switched by relatively long pulses.

5.8. EXPERIMENTAL RESULTS

As a result of the discovery of giant refractive nonlinearities in InSb experiments were undertaken with plane-parallel-sided samples, using the natural material reflectivity (36%) to produce a cavity with the passive response shown in Figure 5.4(a). The feedback from such a cavity was sufficient to give the nonlinear response of Figure 5.14, the first observation of all optical bistability in a semiconductor using continuous-wave lasers (Miller *et al.*, 1979). Nonhysteretic power nonlinearities are detected at below 100 mW of incident power and two bistable regions exist near 130 and 240 mW. The latter are associated with the movement of two adjacent Fabry–Pérot peaks through the operating wavelength. This result was obtained in 1979 at the same time as bistability was reported for pulsed laser operation in GaAs (Gibbs *et al.*, 1979).

Under pulsed conditions one measures the transmitted vs. incident power as a function of time. If switching occurs there should be a sharp step in the transmission at some power level P_u; when the input power decreases after the peak of the pulse a corresponding step down is expected but at a lower power level P_d (Figure 5.15). The step may of course be smeared out by time-averaged detection but one is left with different transmissions for the range of power levels between P_u and P_d on the rise and decay sides of the pulses. Plotted in x–y form (P_t vs. P_i) bistability is observed. It is necessary, however, to know that the excitation recovery time of the sample is very short compared with the pulse duration. If this is not the case then x–y hysteresis is always expected, because the latter part of the pulse experiences the increased excitation level generated by the front part. For true bistability the transmission should be 'stable' at each instantaneous irradiation level.

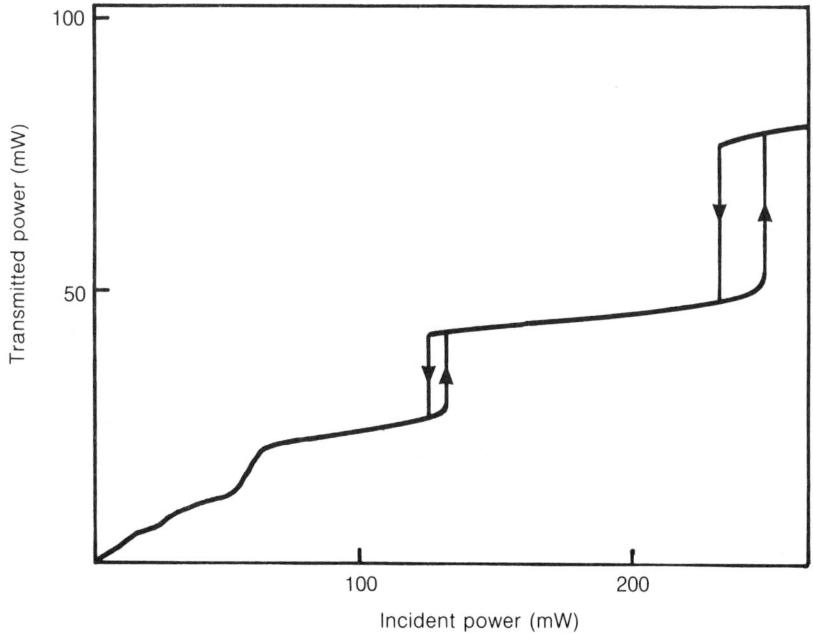

Figure 5.14 First experimental observation of CW optical hysteresis in InSb (5 μm; 5 K); after Miller *et al.* (1979).

The 'Optical Bistability' conference series (Bowden *et al.*, 1981, 1984; Gibbs *et al.*, 1986) provides useful reference for systems in which bistability has now been observed, also the text by Gibbs (1985), the review article by Abraham and Smith (1982) and the project report of Mandel *et al.* (1987).

The majority of systems in which hysteresis associated with electronic excitation has been observed have employed pulsed radiation. This is either because the nonlinearity is too weak to allow CW operation, or because of the effect of unwanted laser heating either occurring as the electronic excitations decay or because of parasitic absorption. The thermal systems make use of the otherwise detrimental heating effect. The major sample configuration used for CW optothermal bistability is the nonlinear interference filter. Figure 5.16 shows the structure of a narrow-band-pass filter in which a central spacer region, typically two optical wavelengths thick, is surrounded by high reflectivity stacks made up of alternate low and high index dielectric materials of $\lambda/4$ optical thicknesses. This forms a high finesse Fabry–Pérot cavity. Karpushko and Sinitsyn (1978) first reported CW bistability for such a structure, using ZnSe as the spacer material. This work was taken up by a number of other groups, in particular by Smith *et al.*, and forms the basis for prototype circuitry. The 0.5 μm band gap of ZnSe means that operation is possible in the visible regime (argon ion lasers have

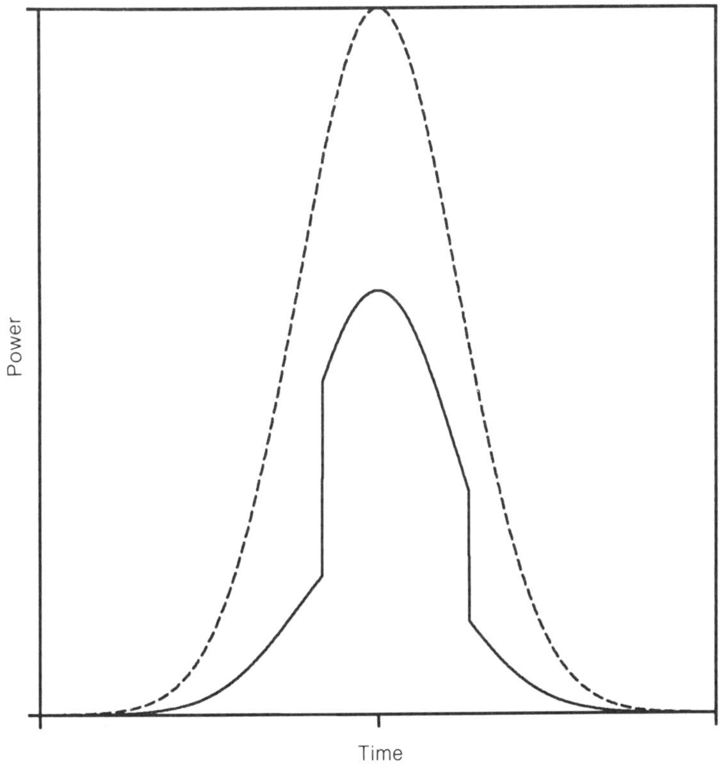

Figure 5.15 Schematic of optical switching of pulsed radiation: symmetric smooth input pulse (– – –) and asymmetric, switching transmitted pulse (———).

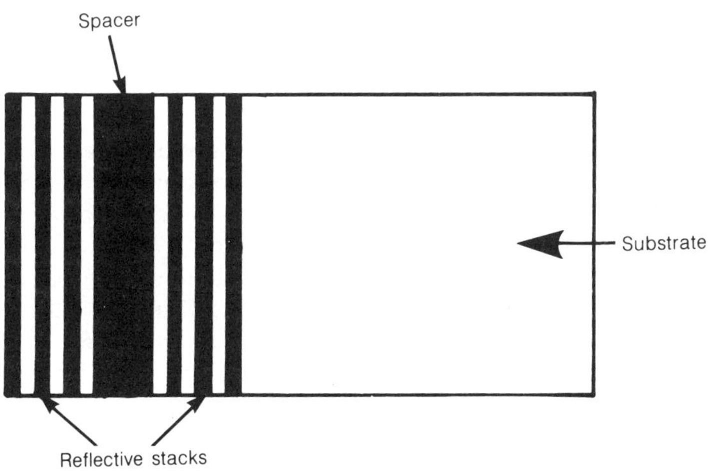

Figure 5.16 Structure of a narrow-bandpass dielectric multilayer interference filter.

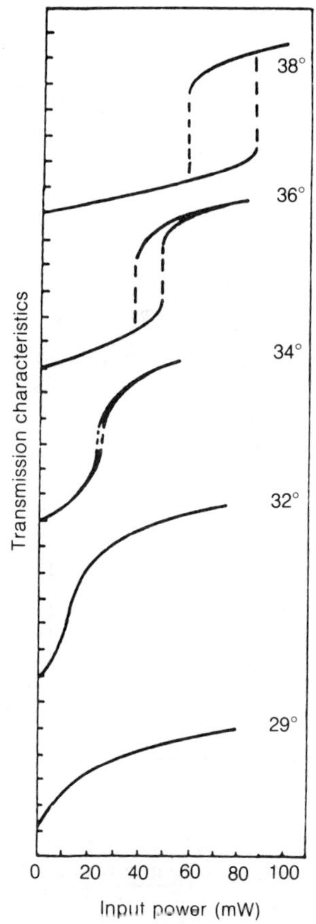

Figure 5.17 Nonlinear and bistable transmission response curves achievable using ZnSe interference filters; after Smith *et al.* (1984).

a powerful CW line at 514 nm) and at room temperature (compared with the 5 K or 77 K operation of InSb). In the larger gap ZnSe the electronic nonlinearity is extremely small and the thermal effect dominates, allowing bistability in the 1–100 mW regime to date (Figure 5.17) (Smith *et al.*, 1984). The filter system has been used to confirm the optimization principles set out in sections 5.5 and 5.6. For example, Figure 5.18(a) shows the spot size scaling of the switching powers, in agreement with section 5.6.4, and Figure 5.18(b) demonstrates both the reduction of switching times with spot size and the critical slowing down that occurs for marginal switching (Smith *et al.*, 1986; Mathew *et al.*, 1986).

Bistable studies have not been restricted to semiconductors (Table 5.3), for example a variety of mechanisms lead to nonlinearities in liquids. An

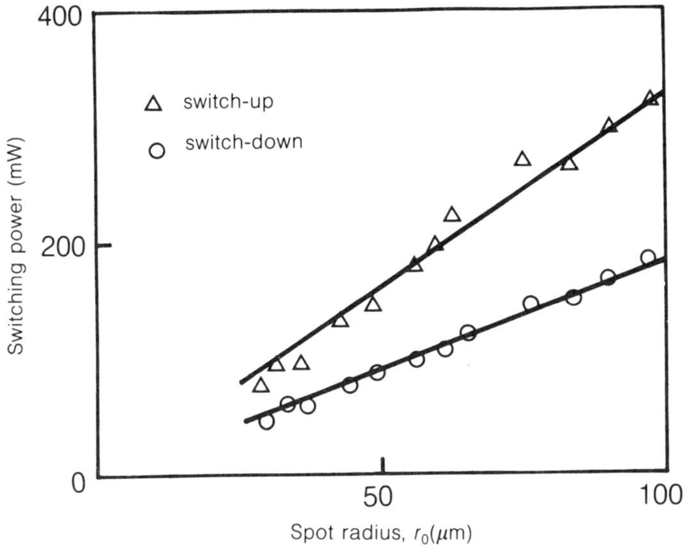

(a)

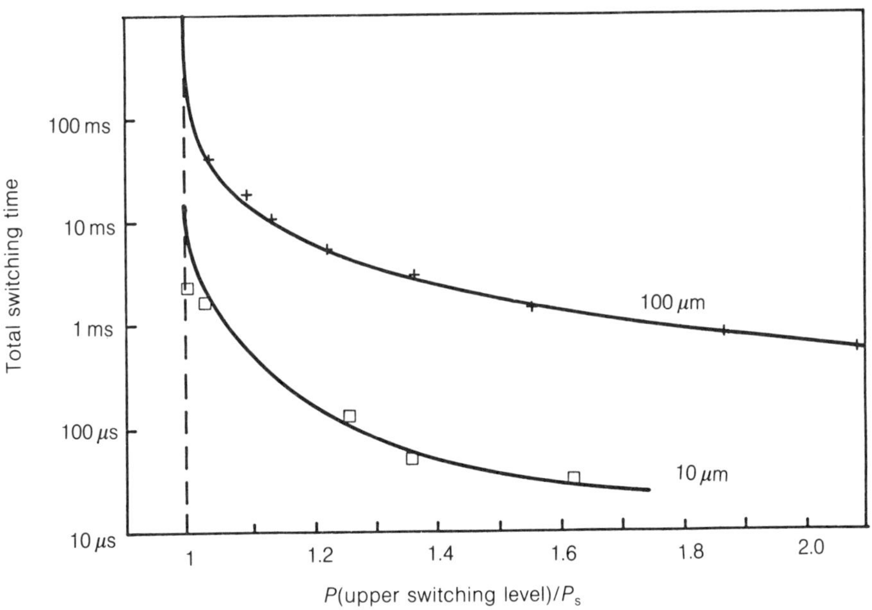

Figure 5.18 (a) Linear scaling of switch powers with spot radius; after Smith *et al.* (1986). (b) Spot size and over-switch dependence of filter switching times; after Mathew *et al.* (1986).

Table 5.3 Optothermal bistability using various liquids

Material	*Critical switching power* $P_c(mW)$	*Thermo-optic coefficient* $\partial n/\partial T(K^{-1})$
Aluminium/aluminium cells		
Water	6	-1.5×10^{-4}
Isopropanol	2.5	-4×10^{-4}
Ethanol	1.7	-4×10^{-4}
Champagne	4	
Glenfiddich	2.2	
Carbon disulphide	1.2	-8×10^{-4}
K15 (ordinary ray)	2.6	4×10^{-4}
K15 (extraordinary ray)	0.38	-2×10^{-3}
Dielectric/gold cell (room temperature)		
K15 (ordinary ray)	0.98	4×10^{-4}
K15 (extraordinary ray)	0.19	-2×10^{-3}

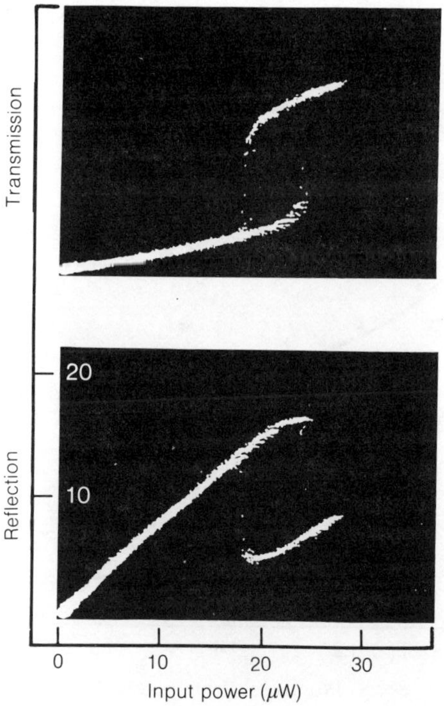

Figure 5.19 Very low power CW optothermal bistability in a Fabry–Pérot cell containing cyanobiphenyl; after Lloyd and Wherrett (1988).

example of extremely low power bistability ($P_c = 14\,\mu W$ for a $25\,\mu m$ spot size) in the nematic liquid crystal cyanobiphenyl (K15) is shown in Figure 5.19 (Lloyd and Wherrett, 1988). This material has a thermo-optic coefficient well above the semiconductor values, quoted in Table 5.2, rising to around $0.1\,K^{-1}$ at temperatures very close to the nematic-to-isotropic phase transition. Because the material is essentially transparent in the 633 nm wavelength range used for this result the cavity construction was slightly different from the conventional filters, with the rear reflector consisting of an absorbing, metallic mirror. Such a structure removes the wavelength selectivity of the material absorption and can therefore be used over a range of wavelengths determined only by the cavity Fabry–Pérot peak positions and by the metal properties (Hutchings *et al.*, 1987). Figure 5.20 demonstrates, for such cavities, the dynamics of switching, as described in section 5.7.1.

An example of fast, low energy pulsed operation is the low temperature CdS work of Dagenais and Sharfin (1985) (Figure 5.21). Here the appropriate material excitation is a discrete, bound exciton peak. Similar excitonic nonlinearities have been studied in detail for GaAs and for the multiple quantum well GaAs/AlGaAs structures (Peyghambarian and Gibbs, 1985; Gibbs *et al.*, 1982). Whilst the value of σ_n is high for such systems, because of the screening effect of general carriers, the requirement on the total index change (equation 5.34) means that the full excitonic oscillator strength is

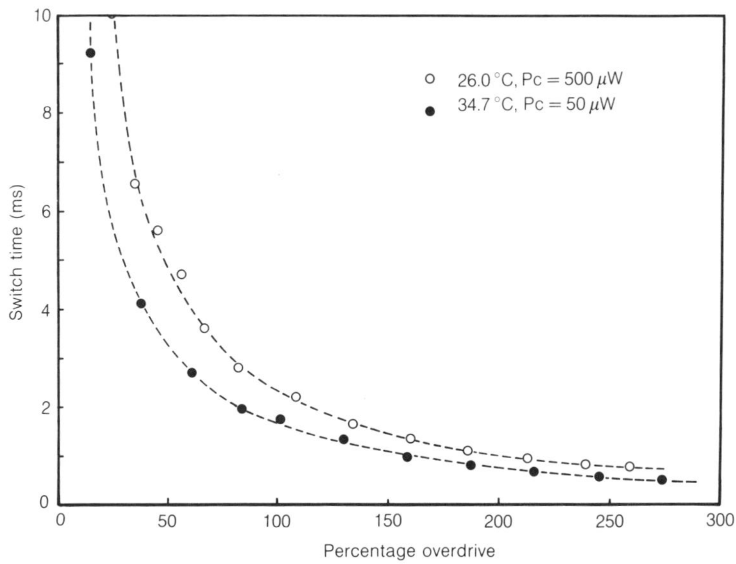

Figure 5.20 Dynamics of switching times for a liquid crystal cell; after Lloyd and Wherrett (1988).

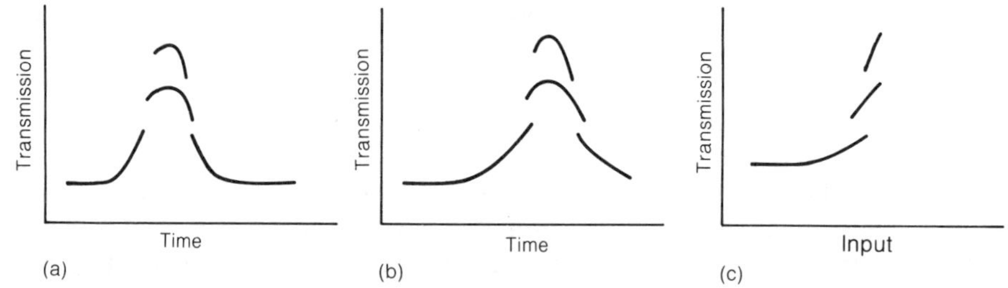

Figure 5.21 Nonlinear transmission in CdS with optical pulses of (a) 100 ns and (b) 500 ns and (c) corresponding hysteresis in input/output characteristics; after Dagenais and Sharfin (1985).

barely sufficient to produce bistability without the additional contribution from interband carrier blocking, unless background absorption can be reduced.

5.9. ALTERNATIVE BISTABILITY MECHANISMS

We have concentrated in this article on refractive bistability in Fabry–Pérot cavities with reflective feedback. For completeness brief comments on a number of the alternative schemes that have been studied are included here; these involve either refractive or absorptive nonlinearities with feedback provided by cavity reflection, or intrinsically in the nonlinear material itself, or by an external nonoptical mechanism.

5.9.1. Bistability by increasing absorption

Consider a semiconductor fundamental absorption edge. For most semiconductors the band gap energy decreases as the sample temperature is raised. Hence starting at a moderate absorption level at low incident power in a poorly heat sunk sample as the power is gradually increased the sample temperature will increase and hence so will the absorption coefficient. There is therefore the possibility of a thermal runaway, with avalanching from low absorption and low temperature to a high temperature and high absorption. Under appropriate initial conditions there is a hysteresis associated with the effect and bistability of the form shown in Figure 5.22 occurs (Wherrett *et al.*, 1984). The control parameters in this example are the incident power level and the laser frequency (and hence the initial absorption coefficient). This is just one example of how a coupling between the intrinsic excitation level and the optical power levels can lead to bistability in the absence of cavity feedback (Hajto and Janossy, 1983; Miller, 1984).

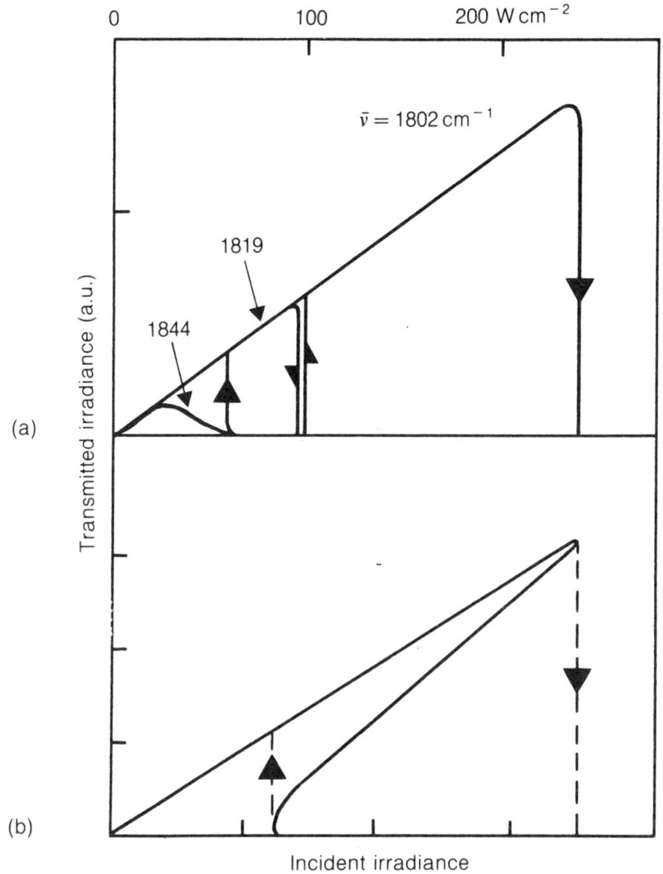

Figure 5.22 (a) Experimental and (b) theoretical plots of bistability by increasing absorption; after Wherrett *et al.* (1984).

5.9.2. Laser diode amplifier bistability

Given that an absorbing nonlinear Fabry–Pérot results in hysteresis it is not surprising that a cavity with material gain is a more effective device in terms of optical power alone. Consider a laser diode held just below the lasing threshold for some electric current input. An incident radiation beam near the notional oscillation frequency experiences amplification inside the cavity, but insufficient amplification to give net round-trip gain. The amplification is caused by the stimulated emission associated with conduction-to-valence transitions of the injected carriers. Hence as the incident power level is increased the population inversion alters very slightly. In turn there is an associated refractive index change and for the current initial condition the cavity can be pulled onto resonance, which is essentially a very high finesse Fabry–Pérot resonance. The bias power (essentially of order 1 mW) is now provided by the electric current, but the optical power for such a device can easily be reduced to the microwatt level (Adams, 1987; Sharfin and Dagenais, 1986).

5.9.3. Laser diode bistability

In contrast to the previous example one now considers a diode laser biased above the threshold current and therefore lasing. The first proposals for bistable systems involved the use of diode lasers with intracavity saturable absorbers (Kosomosky, 1964; Lasker, 1964). In the bistable regime the two self-consistent states are (i) low light level in the cavity and hence a consistent absorption and (ii) a high light level and saturation of the absorption. Switching from one level to the other is achieved by temporary input signals either onto the absorber alone (to saturate the absorption temporarily until the laser level has built up sufficiently to sustain the saturation) or onto the emitter region (such as to quench the emission, to reduce the cavity irradiance and to allow the absorption saturation to recover). A number of multicavity and cleaved cavity structures have been fabricated to enable such bistability. For example, optical switching in 5 ns using 10 μW pulses has been achieved in a two-contact, 1.3 μm structure (Odagiri *et al.*, 1984). Refractively induced laser bistability is also possible, for example by use of an external grating for feedback (Kawaguchi, 1987).

5.9.4. Modal bistability

In parallel stripe waveguides radiation coupling exists if the guides are physically close. The energy of a beam injected into one guide will oscillate between the two as it propagates, being entirely in the second guide after a specific distance. The strength of the coupling depends critically on the guide refractive indices. Hence a beam that has crossed over the guides at low light levels will in principle switch and be propagated entirely within one guide if the level is increased and if the guide material is nonlinear (Jensen, 1982).

A similar effect occurs for twin-stripe lasers, where the spatial mode of the system depends strongly on the details of the index profile in the stripe region. The competition between the gain guiding within each stripe and the index guiding has been used to demonstrate laser modal switching in GaAs/AlGaAs structures for, typically, milliwatt power levels (White and Carroll, 1983).

5.9.5. Polarization bistability

Fabry–Pérot cavities containing birefringent spacer material exhibit different resonant wavelengths for the ordinary and extraordinary rays; indeed TE and TM beams incident nonnormal on isotropic, dielectric-coated cavities also show birefringence as a result of the different phase changes on reflection for the two beams. By arranging for one beam to be initially slightly off resonance, and the other on resonance, it is in principle possible to switch between polarizations by selective transmission or reflection (Lloyd, 1987).

Similar effects have been used to achieve polarization switching in double-heterostructure lasers (Chen and Liu, 1985).

5.9.6. Bistability at nonlinear interfaces

Optical waveguiding and input coupling into waveguides is sensitive to small refractive index differences between the guiding and cladding materials or between for example input prism coupler indices and the guide indices. It is therefore expected that optical nonlinearities in any of the relevant media will lead to highly nonlinear responses. A simple case is frustrated internal reflection: here a beam inside one medium is incident on the second at an angle just above the critical total internal reflection angle and so is almost totally reflected. If the second medium has a positive nonlinear refractive index then as the beam power is increased the evanescent wave causes the index to increase and the critical angle increases. Hence the beam becomes refracted rather than reflected—the irradiance in the second medium becomes even higher and one has a self-consistent condition with no internal reflection. This effect will have an associated hysteresis (Khou and Hou, 1982).

5.9.7. Hybrid bistable schemes and SEED devices

Early schemes used feedback via electronics to demonstrate the principles of optical bistability. Figure 5.23 is a schematic of a dispersive bistable cavity

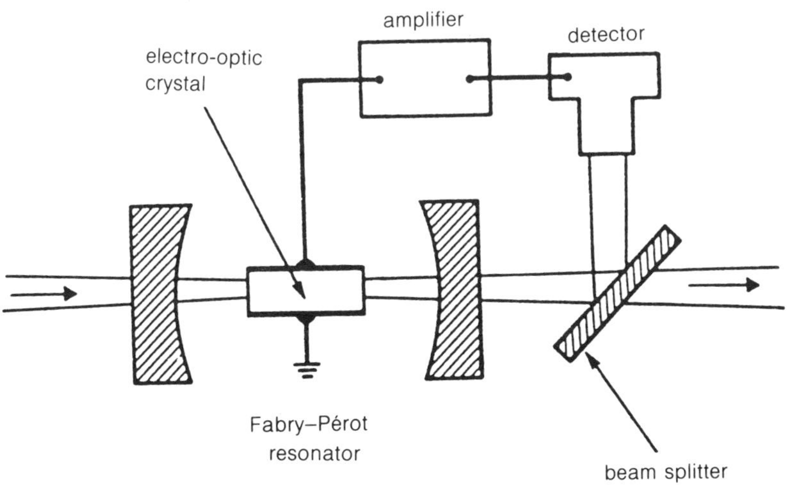

Figure 5.23 Schematic of bistability induced by hybrid feedback; after Smith and Turner (1977).

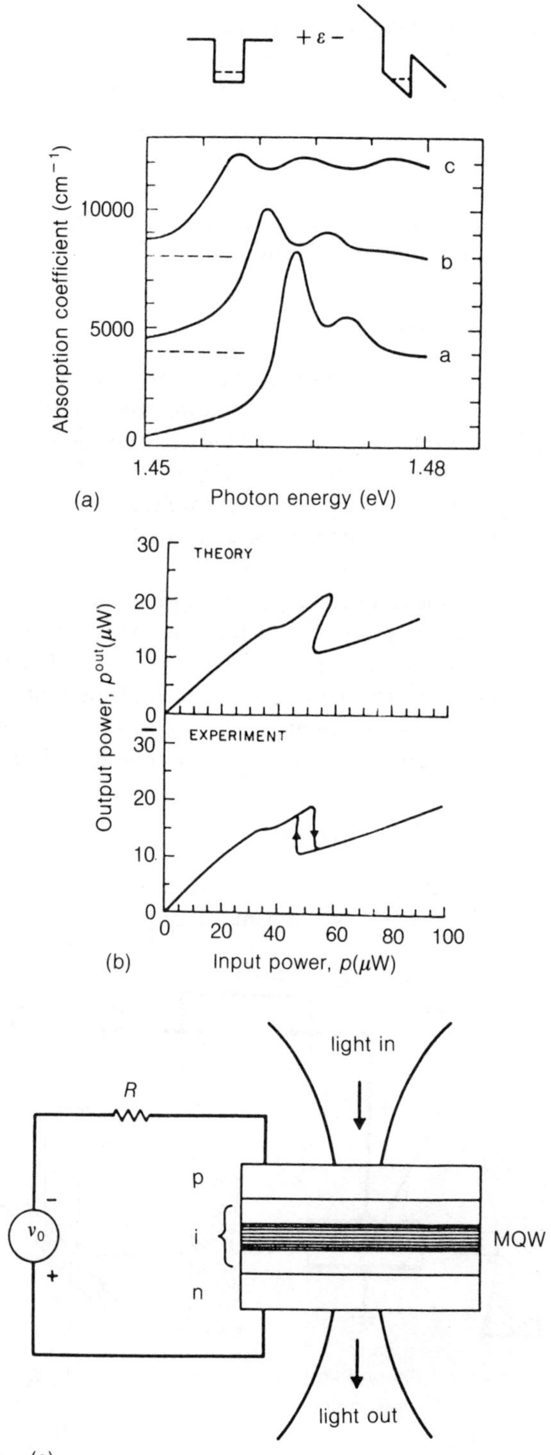

+ε−
Absorption coefficient (cm^{-1})
0
5000
10000
c
b
a
1.45
1.48
Photon energy (eV)
(a)
THEORY
EXPERIMENT
Output power, p^{out}(μW)
0
10
20
30
0
20
40
60
80
100
Input power, p(μW)
(b)
light in
light out
R
v_0
p
i
n
MQW
(c)

with an electro-optic spacer material. Initially it is off resonance and the transmission low. Part of the transmitted power reaches a photodetector and the electronic signal is used to produce a voltage, proportional to the transmitted power, across the electro-optic material. The refractive index is thereby altered, the cavity is brought towards resonance and switching, with bistability, can be achieved in precisely the manner of a nonlinearly refractive cavity (Smith and Turner, 1977).

A similar voltage feedback can be used to produce bistability by increasing absorption, provided that the absorption coefficient is a function of electric field. Excitonic absorption in multiple quantum well structures has attracted particular attention in this respect. Figure 5.24(a) shows the effect of applied field on the excitonic level and Figure 5.24(b) the resulting bistability—of similar form to Figure 5.22 (Chemla *et al.*, 1985). Because the thickness of quantum well material can be extremely small, field strengths of 10^4–10^5 V cm^{-1} are achievable for only a few volts; this is achievable directly from the electrical response of the structure to the radiation field. As a result it is possible to fabricate a self-electroabsorptive device with no electrical amplification; Figure 5.24(c) (Miller *et al.*, 1984). The constant voltage supply produces a field across the MQW device. As the light level increases the device produces a photocurrent so that the voltage drop across the MQW is reduced. The initial conditions are set so that the absorption increases as a consequence. Hence the photocurrent increases and one has the avalanching that forms the basis for switching and bistability. Named the SEED (self-electro-optic effect device), switching energies below nanojoules have been observed. A number of closely similar devices based on the self-electro-optic effect have been designed. Most recently a symmetric SEED structure in which two devices are coupled together has been investigated. Bistability occurs in the ratio of the two input levels and hence can be driven to very low powers. Switching times down to 40 ns (8 mW) and switching powers as low as 10 nW have been observed (Lentine *et al.*, 1988).

5.10. SUMMARY

A vast number of materials and techniques have now been used to achieve optical bistability. No attempt has been made to give a comprehensive list in this article. A selection for the power–time results is shown in Figure 5.25, which follows the style originally presented by Smith (1982). For comparison the operating regions of conventional and Josephson junction electronic

---

Figure 5.24 (a) Effect of applied electric field in the excitonic absorption of GaAs/-AlGaAs MQW. (b) Bistability in GaAs/AlGaAs MQW. (c) Schematic of SEED device; Chemla *et al.* (1985).

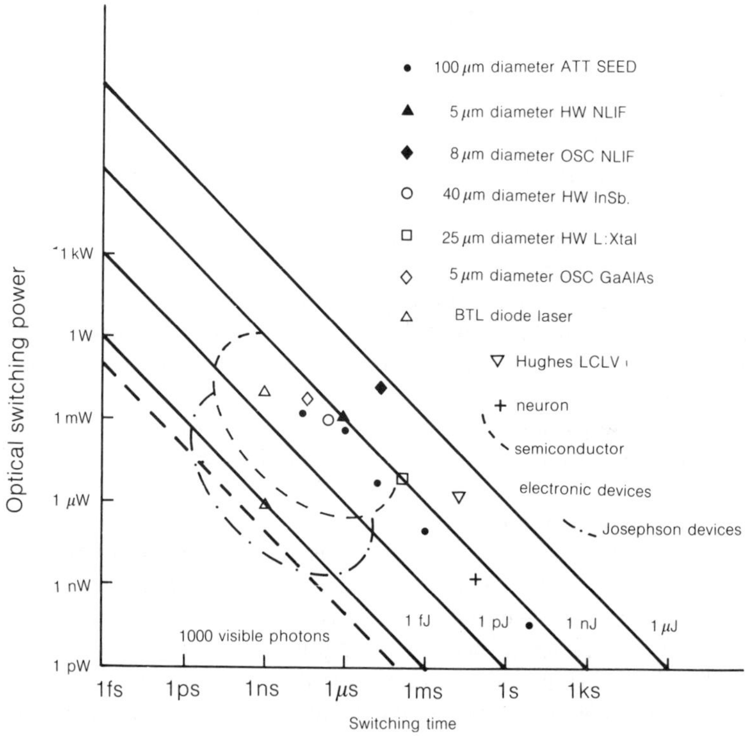

Figure 5.25 Power–time results for various optically bistable devices. The operating regions for conventional and Josephson semiconductor devices and neurons are indicated for comparison.

switches are indicated, and of neurons. Whilst the achievement of low power–time products is clearly important it should be noted that low power and, it follows, relatively slow devices have an important role to play in highly parallel switching schemes such as are mooted for image processing and neural networks. Equally there is still a possibility that ultrafast, and by implication high power, optical switches will have a role in high bandwidth data analysis where the duration and repetition rate of optical pulses prevents the use of direct electronic processing.

REFERENCES

Abraham, E. and Smith, S. D. (1982) *Rep. Prog. Phys.*, **4**, 815.

Adams, M. J., Westlake, H. J., O'Mahony, M. J. and Hendry, I. D. (1985) *IEEE J. Quantum Electron.*, **21**, 1498.

Adams, M. J. (1987) *Solid State Electron.*, **30**, 43.
Al-Attar, H. A., Hutchings, D., Russell, D. *et al.* (1986) *Opt. Commun.*, **58**, 433.
Antonioli, G., Bianchi, D. and Franzosi, P. (1979) *Appl. Opt.*, **18**, 3847.
Ayrault, B., Langlois, H., Lecocq-Mayer, M. C. and Lefin, F. (1973) *Phys. Status Solidi A*, **17**, 665.
Bakiev, A. M., Dneprovskii, V. S., Kovalyuk, Z. D. and Stadnik, V. A. (1983) *Sov. Phys. Dokl.* **28**, 579.
Bowden, C. M., Ciftan, M. and Robl, H. R. (eds) (1981) *Optical Bistability*, Plenum, New York.
Bowden, C. M., Gibbs, H. M. and McCall, S. L. (eds) (1984) *Optical Bistability II*, Plenum, New York.
Chemla, D. S., Miller, D. A. B. and Smith, P. W. (1985) *Opt. Eng.*, **24**, 556.
Chen, Y. C. and Liu, J. M. (1985) *Appl. Phys. Lett.*, **46**, 16.
Craig, D. and Miller, A. (1986) *Opt. Acta*, **33**, 397.
Craig, D., Dyball, M. R. and Miller, A. (1985) *Opt. Commun.*, **54**, 383.
Dagenais, M. and Sharfin, W. F. (1985) *Appl. Phys. Lett.*, **46**, 230.
Dagenais, M. and Sharfin, W. F. (1986) *Springer Proc. Phys.*, **8**, 122.
Eichler, H. J. (1983) *Opt. Commun.*, **45**, 62.
Firth, W. J., Galbraith, I. and Wright, E. M. (1985) *J. Opt. Soc. Am. B*, **2**, 1005.
Firth, W., Peyghambarian, N. and Tauet, P. (eds) (1988) *Optical Bistability IV*, in Les éditions de physique (Los'illis, France). *J. Phys. (Paris), Collog. C2*, Suppl. 6.
Frank, D. and Wherrett, B. S. (1987) *J. Opt. Soc. Am. B*, **4**, 25.
Franken, P. A. and Ward, J. F. (1963) *Rev. Mod. Phys.*, **35**, 23.
Gibbs, H. M. (1985) *Optical Bistability*, Academic Press, New York, pp. 471.
Gibbs, H. M., McCall, S. L. and Venkatesan, T. N. C. (1976) *Phys. Rev. Lett.*, **36**, 1135.
Gibbs, H. M., McCall, S. L., Venkatesan, T. N. C. *et al.* (1979) *Appl. Phys. Lett.*, **36**, 6.
Gibbs, H. M., Venkatesan, T. N. C., McCall, S. L. *et al.* (1979b) *Appl. Phys. Lett.* **34**, 511.
Gibbs, H. M., Tarng, S. S., Jewell, J. L. *et al.* (1982) *Appl. Phys. Lett.*, **41**, 222.
Gibbs, H. M., Mandel, P., Peyghambarian, N. and Smith, S. D. (eds) (1986) *Optical Bistability III*, in *Springer Proc. Phys.*, **8**.
Golik, L. L., Grigor'yants, A. V., Elinson, M. I. and Balkarei, Yu. I. (1983) *Opt. Commun.* **46**, 51.
Guha, S., Van Stryland, E. W., and Soileau, M. J. (1985) *Opt. Lett.*, **10**, 285
Harris, R. J., Johnson, G. T., Kepple, G. A. *et al.* (1977) *Appl. Opt.*, **16**, 436.
Hajto, J. and Janossy, I. (1983) *Philos. Mag. B*, **47**, 347.
Henneberger, F., Puls, J., Rossmann, H. *et al.* (1987) *Physica Scripta*, **T13**, 195.
Hutchings, D., Lloyd, A. D., Janossy, I. and Wherrett, B. S. (1987) *Opt. Commun.*, **61**, 345.
Jain, R. K. and Klein, M. B. (1983) *Optical Phase Conjugation* (ed. R. A. Fisher), (10), 369. Academic Press, New York.
Jain, R. K. and Lind, R. C. (1983) *JOSA* **72**, 647.
Jensen, S. M. (1982) *IEEE J. Quantum Electron.*, **18**, 1580.
Kar, A. K. and Wherrett, B. S. (1986) *J. Opt. Soc. Am. B*, **3**, 345.
Kar, A. K., MacKenzie, H. A., Wei Ji, Reid, J. J. E. *et al.* (1986) *Proc. Phys.*, **8**, 144.
Karpushko, F. V. and Sinitsyn, G. V. (1978) *Zh. Prikl. Spektrosk.*, **29**, 820. (Karpushko, F. V. and Sinitsyn, G. V. (1978) *J. Appl. Spectrosc. USSR*, **29**, 1323.)
Kawaguchi, H. (1987) *Opt. Quantum Electron.*, **19**, S1.

Khou, I. C. and Hou, J. Y. (1982) *J. Opt. Soc. Am.*, **72**, 1761.
Kosomosky, W. F. (1964) In *Proc. Symp. Optical and Electro-optical Information Processing, Boston, MA* (eds J. Tippett, D. Berkowitz, L. Clapp *et al.*) MIT Press, Boston, p. 269.
Kremenitskii, V., Odoulov, S. and Soskin, M. (1979) *Phys. Status Solidi A* **51**, K63.
Lasker, G. J. (1964) *Solid State Electron.*, **7**, 707.
Lee, Y. H., Chavely-Pirson, A., Rhee, B. *et al.* (1986) *Appl. Phys. Lett.* **49**, 1505.
Lentine, A. L., Hinton, H. S., Miller, D. A. B. *et al.* (1988) *Appl. Phys. Lett.*, **52**, 1419.
Lloyd, A. D. (1987) *Opt. Commun.*, **64**, 302.
Lloyd, A. D., Janossy, I., Mackenzie, H. A., Wherrett, B. S. (1987) *Opt. Commun.* **61**, 339.
Lloyd, A. D. and Wherrett, B. S. (1988), Proc. Optical Bistability IV, Aussois in *J. Phys.* (Paris), Collog. C2, Suppl. 6. p. 141.
Maker, P. D., Terhune, R. W. and Savage, C. M. (1964) *Phys. Rev. Lett.*, **12**, 507.
Mandel, P., Smith, S. D. and Wherrett, B. S. (eds) (1987) *From Optical Bistability Towards Optical Computing*, Elsevier, Amsterdam.
Marburger, J. H. and Felber, F. S. (1978) *Phys. Rev. A*, **17**, 335.
Mathew, J. G. H., Kar, A. K., Heckenberg, N. R. and Galbraith, I. (1985) *IEEE J. Quantum Electron.*, **QE-21**, 94.
Mathew, J. G. H., Taghizadeh, M. R., Abraham, E. *et al.* (1986) *Springer Proc. Phys.*, **8**, 57.
Mayer, G. and Gires, F. (1964) *C. R. Hebd. Seances Acid. Sci. Paris*, **258**, 2039.
Miller, D. A. B. (1981) *IEEE J. Quantum Electron.*, **17**, 306.
Miller, D. A. B. (1984) *J. Opt. Soc. Am. B*, **1**, 857.
Miller, A., Parry, G. and Daley, R. (1984) *IEEE J. Quantum Electron.*, **QE-20**, 710.
Miller, D. A. B., Smith, S. D. and Johnson, A. M. (1979) *Appl. Phys. Lett.*, **35**, 658.
Miller, D. A. B., Seaton, C. T. and Smith, S. D. (1981) *Phys. Rev. Lett.*, **47**, 197.
Miller, D. A. B., Chemla, D. S., Eilenberger, D. J. *et al.* (1983) *Appl. Phys. Lett.* **42**, 925.
Miller, D. A. B., Chemla, D. S., Damen, T. C. *et al.* (1984) *Appl. Phys. Lett.*, **45**, 13.
Miller, A., Steward, G., Blood, P. and Woodridge, K. (1986) *Opt. Acta*, **32**, 387.
Odagiri, Y., Komatsu, K. and Suzuki, S. (1984) *CLEO Conf. Digest, J. Opt. Soc. Am.* THJ3.
Oudar, J. L., Abram, I. and Minot, C. (1984) *Appl. Phys. Lett.*, **44**, 689.
Ovadia, S., Gibbs, H. M., Jewell, J. L. *et al.* (1985) *Opt. Eng.*, **24**, 565.
Peyghambarian, N., Gibbs, H. M., Rushford, M. C. and Weinberger, D. A. (1983) *Phys. Rev. Lett.*, **51**, 1692.
Peyghambarian, N. and Gibbs, H. M. (1985) *J. Opt. Soc, Am. B*, **2**, 1215.
Poole, C. D. and Garmire, E. (1984) *Opt. Lett.*, **9**, 356.
Sharfin, W. F. and Dagenais, M. (1986) *Appl. Phys. Lett.*, **48**, 321.
Sharfin, W. F. and Dagenais, M. (1985) *Appl. Phys. Lett.*, **46**, 819.
Smith, P. W. (1982) *Bell Syst. Tech. J.*, **61**, 197.
Smith, P. W. and Turner, E. H. (1977) *Appl. Phys. Lett.*, **30**, 280.
Smith, S. D., Mathew, J. G. H., Taghizadeh, M. R. *et al.* (1984) *Opt. Commun.*, **51**, 357.
Smith, S. D., Walker, A. C., Wherrett, B. S. *et al.* (1986) *Appl. Opt.*, **25**, 1586.
Staupendahl, G. and Schindler, K. (1982) *Opt. Quantum Electron.*, **14**, 157.
Szöke, A., Daneu, V., Goldhar, J. and Kumit, N. A. (1969) *Appl. Phys. Lett.*, **15**, 376.
Tooley, F. A. P., Walker, A. C. and Smith, S. D. (1985) *IEEE J. Quantum Electron.*, **QE-21**, 1340.

Tooley, F. A. P., Firth, W. J., Walker, A. C. *et al.* (1985) *IEEE J. Quantum Electron.*, **21**, 1356.

Walker, A. C. and Hunter, J. J. (1985) Heriot-Watt University, internal report (unpublished).

Watkins, D. E., Phipps, C. R. and Thomas, S. J. (1981) *Opt. Lett.*, **6**, 76.

Wherrett, B. S. (1984a) *J. Opt, Soc. Am. B* **1**, 67.

Wherrett, B. S. (1984b) *Philos. Trans. R. Soc. London, Ser. A*, **313**, 213.

Wherrett, B. S., Tooley, F. A. P. and Smith, S. D. (1984) *Opt. Commun.*, **52**, 301.

Wherrett, B. S., Hutchings, D. and Russell, D. (1986) *J. Opt. Soc. Am. B*, **3**, 35.

Wherrett, B. S., Walker, A. C. and Tooley, F. A. P. (1988) In *Optical Nonlinearities and Instabilities in Semiconductors* (ed. H. Haug), Academic Press, New York.

White, I. H. and Carroll, J. E. (1983) *Electron. Lett.*, **19**, 337.

6 Optical processing using phase conjugation

Robert W. Eason

6.1. INTRODUCTION

Optical phase conjugation (OPC or PC) is a technique that has attracted a great deal of attention during the past ten years or so from those working in the field of optics, lasers, nonlinear materials and optical information processing. In essence, phase conjugation is a process that uses a wide variety of nonlinear optical effects to generate in real time a wavefront that is an exact replica of any arbitrary incident wavefront apart from a reversal of both its direction of propagation and its overall phase.

Although the principle of PC can be traced back to early experiments in holography or wavefront reconstruction, it was not until 1972 that experiments were first performed in media rather more versatile than photographic emulsions. Since then, there has been a remarkable development in activity in this area. Because of the large variety of media currently available for producing phase conjugate waves, particularly those that respond to powers in the μW–mW range, the subject has undergone rapid expansion, and now embraces a wide range of applications.

While a few of these remain purely speculative, the majority are interesting experimental demonstrations that address areas such as optical tracking, high resolution imaging and lithography, a range of optical image processing applications, and, more recently, iterative schemes involving 'intelligent' optical systems.

This chapter will discuss the physics of OPC, while constantly keeping in mind the applications areas that stem from the nonlinear interactions responsible for PC generation. Section 6.2 will give a brief background to PC optics, and will discuss the formalism of the wave equation and its interpretation under phase conjugate reversal. Section 6.3 describes mechanisms for PC, and discusses in some detail the most widely adopted scheme of degenerate four-wave mixing. This area will have been mentioned in several other chapters in the context of interactions involving the third-order nonlinear susceptibility tensor $\chi^{(3)}$. In particular, the analogy between four-wave mixing and real-time holography will be drawn here, as the concept of 'gratings', familiar in conventional holographic optics, also serves a very useful purpose in PC optical systems.

Section 6.4 covers media that are useful for PC generation. In fact, if the four-wave mixing arrangement is adopted, then any material of suitable transparency, optical quality etc. can be used, as there are no restrictions imposed via $\chi^{(3)}$ (apart from the absolute material specific value), as there are for $\chi^{(2)}$ interactions which have the usual centre of symmetry restrictions (see Chapter 1). In particular the area of photorefraction will be discussed in section 6.4. The reason for this is two-fold. Firstly, although analogous to conventional $\chi^{(3)}$ type four-wave mixing, photorefractive phase conjugate interactions are subtly different, but can very easily be thought of in terms of 'gratings' induced in crystalline materials via the linear electro-optic effect. By suitable choice of interaction geometry, and coherence relationship between the three interacting beams, these gratings can readily be interpreted in terms of the relevant material parameters, and all the resultant effects are elegantly described theoretically (Chapter 4). More importantly, however, the photorefractive effect is uniquely able to generate PC waves at exceedingly low intensities. CW powers in the ~ mW range are routinely adequate to produce faithful conjugation, the only drawback being that, in these cases, the time response can be comparatively slow. The ability to use CW rather than pulsed lasers, however, at essentially any wavelength in the visible to near-IR region, does provide the experimenter with a generous dividend in terms of flexibility, and facilitates the optical processing implementation considerably (the objective after all).

Section 6.5 discusses a range of applications of PC in 'spatial' and 'temporal' schemes. This division, while somewhat arbitrary, does attempt to address the processing side from two directions. In practice an area such as PC interferometry would happily fit in either, so the choice here reflects the author's bias rather than any objective decision. It must also be stated that the range of applications discussed is by no means exhaustive. What has been done here is to highlight particular applications that fall under the heading of 'optical processing', in the context of this book, and exclusion of some excellent work is merely due to limited space. Section 6.6 will briefly conclude and point to current and future work in the areas of waveguides, fibres and planar geometries that are rapidly developing in line with the increasingly technologically important diode-pumping planar geometries.

Several excellent texts have appeared on OPC in the last ten years, for example Fisher (1983) for background on all PC generation, Zel'dovich *et al.* (1985) for predominantly stimulated scattering, Günter and Huignard (1988, 1989) for photorefractive PC and its applications, as well as numerous single chapters in graduate level texts such as Yariv (1985) and Shen (1984).

6.2. BACKGROUND AND FORMALISM OF PHASE CONJUGATION

6.2.1. Background

The ideas underlying phase conjugation can be traced back to 1948, when Gabor (1948) discussed the principle behind wavefront reconstruction. It was not until 1965, however, five years after the laser made its first appearance, that Kogelnik (1965) subsequently realized that holography could be used to compensate for static inhomogeneities in a medium. Some further time elapsed before the connection between this compensation process and the act of complex conjugation was appreciated. The concept of real-time holography, in which a grating can be recorded in a medium, and simultaneously read out, provided the next important step in phase conjugate research. (e.g. Gerritsen, 1967; Stepanov *et al.*, 1971).

Independently of this dynamic holographic research, work by Zel'dovich *et al.* (1972) showed that phase distortions could also be corrected through the process of stimulated Brillouin scattering (SBS). It was not until several years later, however, that these parallel but formally unrelated research directions were established on a more rigorous basis by the application of the formalism of nonlinear optics to PC wavefront generation. Notable here are the work of Yariv (1976) and Hellwarth (1977), and the use of a coupled-wave analysis in four-wave mixing by Yariv and Pepper (1977) and Bloom and Bjorklund (1977). Since then, these first analyses have been extended to include the effects of pump depletion, resonant effects and the application of the generalized theory to materials other than Kerr-like media.

In the past ten years, almost the entire catalogue of nonlinear optical processes has been used to produce phase conjugate waves. The mechanisms involved are many and various, and span a very wide range of wavelengths, response times, efficiencies, etc. Photorefractive materials have gained a particularly large following and probably feature most often in current literature on PC optical systems.

Let us therefore consider at this point what effect a phase conjugating device (or phase conjugating mirror (PCM)) has on an incident beam of light, irrespective of the precise mechanism used for PC generation. In Figure 6.1(a), for example, we show light reflected from an ordinary mirror in which it is seen that only the **k** vector component normal to the surface undergoes reflection. For an incident beam with wavevector $\mathbf{k}_{\text{in}}$ given by

$$\mathbf{k}_{\text{in}} = k_x\hat{\mathbf{x}} + k_y\hat{\mathbf{y}} + k_z\hat{\mathbf{z}} \tag{6.1}$$

the wavevector upon reflection $\mathbf{k}_{\text{out}}$ will be given by

$$\mathbf{k}_{\text{out}} = k_x\hat{\mathbf{x}} + k_y\hat{\mathbf{y}} - k_z\hat{\mathbf{z}} \tag{6.2}$$

that is, only the z component has changed sign.

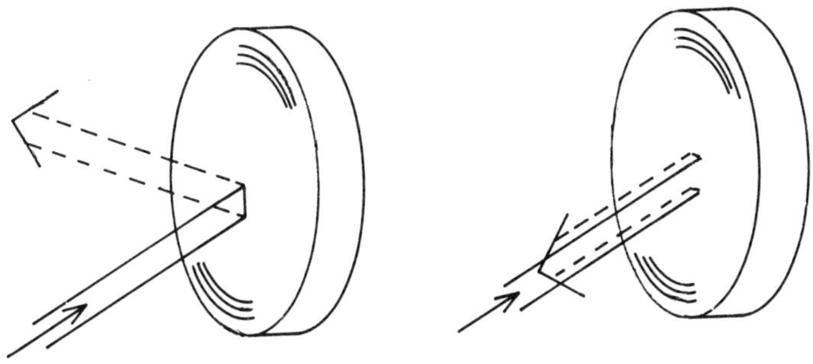

(a) (b)

Figure 6.1 Schematic demonstration of reflection from (a) an ordinary mirror and (b) a phase conjugate mirror (PCM). Note the difference here between specular reflection in (a), and exact retroreflection in (b).

For a phase conjugate mirror, however, shown in Figure 6.1(b), all three wavevector components of the incident wave undergo reflection, so that we write

$$\mathbf{k}_{\text{out}} = -k_x\hat{\mathbf{x}} - k_y\hat{\mathbf{y}} - k_z\hat{\mathbf{z}} = -\mathbf{k}_{\text{in}} \tag{6.3}$$

The incident wave therefore has undergone complete retroreflection. Notice too that Figure 6.1(b) shows the PCM in an arbitrary position with respect to the incident beam. Tilting the mirror to another (different) arbitrary angle has no effect on this property of retroreflection. Extending this argument to an incident converging beam as shown in Figure 6.2, or a beam that has undergone random scattering from, for example, traversing a distorter, it is

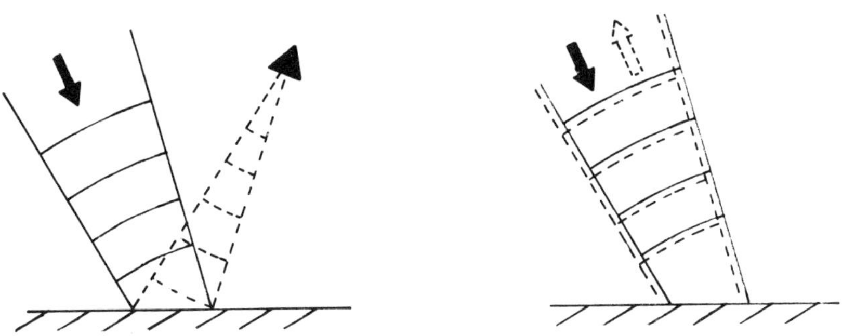

(a) (b)

Figure 6.2 Retroreflection properties of (b) a PCM vs. (a) a conventional mirror. In (b) any arbitrary phase front will be reflected with equal (but opposite) curvature, so that the 'reflected' beam will retrace the exact path of the incident beam.

clear that the PCM will perform the operation of 'healing' the convergent, distorted or aberrated beam. A plane wave for example, traversing a distorter, spreads out, and is incident on the PCM. Each portion of the distorted beam, however, undergoes exact retroreflection, and is incident on the distorter a second time, to emerge in its original undistorted form as a plane wave.

This seemingly magic property of healing has been the basis for much of the recent appeal of phase conjugation. The possibility of generating aberration-free waves for imaging with very high resolution while using cheap and possibly distorted optics is clearly a tantalizing goal, and several experimental arrangements have successfully demonstrated this property for high resolution lithographic systems. (See section 6.5, and Levenson *et al.* (1981), Gower (1985) and Gower *et al.* (1989).)

6.2.2. Definitions and formalism

In its simplest form, the process can be described as converting an optical field given by

$$E_1(r,t) = \mathrm{Re}[\mathscr{E}(r)\mathrm{e}^{\mathrm{i}(\omega t - kz)}] \tag{6.4}$$

(where $\mathscr{E}(r)$ is a complex amplitude, i.e. $\mathscr{E}(r) = |\mathscr{E}(r)|\mathrm{e}^{\mathrm{i}\Phi(r)}$) to a field described by

$$E_2(r,t) = \mathrm{Re}[\mathscr{E}^*(r)\mathrm{e}^{\mathrm{i}(\omega t + kz)}] \tag{6.5}$$

(where $\mathscr{E}^*(r)$ is the complex conjugate of $\mathscr{E}(r)$).

We shall delay for the moment any discussion about how this is achieved, but merely look at the implications that complex conjugation has, and the properties that such a conjugated wave possesses.

The field described by

$$E_2(r,t) = \mathrm{Re}[\mathscr{E}^*(r)\mathrm{e}^{\mathrm{i}(\omega t + kz)}] \tag{6.6}$$

is one which has had the operation of complex conjugation performed on its spatial part only. The time factor $\mathrm{e}^{\mathrm{i}\omega t}$ has been left alone. Complex conjugation, therefore, reverses the direction $(-\mathbf{k} \longrightarrow +\mathbf{k})$ of the wave, and also its phase $(\mathscr{E}(r) \longrightarrow \mathscr{E}^*(r))$.

Because the cosine function is an even one, i.e. $\cos(\theta) = \cos(-\theta)$, we can rewrite the above field as

$$E_2(r,t) = \mathrm{Re}[\mathscr{E}^*(r)\mathrm{e}^{\mathrm{i}[\omega(-t) - kz]}] \tag{6.7}$$

In effect equation 6.7 says that $E_1(r,t)$ and $E_2(r,t)$ are related by 'time reversal', and it is this concept that has attracted the attention of many people over the past ten years, and that forms the basis for many of the imaging and distortion correcting applications discussed in section 6.5.

We can look at the properties of such a PCM therefore and reconsider the action of reflection. As shown in Figure 6.2(a), a conventional mirror has no overall phase reversing properties. The PCM, however, of Figure 6.2(b) can achieve exact retroreflection, as the complex amplitude of wave E_2 is everywhere the complex conjugate of E_1, therefore its wavefronts coincide everywhere with those of E_1.

This result is expressed via a 'distortion correction theorem' (Yariv, 1977) which states:

> If a (scalar) wave $E_1(r)$ propagates from left to right through an arbitrary dielectric (but lossless) medium, then if we generate in some region of space (say near $z = 0$) its phase conjugate replica $E_2(r)$, then E_2 will propagate backward from right to left, through the dielectric medium, remaining *everywhere* the phase conjugate of E_1.

In terms of the wave equation, we can consider this healing properly via equation 6.4. The wave $E_1(r, t)$, propagating through a lossless medium ($\varepsilon(r) = \text{real}$), obeys the following wave equation

$$\nabla^2 E_1 + \omega^2 \mu \varepsilon(r) E_1 = 0 \tag{6.8}$$

Applying this to $E_1(r, t)$ yields (in the paraxial case)

$$\nabla^2 \mathscr{E} + [\omega^2 \mu \varepsilon(r) - k^2]\mathscr{E} - 2\text{i}k\frac{\partial \mathscr{E}}{\partial z} = 0 \tag{6.9}$$

Now, if we apply the same equation to

$$E_2(r, t) = \text{Re}[\mathscr{E}^*(r)\text{e}^{\text{i}(\omega t + kz)}] \tag{6.10}$$

we obtain

$$\nabla^2 \mathscr{E}^* + [\omega^2 \mu \varepsilon(r) - k^2]\mathscr{E}^* + 2\text{i}k\frac{\partial \mathscr{E}^*}{\partial z} = 0 \tag{6.11}$$

(recall $\varepsilon(r) = \varepsilon^*(r)$).

However, equation 6.11 is the same as equation 6.9, apart from the act of complex conjugation. In other words $E_1(r, t)$ and $E_2(r, t)$ obey the same wave equation. This would equally apply to the case where $\varepsilon(r) = \alpha\varepsilon^*(r)$, where α is an arbitrary constant.

The one proviso in the above is that the distortion being healed on the return path cannot change at any (arbitrarily fast) rate. Phase conjugation, although it is extremely impressive in its nonlinear behaviour, still obeys physical laws. As we will see later, some PC media respond agonizingly slowly, so that the rate at which dynamic phase distortions can be compensated is unfortunately equally slow.

6.3. MECHANISMS FOR PC GENERATION

The wide range of nonlinear processes that can be, or have been, used to generate phase conjugate waves fall into two broad classes; parametric interactions which involve elastic photon scattering, and stimulated processes in which energy is exchanged between the medium and the incident light via inelastic scattering. The almost universally preferred route for signal processing is an example of the first class, and is termed four-wave mixing (FWM) or degenerate four-wave mixing (DFWM or D4WM). We will discuss this process in most detail. In DFWM it is the inherent field multiplication that occurs via three waves interacting to generate a fourth that offers numerous possibilities in areas such as correlation and logic, particularly when the fields are optical Fourier transforms. This point will be addressed again later.

Before we look at DFWM, however, we will review a 'precursor' of four-wave mixing namely three-wave mixing, and discuss its limitations for PC generation.

6.3.1. Three wave mixing

Chapter 1 will have discussed second harmonic generation (SHG) which is a specific example of three-wave mixing much used today for laser wavelength conversion from IR to visible for example. In this particular case two waves derive from the same incident laser beam, so that two photons at frequency ω_1 combine to produce one photon at frequency ω_2. The process involves $\chi^{(2)}$, the second-order non-linear optical susceptibility tensor of a material.

In the general case of difference frequency mixing between three waves interacting under the conditions $\omega_1 < \omega_2 < \omega_3$, the simultaneous introduction of waves at ω_3 and ω_2 generates a nonlinear polarization source term of the form

$$P_{\mathrm{nl}} = \tfrac{1}{2}\chi^{(2)}\mathscr{E}_3(r)\mathscr{E}_2^*(r)\exp\{\mathrm{i}[(\omega_3 - \omega_2)t - (k_3 - k_2)z]\} + \text{c.c.} \tag{6.12}$$

where the symbols used are the same as in equation 6.4, and c.c. denotes the complex conjugate. For $\omega_1 + \omega_2 = \omega_3$, equation 6.12 becomes

$$P_{\mathrm{nl}} = \tfrac{1}{2}\chi^{(2)}\mathscr{E}_3(r)\mathscr{E}_2^*(r)\exp\{\mathrm{i}[\omega_1 t - (k_3 - k_2)z]\} + \text{c.c.} \tag{6.13}$$

This nonlinear polarization will only efficiently radiate a new wave at frequency ω_1 when the phase-matching condition

$$k_3 - k_2 = k_1 \tag{6.14}$$

is met (i.e. $k_1 + k_2 = k_3$). This is the fundamental problem of such three-wave mixing interactions. The condition imposed by equation 6.14 is often met via birefringence, and thus the direction-dependent value of the extraordinary refractive index n_e of the crystal used for the parametric interaction allows efficient three-wave mixing, but only for a very restricted angular range of the interacting beams. This restriction therefore places severe constraints on the most sought-after property of PC, which is the ability to compensate for distortion (i.e. deviation) of the input beam away from the very restrictive phase-matching direction.

The above scheme has indeed been implemented for PC generation by Avizonis *et al.* (1977) who used a double three-wave mixing geometry using an Nd:YAG laser. Firstly the second harmonic frequency was generated at frequency ω_2 in a crystal of $LiIO_3$, from the fundamental at frequency ω_1. Subsequently a difference frequency mixing occurred in a crystal of lithium formate, via the interaction $\omega_c = \omega_2 - \omega_1$. The conjugate output wave therefore was at frequency $\omega_c = \omega_1$, and compensation was achieved for a deliberate phase distortion introduced onto the input wave at frequency ω_1. As expected, however, the usual phase-matching restrictions applied, and hence this three-wave mixing interaction does not show great utility in PC generation.

6.3.2. Degenerate four-wave mixing

The phase-matching constraint of three-wave mixing is relaxed by using a fourth beam in the mixing process. In the simplest case, the frequencies of all the four beams are the same, and this is termed degenerate four-wave mixing. Extension of equation 6.12 to the interaction of three waves yields (for a full derivation of equations 6.12 and 6.15, consult e.g. Yariv (1985))

$$P_{\text{nl}}(r,t) \propto \chi^{(3)} \mathscr{E}_1(r) \mathscr{E}_2(r) \mathscr{E}_3^*(r) \cdot \exp\{i[(\omega_1 + \omega_2 - \omega_3)t - (\mathbf{k}_1 + \mathbf{k}_2 - \mathbf{k}_3).\mathbf{r}]\} + \text{c.c.} \tag{6.15}$$

for three waves, E_1, E_2, E_3, incident simultaneously on the $\chi^{(3)}$ medium. The resultant wave that is generated by this nonlinear polarization is at frequency ω, as $(\omega_1 + \omega_2 - \omega_3 = \omega + \omega - \omega = \omega)$. By adopting a rather simple geometrical arrangement, in which beams E_1 and E_2 are counter propagating, as shown in Figure 6.3, the phase term $(\mathbf{k}_1 + \mathbf{k}_2 - \mathbf{k}_3).\mathbf{r}$ is greatly simplified.

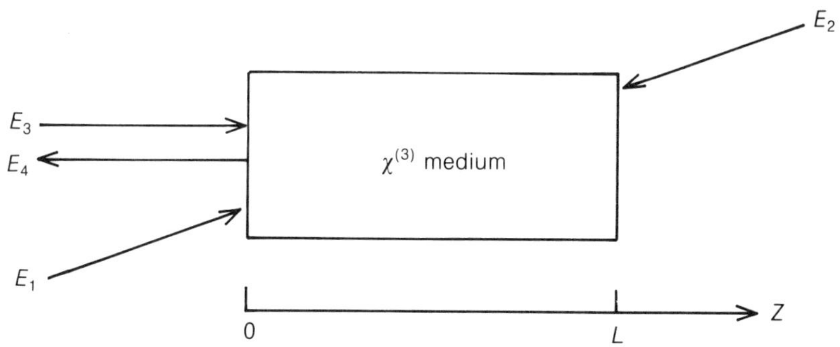

Figure 6.3 Schematic four-wave mixing (FWM) geometry involving interaction between optical fields E_1, E_2 and E_3 to generate a backward-going field E_4. The nonlinearity here involves a $\chi^{(3)}$ active medium. For degenerate four-wave mixing (DFWM) all optical fields are at the same optical frequency ω.

As $\mathbf{k}_1 = -\mathbf{k}_2$, the above term becomes $-\mathbf{k}_3.\mathbf{r}$, or if E_3 is propagating in the z direction, it becomes $-k_3 z$.

The result is a wave, E_4, generated by the interaction of the other three waves, which has the form

$$E_4(r,t) = \tfrac{1}{2}\mathscr{E}_4(z)e^{i(\omega t + k_3 z)} + \text{c.c.} \tag{6.16}$$

where

$$\mathscr{E}_4 \propto \mathscr{E}_1 \mathscr{E}_2 \mathscr{E}_3^* \tag{6.17}$$

In DWFM, E_1 and E_2 are usually called pump waves, E_3 is called the signal (or probe) wave, and E_4 is the phase conjugate output (conjugate return). (Please note, however, that this nomenclature has not been adopted as standard in the literature. Other terminology interchange E_3 and E_4 (e.g. Yariv and Pepper, 1977) E_2 and E_3 (Feinberg, 1983), replaces E_3 by E_p (probe) or E_s (signal), and numerous other variants.)

Of fundamental importance here is the resultant amplitude of E_4 for given values of E_i, $i = 1$–3. With reference to Figure 6.3, the coupled wave formalism of nonlinear optics yields (Yariv and Pepper, 1977)

$$\frac{d\mathscr{E}_3^*}{dz} = i\kappa\mathscr{E}_4 \tag{6.18}$$

and

$$\frac{d\mathscr{E}_4}{dz} = i\kappa^*\mathscr{E}_3^* \tag{6.19}$$

where

$$\kappa = \frac{\omega}{2}\left(\frac{\mu}{\varepsilon}\right)^{1/2} \chi^{(3)}\mathscr{E}_1^*\mathscr{E}_2^* \tag{6.20}$$

By applying boundary conditions at $z = 0$ and $z = L$ as in Figure 6.3, the general solution for $\mathscr{E}_3$ and $\mathscr{E}_4$ is given by

$$\mathscr{E}_4(0) = -i[(\kappa^*/|\kappa|)\tan|\kappa|l]\mathscr{E}_3^*(0) \tag{6.21}$$

and

$$\mathscr{E}_3(L) = \mathscr{E}_3(0)/\cos(|\kappa|L) \tag{6.22}$$

Inspection of equations 6.21 and 6.22 reveals the following interesting facts.

1. As $|\cos(|\kappa|L)| \leqslant 1$, $|\mathscr{E}_3(L)|$ is greater than $\mathscr{E}_3(0)$.
2. For values of $\pi/4 < |\kappa|L < 3\pi/4$, $|\mathscr{E}_4(0)|$ is greater than $|\mathscr{E}_3(0)|$. In other words, the PCM exhibits a net gain, or a reflectivity $> 100\%$.
3. Finally for the value $|\kappa|L = \pi/2$, we have the unique situation that both

$$\frac{\mathscr{E}_4(0)}{\mathscr{E}_3(0)} = \infty \quad \text{and} \quad \frac{\mathscr{E}_3(L)}{\mathscr{E}_3(0)} = \infty \tag{6.23}$$

This situation corresponds to oscillation, which occurs without the normal feedback requirement. The four-wave mixing geometry therefore is seen to be extremely adaptable and versatile, and it is clear that the freedom from phase-matching constraints (provided that $k_1 = -k_2$) is a very useful prize.

6.3.3. Analogy between DWFM and real-time holography

The arrangement of Figure 6.3 can readily be interpreted as a simultaneous version of the normally distinct processes involved in holographic recording and subsequent reconstruction. In holographic terms fields E_1 and E_3 interfere to write a grating whose spatial period Λ, is given by $\Lambda_g = \lambda/2\sin\theta$, where λ is the recording wavelength, and 2θ is the external angle between the two recording beams. This grating can subsequently scatter a reading beam E_2 that is counter propagating with respect to E_1, to generate a field E_4, that is the PC of E_3. The medium in Figure 6.3 is 'thick' in the holographic context, so Bragg scattering would occur. This grating analogy is very useful in visualizing the interactions involved here, but as it stands does not provide a complete picture. Because the actual PC interaction is simultaneous, not consecutive as in holography, fields E_3 and E_2 can also interfere (given the requisite coherence relationship) and the resultant grating can additionally scatter E_1 into the direction of E_4. Figure 6.4 illustrates these processes firstly as discrete events, and subsequently as a DFWM interaction, via real-time holography.

If we take the basic arrangement of Figure 6.4(a), then the interference pattern generated via beams 1 and 3 with complex amplitudes $\mathscr{E}_1(x, y)$ and $\mathscr{E}_3(x, y)$ respectively yields a transmission function, T, given by

$$T \propto (\mathscr{E}_1 + \mathscr{E}_3)(\mathscr{E}_1^* + \mathscr{E}_3^*) \tag{6.24}$$

in the hologram plane.

To view the hologram, and to reconstruct the original object, a third beam, beam 2, is used, which is directed at the hologram from the opposite direction to beam 1, so that $\mathscr{E}_2 = \mathscr{E}_1^*$. The reconstructed image field $\mathscr{E}_4$ therefore is given by

$$\begin{aligned}\mathscr{E}_4 &= T\mathscr{E}_2 = T\mathscr{E}_1^* \\ &= (|\mathscr{E}_1|^2 + |\mathscr{E}_3|^2 + \mathscr{E}_1^*\mathscr{E}_3 + \mathscr{E}_1\mathscr{E}_3^*)\mathscr{E}_1^*\end{aligned} \tag{6.25}$$

or

$$\mathscr{E}_4 = (|\mathscr{E}_1|^2 + |\mathscr{E}_3|^2)\mathscr{E}_1^* + (\mathscr{E}_1^*)^2\mathscr{E}_3 + |\mathscr{E}_1|^2\mathscr{E}_3^* \tag{6.26}$$

Taking this term by term, we have

1. a term proportional to the incident field

   $(|\mathscr{E}_1|^2 + |\mathscr{E}_3|^2)\mathscr{E}_1^* = (|\mathscr{E}_1|^2 + |\mathscr{E}_3|^2)\mathscr{E}_2$

2. a term $(\mathscr{E}_1^*)^2\mathscr{E}_3$ which is not phase-matched, and hence will not radiate, and

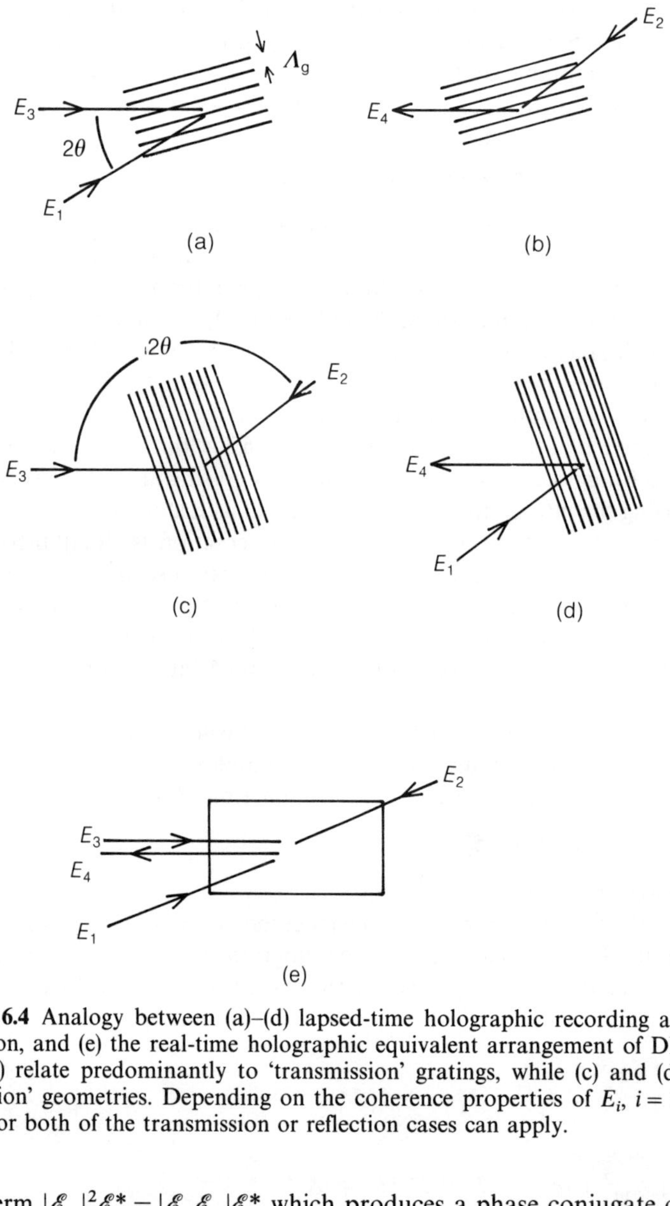

Figure 6.4 Analogy between (a)–(d) lapsed-time holographic recording and reconstruction, and (e) the real-time holographic equivalent arrangement of DFWM. (a) and (b) relate predominantly to 'transmission' gratings, while (c) and (d) refer to 'reflection' geometries. Depending on the coherence properties of E_i, $i = 1–3$, in (e) either or both of the transmission or reflection cases can apply.

3. a term $|\mathscr{E}_1|^2\mathscr{E}_3^* = |\mathscr{E}_1\mathscr{E}_2|\mathscr{E}_3^*$ which produces a phase conjugate output of beam E_3, the subject wave.

This last term $\mathscr{E}_1\mathscr{E}_2\mathscr{E}_3^*$ is identical to that in equation 6.17 which was for the case of a $\chi^{(3)}$ DFWM interaction. The analogy here between DFWM and real-time holography is useful, as it permits the updating of various established holographic techniques, such as interferometry, into their real-time DFWM counterparts. We shall discuss some of these in the sections on applications.

6.4. MEDIA FOR PC GENERATION

For DFWM, the material requirements can be rather forgiving. As phase-matching is not required, then in principle any suitably transparent media can be used, as the $\chi^{(3)}$ nonlinearity is not governed by restrictive symmetry considerations. The parameters of concern therefore in selecting a DFWM material would reflect the particular application, and might well cover aspects such as response time, damage threshold, efficiency and wavelength reponse, as well as cost and availability. Media in which phase conjugation has so far been demonstrated include solids, liquids (and liquid crystals), gases, plasmas and aerosols. As an indication, an early review by Pepper (1982) lists a variety of such experiments, in which he itemizes some of the more important parameters of each interaction. Similar reviews, ten years on, would have a much longer list of such experiments.

One particular parameter that clearly affects the viability of any particular experiment is the power density required to induce the nonlinear interaction with the required efficiency. As an example, early work using CS_2 (Pepper *et al.*, 1978) achieved phase conjugate reflectivities of the order of unity, but this required power densities in the $MW\,cm^{-2}$ range, and interaction lengths of $\sim 0.5\,m$. Although the response time here was fast (ps) it is an often-observed result that the product of response time and required power density is approximately constant irrespective of the precise interaction mechanism involved. This can be seen at the other extremes of power and response time, for photorefractive nonlinearities (Chapter 4), where, in extreme cases, response times of $\sim 3\,h$ have been reported. Normally, at the $1\,W\,cm^{-2}$ level, photorefractive response times vary, however, in the μs–s range, depending on the material involved.

6.4.1. Kerr-type nonlinearities

Kerr-type effects such as molecular re-orientation, or band structure effects, rely on a refractive index change produced by the incident light, given by

$$n = n_0 + n_2 |E|^2 \tag{6.27}$$

The effect is local, and is directly proportional to the local intensity. We will list some of the effects that come under this heading, with references indicating earlier work rather than specific improvements reported more recently.

Off-resonance excitation in Na vapour has been used (Bloom *et al.*, 1978) to produce phase conjugate gains of ~ 60 via refractive index changes produced by light-induced shifts in the dispersion curve.

Molecular orientation in liquids such as CS_2 has produced phase conjugation which shows gain and even oscillation (Pepper *et al.*, 1978). CS_2 is a well-known nonlinear liquid, and the large anisotropic polarizability has been utilized in geometries such as waveguides and multipass transverse schemes to produce reflectivities greater than unity (Jensen and Hellwarth, 1978).

Saturable absorption in dyes has been studied (Tocho *et al.*, 1980; Lind *et al.*, 1982) for which reflectivities up to $\times$ 10 have been produced. Both absorptive and dispersive nonlinearities can be used in this case.

Thermal effects produced by direct heating via nonradiative transitions can also write gratings (Martin and Hellwarth, 1979). The local heating produces a dn/dt grating, which because the medium is usually in the liquid state will tend to spread spatially, and hence wash out. In fact the thermal gratings caused via saturable absorption also contribute a slower, longer-lived mechanisms for phase conjugation which may interfere with the intended absorptive nonlinear behaviour (Tocho *et al.*, 1981).

Anharmonic oscillations of bound electrons and nonlinear motion of free carriers in semiconductors provides a means for achieving efficient FWM reflectivities. Resonance enhancement is also possible for producing extremely strong nonlinear behaviour. Semiconductor doped glasses (colour glass filters) have also been used as a convenient form for the semiconductor required (Jain and Lind, 1983), achieving reflectivities of $\sim 10\%$ in CdS_xSe_{1-x} microcrystal glasses for example.

6.4.2. Nonlocal effects

Electrostriction in e.g. aerosols (Palmer, 1980) has been used to produce a grating via density changes resulting from particle movement either into or away from regions of high field. Perhaps the best known non-local effect, however, is photorefraction (Günter and Huignard, 1988), a very versatile technique which is discussed in Chapter 4. One point that stands photorefractive nonlinear optics apart from other nonlinear physics discussed in other chapters is that the interaction does not depend on the absolute intensity of the interacting beams in, for example, a DFWM set-up, but on their relative intensity. The trade-off here is between speed and intensity, rather than the more usual efficiency and intensity.

A particularly novel and appealing feature of photorefractives is a technique known as self-pumping. For a wide range of input geometries certain crystals such as $BaTiO_3$, and $Sr_xBa_{1-x}Nb_2O_6$(SBN) can self-generate the beams normally required as separate inputs in a DFWM arrangement. The exact processes that contribute to such an apparently unlikely process are involved; photographic evidence obtained from examining the beam paths within the crystals undergoing self-pumping indicates that more than one mechanism can simultaneously contribute to the overall effect.

The first reports of such self-pumping by White *et al.* (1982), Feinberg (1982) and Cronin-Golomb *et al.* (1982) made use of strategically placed external mirrors, or total internal reflection within the crystal itself, to provide self-induced optical beams via directionally dependent two-wave mixing gain. These first demonstrations have generated enormous subsequent research into all aspects of such behaviour. Hundreds of publications have since reported reflectivity data, response times, wavelength selectivity, instabilities of operation, uses in image processing, and theoretical of analyses of the coupled-wave nature of this intriguing phenomenon.

Equally fascinating are geometries that have currently become known as mutually pumped phase conjugators (MPPC). These schemes involve coupling within a crystal of e.g. $BaTiO_3$ between two input beams that are mutually incoherent. This incoherence can result from either a long path length difference, using different lasers operating at the same nominal wavelength (e.g. two Ar^+ lasers at $\lambda = 514.5\,\text{nm}$) or perhaps even more surprisingly, two lasers operating at different wavelengths (e.g. He–Ne and Ar^+).

The first reports of such geometries by Sternklar *et al.* (1986), Eason and Smout (1987) and Ewbank (1988) have again stimulated extensive research and interest. Their use in optical processing has so far been limited compared with the effort that has gone into understanding the physics behind their operation.

6.4.3. Inelastic scattering processes

The other major area for PC generation is the inelastic scattering process known as stimulated Brillouin scattering. This is one of several processes which also includes stimulated Raman, and Rayleigh, scattering in which the medium takes some of the photon energy. The conjugate wave produced therefore is not exactly degenerate. In the case of SBS, for example, there is a difference in frequency between incident and conjugated waves of roughly 1 part in 10^5. This clearly may have consequences in terms of fidelity of conjugation or aberration correction.

Unlike some FWM processes, the stimulated nature of these inelastic processes implies that a high intensity is required, and there may also exist an intensity threshold which must be exceeded for the effect to occur. SBS does have the great attraction, however, of being extremely simple. A high intensity beam is focused into a cell containing e.g. liquid or gas, and that is all that is requried. This passive nature of SBS is clearly very attractive, but for signal processing applications it is somewhat restrictive in terms of its requirement for a threshold gain–length product. The length aspect may not be so much of a problem (Chapter 9) when nonlinearities in fibres are considered, but in general SBS offers limited scope for many signal processing architectures compared with the rapidly developing areas of organic nonlinear materials and customized quantum well structures.

6.5. APPLICATIONS OF PC IN SIGNAL PROCESSING

We can divide applications into two broad classes, spatial and temporal.

Spatial includes such topics as aberration correction (for precise focusing), imaging devices, lensless imaging, transmission in fibres, pointing, tracking three-dimensional projection, photolithography, spatial convolution and correlation, edge enhancement, pattern recognition, interferometry, and iterative feedback techniques.

Temporal (includes frequency) effects cover temporal signal processing (pulse encoding), time domain optical filters, optical computing and logic, PC resonators, thresholding, laser beam combining and distributed feedback geometries. We will consider a representative sample of these in the context of signal processing.

6.5.1. Spatial applications

6.5.1.1. Aberration correction

This is perhaps the most direct application of a PCM in correcting for aberrations induced via, for example, turbulence, classical optical aberrations, thermal lensing and other similar 'defects'. The aberrations here are phase aberrations, i.e. distortions of the quality of the optical wavefront. If we retroreflect with phase conjugation, then retraversing induces an equal and opposite effect, to 'heal' or correct the damaged wavefront, as shown schematically in Figure 6.5. The alternative to this automatic aberration correction involves adaptive optical elements such as deformable telescope mirrors that are coupled to wavefront sensing devices. While adaptive optics certainly

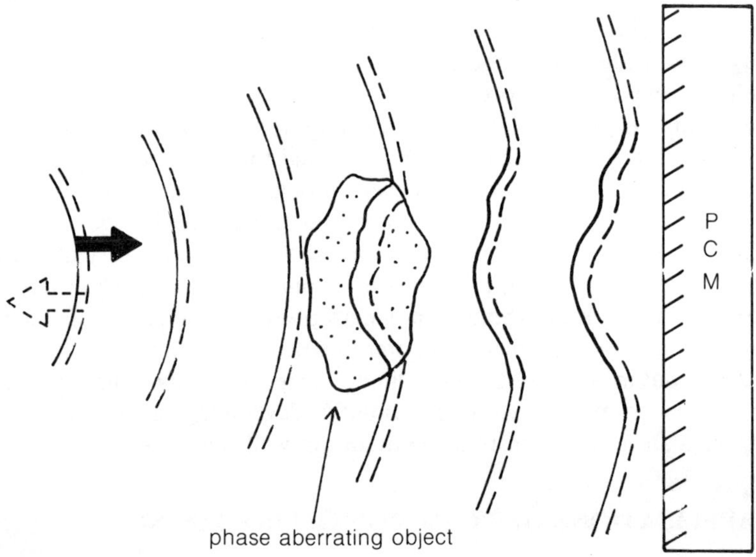

Figure 6.5 Wavefront healing on double passing a phase aberration with reflection from a PCM.

has a role to play, it is undoubtedly more involved than PC mirror schemes, as for large optical systems (telescopes) the number of actuators positioned on the mirror rear surface can be very large. The advent, however, of parallel computing technology that can address in real time such actuator arrays clearly makes adaptive optics schemes more attractive. PC mirrors, however, do have a unique advantage here in possessing gain.

The natural application of such automatic PC involves secure optical communications schemes. The transmitting station receives all the returning conjugate signal, with no 'spillage'. Furthermore, any attempt to intercept the outward transmission would immediately be noticed, as the DFWM operates as a logical AND gate. One last point is that the PCM can be tuned to operate as an extremely narrowband filter, so only a previously determined and very specific frequency will generate any conjugate return (Pepper and Abrams, 1978).

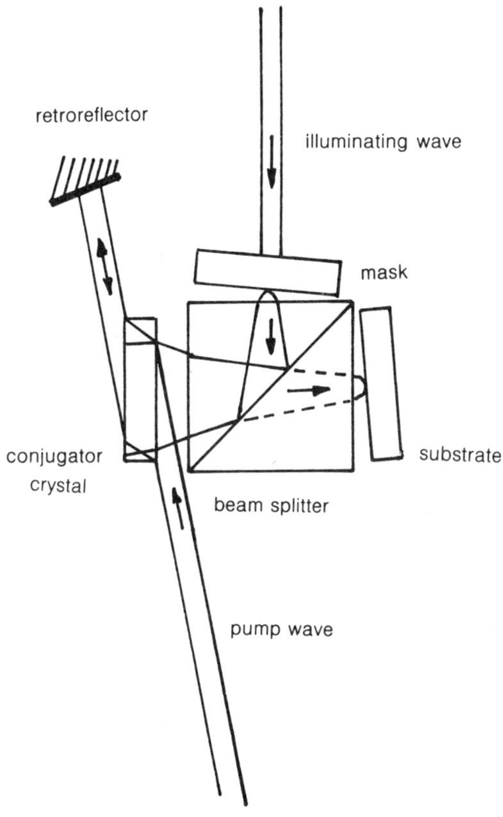

Figure 6.6 Experimental arrangement for lensless imaging using a $LiNbO_3$ phase conjugate reflector. The mask and photoresist-coated substrate were 0.5 mm from the beam splitter faces. (After Levenson *et al.* (1981).)

6.5.1.2. Lensless Imaging

Several schemes have demonstrated improvements to microlithographic techniques that formerly have relied on direct contact printing, or the use of multi-element, computer-optimized, aberration-free large area UV imaging lenses, produced at enormous cost. The schematic arrangement is shown in Figure 6.6 (Levenson *et al.*, 1981) in which the mask to be replicated and substrate are at equivalent planes on different faces of a (high quality) beam splitter. In this particular case, the normal PC aberration correction would only apply to aberrations induced via common path, so some aberrations present in the beam splitter, for example curvature of reflecting surfaces, remain uncorrected. Results of this work are shown in Figure 6.7.

The important features here are clearly that there is no contact involved between the (usually extremely costly) mask and the substrate, as there would have to be for the more common technique of direct contact printing. For high resolution it is advantageous to use as short a wavelength as possible.

More recently several schemes have been demonstrated in which both CW Ar^+ and pulsed excimer pumped dye lasers were used with self-pumped crystals of $BaTiO_3$, or a cell containing rhodamine dye, to replicate submicron features (Gower *et al.*, 1985, 1989). Results from his work are shown in Figures 6.8 and 6.9.

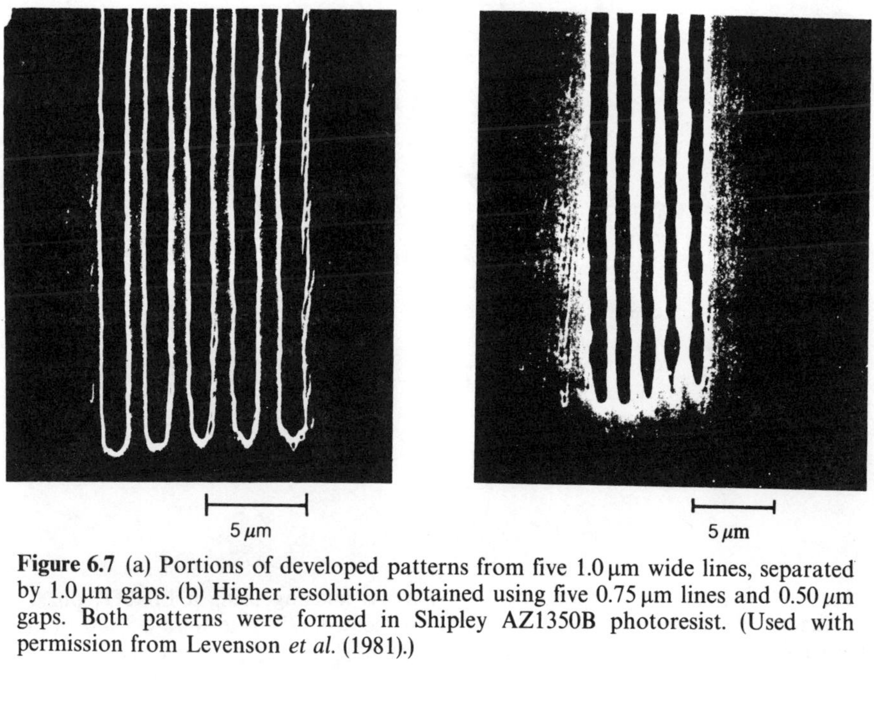

Figure 6.7 (a) Portions of developed patterns from five 1.0 μm wide lines, separated by 1.0 μm gaps. (b) Higher resolution obtained using five 0.75 μm lines and 0.50 μm gaps. Both patterns were formed in Shipley AZ1350B photoresist. (Used with permission from Levenson *et al.* (1981).)

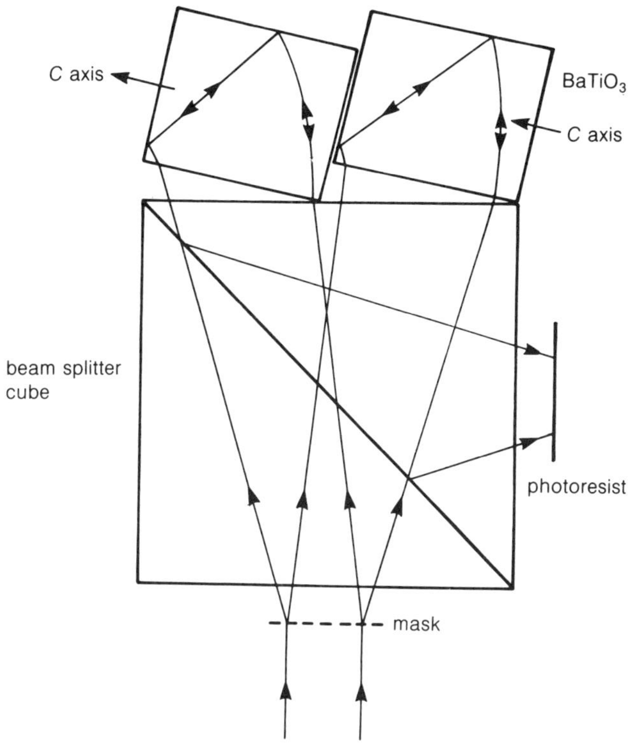

Figure 6.8 Large area double-crystal self-pumped PCM image projection arrangement. (After Gower *et al.* (1989).)

Correction has also been achieved for aberrations resulting from modal scrambling when propagating an image through a multimode optical fibre. The distortions occur because the spatial image travels down the fibre not in a single mode, but (in the geometrical analogy) in many zig-zag paths each of which takes a different time, and hence the image is split up into an unrecognizable pattern after only a short distance travelled inside the fibre. This modal dispersion is an example in which the image has undergone phase aberration. If the image is incident on a phase conjugator, however, and then sent back in the same fibre, then this modal distortion will be corrected and the image quality restored. This has been achieved experimentally by several groups using this double-pass geometry, for example Dunning and Lind (1982), shown in Figure 6.10 and Figure 6.11, and also for multimode fibre geometries for polarization-preserving applications (McMichael *et al.*, 1987). This geometry is also ideal for situations involving pumped (active) multimode fibres.

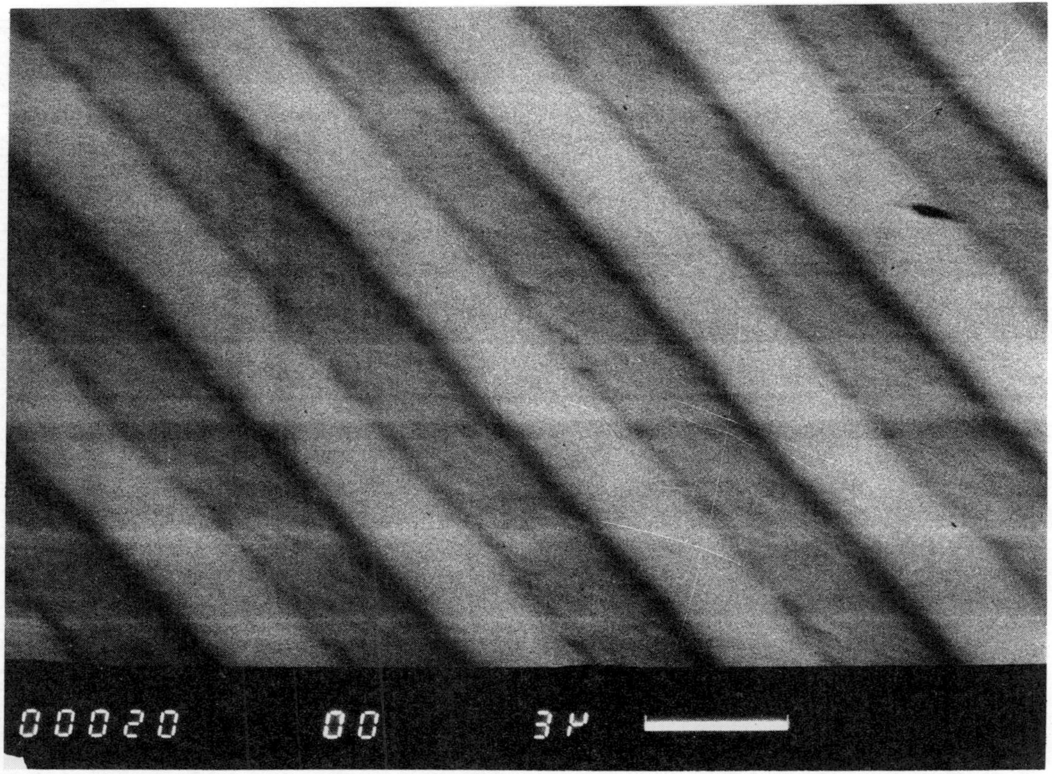

Figure 6.9 Scanning electron microscope pictures of 2.5 μm wide lines and spaces recorded in photoresist using two-crystal projector of Figure 6.8. The laser source was an excimer-pumped dye laser operating at 460 nm, at an average intensity of 5 W cm^{-2}. Exposure time required was 5 s.

Figure 6.10 Results of image transmission though an optical fibre. (a) Input image resolution test pattern; (b) output after a single pass; (c) result of double pass with intermediate phase conjugate reflection. (Used with permission from Dunning and Lind (1982).)

6.5.1.3. Pointing and tracking

Several authors have suggested and experimentally verified schemes for automatic alignment and tracking of objects via PCM (Guiliano, 1981; Gower, 1982). Figure 6.12 shows one such experimental verification, which has obvious applications in inertial confinement fusion. The process is started by a weak illuminating laser beam striking a target, for example, a laser fusion pellet. The glint of light that reflects off the pellet is incident on the nearby amplifier, and undergoes amplification in the (aberrating) amplifying medium. On retroreflection by the PCM, in this case an SBS cell (benzene), an amplified pulse is incident on the pellet with essentially perfect alignment. The scheme in Figure 6.12 achieved considerable success, destroying targets that were allowed to free fall in the vicinity of the nominal focus of lens L_1. The SBS cell achieved high efficiencies ($R \sim 100\%$), even under optically extreme conditions such as being inadvertently set on fire through careless focusing on the liquid surface.

Such automatic targeting schemes have clear military potential. To the best of this author's knowledge any successful trials on other than laboratory-scale experiments probably remain classified.

6.5.1.4. Correlation and pattern recognition

This application is a very good example of optical processing that uses the principle of field multiplication described in equations 6.17 and 6.26. With reference to Figure 6.13 (White and Yariv, 1980) the DFWM occurs between three fields whose amplitudes $u_1(x, y)$, $u_4(x, y)$ and $u_2(x,y)$ undergo Fourier transformation by two lenses of common focal length $f = 30$ cm, to a common focal plane situated inside the BSO crystal. The DFWM interaction therefore occurs between the three fields given by

$$U_i = \mathrm{FT}[u_i(x, y)],\ i = 1, 2, 4 \tag{6.28}$$

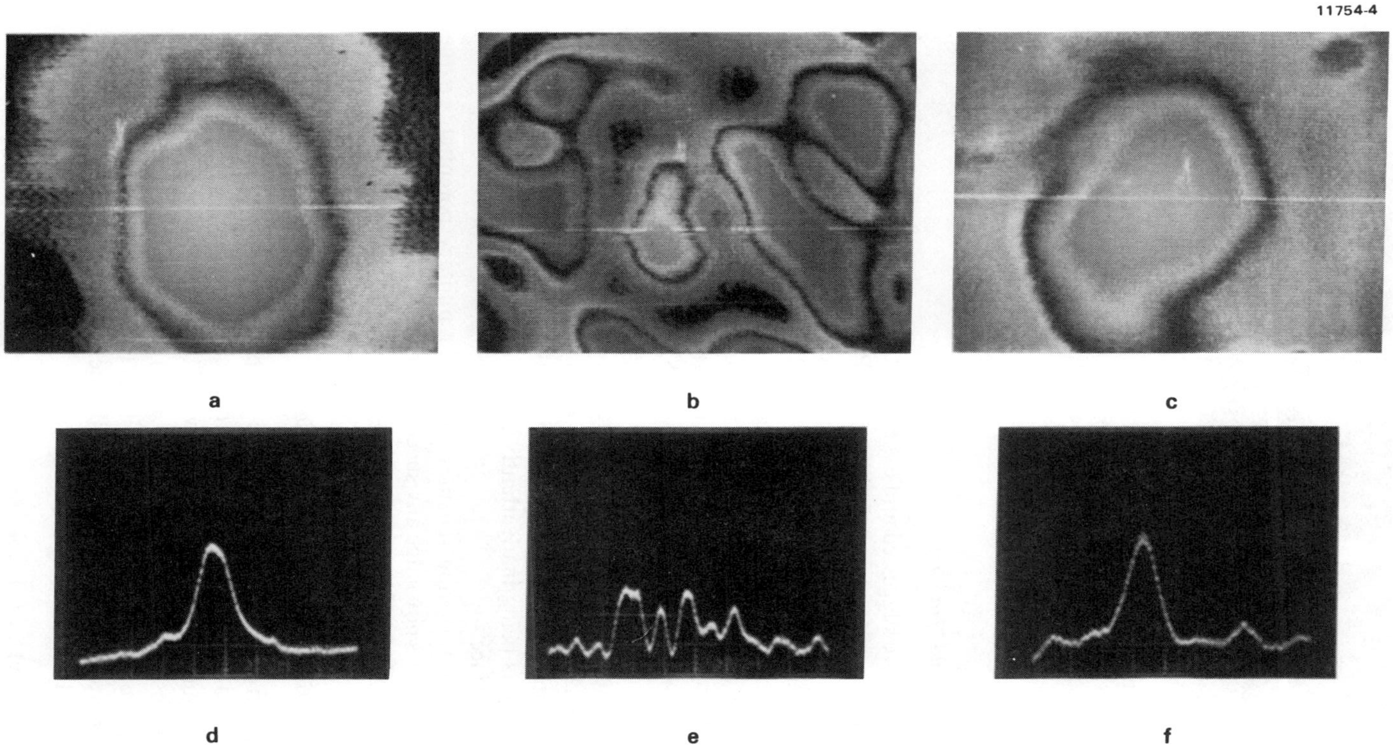

Figure 6.11 Fidelity test for phase conjugation through a fibre. (a) shows a Gaussian input profile, (b) is the far-field intensity distribution after a single pass through the fibre, while (c) shows the restored profile after a double pass and intermediate phase conjugate reflection. Note the faithful reconstruction of the input profile. (d)–(f) show video line-outs through representative portions of the images above. (Used with permission from Dunning and Lind (1982).)

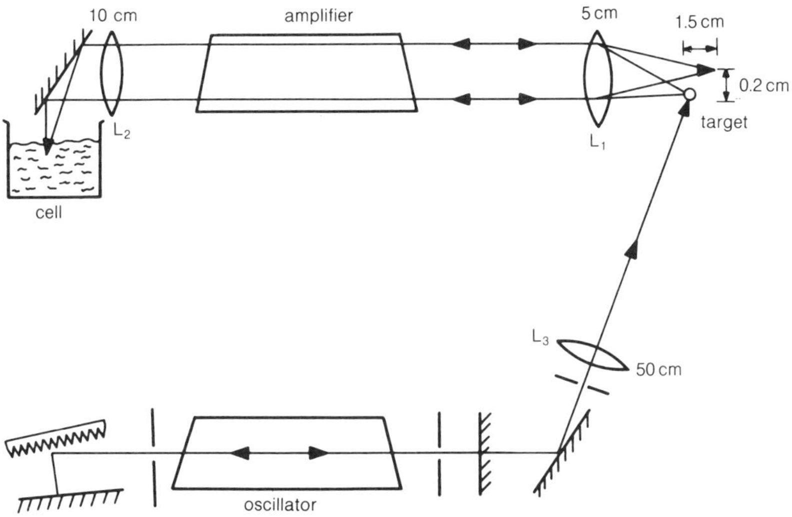

Figure 6.12 Experimental arrangement for demonstrating automatic targeting and destruction of laser fusion target pellet, using KrF oscillator–amplifier configuration. Targeting still occurred even when the pellet was purposely misplaced with respect to the nominal focus of lens L_1 by the amount shown. The phase conjugator used was an SBS cell containing benzene. (Used with permission from Gower (1982).)

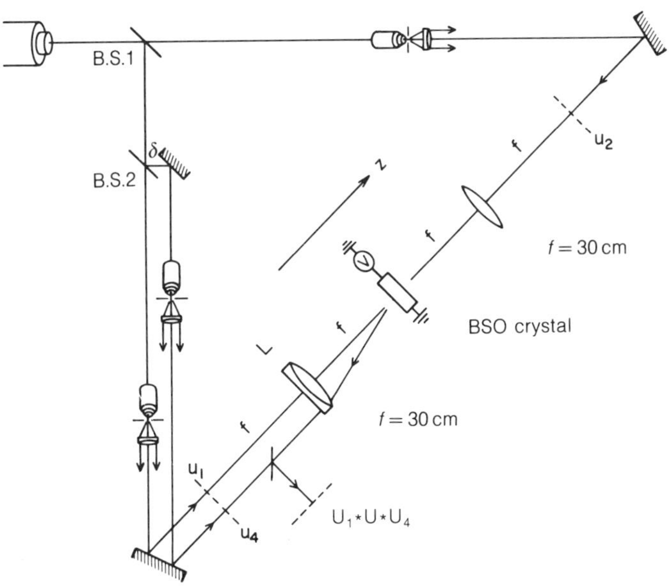

Figure 6.13 Arrangement used to perform spatial convolution and correlation using BSO. – – –, input and output planes. (After White and Yariv (1980).)

so that the wave mixing process generates a wave which, at the equivalent plane shown in Figure 6.13, has the form

$$u_3 \propto u_1 * u_2 \bigstar u_4 \tag{6.29}$$

where * and ★ stand for convolution and correlation respectively. For the arrangement shown a spatial delta function distribution for field u_2 will generate a plane wave at the BSO crystal plane, so that equation 6.29 simplifies to

$$u_3 \propto u_1 \bigstar u_4 \tag{6.30}$$

In other words, the generated wave when back-Fourier-transformed to the equivalent plane of u_1 and u_4 appears as the spatial correlation of these two original fields.

Results of this experimentally simple, but extremely versatile, arrangement are shown in Figure 6.14 (White and Yariv, 1980), in which for various input field distributions we see correlation and also convolution. Figure 6.14(d) clearly illustrates the process of pattern recognition in which the letter 'C'

	u_1	u_2	u_4	u_3
(a)	· · ·	Delta function	: :	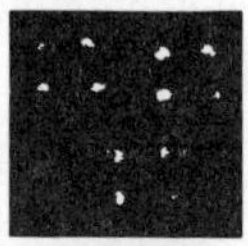
(b)	· · ·	Delta function	E	
(c)	Delta function	Delta function	Resolutior chart	
(d)	C	Delta function	CAL TECH	
(e)	C	: :	Delta funcion	

Figure 6.14 Demonstration of real-time spatial convolution and correlation of objects placed at positions u_1, u_2 and u_4. (Used with permission from White and Yariv (1980).)

present in both input fields has been selected as a common feature. The appealing features of this demonstration are clear. Firstly, with the inclusion of real-time updatable input, via spatial light modulators, and sufficiently fast response from the crystal achieved for example by using pulsed lasers, and/or faster crystals such as GaAs, or InP, a compact, optical correlator can be used to recognize features such as fingerprints, signatures, and other essentially optical format input. With the advent of high power diode lasers, such a correlator package can be rather compact. The author is aware of several companies who have had projects on such miniature pattern recognition systems, and has seen devices at for example GEC, UK, and Thomson CSF, France, that even without diode laser sources are already fairly small and portable.

One obvious immediate drawback that must be mentioned, however, is that the accuracy of correlation is very dependent on similarity in the scale, and rotational position (but not translational position) between the two input objects. Schemes that are insensitive to all three parameters are being developed, and transforms exist that can cope with such ambiguities.

Finally, the process of simultaneous multiplexing of several holograms is also easy to achieve in materials such as BSO, so that simultaneous correlation between, for example, one pattern and two different test objects has been achieved with little or no cross-talk (Vainos and Eason, 1987).

6.5.1.5. Edge enhancement and contrast manipulation

Associated techniques involving Fourier transform manipulation also allow phase conjugation of Fourier transforms that have had modifications made on some parts of their spatial frequency components. When a simple lens is used to Fourier transform an image, the spatial frequencies present in the image are placed at different radially varying positions in the focal plane of the transform lens. It is an easy matter to use an additional 'erase' beam either to destroy completely, via redistribution of the previously localized charge distribution, or merely to modify, via a reduction of the modulation ratio between the two writing beams, any portion of a hologram written between a reference wave and a Fourier transform distribution in a photorefractive crystal. This erasure can be controlled, dynamic, iterative, or even 'intelligent' depending on what is required.

For example, preferential erasure of low spatial frequencies in the hologram plane yields a phase conjugate that shows reduced fidelity, and appears edge enhanced. The effect is to weight spatially the reconstructed image in any desired manner. Such weighted Fourier transforms have been used for example in associative memory schemes (White *et al.*, 1988), and a dynamic weighting procedure may prove useful here.

Results on edge enhancement using essentially 'passive' schemes have been generated in several ways. Firstly, if the hologram is recorded between reference and object beams of very unequal intensities, then the reconstruction process is inefficient for all areas of the object, except at edge features, where the step function in intensity will match the reference beam intensity much better.

Secondly, a recording between a Fourier transform and a plane wave will also demonstrate such an imbalance of intensities, as the modulation ratio will be a radially varying function. These effects have been demonstrated by several authors, for example Huignard and Herriau (1978) and Feinberg (1980). An active scheme in which an erase beam is used has been reported by Vainos and Eason (1986).

Manipulation of image fidelity in this manner can be effectively used, but the preferential reconstruction of high spatial frequencies does have the disadvantage that both image brightness and contrast are also affected, as the zero-order (or d.c.) term in the Fourier transform is responsible for the overall illumination of the image. Poor reconstruction of this term therefore generates a less bright image. A technique that overcomes this has been demonstrated by Khoury *et al.* (1989a) for improving the contrast of an object by selective amplification of higher spatial frequencies in an object while maintaining the d.c. component at essentially unity gain. Their technique relied on two-wave mixing (two-beam coupling) in $BaTiO_3$, between a Fourier transform, and a plane wave. The dynamics of two-beam coupling yields an intensity dependent gain for features in the object Fourier transform, such that for example high spatial frequencies see a gain of $\times$ 100, while the d.c. term has unity gain. The net effect is to improve the contrast of features that would otherwise be swamped by the overall illumination of the whole image.

6.5.1.6. *Interferometry and vibration studies*

So far we have said little about the real-time holography aspects of phase conjugation, but of course these are perhaps one of the most appealing features of the technique.

An early demonstration by Marrakchi *et al.* (1980), using BSO, was of visualization of moving (vibrating) objects. Figure 6.15 shows the results of the image of a vibrating loudspeaker cone after four-wave mixing. The points which are visualized clearly are those regions of the cone that are stationary, and hence are able to write a grating within the crystal. Any point on the object that suffered motion at a speed that was greater than the recording time (response time) of the crystal would not be recorded. On reconstruction therefore only the stationary, or slowly varying, points would appear bright, and hence the image is a nodal map of the vibrating structure.

Double exposure or holographic interferometry has also been demonstrated via various routes. The simplest scheme involves DFWM via a two-step recording sequence in which one exposure is recorded, the object is deformed or moved, and a second exposure is taken so that the resultant PC output

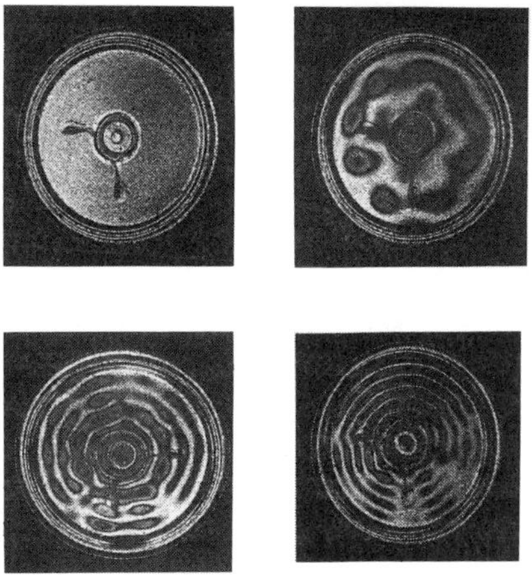

Figure 6.15 Results of using a vibrating object in a four-wave mixing experiment using BSO:(a) shows the static loudspeaker cone ($f = 0\,\text{kHz}$), while the other three figures show nodes at frequencies of (b) 4 kHz, (c) 8 kHz and (d) 12 kHz. (Used with permission from Marrakchi *et al.* (1980).)

from the crystal shows a transient interferogram, that lasts for the decay time of the holographic grating, and can be captured on film or other memory device. This recording technique was first demonstrated by Huignard and Herriau (1977).

If the interferogram is required between a permanent reference object and a dynamic test object, than various fixing procedures are required to store one image, while still allowing the crystal to generate a real-time hologram for dynamic comparison. A novel approach to achieve this has been reported by Vainos *et al.* (1989a) in which photochromic recording via DFWM in BSO has permitted permanent storage of one image, while still allowing efficient photorefractive recording within the same crystal volume. By adjusting the initial level of photochromic recording, an intensity ratio of unity between the permanent and dynamic images yields interferometric results of good contrast. This scheme has also been used for a range of other applications including optical logic and differentiation (Vainos *et al.*, 1989b), and spatial light modulation techniques (Clapham *et al.*, 1990). Figure 6.16 shows regions of permanent holographic storage, and recording characteristics.

6.5.1.7. Analogue processing, mathematical operations and optical logic

There have been a range of schemes discussed in the literature for operations involving analogue processes such as division and multiplication, as well as both analogue and digital implementations of addition, weighted addition and subtraction. All of these are low power applications involving a range

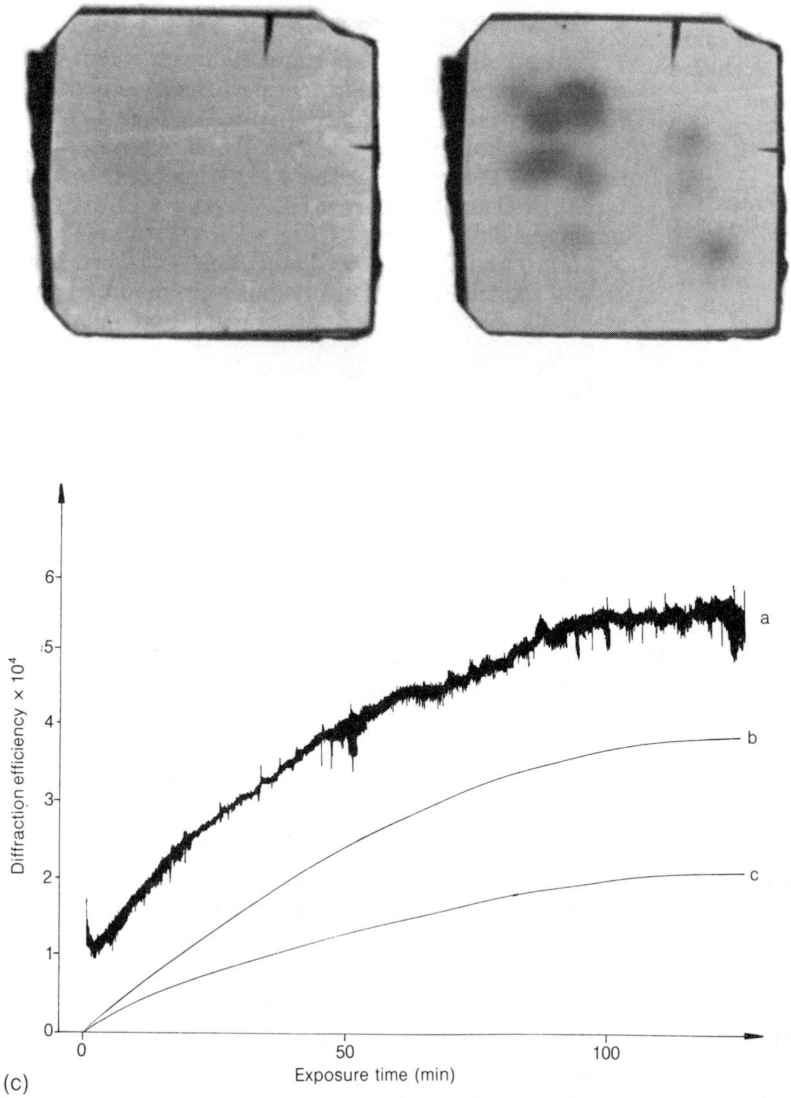

(c)

Figure 6.16 Pictures of a $10 \times 10 \times 2\,\text{mm}^3$ crystal of BSO (a) before and (b) after recording of permanent holograms in a DFWM arrangement using $\lambda = 488\,\text{nm}$ from an Ar^+ laser. The dark areas in (b) show ~ 10 separate stored holograms. The recording characteristics are shown in (c), as a plot of diffraction efficiency (measured via a Bragg-matched He–Ne laser), vs. exposure time. Curve a shows the effect of combined permanent and photorefractive holograms, curve b the permanent hologram in the presence of one of the writing beams, and curve c the permanent hologram only.

of photorefractive crystals. By slight rearrangement of the normal phase-matched DFWM scheme, additional effects such as logic, performed on two-dimensional pictures or matrix arrays, and differentiation have also been demonstrated.

Image division has been performed using BSO, via the nonlinear response induced from unequal beam intensity ratios used in FWM (Ja, 1982). This effect in which the diffraction efficiency is proportional to the inverse of the object field distribution can also be used for deconvolution schemes (Khoury and Eason, 1989), although the fidelity of the division process must be high over a wide dynamic range to achieve for example image deblurring schemes. Multiplication serves a less obvious requirement in image processing schemes, but clearly can be achieved via the basic equations of DFWM as shown in equation 6.17.

Image addition and subtraction schemes have an obvious application to optical logic. The configuration used here is often one of double (or two simultaneous) phase conjugate inputs, and hence two outputs. If the two inputs are derived from a common beam initially, via a beam splitter for example, then under perfect phase conjugation the output port of the beam splitter should yield nothing (nothing went in from that direction, hence nothing should come out). This scheme has also been used as a novelty filter, to detect dynamic change in an input picture or scene (see section 6.5.2.5). The process of image subtraction is equivalent to the operation of the Boolean exclusive OR (XOR), and various schemes have been used to demonstrate this. The first reported method involved a double recording technique of one grating followed by a second complementary grating that had an additional π phase delay imposed on the reference beam (Huignard *et al.*, 1976). Subsequently both Kwong *et al.* (1986) and Chiou and Yeh (1986) have used self-pumped $BaTiO_3$ to construct a phase conjugate mirror interferometer that automatically generates image subtraction.

Another scheme reported by Vainos *et al.* (1988) uses BSO in which the DFWM has slightly misaligned (non phase conjugate) pump beams. In this case, complete fidelity of phase reversal is not achieved so that an output is observed at the beamsplitter output port, which may be for example an operation of coherent addition, subtraction, or differentiation. A truth table results for various degrees of misalignment, generating the logic functions, AND, OR, XOR, NOT (and implicitly NOR). These results are shown in Figure 6.17.

Several other schemes for optical logic have also been devised, based on DFWM and also two-beam coupling (Fainman *et al.*, 1986). Although many of these schemes are somewhat cumbersome or perhaps inapplicable for general-purpose computing, it is important to stress that the architectures that they illustrate, and the concepts involved, are the important points. The parallel nature of the operations being demonstrated, and the intrinsic speed for a single processing step (e.g. $10^3 \times 10^3$ pixels in 1 ms) are the important features here.

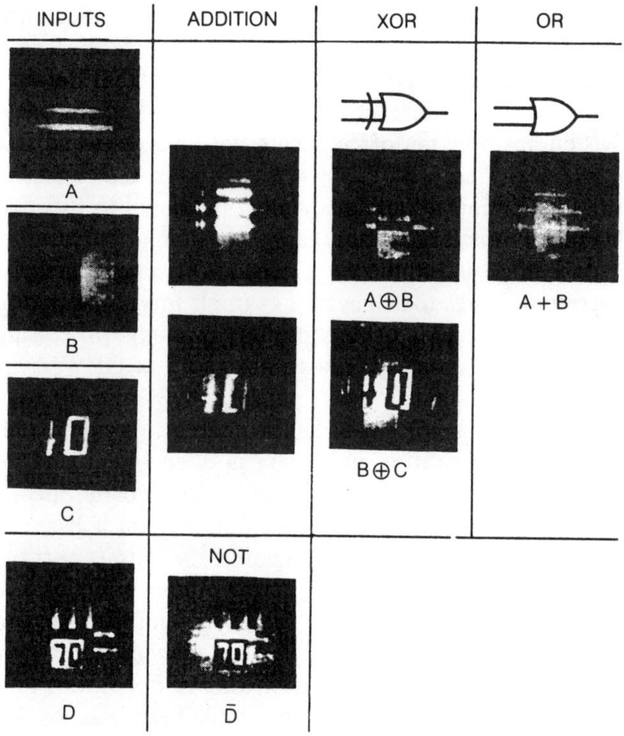

Figure 6.17 Examples of computing functions performed between two simultaneous inputs to a DFWM arrangement in BSO, under conditions of intentional slight misalignment of pump beams. Coherent addition operations $(A + B)^2$ and $(B + C)^2$ are shown in the second column. In the third column, the exclusive OR operations *A* XOR *B* and *B* XOR *C* are shown. In the last row, the negation (NOT) of *D* is shown, while in the last column the OR operation *A* OR *B* appears. (From Vainos *et al.* (1988).)

6.5.2. Temporal applications

Under the heading of temporal applications we shall also include frequency domain applications, and mention laser resonators that incorporate a PCM. There are also a wide range of both potential and experimentally demonstrated techniques concerning short pulse generation, for example, *Q*-switching and mode-locking, that fall outside the scope of this chapter. The concepts of coherent beam combination will briefly be discussed also, although several such schemes rely on two-beam coupling rather than PC in the true sense. Finally, 'intelligent' systems covering novelty filters (systems that detect change only) and associative memory implementations will be described.

6.5.2.1. Frequency Filtering

For the DFWM case considered earlier, the situation was quite straightforward. The degenerate situation was one in which both pump beam and probe (or signal) beam were at the frequency ω. What happens now, if you allow the frequency of one of these beams to vary, so that $\omega_1 = \omega_2 \neq \omega_p$, i.e. pump beams degenerate, but the signal beam is at $\omega \pm \delta\omega$ (where $\delta\omega \ll \omega$)?

The situation now is such that the conjugate return wave is radiated at a frequency given by $\omega_1 + \omega_2 - \omega_p$, so that, for $\omega_p = \omega + \delta\omega$, the return wave has a frequency $\omega_c = \omega - \delta\omega$. In other words the input wave has been 'frequency flipped' about its centre frequency ω.

In physical terms, the grating written between the probe wave and the first pump (the writing beam) is no longer stationary in space in this nondegenerate case, but travels through the phase conjugating medium. The other pump beam (reading beam) that scatters of this travelling grating is now Doppler shifted, to produce the conjugate output with the appropriately flipped

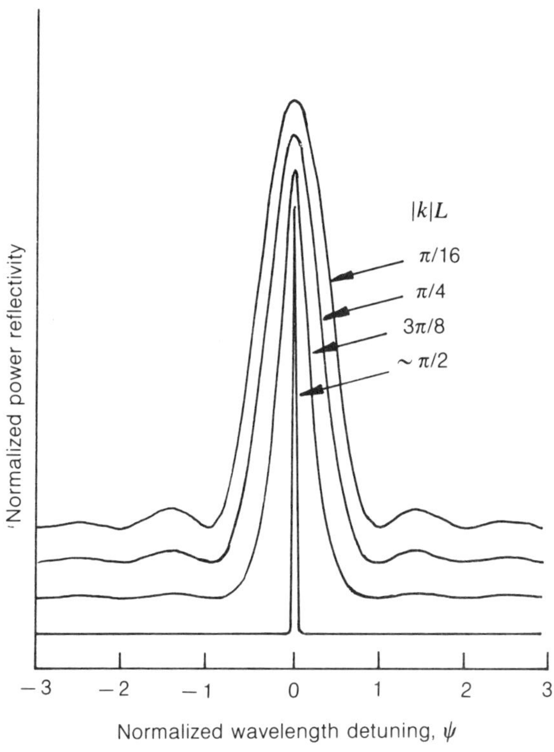

Figure 6.18 Power reflectivity vs. normalized wavelength detuning ψ for several values of nonlinear gain $|\kappa|\ L$. (After Pepper and Abrams (1978).)

frequency. The presence of this nondegeneracy, however, results in a phase mismatch $|\Delta k|$ given by (Pepper and Abrams, 1978).

$$|\Delta k| = 2n\pi(|\Delta\lambda|/\lambda^2) = 2n|\delta|/c \tag{6.31}$$

The effect of this is that the reflectivity of the phase conjugate mirror is not a constant, but depends intimately on the frequency detuning parameter δ. It is clear then that what we have here is a narrow bandpass filter. Work using sodium vapour as the (resonant) phase conjugate medium has achieved a fractional bandpass of 10^{-8}.

Figure 6.18 (Pepper and Abrams, 1978) gives a numerical calculation of the power reflectivity from such a nondegenerate mirror, as a function of the normalized wavelength detuning ψ. In this case, ψ is $\Delta k\,L/2\pi$, or the phase mismatch divided by 2π. The shape of these curves is the familiar sinc^2 function observed in SHG under conditions of slight phase mismatch.

6.5.2.2. *Logic functions*

As well as spatial logic functions described in section 6.5.1.7, we can also consider temporal logic, or optical gating. The basic four-wave mixer already described has the precise properties that you would require for either an AND gate or an OR gate. Considering Figure 6.19(a) (O'Meara *et al.*, 1983), we can see that the probe beam will only be reflected as a conjugate return if both A and B are simultaneously present. If the beams A and B are actually short duration optical pulses (0.5 ps) then the FWM optical gate can perform a logic cycle of <0.5 ps, giving the answer 0 for nontemporal coincidence of A and B, and 1 for their temporal simultaneity. This idea can be extended to perform the OR function, as shown in Figure 6.19(b) where either A or B will generate the FWM conjugate return.

This idea is rather similar to the technique of SHG autocorrelation for ultrashort pulse duration measurement. Indeed, a recent elegant experiment by Dominic *et al.* (1990) has achieved precisely this using transient two-beam coupling in $BaTiO_3$, to display the autocorrelation function of mode-locked laser pulses.

6.5.2.3. *Phase conjugate laser resonators*

Laser cavities that have either one mirror or both mirrors replaced by a PCM are extremely interesting when one considers their implications for aspects familiar from 'conventional' resonators, such as transverse mode structure, longitudinal mode spacing, stability criteria, and intracavity aberration. Indeed, it is easy to construct resonators that have gain, but are empty. For PC resonators the mirrors themselves can possess the gain, so a laser medium is not required. Although PC resonators do not fall easily under the heading of signal processing, their brief inclusion in this chapter is because they can readily be used in a range of schemes involving *Q*-switching, mode locking,

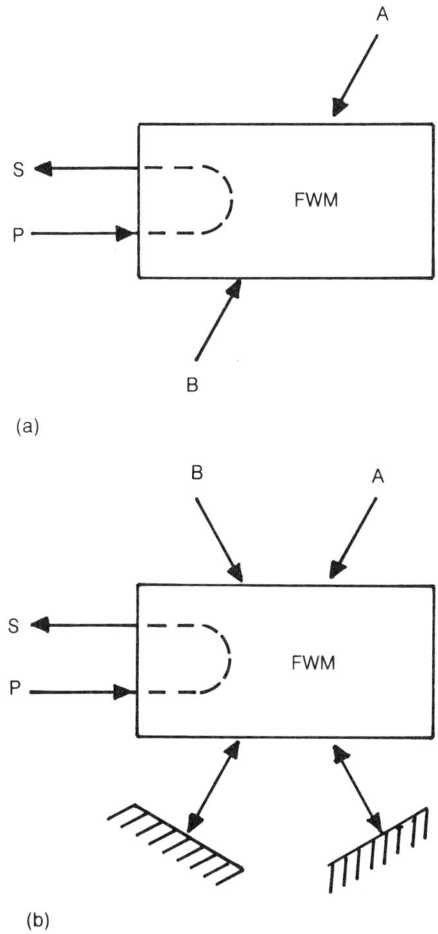

Figure 6.19 Schematic optical logic devices based on (a) AND gate and (b) OR gate (inclusive OR). (After O'Meara *et al.* (1983).)

image processing within the resonator (iterative techniques), and phase locking of several lasers via a common set of pump beams within a resonator geometry. A very good introductory chapter by Siegman *et al.* (1983) exists.

PC resonators are one of the few areas in which SBS techniques play an important role, as they are ideally suited to many high power pulsed laser systems.

6.5.2.4. Coherent beam combination

Beam combining, via phase locking of two or more separate lasers using a nonlinear coupling medium such as a photorefractive crystal, has been demonstrated several times already (Feinberg and Bacher, 1986; Cronin-

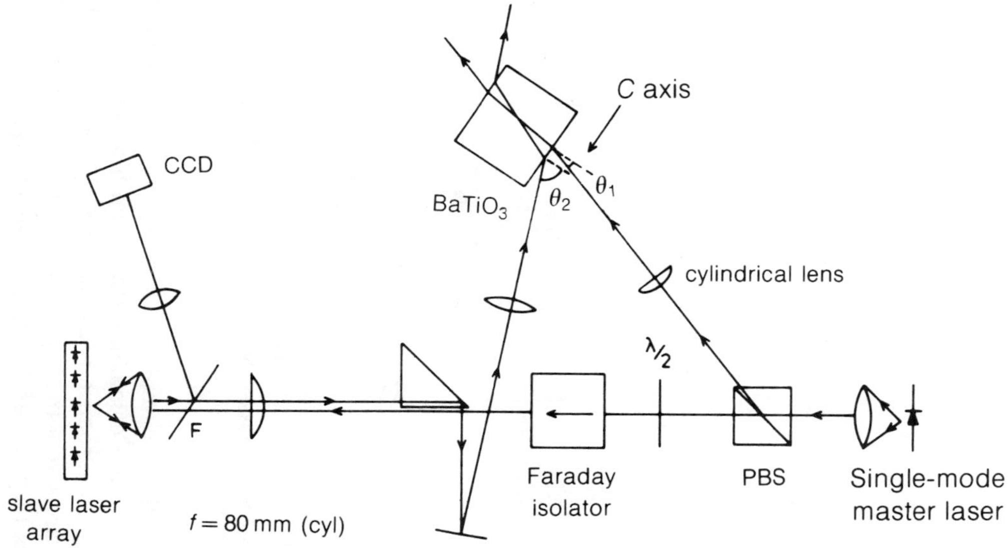

Figure 6.20 Experimental arrangement for injection locking a 1 W 20 stripe semiconductor laser array (SLA) by a 50 mW single longitudinal mode laser (ML). After injection locking, the mutually coherent outputs from SLA and ML are two beam coupled in the $BaTiO_3$ crystal. The net output is now diffraction limited with ideally all the power of the array. (After MacCormack and Eason (1990b).)

Golomb *et al.*, 1986). Of particular relevance here, though, is the conversion of many separate outputs, e.g. from a semiconductor diode laser array (laser 'bar') into one diffraction-limited beam, with ideally all the output power of the total array outputs.

Several schemes for achieving this have recently been proposed (Yeh *et al.*, 1986) and subsequently demonstrated. MacCormack and Eason (1990a) have used sequential two-beam coupling between successive stripes of a 100 mW CW ten-stripe diode laser array operating at ~800 nm, to yield an output that showed ~50% power transfer into the selected 'signal' stripe by the remaining nine 'pump' stripes. This geometry was extremely simple and involved no pre-injection of the array to induce coherence between the otherwise poorly coherent neighbouring stripes. In this experiment, the results were not power scalable to allow the array to operate at its full output power.

A much better scheme has been demonstrated by Verdeill *et al.* (1990) and MacCormack and Eason (1990b), shown in Figure 6.20, in which two-beam coupling occurs between an injection-locked array, e.g. 500 mW or 1 W device, and a portion of the injecting beam. After locking, the two mutually coherent outputs can couple very efficiently, so that outputs of 20–50% of the total array power can be achieved in a good diffraction-limited output beam. The main drawback to all these schemes is, however, the requirement that, for efficient injection locking to occur, then an antireflection coated (or reflection modified) laser array is preferred, as then the array can be operated at higher drive currents before the effects of internal array modes become a problem.

Cascading of such arrays is also possible (Pang *et al.*, 1990), so in principle the technique can be extended to incorporate arbitrarily large numbers of arrays. Practical problems (predominantly cost at present) and possible instabilities may limit this to a more realistic figure of a few cascaded stages only (perhaps three to five).

6.5.2.5. *Novelty filters and motion detection*

An experimentally simple but potentially useful class of processor known as novelty filters allows the visualization of dynamic change in an input image or scene. Several different schemes using either $BaTiO_3$ or BSO have been reported, all of which use the properties of a phase conjugate Michelson interferometer.

In the general case, if two signals generated via a reflection–transmission beam splitter are simultaneously input into a phase conjugator, then their PC returns will also undergo the same transmission–reflection effects, including phase shifts, path lengths and so on, such that there will be no output from the beam splitter under the conditions of perfect phase conjugation. If, however, aspects such as rise time for PC generation after a disturbance in the input (induced by motion for example) are considered, then the 'exclusive OR' type of operation that would occur under ideal conditions is spoiled, and a transient output is observed, lasting only for the duration of the induced movement.

This idea is embodied in the Stokes principle for time reversibility under which the equations for reflection and transmission at an interface between two dielectric media can be deduced merely by considering what must occur on running time backwards (i.e. PC generation) to satisfy conservation of energy.

Novelty filtering devices were first reported by Anderson *et al.* (1987) using a spatial light modulator (SLM) as an input device and $BaTiO_3$ as the PC medium. Subsequently rather dramatic results have been shown for a device called a transient detection microscope (Cudney *et al.*, 1988) in which two-beam coupling allows visualization of moving microscopic organisms (termed 'dancing bugs' in a subsequent video of the effect!).

Alternative schemes using simultaneous inputs to BSO have allowed much faster visualization of change, as the typical speed of response of BSO is $\sim$ 100–1000 times faster than that of $BaTiO_3$ at equivalent input intensities. Khoury *et al.* (1989b) have used the grating period response time dependence of BSO to tune the desired characteristics of a motion detector scheme. A schematic of this is shown in Figure 6.21. The technique of complementary grating recording can also be adopted to achieve cancellation in a stationary image, but this particular scheme has added possibilities in terms of its velocity filtering capabilities, and not merely arbitrary change (Hussain and Eason, 1991).

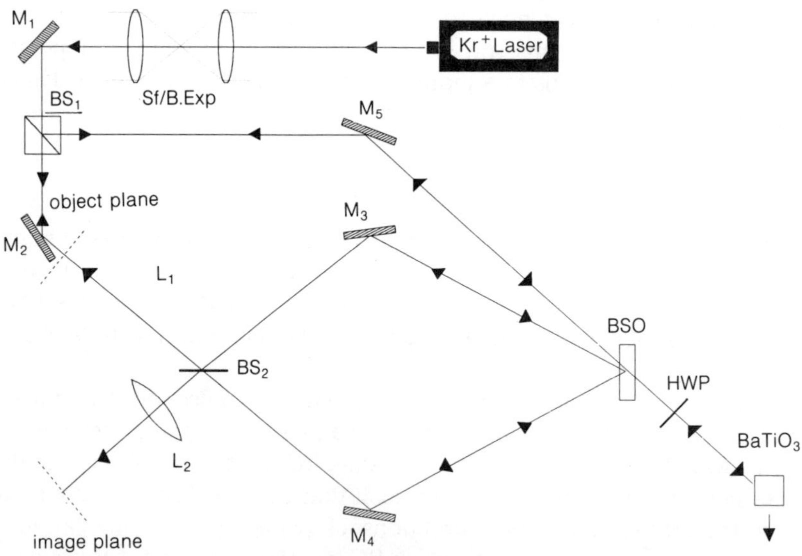

Figure 6.21 Schematic for motion detection and novelty filtering using two angularly multiplexed gratings (with different response times) using BSO. (After Khoury *et al.* (1989b).)

6.5.2.6. *Associative memories and intelligent systems*

Perhaps the most promising aspect of phase conjugate optical systems involves the use of PC mirrors in schemes for data retrieval via association between a given (perhaps incomplete) data input, and previously stored and sorted data. In this respect, the ability to store, sort, compare, decide and finally output a result is very similar to the operation of learning, thought and human decision that guides every aspect of all our lives. It would appear that a comparison can be made between the 'networks' involved in such an optical learning machine, which involves training and modification of the decision-making circuits (the stored holograms within the nonlinear crystal), and the substantially more complex, but perhaps fundamentally similar, myriad pathways inside the human brain.

So far, the experimental demonstrations of associative recall of complete data using only partial input have had exciting but limited results (Soffer *et al.*, 1986). The overwhelming problem appears to be extending these initial promising results which have stored e.g. four discrete test patterns and achieved successful recall of one to the much more desirable goal of recall of one input test object, e.g. your uncle's face, among a possible total of $\sim 5 \times 10^9$ other stored images (the rest of our present population). Nevertheless, these 'intelligent' systems are the first optical implementations of the area known as neural networks (Chapter 7), which are making great progress in tackling problems of a more human kind such as recognition and learning, which are extremely hard concepts for even the most advanced digital computer to tackle. It is to be hoped that progress along this line of research involving reconfiguration of holographic systems and stored data will advance successfully.

6.6. CONCLUSIONS

This chapter has addressed some of the background of optical phase conjugation and discussed a range of applications in the area of optical information processing. Clearly the list of devices and techniques discussed is not exhaustive or complete. It should also be stated that, quite apart from these applications-oriented results, there is a large research effort into new materials, optimized performance, improved modelling of PC behaviour, and so forth. The techniques discussed in this chapter are often proof of principle only, and as such do not rely on such optimization to work. However, for such schemes to yield commercially feasible devices successfully does require serious questions concerning, speed, size, scalability etc. to be addressed.

It might also appear that, on the applications side, the materials that are predominantly used are photorefractive crystals. This is currently true, but only because they are ideally suited to low power (e.g. diode laser compatible) intensities. There is no fundamental reason for this to remain the case, and, given the necessary research investment, novel materials such as organics and polymeric compounds could equally well come to the fore.

ACKNOWLEDGEMENTS

The auther would like to thank numerous people for the provision of original figures for inclusion in this chapter. Thanks also go to previous research students, and to the Science and Engineering Research Council (SERC) for research funding under the Physics Committee, NLO initiatives and the IRC programme.

REFERENCES

Anderson, D. Z., Liniger, D. M. and Feinberg, J. (1987) *Opt. Lett.*, **12**, 123.

Avizonis, P., Hope, P. A., Bomberger, W. D., Jacobs, S. F., Tomita, A. and Womack, K. H. (1977) *Appl. Phys. Lett.*, **31**, 435.

Bloom, D. M. and Bjorklund, G. C. (1977) *Appl. Phys. Lett.* **31**, 592.

Bloom, D. M., Liao, P. F. and Economou, N. P. (1978) *Opt. Lett.*, **2**, 58.

Chiou, A. E. and Yeh, P. (1986) *Opt. Lett.*, **11**, 306.

Clapham, S. L., Eason, R. W. and Vainos, N. A. (1990) *Opt. Commun.*, **74**, 290.

Cronin-Golomb, M., Fischer, B., White, J. O. and Yariv, A. (1982) *Appl. Phys. Lett.*, **41**, 689.

Cronin-Golomb, M., Yariv, A. and Ury, O. (1986) *Appl. Phys. Lett.*, **48**, 1240

Cudney, R. S., Pierce, R. M. and Feinberg, J. (1988) *Nature* (*London*), **332**, 124.

Dominic, V., Yao, X. S., Pierce, R. M. and Feinberg, J. (1990) *Appl. Phys. Lett.*, **56**, 521.

Dunning, G. J. and Lind, R. C. (1982) *Opt. Lett.*, **7**, 558.

Eason, R. W. and Smout, A. M. C. (1987) *Opt. Lett.*, **11**, 51.

Ewbank, M. D. (1988) *Opt. Lett.*, **13**, 47.

Fainman, Y., Guest, C. and Lee, S. (1986) *Appl. Opt.*, **25**, p. 1598.

Feinberg, J. (1980) *Opt. Lett.*, **5**, 330.

Feinberg, J. (1982) *Opt. Lett.*, **7**, 486.

Feinberg, J. (1983) In *Optical Phase Conjugation* (ed. R. A. Fisher), Academic Press, New York, Chapter 11.

Feinberg, J. and Bacher, G. D. (1986) *Appl. Phys. Lett.*, **48**, 570.

Fisher, R. A. (ed.) (1983) *Optical Phase Conjugation*, Academic Press, New York.

Gabor, D. (1984) *Nature* (*London*), **161**, 777.

Gerritsen, H. J. (1967) *Appl. Phys. Lett.*, **10**, 237.

Gower, M. C. (1982) *Opt. Lett.*, **7**, 423.

Gower, M. C. (1985) *Conf. on Lasers and Electro-optics, CLEO, Baltimore, MD, 1985*, Paper Th T2.

Gower, M. C., Davis, G. M. and Gibson, A. F. (1989) *Int. J. Optoelectron.*, **4**, 501.

Guiliano, C. R. (1981) *Phys. Today*, **34**, 27.

Günter, P. and Huignard, J.-P. (eds) (1988) *Photorefractive Materials and their Applications I*, Topics in Applied Physics, Vol. 61, Springer, Heidelberg.

Günter, P. and Huignard, J.-P. (eds) (1989) *Photorefractive Materials and their Applications II*, Topics in Applied Physics, Vol. 62, Springer, Heidelberg.
Hellwarth, R. W. (1977) *J. Opt. Soc. Am.*, **67**, 1.
Huignard, J.-P., and Herriau, J.-P. (1977) *Appl. Opt.*, **16**, 1807.
Huignard, J.-P., and Herriau, J.-P. (1978) *Appl. Opt.*, **17**, 2671.
Huignard, J.-P., Herriau, J.-P. and Micheron, F. (1976) *Ferroelectrics*, **11**, 393.
Hussain, G. and Eason, R. W. (1991) *Opt. Commun.*, **86**, 106.
Ja, Y. H. (1982) *Opt. Commun.*, **44**, 24.
Jain, R. K. and Lind, R. C. (1983) *J. Opt. Soc, Am.*, **73**, 647.
Jensen, S. M. and Hellwarth, R. W. (1978) *Appl. Phys. Lett.*, **35**, 404.
Khoury, J. A. and Eason, R. W. (1989) *J. Mod. Opt.*, **36**, 369.
Khoury, J. A., Hussain, G. and Eason, R. W. (1989a) *Opt. Commun.*, **70**, 272.
Khoury, J. A., Hussain, G. and Eason, R. W. (1989b) *Opt. Commun.*, **71**, 138.
Kogelnik, H. (1965) *Bell. Syst. Tech. J.*, **44**, 2451.
Kwong, S. K., Rakuljic, G. A. and Yariv A. (1986) *Appl. Phys. Lett.*, **48**, 201.
Levenson, M. D., Johnson, K. M., Hanchett, V. C. and Chiang, K. (1981) *J. Opt. Soc. Am.*, **71**, 737.
Lind, R. C., Steel, D. G. and Dunning, G. J. (1982) *Opt. Eng.*, **21**, 190.
MacCormack, S. and Eason, R. W. (1990a) *J. Appl. Phys.*, **67**, 7160.
MacCormack, S. and Eason, R. W. (1990b) *Opt. Lett.*, **15**, 1212.
Marrakchi, A., Huignard, J.-P. and Herriau, J.-P. (1980) *Opt. Commun.*, **34**, 15.
Martin, G. and Hellwarth, R. W. (1979) *Appl. Phys. Lett.*, **34**, 371.
McMichael, I., Yeh, P. and Beckwith, P. (1987) *Opt. Lett.*, **12**, 507.
O'Meara, T. R., Pepper, D. M. and White, J. O. (1983) In *Optical Phase Conjugation* (ed. R. A. Fisher), Academic Press, New York, Chapter 14.
Palmer, A. J. (1980) *Opt. Lett.*, **5**, 54.
Pang, L. Y., Kintzer, E. S. and Fujimoto, J. G. (1990) *Opt. Lett*, **15**, 728.
Pepper, D. M. (1982) *Opt. Eng.*, **21**, 156.
Pepper, D. M. and Abrams, R. L. (1978) *Opt. Lett.*, **3**, 212.
Pepper, D. M., Fekete, D. and Yariv, A. (1978) *Appl. Phys. Lett.*, **33**, 41.
Shen, Y. R. (1984) *The Principles of Nonlinear Optics*, Wiley, New York.
Siegman, A. E., Belanger, P. A. and Hardy, A. (1983) In *Optical Phase Conjugation* (ed. R. A. Fisher), Academic Press, New York, Chapter 13.
Soffer, B. H., Dunning, G. J., Owechko, Y. and Marom, E. (1986) *Opt. Lett.*, **11**, 118.
Stepanov, B. I., Ivakin, E. V. and Rubanov, A. S. (1971) *Sov. Phys. Dokl. Tech. Phys.*, **16**, 46.
Sternklar, S., Weiss, S., Segev, M. and Fischer, B. (1986) *Opt. Lett.*, **11**, 528.
Tocho, J. O., Sibbett, W. and Bradley, D. J. (1980) *Opt. Commun.*, **34**, 122.
Tocho, J. O., Sibbett, W. and Bradley, D. J. (1981) *Opt. Commun.*, **37**, 67.
Vainos, N. A. and Eason, R. W. (1986) *Opt. Commun.*, **59**, 167.
Vainos, N. A. and Eason, R. W. (1987) *Opt. Commun.*, **62**, 311.
Vainos, N. A., Khoury, J. A. and Eason, R. W. (1988) *Opt. Lett.*, **13**, 503.
Vainos, N. A., Clapham, S. L. and Eason, R. W. (1989a) *Appl. Opt.*, **28**, 4381.
Vainos, N. A., Clapham, S. L. and Eason, R. W. (1989b) *Appl. Opt.*, **28**, 4386.
Verdeill, J. M., Rajbenbach, H. and Huignard, J.-P. (1990) In *Digest of Meeting on Photorefractive Materials, Effects and Devices II*, Société Française d'optique, Aussois, France, p. 381.
White, H. J., Aldridge, N. B. and Lindsay, I. (1988) *Opt. Eng.*, **27**, 30.

White, J. O. and Yariv, A. (1980) *Appl. Phys. Lett.*, **37**, 5.

White, J. O., Cronin-Golomb, M., Fischer, B. and Yariv, A. (1982) *Appl. Phys. Lett.*, **40**, 450.

Yariv, A. (1976) *Appl. Phys. Lett.*, **28**, 88.

Yariv, A. (1977) *Opt. Commun.*, **21**, 49.

Yariv, A. (1985) *Optical Electronics*, 3rd edn, Holt, Rinehart and Winston, New York.

Yariv, A. and Pepper, D. M. (1977) *Opt. Lett.*, **1**, 16.

Yeh, P., Ewbank, M. D. and Khoshnevisan, M. (1986) *US Patent 4,761,059.*

Zel'dovich, B. Ya., Popovichev, V. I., Ragul'skiy, V. V. and Faizullov, F. S. (1972) *Sov. Phys. JETP*, **15**, 109.

Zel'dovich, B. Ya., Pilipetsky, N. F. and Shkunov, V. V. (1985) *Principles of Phase Conjugation*, Springer Series in Optical Sciences, Vol. 42, Springer, Berlin.

Architectures for Optical Neural Networks

7

ASHLEY J. BOSTEL, ANDREW K. POWELL
and TREVOR J. HALL

7.1. INTRODUCTION

Artificial neural networks are parallel processing systems which have applications in speech and pattern recognition (Rumelhart and McCelland, 1986; Prager *et al.*, 1986; Lippmann, 1987; Szu, 1986; Geman and Geman, 1984; Luttrell, 1985; Widrow *et al.* 1988), function optimization (Geman and Geman, 1984; Barhen *et al.*, 1987; Hinton *et al.*, 1984; Kirkpatrick *et al.*, 1983; Hopfield and Tank, 1986), robotics (Barhen *et al.*, 1987), and control (Psaltis *et al.*, 1988a). They consist of a set of identical, nonlinear processing elements, generically known as neurons, which are linked together to form a highly interconnected network. Information is represented by the pattern of activity of the neurons and data are stored by distributing them throughout the network's connections. This is done by weighting the links with positive and negative values to indicate the effect that one neuron has on another. This makes the system fault tolerant since each neuron is connected to many others and each weight represents the 'average' stimulus over the data set, so the loss of a few connections does not drastically affect the operation of the network.

These systems function as associative memories (Kohonen, 1984) since the weights force the rest of the network to assume specific patterns when some of the neurons have their values fixed by an external input. The information represented by the rest of the network has been 'associated' with the input. The particular pattern formed by the network is dictated by the values of the weights. This process is extremely fast since the system operates in parallel with the various alternative states competing with each other simultaneously. The state which wins is the one which is the most closely associated with the input.

The desired states of the network can be fixed by altering the weight values to reinforce those states. The current interest in neural networks derives from the fact that there are simple training or learning rules (Rumelhart and McCelland, 1986a; Hinton *et al.*, 1984; Rumelhart *et al.*, 1986; Derthick, 1984) which enable the weight values to be modified to improve the performance of the system or fix desired configurations.

There is currently a great deal of interest in implementing neural network models to exploit their parallelism (Jang *et al.*, 1988a, Lee *et al.*, 1989; Wagner and Psaltis, 1987; Psaltis *et al.*, 1987, 1990a; Farhat, 1987; Marks *et al.*, 1988; Guest and Tekolste, 1987; White *et al.*, 1988; Shamir *et al.*, 1989; Farhat, 1989; Ticknor and Barrett, 1987; White and Wright, 1988; Psaltis and Farhat, 1985; Farhat *et al.*, 1985). The burden of this implementation falls on the connections since the number of connections in a neural network increases with the square of the number of neurons. Optical systems can form connections in free space (Psaltis *et al.*, 1990a; Feldman and Guest, 1987) and this allows three dimensions to be exploited. This in turn means that processor densities can be increased and larger networks constructed. In addition, real-time holographic media (Peltier and Micheron, 1977; Staebler and Amodei, 1972; Hall *et al.*, 1985) and dynamic spatial light modulators (Collins, 1988; Johnson and Moddel, 1990; Collins *et al.*, 1989; Powell *et al.*, 1988; Williams *et al.*, 1988; Collins *et al.*, 1989) exist which can be used for reconfigurable interconnects in learning systems.

In this chapter some of the techniques used in implementing neural networks are described. In section 7.2, the most commonly implemented network models are briefly described. Section 7.3 reviews some of the components that can be used for their optical implementation. The various connection architectures that can be employed are then discussed in section 7.4. Mechanisms for optical learning are described in section 7.5. Finally, some conclusions are drawn in section 7.6.

7.2. NEURAL NETWORK MODELS

In most neural networks, the neuron algebraically sums its inputs and passes this value through a nonlinear function to determine its next output. The neuron will generally have an internal threshold which offsets this input in order to produce a desired response. This behaviour is shown schematically in Figure 7.1 and can be described by the following equation:

$$y_i = f(u_i)$$

with

$$u_i = \sum_{j=1}^{N} W_{ij} y_j - \theta_i \tag{7.1}$$

In equation 7.1 y_i is the output from the ith neuron and $f(u_i)$ is a nonlinear function. W_{ij} is the weight on the connection between neurons i and j, and

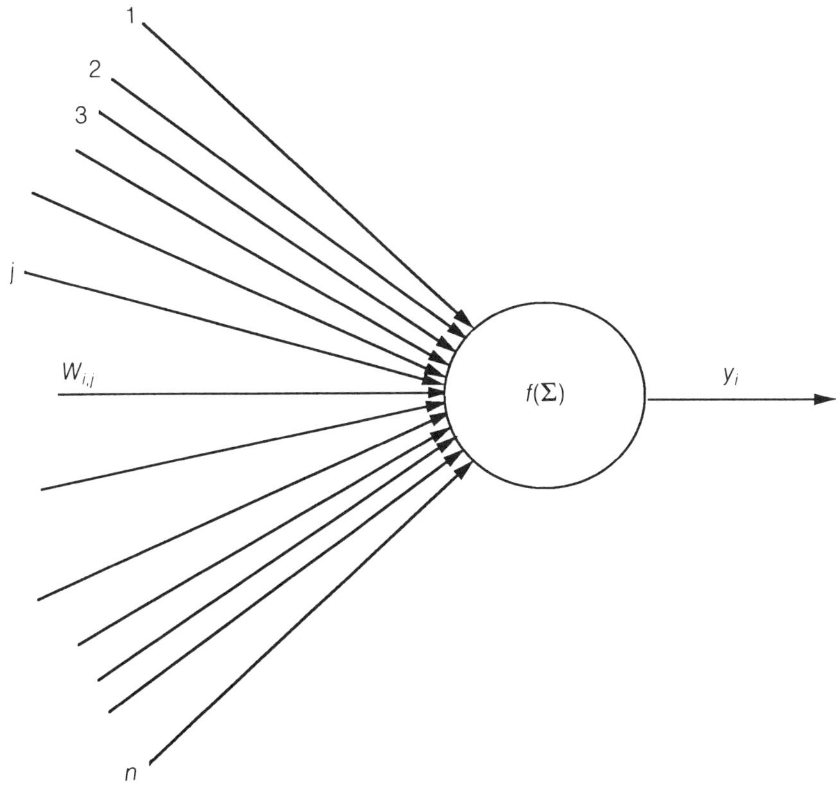

Figure 7.1 The neuron.

θ_i is the threshold of the ith neuron. In this case there are N neurons connected to the ith one. The most common types of nonlinear functions are shown in Figure 7.2. Since all the neurons perform the same function, they can all be implemented by a single device with a transfer characteristic similar to one of the curves shown.

The two neural models most often implemented are the Hopfield model (Hopfield, 1982) and the perceptron (Rumelhart and McCelland, 1986). Because of their popularity these two models will be described in more detail.

7.2.1. The Hopfield model

The Hopfield model is shown schematically in Figure 7.3. This network employs either two-state or binary neurons (Hopfield, 1982) or continuous, sigmoidal nonlinearities (Hopfield, 1984). With binary nodes the system functions as a content-addressable memory, while with continuous nonlinearities it can also be used to solve optimization problems, and may be implemented by an analogue electrical circuit (Hopfield and Tank, 1986).

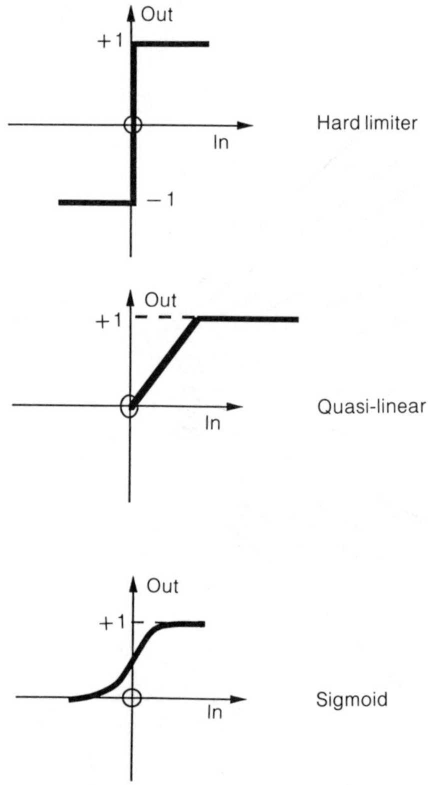

Figure 7.2 Node characteristics.

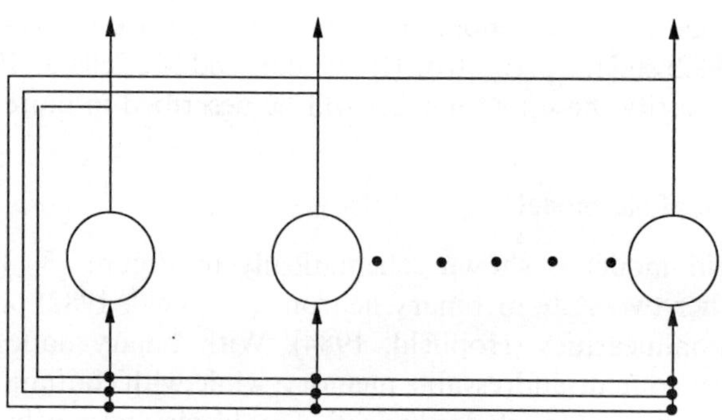

Figure 7.3 The Hopfield model.

The network consists of N nodes with weighted connections linking any particular neuron to every other one. If it is to act as an associative or content-addressable memory, the various patterns of nodal activity which constitute the memories must be defined as stable states. This can be done by defining the weights by the rule

$$W_{ij} = \begin{cases} \sum_{s=0}^{M-1} V_i^s V_j^s, & \text{if } i \neq j \\ 0, & \text{if } i = j \end{cases} \quad \text{with } 0 \leqslant i,j \leqslant N \tag{7.2}$$

where $\mathbf{V} = (V_1, V_2, \ldots, V_N)$ is a desired stable state and V_i is the value of the ith node in this pattern and can assume the values $+1$ or -1. M such states are stored by equation 7.2. An alternative representation for equation 7.2 when the nodes are binary is

$$W_{ij} = \sum_{s=0}^{M-1} (2X_i - 1)^s (2X_j - 1)^s \quad i \neq j \tag{7.3}$$

where X_i is either 1 or 0. This relation generates negative connection weights from the nonnegative neuron values.

Stability of the network is ensured by the fact that the connection weights are symmetric, and there is no feedback from a neuron to itself. These conditions are sufficient to guarantee that the network will not oscillate. As a further simplification, the weights can be clipped to produce weight values that are $+1$, -1 or 0.

In operation, an input forces the neurons to assume specific values. If this state is near to one of the stable patterns, this initial configuration will be unstable and the weights will cause the network to change until the nearest stable state is reached. The final pattern will tend to be the closest match to the input and will only differ from it by a few node values. In this case the input can be regarded as a copy of the stored pattern which has been corrupted by noise and the action of the network is to restore the fidelity of the information. The Hopfield model can thus be applied to pattern recognition and image restoration problems.

The binary Hopfield model can be guaranteed to converge to a stable state providing that the neurons change asynchronously. The reason for this is that this type of network can be described by the following function (Hopfield, 1982):

$$E = -\frac{1}{2} \sum_{i \neq j} \sum_{j \neq i} W_{ij} V_i V_j + \sum_i \theta_i V_i \tag{7.4}$$

This resembles the Hamiltonian function of a physical system such as a spin-glass ferromagnet (McKenzie and Sagi, 1986). This function is called the 'computational energy' of the network and describes a surface of possible network states. The action of a network obeying the above dynamics is always to reduce its value of E. The energy function is a nonnegative quantity which means that there will be minima in the surface and these correspond to the stable states of the network. In practice the restriction of asynchronous updating can be relaxed to allow all the nodes to change simultaneously, providing a parallel dynamical operation.

A similar function can be defined for a network of continuous neurons (Hopfield and Tank, 1986). This is analogous to a Lypanov function which describes the stability of continuous systems. In this format the Hopfield network can be used to solve optimization problems. The weights are used to define constraints and the neurons represent variables in the problem. Minima in E correspond to the satisfaction of as many constraints as possible given the current input. This represents the optimum decision based on the current input and the underlying constraints. The Hopfield optimization network has been applied to analogue-to-digital conversion, signal decomposition and linear and nonlinear programming problems.

Other networks can also be described by energy functions. Most notable of these are the Boltzmann machine (Hinton *et al.*, 1984; Ticknor and Barrett, 1987; Bounds, 1987; Derthick, 1984) and the bidirectional associative memory (BAM) (Kosko, 1988). These structures are all very similar to the Hopfield model but are generally partitioned into two or more layers and operate using more complicated dynamics.

7.2.2. The perceptron

The other type of network most often discussed is the perceptron. This is a feedforward system that acts as a pattern classifier (Rumelhart and McCelland, 1986; Widrow *et al.*, 1988; Rumelhart *et al.*, 1986). They generally consist of two or more layers with connections between them but no connections within a layer. The structure of a multilayer perceptron is shown in Figure 7.4.

The weights in a perceptron are set by an iterative training procedure which is driven by the error between the output produced by the network and the desired output. This error is used to alter the weights in order to improve the performance of the system. Many iterations are required in order to get the system to learn a complete set of classifications.

There are several forms of the training algorithm. The particular algorithm chosen depends on the task to be performed and the exact structure of the network. Most are only appropriate for two-layer networks; however, one of the most interesting procedures, known as the back-propagation algorithm (Rumelhart *et al.*, 1986), can be used to train large multilayer structures having nonlinear nodes.

The back-propagation procedure requires neurons with continuously differentiable nonlinearities and is designed to minimize the mean-squared error between the actual output of the system and the desired output. Nodes

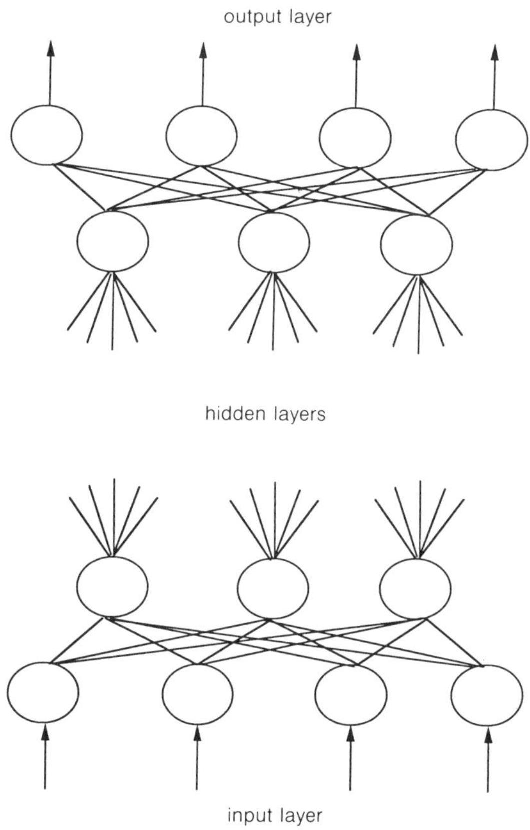

Figure 7.4 Multilayer perceptron.

with a sigmoidal transfer characteristic of the form

$$y = f(x) = \frac{1}{1 + \exp[-(x - \theta)]} \tag{7.5}$$

are usually used, although any differentiable function will suffice.

The procedure begins by initializing the weights with small, nonzero, random values. The function of the network is to associate a given input pattern, **A** with a desired classification pattern **B**. The pattern **A** is presented to the input layer and the pattern on the output nodes is compared with **B**. The error in the output is used to adapt all the weights in the system, starting with the links to the output layer, and working back to the connections between the input and first hidden layer. The new weight values are given by

$$W_{ij}(t + 1) = W_{ij}(t) + v\delta_j o_i \tag{7.6}$$

where, at time t, $W_{ij}(t)$ is the connection weight between the ith node in one layer and the jth node in the next layer. The output of the ith node is o_i and δ_i is the error; v is a gain term which regulates the size of the next weight step. If the node is in the output layer then (using nodes with sigmoidal characteristics),

$$\delta_i = y_i(1 - y_i)(d_i - y_i) \tag{7.7}$$

where d_i is the desired output of node i, y_i is the actual output, and $y_i(1 - y_i)$ is the differential of equation 7.5 with respect to x. If the ith node is a hidden node then

$$\delta_i = y_i(1 - y_i)\sum_k \delta_k W_{ik} \tag{7.8}$$

where k refers to nodes in the layer above the one being considered. The thresholds at individual nodes can also be adapted using this procedure by assuming that they are connections from nodes that are permanently on.

Once all the weights have been adjusted, the input is presented again and the adaptation procedure repeated. This process may have to be repeated several thousand times before all the errors have been reduced to zero. The convergence rate can sometimes be increased if a momentum term that depends on the weight change in the previous iteration is included (Rumelhart *et al.*, 1986). Several pairs of patterns can be memorized using this algorithm. The actual number depends on the size of the network. Networks implementing simplified back-propagation procedures are frequently considered for optical learning systems (Szu, 1986; Wagner and Psaltis, 1987; Psaltis *et al.*, 1987, 1988b, 1990a; Pack *et al.*, 1989).

There are several other algorithms that have been employed to train neural networks. some of these are 'self-organizing' and do not require the presence of training patterns. Many have yet to be implemented, mainly because of their increased complexity. The methods mentioned above are among the simplest but the back-propagation algorithm in particular frequently performs surprisingly well with highly complex problems (Rumelhart and McCelland, 1986).

7.3. COMPONENTS FOR THE OPTICAL SYSTEM

The first requirement for the implementation of a neural network is an array of neurons. This requires devices capable of a nonlinear response. The recent interest in optical processing has stimulated the development of several real-time optical modulators. Many of these can be used as neuron arrays in optical implementations. The following section reviews some of the basic types of devices that are available.

7.3.1. Spatial light modulators

Spatial light modulators (SLMs) are the most common type of real-time optical device. This generic title covers almost everything from the commercially available liquid crystal television, to the more experimental deformable mirror devices (Collins *et al.*, 1989). Spatial light modulators come in two classes (Johnson and Moddel, 1990): optically addressed and electronically addressed. Optically addressed devices use one light beam to control another. The original motivation behind this type of addressing was for the incoherent-to-coherent conversion of light so that incoherent optical information from the environment could interface directly with coherent processing systems. The advantage of optical addressing is the high speed obtainable through the parallelism in both input and output. From the point of view of all-optical networks an optically addressed light modulator is the ideal method of implementing the neurons.

Most optically addressed devices work in reflection (Collins, 1988; Johnson and Moddel, 1990). The basic structure of the most common form of photo-addressed SLM is shown in Figure 7.5. The image on the optical 'write'

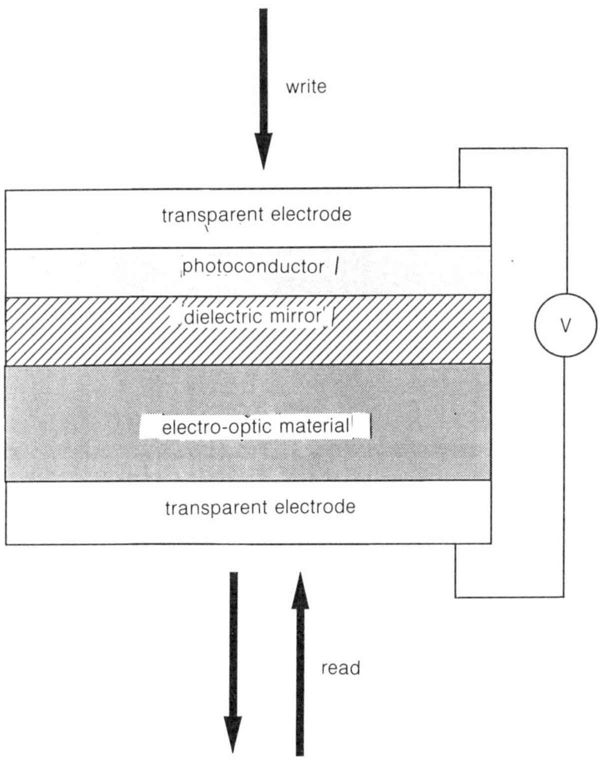

Figure 7.5 Optically addressed spatial light modulator (SLM).

signal activates the photoconductor and produces a charge distribution identical to the intensity distribution in the image. This charge pattern modifies an electric field, applied across the electro-optic material by an external voltage, V, and produces a change in the material which depends on the intensity at each point in the input image. This change is then detected by the optical read signal. Most devices employ media with electrically controlled birefringence so the light has its polarization modified as it passes through the medium. This change in polarization is measured by an analyser placed in the path of the reflected beam. The read beam can be much brighter than the writing signal since the two are isolated by the mirror and any other layers between the photoconductor and the electro-optic material. This means that some SLMs can act as amplifiers (Powell *et al.*, 1988).

The first photoconductors were selenium or cadmium sulphide (Johnson and Moddel, 1990). Recently other materials with better performance such as the amorphous-silicon–hydrogen alloys and GaAs have been investigated. The electro-optic material often employed is liquid crystal (Collins *et al.*, 1988; Johnson and Moddel, 1990; Powell *et al.*, 1988; Williams *et al.*, 1988). This has the advantage of low voltage requirement, sensitivity and high resolution. Typical input vs. output curves for a liquid crystal SLM are shown in Figure 7.6.

Apart from these liquid crystal devices, there are some alternative structures which have also been employed in optically addressed SLMs (Collins *et al.*, 1988). For example, the microchannel spatial light modulator (MSLM) uses a standard photocathode to produce electrons. These electrons are multiplied by a microchannel plate and deposited on the surface of a piece of birefringent

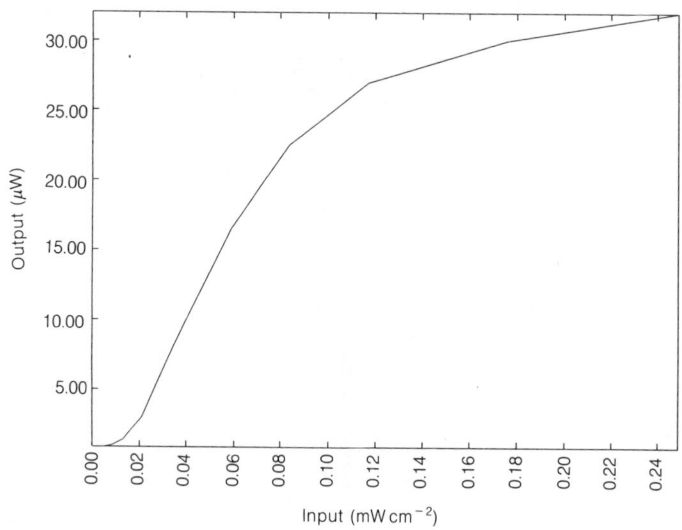

Figure 7.6 Input vs. output characteristic of a spatial light modulator.

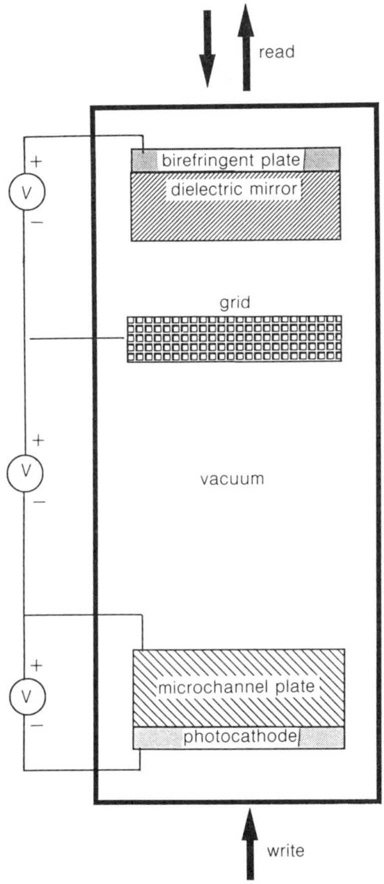

Figure 7.7 Microchannel spatial light modulator (MSLM).

material. This charge layer controls the voltage across the material and hence its birefringence. The read light is modulated by reflection through the material. The device is shown schematically in Figure 7.7.

The other type of photo-addressed SLM is the Pockel's readout optical memory (PROM). The main feature of this device is that images can be stored and erased again. The device consists of a thin crystal of bismuth silicon oxide ($Bi_{12}SiO_{20}$ or BSO). This is both photoconductive and birefringent. The device is shown in Figure 7.8. A strong electric field is initially applied across the BSO to make it birefringent. Images are written into the device using blue light which makes the BSO photoconductive in the illuminated areas. This causes the initial charge to drain away, reducing the electric field across the BSO. This in turn reduces the birefringence in the illuminated regions. The pattern can be read out using red light polarized at 45° to the

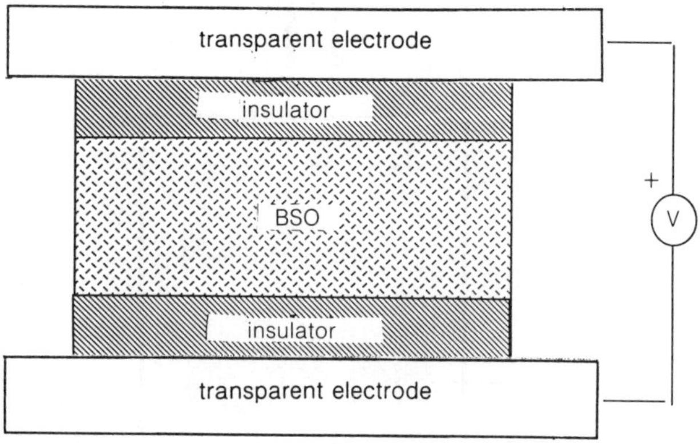

Figure 7.8 Pockel's readout optical memory (PROM).

fast and slow birefringent axes. A larger effect for a given source voltage can be obtained by reversing the direction of the applied voltage. This effectively doubles the voltage across the BSO. This device requires light of different wavelengths for the reading and writing operations. This means that the operations can not be cascaded very easily and makes iterative systems difficult to build. High voltages ($\sim$ 1000 V) are also required and its resolution is low. The storage properties could be used to store connection weight values, however.

The above devices all function by using one beam of light to modulate the polarization state of another. Most are capable of nonlinear thresholding operations by tuning the applied voltage. The liquid crystal devices do not have the storage abilities of the other types of SLMs, but they do have the advantage of high sensitivity coupled with high resolution and low voltage requirements. Large devices are also feasible. The MSLM has several desirable features. It has a broad spectral response, coupled with low optical switching energy requirements. It can be read out with a higher intensity beam and can be either operated in a real-time mode or used to store information. By proper adjustment of the grid voltage, thresholding or hard clipping operations are possible, thus logical AND or OR functions can be obtained. The main drawbacks of this device are its relatively low resolution, small active area and slow temporal response. Table II of Johnson and Moddel (1990) gives a summary of the important operating characteristics for a range of optically addressed devices.

The other type of SLM is addressed electronically. These devices generally operate in transmission and need to be interfaced to a TV camera or photodetector to be controlled by another optical signal. The read beam is affected by directly controlling the voltage across the modulating mechanism. This can be done actively, by the inclusion of transistor and photodiode elements with each pixel, thus forming active backplane devices. They are generally pixellated with the voltage applied to each pixel supervised by a microcomputer. Electrically addressed modulators are often binary with only transparent or opaque pixels. Although not specifically designed for use in optical systems the commercially available liquid crystal television (LCTV) can provide a grey level display capability at relatively low cost. Better quality analogue grey level devices are currently under development (Johnson and Moddel, 1990). A range of birefringent materials have been employed; these include $LiNbO_3$, PLZT, nematic liquid crystals, magneto-optic media such as gadolinium gallium garnet and ferroelectric liquid crystals. Apart from these birefringent devices, there is also a novel SLM called a deformable mirror device (DMD) (Collins *et al.*, 1989). It consists of a 2D array of thousands of small metal mirrors ($\sim 15\,\mu m \times 15\,\mu m$) which can be tilted electromechanically. Another development on the horizon is the inclusion of active circuitry on each pixel (Collings *et al.*, 1989). This will allow some elementary electronic processing to take place there. A simple version of an electrically addressed SLM is the commercially available liquid crystal television (Johnson and Moddel, 1990). These devices can be adapted to operate as relatively inexpensive controllable light displays.

The performances of several currently available electrically addressed SLMs are compared in Table I of Johnson and Moddel (1990). Electrically addressed devices are controllable displays. The desired nonlinear function is implemented electronically (via the microcomputer) on the signal from the photodetector, and the results are then displayed on the SLM. This approach is very flexible as it allows several types of neurons to be implemented but introduces a serial bottleneck.

Apart from acting as nonlinear elements for thresholding purposes the main rolé of SLMs in optical neural network architectures is as 2D image multipliers. These can be used as updateable connection matrices and can form the basis of real-time learning systems (Farhat, 1987; Fisher *et al.*, 1987).

7.3.2. Nonlinear Fabry–Pérot resonators

Nonlinear Fabry–Pérots (Smith, 1988) can be used as neurons in optical neural network implementations. A Fabry–Pérot resonator consists of a pair of plane, partially reflecting mirrors. Under normal conditions, when light is injected into the cavity, the transmitted light will oscillate between a maximum and minimum as the separation of the mirrors is altered. This is due to conditions of constructive and destructive interference within the cavity. The same effect can be obtained if the mirrors are held fixed and the refractive index of the material between the mirrors altered. A nonlinear Fabry–Pérot (NLFP) contains a material whose index of refraction depends on the light intensity within it. Such a device is shown schematically in Figure 7.9. Initially, the device is arranged to be in a state where the transmission I_{trans} is slightly off maximum. As the input intensity I_{in} increases,

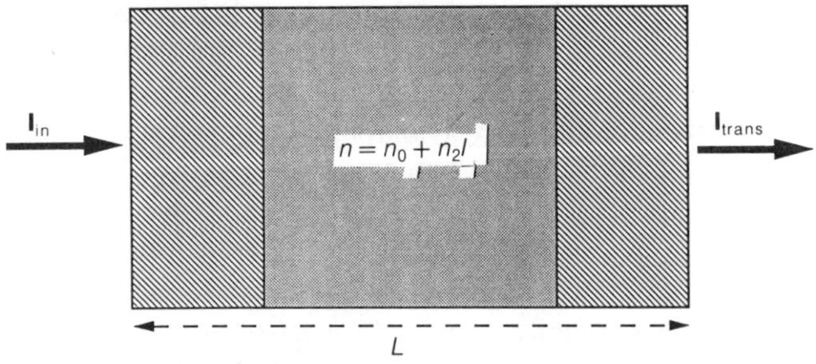

Figure 7.9 Nonlinear Fabry–Pérot.

the intensity inside the resonator increases. This changes the refractive index, which draws the cavity closer to resonance. The intensity in the cavity is magnified, giving a further change in refractive index, which further increases the intensity, and so on. The positive feedback eventually switches the device into a resonant state, where it will be forced to stay, even if I_{in} is reduced. When I_{in} is reduced too much, however, the device suddenly falls into a lower transmitting state since the cavity has been moved too far off resonance. Once this has happened the whole process has to be repeated in order to reach the highly transmitting state. The system is thus bistable, since there are conditions under which the same input intensity will produce either a high or a low transmission. Such a bistable system can display both this hysteresis effect and an abrupt switching from one state to another depending on the tuning of the cavity. The different forms of characteristics are shown in Figure 7.10.

Many different nonlinear materials can be employed, ranging from photorefractive crystals such as barium titanate ($BaTiO_3$) or BSO, to liquid crystals and saturable absorbers. The advantage of NLFPs are the low optical switching energies that can be employed. The main disadvantages are the external bias beams which are required to set the device near resonance initially, and the precision with which the device must be fabricated.

7.3.3. Opto-electronic devices

For present purposes opto-electronic devices refers to LED or laser diode arrays connected to optical detectors or devices fabricated directly from semiconductor materials. An LED provides a cheap incoherent source of light and can be used in conjunction with a photodetector to form an elementary 'neuron' by using the thresholded output from the photodetector to control the LED. This approach is one of the simplest and commonest ways of building a neural network system and many implementations favour them because of the availability of off-the-shelf components.

Laser diode arrays are promising for future optical implementations. They

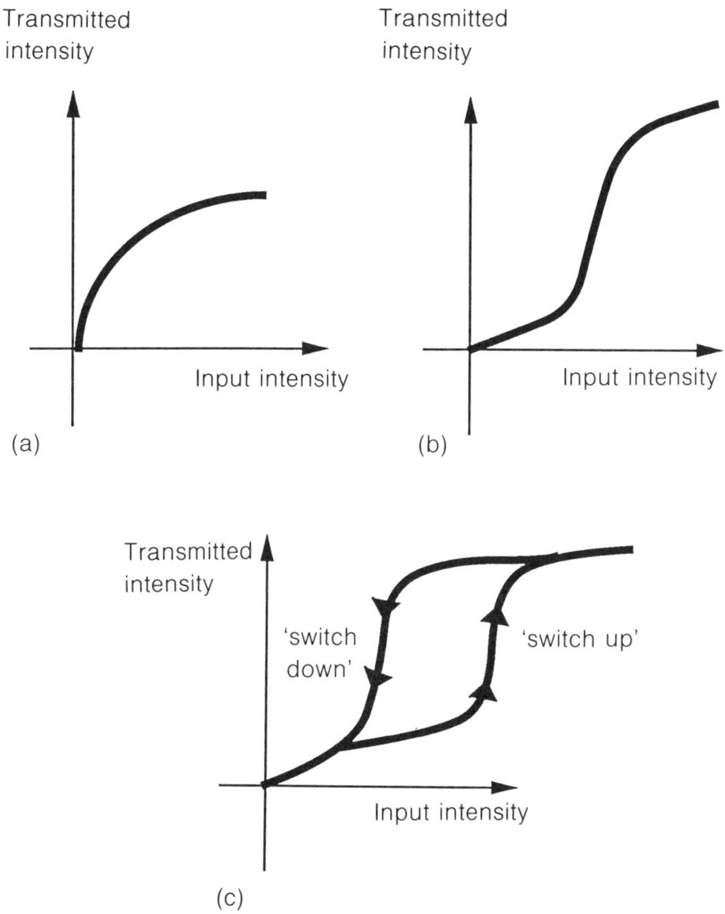

Figure 7.10 Nonlinear Fabry–Pérot characteristics. The following input vs. output characteristics can be obtained from an NLFP by tuning the cavity parameters: (a) optical limiter; (b) optical switch; (c) hysteresis.

are capable of providing small, compact sources of coherent light and are ideal for holographically connected networks. Commercial diode lasers (Bradley *et al.*, 1989) can emit at wavelengths from 650 to around 980 nm. Laser diodes can be obtained in one- and two-dimensional arrays with optical powers in the region of 100 mW (single stripe) or 10 W (bars). Recently, a two-dimensional array of micro-laser diodes with diameters around 5 μm has been fabricated (Pack *et al.*, 1990). This gives output powers of around 100 mW at 960 nm. The lasers in this array are capable of high speed switching (< 1 ns) and can be driven by currents of around 1 mA.

Another opto-electronic device which is actively being investigated is the self-electro-optic effect device (SEED) (Miller, 1988). The SEED consists of

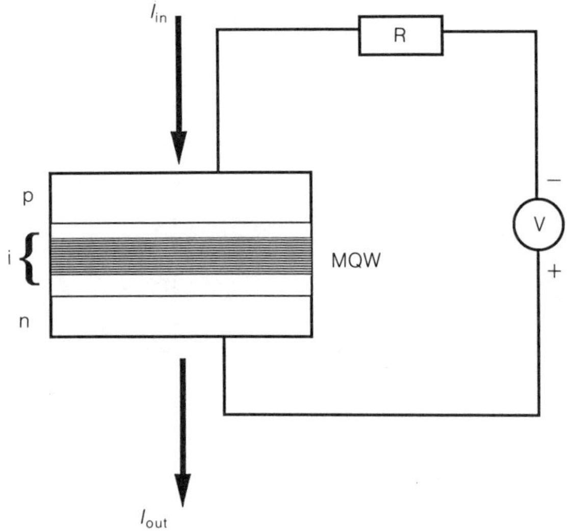

Figure 7.11 Self-electro-optic effect device (SEED).

layers of different semiconductors which form a multiple quantum well (MQW) structure, i.e. the layers are thin enough that the charge carriers are quantum-mechanically 'confined' and act like particles in a box in one dimension. These devices use an effect called the quantum-confined Stark effect to modulate light. The details of their function can be found elsewhere (see Miller (1988) and the references therein). Several forms of SEEDs have been proposed. The simplest is shown in Figure 7.11. This is the resistor-biased bistable SEED. This device acts through photoabsorption. Initially, all of the voltage supply is applied as a reverse bias across the quantum well diode which results in low optical absorption. The device is then illuminated, generating a photocurrent. This causes a voltage drop across the resistor which reduces the bias voltage across the quantum well diode. This, in turn, increases its optical absorption with a corresponding increase in photocurrent. The voltage drop across the resistor grows larger at the expense of the voltage across the quantum wells, giving a still larger absorption in the diode. This positive feedback mechanism can become so strong that the device switches into a highly absorbing state and leads to bistability. The switching power and speed depend on the resistance of the circuit. Switching speeds of a few tens of nanoseconds can be obtained in devices of 100 μm dimensions. Typical switching powers are of the order of milliwatts. This could be reduced to nanowatts but with a corresponding increase in switching time. For a given power, faster switching speeds can be obtained from devices with smaller areas.

An optical neural network can be formed when two optical modulators are linked together, with any point on one modulator influencing several elements on the second. Alternatively the light modulator and light detector of the same device can be linked together in a similar fashion to form a self-contained network. The type of network that is formed depends on the details of the connection scheme. Various connection methods can be employed. These are discussed in the next section.

7.4. CONNECTION SCHEMES

7.4.1. Matrix–vector multipliers

The first optical implementations of neural networks were systems to implement the Hopfield model. The original architecture was based on an incoherent method for performing parallel matrix–vector multiplication developed in the late 1970s by Goodman and his coworkers at Stanford (Goodman *et al.*, 1979). The architecture, sometimes called the Stanford matrix–vector multiplier, is shown schematically in Figure 7.12. The system consists of a column of light sources which represent the elements of the input vector. These sources are usually LEDs or laser diodes. The light from each source is spread into a uniform, horizontal, strip which illuminates a row of a 2D mask which faces the sources. This is accomplished by imaging in the vertical direction, while spreading the light in the horizontal direction using a combination of spherical and cylindrical lenses.

The matrix mask is a transparency consisting of a set of $p \times q$ cells. The transmittance of the cell in the ith row and jth column is proportional to M_{ij}, the (i,j)th element of the matrix to be multiplied. The constant of proportionality is the same for all cells and ensures that all the elements have values between 0 and 1. If the matrix is fixed, it can be implemented photographically with the optical density in each cell defining the transmittance. Another technique uses opaque cells and defines the transmittance

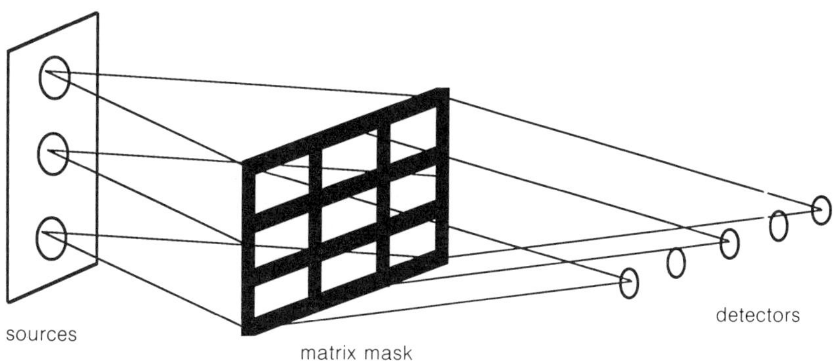

Figure 7.12 Matrix–vector multiplier.

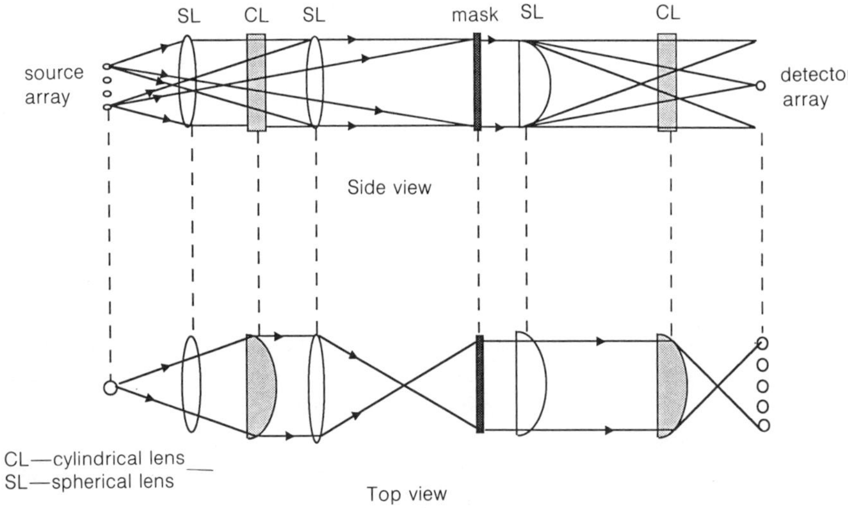

Figure 7.13 Optical system for matrix–vector multiplier. (After Goodman *et al.*, 1979).

by the area of a transparent region within each cell. If the matrix has to change rapidly a spatial light modulator can be used.

The light transmitted by the mask represents $p \times q$ analogue optical multiplications, performed in parallel, between each cell and a source in the array. The elements of the matrix–vector product are formed by summing the light in each column. This is achieved by imaging in the horizontal direction while focusing in the light from each mask element vertically. A combination of spherical and cylindrical lenses similar to the one used in the first step can be employed. One possible optical system is shown in Figure 7.13.

The summed light falls onto a row of detectors oriented perpendicularly to the source array. Each detector produces a current proportional to a different element of the output vector, and all the elements are available in parallel. The complete system performs the function

$$y_i = \sum_{j=1}^{q} M_{ij} x_j \quad \text{for} \quad i = 1, \ldots, p \tag{7.9}$$

where x_j and y_i are the input and output vector elements respectively.

The above system can be used to implement neural networks having fixed, precalculated, connection weights. The matrix elements represent the weights on the links betwen neurons. To form an interconnected set of neurons the output from each photodetector must be thresholded and the result fed back to drive an associated LED in the source array. The LED–photodetector pair forms the neuron.

The weights in a neural network assume both positive and negative values.

The negative elements of the matrix can be represented in several ways. A constant bias can be added to the weight matrix ensuring that all the elements are positive. In other words the system implements a matrix $[A]$ where

$$[A] = [W] + [a] \tag{7.10}$$

The input to the photodetectors is then

$$[A][x] = [W][x] + [a][x] \tag{7.11}$$

where $[x]$ is the vector of neuron values. Thus, a neuron-dependent bias term must be subtracted from the resulting product in order to obtain the correct answer. This can be done by including an extra column in the transmission mask with cells of equal transmission $[a]$. The signal from the photodetector associated with this column can then be electronically subtracted from the signals of the other detectors. An alternative approach is to decompose the connection matrix into a difference of two matrices:

$$[W] = [W^+] - [W^-] \tag{7.12}$$

$[W^-]$ is a matrix with nonnegative elements which represents the magnitudes of the negative elements of the weight matrix. The full weight matrix can be implemented as two matrix–vector multipliers, one for $[W^+]$, the other for $[W^-]$. The outputs from each multiplier can then be subtracted from each other to form the result. The two systems can be multiplexed together into a single crossbar architecture by doubling the number of columns in the matrix, and the number of photodetectors in the array. The light from each source is now spread out to cover two of the columns in the mask. One column represents the elements of $[W^+]$, the other the corresponding elements of $[W^-]$. At the detector array the light from each column is focused onto separate detectors. The signals from adjacent pairs of detectors can then be subtracted to form the input to the corresponding LED. The architecture is shown schematically in Figure 7.14.

The seminal paper on the optical implementation of neural networks dispensed with the optical summation stage, preferring instead to perform this task electronically with a pair of two-dimensional detector arrays (Farhat *et al.*, 1985). This approach reduces some of the problems with light collection efficiency that affect the architecture of Figure 7.14. A 32 neuron network was implemented by this system. Binary connection matrices were also used allowing the construction of the masks to be simplified.

These architectures become updateable in real time if a spatial light modulator is used to implement the connection matrix. Electrically addressed devices are particularly appropriate for this purpose as they also allow additional processing to be incorporated into the system. For example, the memory of the microcomputer controlling the SLM can be used to store several different connection matrices. This allows multilayer feedforward systems to be implemented by changing the connection pattern at each

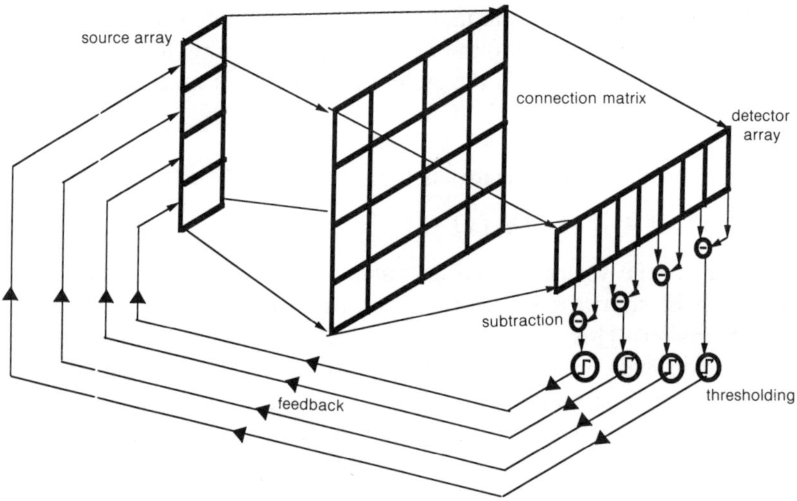

Figure 7.14 Electronic subtraction and feedback.

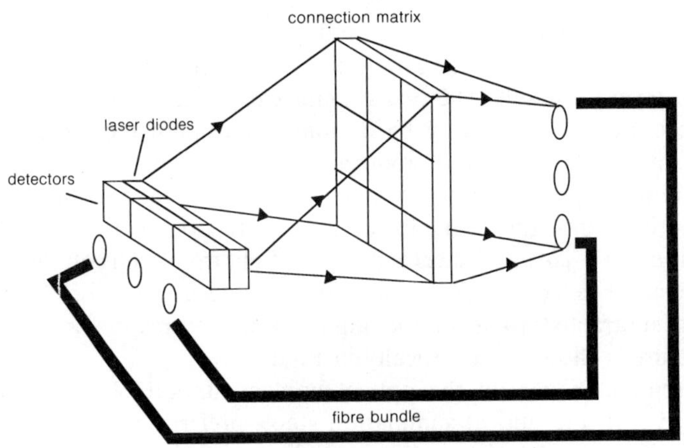

Figure 7.15 Optical fibre feedback system.

iteration. Alternatively, much larger networks can be simulated using a similar time-division multiplexing technique (Oita *et al.*, 1990).

If high power sources are used optical feedback becomes possible. An architecture using a back-to-back array of photodetectors and laser diodes, with optical fibres for light guiding, is shown in Figure 7.15. A similar scheme with free space optical feedback is shown in Figure 7.16. Architectures of this type require the light transmitted by the mask to be rotated through 90° in order to match the orientation of the input vector. In the free space case this can be done with a dove prism. A diffuser is sometimes employed

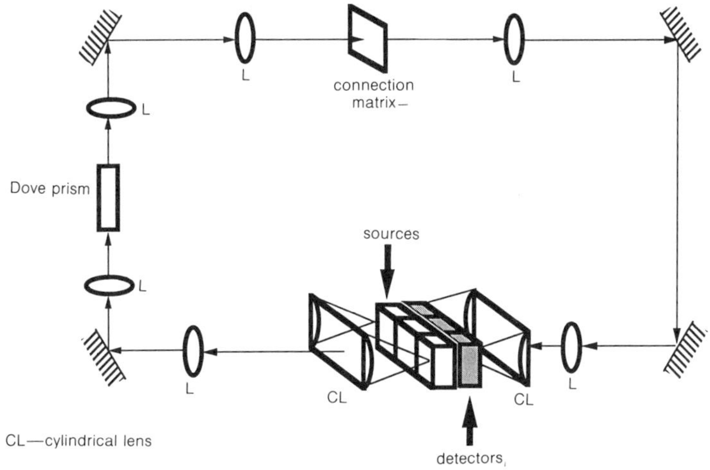

Figure 7.16 Free space optical feedback.

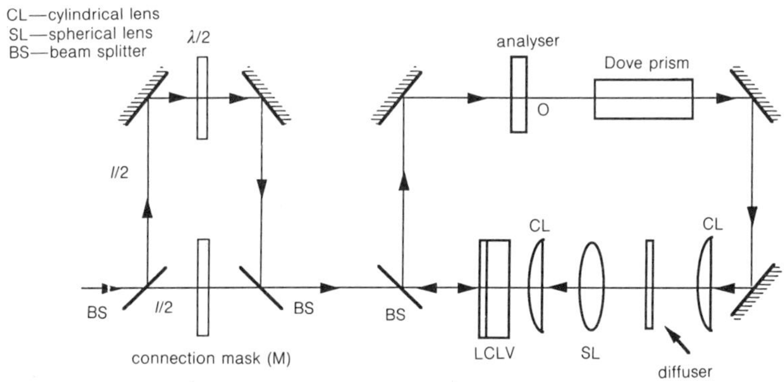

Figure 7.17 All optical feedback architecture with polarization encoding. (After Tekolste and Guest, 1987.)

in order to ensure that the illumination of the photodetectors is proportional to the incoherent summation of the light transmitted by a row of the matrix. The diffuser can be a major source of loss in the system.

A novel all-optical system, based on the free space architecture, is shown in Figure 7.17. This has been implemented using a Hughes liquid crystal light valve (LCLV) to perform both matrix–vector multiplication and thresholding (Tekolste and Guest, 1987). The system employs an ingenious polarization encoding scheme for the negative weight values. The LCLV's polarized read beam is split in two. One beam illuminates the connection mask which is biased to make all elements nonnegative. The polarization of the other beam is rotated by 90° using a half-wave plate. The two beams are recombined to form the total read beam for the SLM. An analyser, oriented parallel to the original polarization state, transmits the sum of two components: (1) the image written on the SLM multiplied by the connection matrix and (2) the inverse of the image. The output $\boldsymbol{O}$ from the analyser with a pattern $\boldsymbol{x}$ illuminating the SLM and a read beam of intensity I is

$$\boldsymbol{O} = \frac{I}{2}\boldsymbol{M}\boldsymbol{x} + \frac{I}{2}(1 - \boldsymbol{x}) \tag{7.13}$$

where $\boldsymbol{M} = \boldsymbol{W} + b$. $\boldsymbol{W}$ is the desired connection matrix and b is a constant bias whose value is set to make the elements of $\boldsymbol{M}$ positive or zero. When the values of $\boldsymbol{W}$ are constrained to lie in the range from -1 to 1, and $b = 1$,

$$\boldsymbol{O} = \frac{I}{2}(\boldsymbol{W}\boldsymbol{x} + 1) \tag{7.14}$$

This is the desired input to the SLM on the next iteration with the SLM threshold set equal to $I/2$.

An alternative polarization encoding technique is to implement the positive and negative weights by polarizing the light in orthogonal directions. This can be done with a birefringent spatial light modulator such as a liquid crystal television. The system is shown in Figure 7.18. LCTV1 acts as an input device introducing one-dimensional vectors as columns of ON or OFF pixels. This column display approach obviates the need for an anamorphic lens system. The light is incident on LCTV2 which displays the connection values. If a positive connection weight is encoded on one polarization and a negative weight as the perpendicular polarization, a polarizing beam splitter

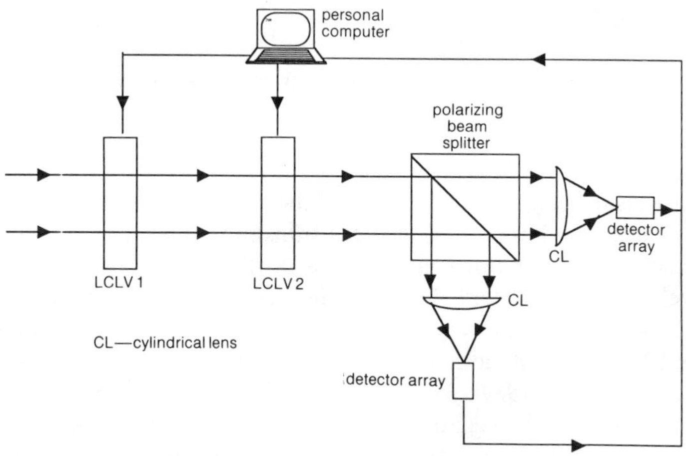

Figure 7.18 Alternative polarization encoding scheme. (After Johnson and Moddel, 1990.)

can be used to separate the two contributions. These components can be individually summed using cylindrical lenses, and the resulting vectors subtracted from each other and thresholded electronically. The result forms the new input to the first LCTV. Analogue weight values can be obtained by using polarizations between the parallel and perpendicular. In this scheme a zero weight occurs when the light is polarized at 45°. This scheme is implemented by the optical connectionist machine found in Johnson and Moddel (1990).

Apart from external feedback systems, reflexive architectures, in which the light is sent back through the same optical system, have also been suggested (Farhat, 1987; Guest and Tekolste, 1987). Figure 7.19 shows two concepts for systems of this type. In the first scheme in Figure 7.19(a), the sources and detectors are in the same plane but oriented at 90° to each other. The sources produce expanding wavefronts which are transmitted by the connection mask before being reflected back to the source–detector plane through a plano-

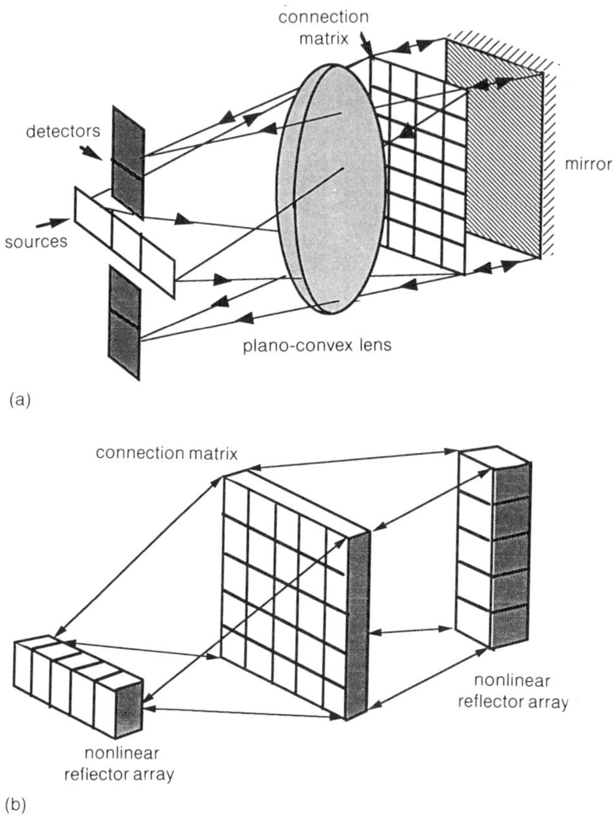

Figure 7.19 Reflexive architectures. (After Guest and Tekolste, 1987.)

convex lens. This images the light in the correct orientation for the detectors. In the second arrangement in Figure 7.19(b), two nonlinear reflectors are employed. These could be bistable optical devices such as nonlinear etalons working in reflection. The reflectors amplify and threshold the incoming light and retroreflect the result. In these systems the light makes a double pass through the connection mask. The transmittance of each cell must therefore be altered accordingly. In addition, the second system, in Figure 7.19(b), generally requires a symmetric connection matrix because of the π phase change on conventional reflection (this may not be necessary if the reflector also phase conjugates the light). Systems of this type have been proposed for implementing bidirectional associative memories (BAM).

Recent advances in SLM technology have produced designs which can eliminate the need for bulk optics and minimize diffraction problems. These electrically addressed devices have bar-shaped modulators that can be used with strip photodetectors to form compact systems in which the connection mask is sandwiched between the two (Farhat, 1989). An exploded view of

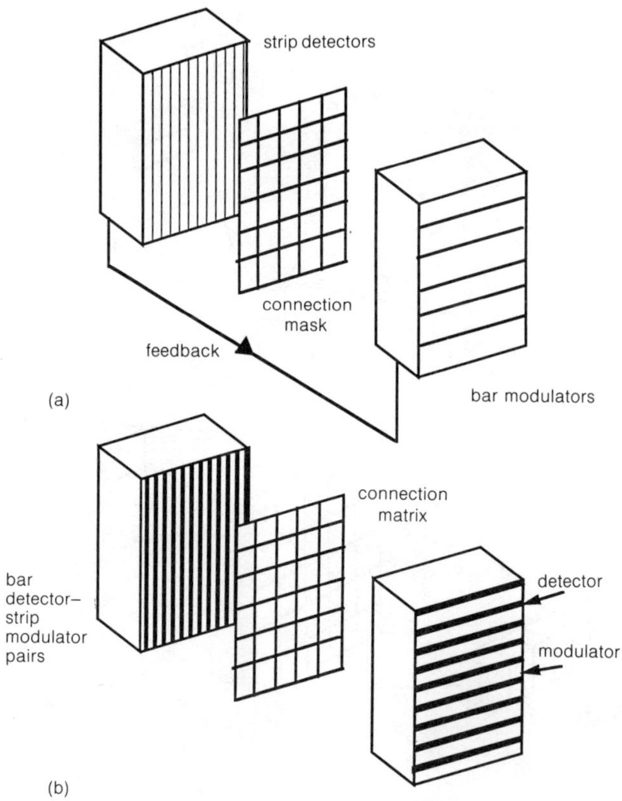

Figure 7.20 Bar modulator architectures.

the design is shown in Figure 7.20. This architecture eliminates the need for fan-out and summation optics by using the extended modulator and spatial integration. Diffraction effects are virtually eliminated by placing the connection mask directly against the detector, and the whole assembly against the modulator. A device of this type employing ambient light has recently been demonstrated (Von Lehman *et al.*, 1990). A reflexive form of this system with contiguous modulator–detector strip pairs (Figure 7.19(b)) could be used for systems such as the BAM which have complicated feedback requirements (Guest and Tekolste, 1987).

7.4.2. Inner product memories

All the systems described so far form networks of fully interconnected neurons. As such they can, in principle, implement most neural network algorithms. However, these systems are implemented by devices which are of limited accuracy. This can limit the choice of implementable algorithms. By examining particular schemes in more detail alternative architectures can sometimes be used which are more suitable for optics and are therefore easier to implement. In the Hopfield model, for instance, each connection weight is given by equation 7.2. The input to each neuron is therefore

$$u_i = \sum_{j=1}^{N} \left(\sum_{s=1}^{M} V_i^s V_j^s \right) V_j^p \tag{7.15}$$

where p refers to a particular input and s is the set of memorized patterns. The order of summation in equation 7.15 can be interchanged giving

$$u_i = \sum_{s=1}^{M} V_i^s \left(\sum_{j=1}^{N} V_j^s V_j^p \right) \tag{7.16}$$

This equation says that the Hopfield model can be implemented by first summing the point-by-point multiplication between the input and a stored pattern (forming the inner product of the patterns) for each of the memories. Each sum then multiplies its corresponding memory so that the input to a neuron is the weighted summation of the set of stored patterns. The thresholding operation then selects the pattern with the largest weight. This generates the correct output for the Hopfield model memory.

The inner-product can be formed optically by cross-correlating the input with the set of memories and selecting the unshifted correlations (Psaltis and Farhat, 1985; Psaltis and Hong, 1987). These correlation peaks can then be used to illuminate the set of memory patterns. The illuminated patterns are summed to form an intensity-weighted input to a thresholding device. The thresholded output yields the memory with the strongest correlation to the input. Two designs for implementing the inner-product scheme based on the matrix–vector multiplier architecture are shown in Figure 7.21.

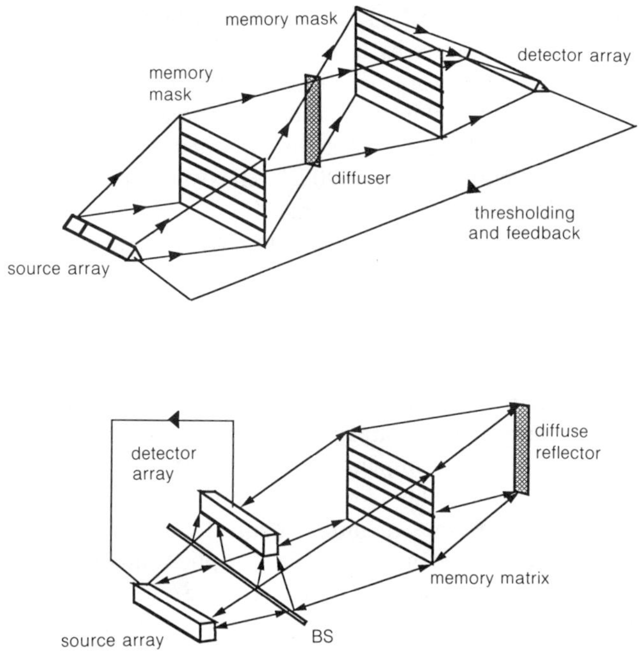

Figure 7.21 Inner product architectures. (After Farhat, 1987.)

The system shown in Figure 7.21(a) consists of two back-to-back matrix–vector multipliers. The first matrix contains the memory vectors arranged in a vertical stack. The output from this multiplier is a vector representing the summation of the point-by-point multiplication of the input pattern with each stored vector. The intensity of each element represents the discrete correlation between the input and each memory. The second multiplier contains the same set of vectors arranged in the same order as in the first. Each memory is illuminated with an intensity that corresponds to its correlation with the input. The total transmitted light is focused vertically and imaged horizontally on to an array of detectors. This represents the sum of the memories, weighted by the intensity of their illumination. Thresholding at the detector selects that memory with the strongest illumination or correlation with the input. If several patterns closely match the input the output will be a mixture of all of them. By thresholding the image and feeding the result back into the system, a slightly better match will be generated on the next pass. If it is allowed to iterate several times the system will eventually stabilize on one of the stored patterns. If the input is not too noisy the output will be the noise-free image, otherwise it will be the pattern in the system which the network 'thinks' is closest to what it sees. In order to function correctly, the output from the first multiplier must uniformly illuminate the second matrix.

This means that a diffuser or SLM must be placed at the output of the first multiplier in order to ensure that the illumination of the second stage is proportional to the magnitude of the correlations. Figure 7.21(b) is a reflexive version of the same system. This is possible since the two memory masks are identical. The arrangement is more compact and only requires the construction of one mask. The use of a diffusing reflector or an array of contiguous detector–emitter pairs is necessary in order to produce the correct reflected illumination.

These matrix–vector multiplier architectures are popular because of their simplicity and because of the availability of off-the-shelf components. The drawback of this architecture is the requirement that the neurons must lie in a one-dimensional plane. This arrangement can make large systems (with more than a few hundred neurons) cumbersome since it uses space inefficiently.

7.4.3. Matrix–matrix multipliers

Large networks can be implemented more efficiently if two-dimensional arrays are used. This approach is also attractive from the applications viewpoint, as many neural network applications can be represented in a two-dimensional format or can easily be arranged in this way.

There are several techniques that can be employed to implement networks with neurons arranged in two dimensions. Two-dimensional analogues of the matrix–vector multiplier can be constructed using arrays of small lenses (Farhat, 1989). Several examples of these matrix–matrix multiplier architectures are shown in Figure 7.22. The salient features of these schemes are (i) a two-dimensional diffuse light display such as a TV monitor or 2D array of LEDs, (ii) a two-dimsnsional array of lenses which generates multiple images of the display, (iii) a partitioned connection mask onto which the multiple images are focused (this mask has N^4 elements arranged in an $N \times N$ array of submasks (one for each node) with $N \times N$ cells in each), (iv) an array of light integrating elements (e.g. a two-dimensional photodetector array), and (v) a thresholding device and a feedback mechanism (e.g. an SLM with gain, or electronic detection and feedback).

As with the matrix–vector multiplier the connection masks can be produced photographically. One of the advantages of these systems is that the mask can be formed using the system itself. A photographic plate can be placed in the image plane of the lenslet array, all but one of the lenses can be blocked, and the appropriate connection pattern formed on the input light display. This light distribution exposes the plate and forms part of the connection mask. The procedure is repeated for each lens in the array with the appropriate connection pattern. The whole connection mask is built up bit by bit, one submatrix at a time. The technique can be easily automated and has the advantage that systematic errors and aberrations are taken into

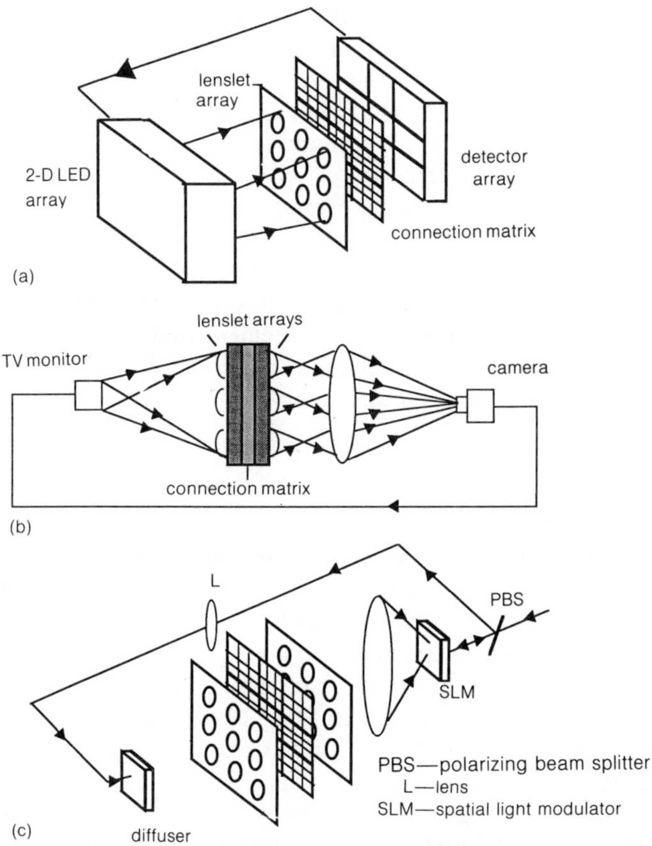

Figure 7.22 Matrix–matrix multiplier architectures. (After Farhat, 1989.)

account. The negative elements of the connection matrix can be handled in exactly the same way as in the one-dimensional systems. In electronic detection and feedback schemes electronic subtraction can be employed. In all-optical systems the negative quantities are more difficult to handle although it may be possible to devise a polarization scheme using an SLM.

Two-dimensional lenslet arrays can be fabricated in both plastic and glass with resolutions of up to 50 line pairs mm^{-1}. Using photolithographic techniques densities between 9 and 25 lenslets mm^{-1} can typically be obtained; Farhat (1989) reports that a 64 neuron network based on the first architecture of Figure 7.22 has been demonstrated using an 8×8 lenslet array.

7.4.4. Holographic systems

Another approach to implementing two-dimensional neural networks is to use holography (Psaltis *et al.*, 1990a; White *et al.*, 1988; Shamir *et al.*, 1989;

Figure 7.23 CGH fanout interconnect producing a 5×5 array of spots of $1\,mm^2$ diameter.

White and Wright, 1988; Shariv and Friesem, 1989; Jang *et al.*, 1988b). This is perhaps the most flexible way of building a complex network. A single hologram can define a two-dimensional light distribution at an arbitrary position. The simplest way of building a compact two-dimensional network is therefore to implement the neuron with a coherent light source and to use a hologram to define its connectivity. Computer-generated holograms (CGHs) (Feldman and Guest, 1987; Huang, 1971) can be used which are not restricted by the physical constraints of a recording stage. In this technique an arbitrary light distribution can be formulated on a computer and the hologram required to produce it calculated. The hologram can then be fabricated directly using a computer-controlled optical plotting facility or electron beam system. The reconstruction from a computer-generated fan-out element is illustrated in Figure 7.23. The images from an array of source–hologram combinations can be made to overlap at an array of light detectors and form a highly compact 2D neural network.

Figure 7.24 shows several architectures for holographic optical neural network systems employing both electronic and optical feedback. Figure 7.24(a) shows an electronic feedback system. This has many advantages. Firstly, electronic amplification at the detectors reduces the optical power requirements; secondly, negative connections can be handled by a simple electronic subtraction stage and, third, the system allows for additional electronic processing to be performed before the signal is fed back. The

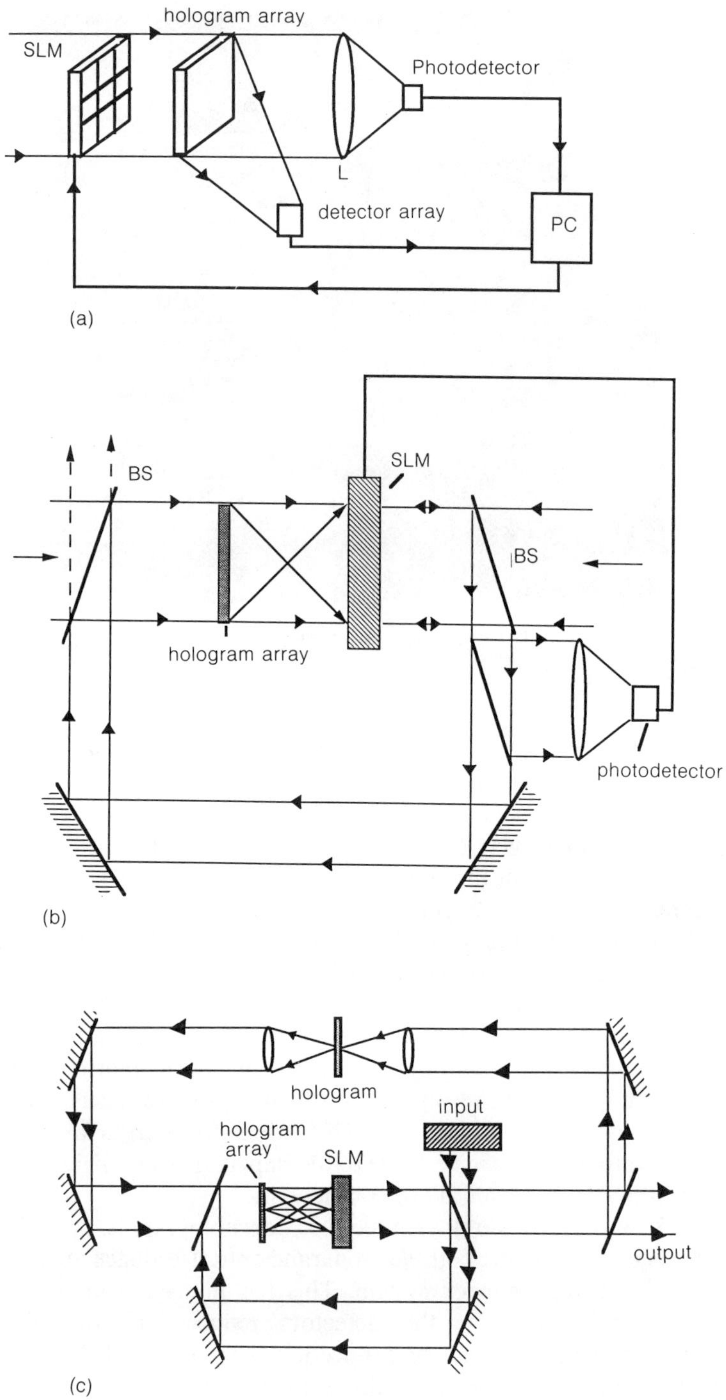
hologram array
SLM
Photodetector
L
detector array
PC
(a)
BS
SLM
BS
hologram array
photodetector
(b)
hologram
input
hologram array
SLM
output
(c)

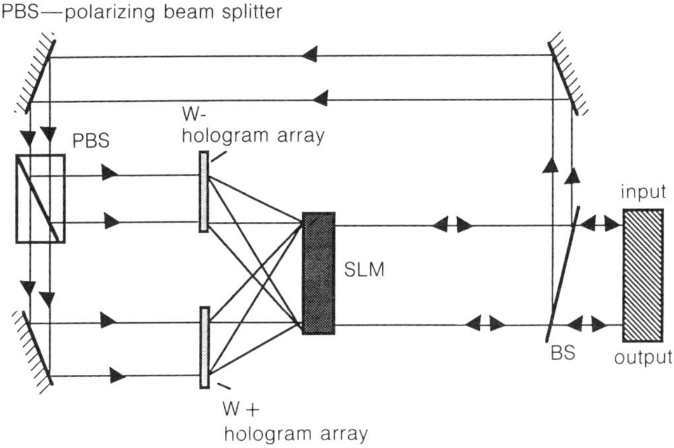

(d)

Figure 7.24 Holographic architectures: (a) after Jang *et al.* (1988a); (b) after White and Wright (1988); (c) after White and Wright (1988); (d).

implementation of negative weights is illustrated in Figure 7.24(a) by the collection and summation of the undiffracted light from the hologram. This can be used to form an activity-dependent threshold level for the purely positive optical connections. This implements the positive bias technique for negative quantities discussed earlier for the matrix–vector multiplier. The same idea is employed in the optical feedback system of Figure 7.24(b). In this case the threshold level of an optically addressed SLM is controlled by the optically modulated external bias voltage. An all-optical version of the dynamic threshold technique is illustrated in Figure 7.24(c). In this proposal, the light is diffracted by the second hologram in the top loop (White and Wright, 1988). This signal will be proportional to the cumulative light intensity of all the active elements of the neuron array. The hologram provides a set of bias beams which optically resets the threshold level of the SLM. In practice, this technique would require an extra amplifying stage in the top loop in order to overcome the inefficiency of the hologram and to provide bias levels of sufficient intensity to switch the SLM. Figure 7.24(d) shows a much simpler method of implementing negative connection weights. Devices such as liquid crystal SLMs or nonlinear etalons can simultaneously generate both the image and its complement. The negative connections can be formed by decomposing the connection matrix into two nonnegative components which are subtracted in order to form the total connectivity matrix. Once this decomposition has been made, two holograms can be fabricated which represent each new matrix. The hologram containing the positive elements of the original matrix is illuminated by the image. The other hologram containing the rest of the elements of the connection matrix is illuminated by its complement. The diffracted output from the two holograms is superimposed to form the desired connection pattern with the addition of a constant bias which can be thresholded out. The process can be described as follows. If $\boldsymbol{W}$ is the original connection matrix containing both positive and negative elements, it can be decomposed as follows:

$$\boldsymbol{W} = \boldsymbol{W}^{+} - \boldsymbol{W}^{-} \tag{7.17}$$

where $\boldsymbol{W}^{+}$ contains the positive elements of $\boldsymbol{W}$ and $\boldsymbol{W}^{-}$ contains the negative elements. If $\boldsymbol{W}^{+}$ is illuminated by a binary image vector $\mathbf{X}$ and $\boldsymbol{W}^{-}$ is illuminated by its complement $\mathbf{U} - \mathbf{X}$, where $\mathbf{U}$ represents a vector with all elements equal to unity, the result of adding the patterns is

$$\boldsymbol{W}^{+}\mathbf{X} + \boldsymbol{W}^{-}(\mathbf{U} - \mathbf{X}) = \boldsymbol{W}\mathbf{X} + \boldsymbol{W}^{-}\mathbf{U} \tag{7.18}$$

The product $\boldsymbol{W}^{-}\mathbf{U}$ is a constant which can be thresholded out to yield the desired connectivity for the system. This technique is attractive because of its simplicity. It implements both the positive and negative elements in one step, and also exploits the optical characteristics of the thresholding device. The geometries of two types of devices are shown in Figure 7.25.

The inner product representation of the Hopfield model is tailor made for holographic implementation since it is formed by two back-to-back correlators. The cross-correlation of two images can be obtained by multiplying their Fourier transforms and inverse transforming the product. The Fourier transform of an image can be recorded as a hologram. If this hologram is placed in the back focal plane of a lens and illuminated by a coherent wave, then the distribution of light across the wavefront at the front focal plane becomes spatially Fourier transformed by diffraction from the lens. The total light distribution at the back plane of the lens is therefore the product of the Fourier transforms of the light and the recorded image. If this product is retransformed by a second lens, we obtain a pseudo-inverse transform, i.e. the inverse transform with the space variables reversed. This

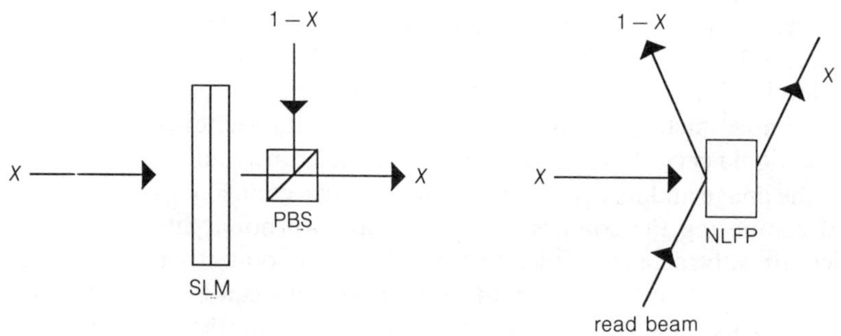

Figure 7.25 Complement image generation.

represents the correlation between the input and the stored image (Huang, 1971). If the cross-correlation pattern were to have originally entered the system, the output would have been the pattern recorded on the hologram because of reciprocity. Several Fourier transforms can be recorded in the same hologram allowing multiple memories to be stored.

The neural network is formed by arranging two correlators back to back. An image enters the first correlator and the stored image with the greatest correlation to the input is generated by the second. The intensity of the reconstruction depends on the match between the input and the stored pattern. The system is depicted in Figure 7.26. The original demonstration of this system is described by Psaltis and Farhat (1985). Pin-holes are used at the output of the first correlator in order to select the central peak of the correlation signal from each stored pattern. The correlations enter the second

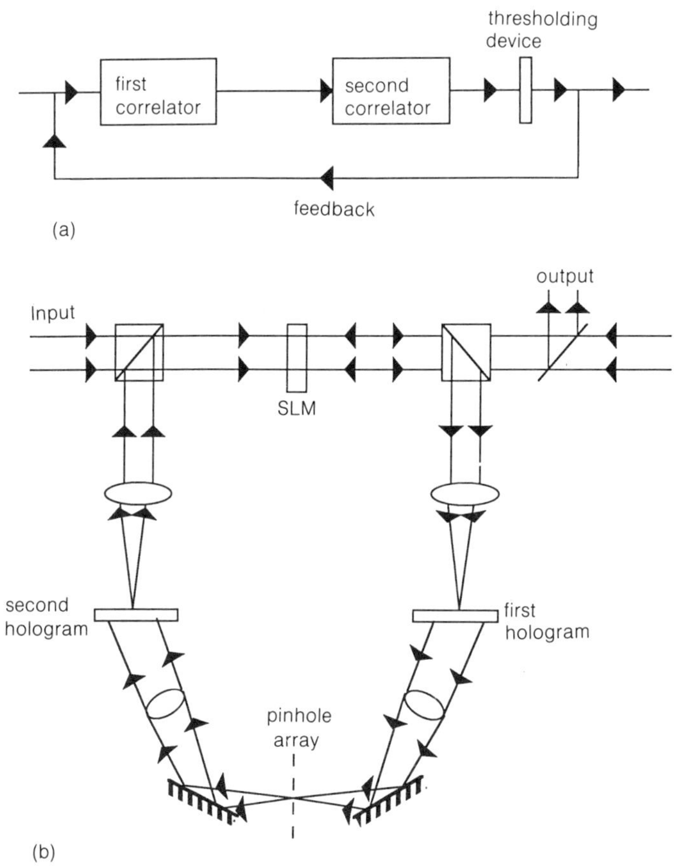

Figure 7.26 Holographic implementation of the inner product memory (After Psaltis and Farhat, 1985.)

correlator and produce a set of reconstructions which the SLM thresholds to generate the output pattern. In the demonstration of this system four images each containing around 100 000 neurons were stored and successfully retrieved. The compatability of this formalism with optical processing systems allows very large networks to be implemented. The main drawback of this approach is the restrictions imposed on the connection matrix, i.e. the connections must be specified by a learning rule similar to equation 7.2 or equation 7.3. This type of rule is called a Hebbian or Hebb learning rule after Hebb (1949) who first suggested this form of learning mechanism in the brain.

The above architectures have all been based on the idea that the output from the thresholding element reconstructs the connection hologram. This technique can be employed regardless of whether the SLM used is optically or electrically addressed. A different architecture can be employed if an electrically addressed transmitting device is used. This architecture will be mentioned briefly since it suffers adversely from diffraction effects, which limits its versatility. The architecture is shown in Figure 7.27 and consists of a hologram array containing the desired connectivity patterns, an electrically addressed SLM and a 2D photodetector array. Instead of selectively reconstructing the array of connection holograms all the holograms are reconstructed at once. The reconstructions are imaged onto the detector array using a double-lens system and pass through the SLM on the way. Every pixel in the SLM receives a contribution from every hologram in the array. Providing that the transmission of the SLM is linear, each of these components separates after passing through the device because they have different phases. The second lens images each component separately and so each detector only

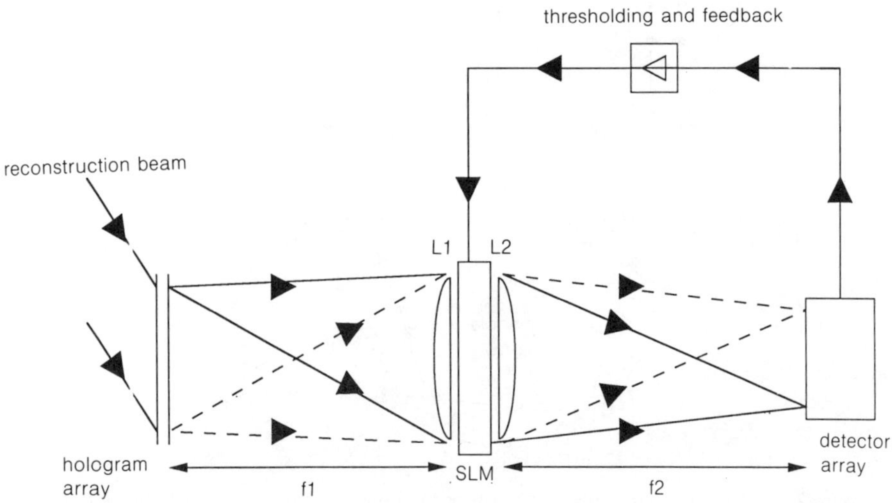

Figure 7.27 Alternative holographic architecture. (After Shamir *et al.*, 1989.)

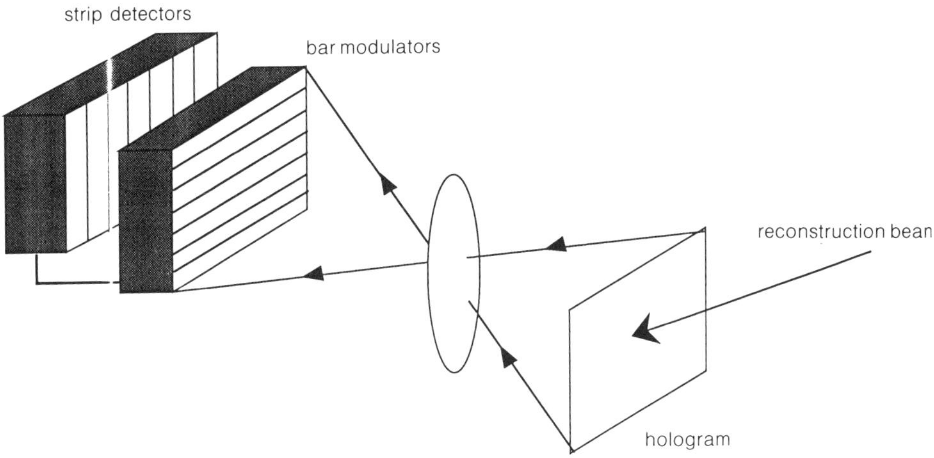

Figure 7.28 One-dimensional holographic architecture.

receives the reconstructions from a single hologram. These are weighted, however, by the spatially modulated transmission of the SLM. Information is input into the system through the SLM and the output from the detector can be fed back to the SLM to allow the system to iterate. This architecture is severely limited by diffractive cross-talk at the detectors and by the angular sensitivity of the SLM. This places serious restrictions on the system and makes it impractical for most applications.

A variation of the architecture can be used to implement one-dimensional systems with two-dimensional modulators. The scheme is shown in Figure 7.28. In this system individual columns (rows) of the SLM assume the same transmittance values. The problems with the angular sensitivity are reduced since summation can be performed electronically by spatially integrating detectors. This allows the light to be incident normally on the device and allows the detector array to be directly attached to the SLM. This system can potentially form a highly compact architecture by using SEED devices for example.

7.4.5. Optical disk systems

One of the more interesting developments of recent years is optical disk technology. This can be used to store the connection matrices for optical neural networks (Lu *et al.*, 1989; Psaltis *et al.*, 1990b). The advantage of using optical disks is that many different connection patterns can be stored on the same disk and extracted optically to update a system very rapidly.

Optical disks store information in a binary format. The patterns can either be stored directly, allowing them to be used as connection matrices in matrix–

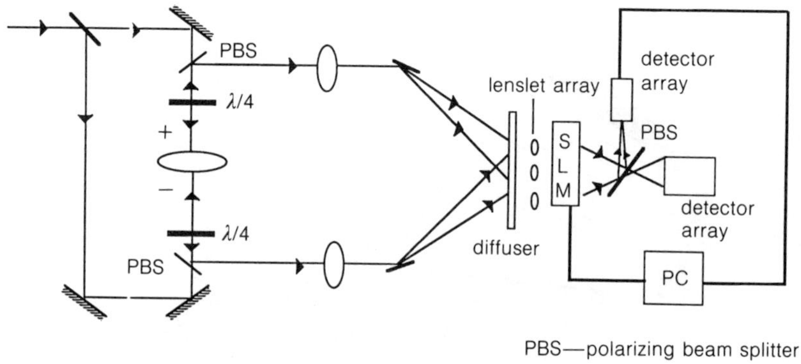

Figure 7.29 Optical disk based neural network. (After Lu *et al.*, 1989.)

vector or matrix–matrix multiplier architectures, or be recorded as binary holograms, which allows two-dimensional networks to be formed. Unfortunately, this technique uses up a lot of the disk's storage capacity and degrades rapidly as the number of connections increases. Analogue weightings can be implemented by using half-toning techniques.

Several different architectures have been proposed using optical disks to store the connection weights. In one architecture the positive and negative connections are stored on opposite sides of the disk and the system implements a two-dimensional network by partitioning the connection matrix into a 2D array of submatrices (Lu *et al.*, 1989). The architecture is shown in Figure 7.29. The disk is read by a pulsed laser which is split into two orthogonally polarized beams. One beam reads the connections on one side of the disk, and the other beam reads the pattern on the reverse. Each beam passes through a polarizing beam splitter followed by a quarter-wave plate which rotates its polarization. The beams are reflected from the disk back through the quarter-wave plates. They are now reflected by the beam splitters since the polarization of each beam has been rotated twice by the quarter-wave plates. The two beams are enlarged and superimposed on a diffuser. A lenslet array images each matrix through an electrically addressed SLM displaying the input pattern. The light transmitted by the SLM is separated by another polarizing beam splitter to form two 2D output displays. These represent the product of the input pattern and the interconnection matrix. The two output patterns are detected and subtracted electronically. The result can be thresholded and displayed on the SLM again to form an iterative loop.

Another development is the implementation of networks on integrated circuits with connection weights specified optically through photodetectors on the links between neurons. An optical disk can be used to store and update rapidly the connection weights for the system (Psaltis *et al.*, 1990b). The scheme is shown schematically in Figure 7.30. This arrangement allows the

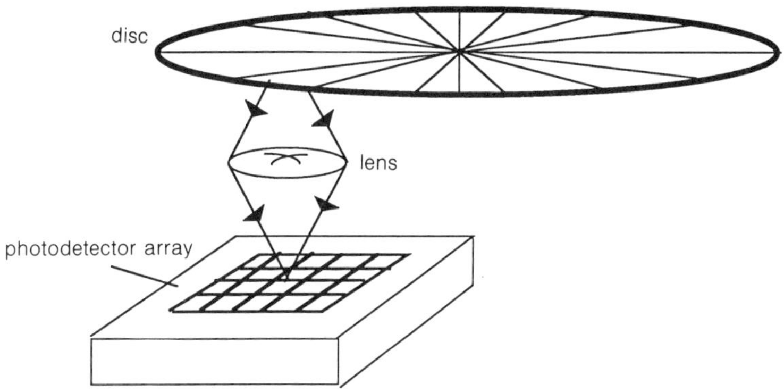

Figure 7.30 Optoelectronic VLSI network.

connection pattern on the disk to be imaged directly onto the chip. Negative weights can be implemented electronically by using separate photodetectors or by using the complement of the input pattern.

Most of the currently available optical disks are of the write-once read many times (WORM) variety. This means that once the information is stored it cannot be altered. An optical disk, however, can store several gigabytes of information on the two sides of the disk. Many different networks can thus be implemented with the same basic system. There is currently much research into producing erasable disks which will allow new information to be recorded *in situ*. This will further improve the attractiveness of optical disk based systems.

Currently, disks can rotate at over 1000 rpm. Assuming each connection matrix has about 500 × 500 bits, around 16 000 matrices can be stored on each side of the disk. It has been estimated that more than 10^9 connections per second can therefore be made using an optical disk system. This is well above most available electronic systems and is very encouraging for neural network implementations as well as optoelectronic processing in general.

7.4.6. Real-time holography

The use of photorefractive materials such as $BaTiO_3$ and $LiNbO_3$ can have a significant impact on the architectures of neural network implementations (Lee *et al.*, 1989; Wagner and Psaltis, 1987; Psaltis *et al.*, 1987 1988b, 1990a; Soffer *et al.*, 1986; Yariv and Kwong, 1986). These materials exhibit optically induced temporary changes in refractive index which allow them to be used as real-time processing materials. The interference of two or more beams within a crystal of $BaTiO_3$, for example, generates a holographic phase grating which persists for as long as the beams remain. This can allow one beam to couple power into another. This ability to amplify optical signals makes them especially useful in iterative optical schemes. In addition, the physical process which gives rise to these effects can be made to exhibit threshold properties (Cronin-Golomb and Yariv, 1986), allowing them to be used as continuous fields of neurons with a single crystal being capable of forming arrays of hundreds or thousands of neurons.

The persistence of the holographic grating in the absence of illumination depends on the material. In $BaTiO_3$ it is of the order of seconds, while in $LiNbO_3$ it can be several hours (Peltier and Micheron, 1977; Staebler and Amodei, 1972; Hall *et al.*, 1985). These features can be combined to produce neural network architectures with interesting capabilities. $LiNbO_3$ can be used to store several holograms simultaneously (Psaltis *et al.*, 1987; White *et al.*, 1988; Soffer *et al.*, 1986; Yariv and Kwong, 1986). These recordings are separated from each other during manufacture by using reference beams with different phases. The recordings can be selectively erased by rewriting over an existing hologram using a reference beam with the same phase as in the original process. This means that in principle new holograms can be stored without having to change the rest of the system.

The inner product architecture discussed previously can be implemented using $LiNbO_3$ holograms to store the memories and $BaTiO_3$ crystals as optical amplifiers and thresholding devices (Lee *et al.*, 1989). The scheme is illustrated in Figure 7.31. An image can be introduced to the system (via the

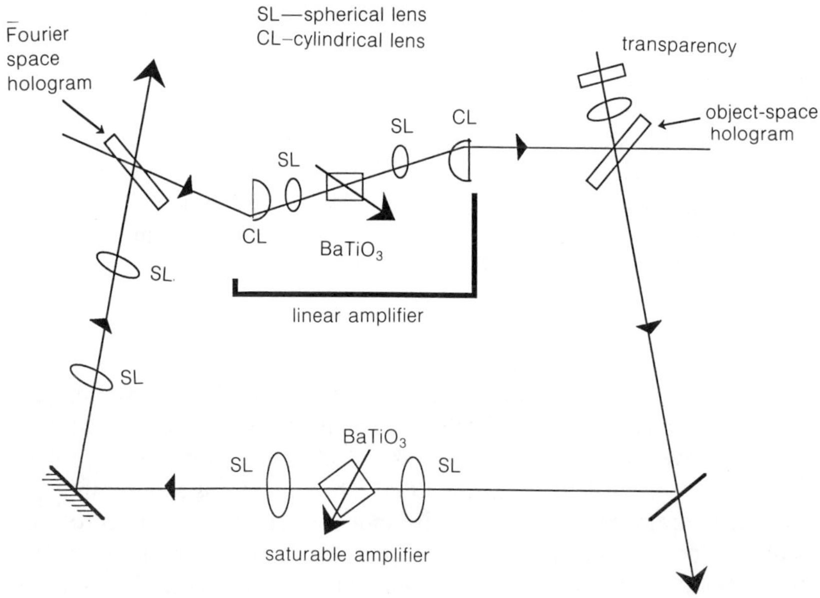

Figure 7.31 Inner product architecture using real-time holograms. (After Lee *et al.*, 1989.)

transparency) as the undiffracted light through the first (object space) hologram. The saturable $BaTiO_3$ amplifier acts as a neuron array, thresholding and amplifying the image. The amplified image then illuminates the second (Fourier space) hologram which generates the correlation signal. This is guided through a second amplifier placed in the correlation plane. The angular selectivity of the holographic process is exploited to enhance preferentially the d.c. components of the correlation peaks. The amplified correlations then illuminate the object space hologram and the diffracted wave introduces the closest matching memory pattern into the system. As before the system iterates until it stabilizes on a particular memory. The amplifier in the correlation plane increases the flexibility of the system through the gain it imparts to the correlation signals. A $BaTiO_3$ crystal with transverse dimensions around 5.5 mm × 6 mm, acting as a saturable amplifier, can implement over 500 neurons in a two-dimensional array. The use of $LiNbO_3$ holograms in principle allows old data to be removed and new information added making this system updateable.

Photorefractive materials such as $BaTiO_3$ can act as real-time phase conjugate mirrors (PCMs) (Hall *et al.*, 1985). A PCM generates a reflection that is the spatial conjugate of the input beam, i.e. each point on the reflected wave propagates back along the same path taken by the corresponding point of the input wave. The distinction between this and an ordinary mirror is shown in Figure 7.32. The important point about phase conjugate reflection is that any phase changes incurred by the incoming light on the course of its journey will be reversed as the conjugate travels back along the path. This means that any phase distortions introduced by the optical system will be corrected and an undistorted image produced at the input to the system.

The PCM is formed when two counterpropagating light fields interact within the photorefractive material along with a third, weaker beam. The counterpropagating light fields can either arise spontaneously from scattering within the material or result from the interaction of two external 'pump'

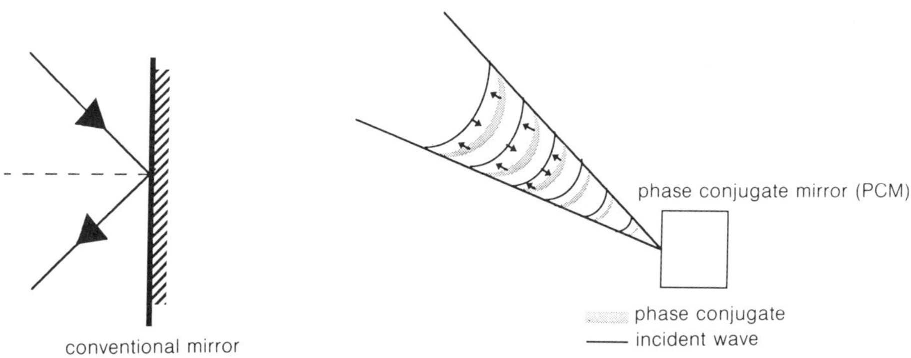

Figure 7.32 Reflection from a phase conjugate mirror (PCM).

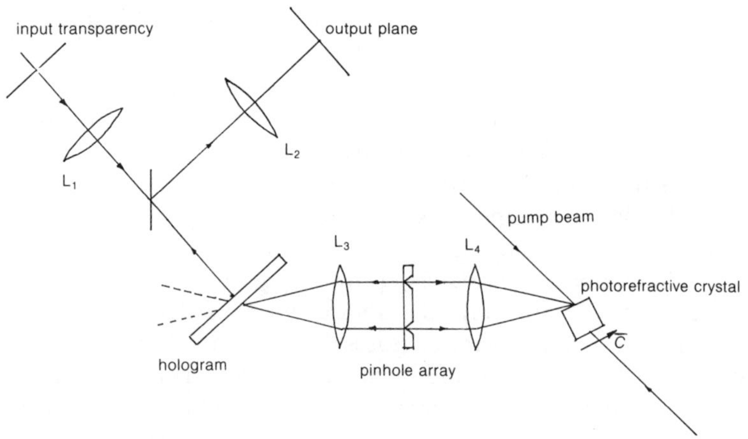

Figure 7.33 Inner product architecture using PCMs. (After White *et al.*, 1988).

beams. In the latter case, power can be coupled from the pump beams into the reflected field, amplifying the conjugate beam. This makes photorefractive materials and PCMs in particular a very powerful tool in the implementation of iterative optical networks. Amplified phase conjugation naturally forms reflexive iterative processing architectures and produces systems which are self-aligning and tolerant to aberration and system error.

The ability to form self-aligning reflexive systems can considerably reduce the complexity of an implementation. For example, the inner product architecture becomes much less complicated if a PCM is used (Soffer *et al.*, 1986; Yariv and Kwong, 1986; Bostel *et al.*, 1989). The system is shown in Figure 7.33. The Fourier transform of the input image illuminates the hologram and the diffracted correlations are filtered through a pin-hole array and focused into a PCM. This preferentially amplifies the strongest correlation and sends the conjugate signal back through the system to reconstruct the hologram with the closest match to the input. This output can be focused into another PCM which amplifies the image and returns it to the system to form the iterative loop. The advantages of this system are clear. Not only are the signals amplified but aberrations introduced by the optical system are corrected and the system self-aligns as a result of the phase conjugation operation. In addition, the retroreflective scheme uses fewer components, i.e. only a single hologram is required, and simplifies the construction. The results of implementing a single pass of the above scheme are shown in Figure 7.34. The perfect recall from the single pass results from the high gain of the optical system (Bostel *et al.*, 1989). The actual configuration employed is shown schematically in Figure 7.35.

The use of nonlinearities in the correlation plane improves the discriminatory capabilities of the system (Athale *et al.*, 1986; Psaltis and Hong,

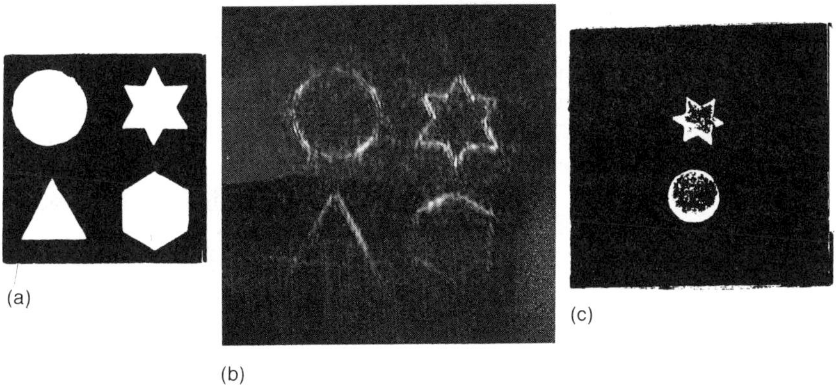

Figure 7.34. Experimental results from the system in Figure 7.35: (a) stored pattern; (b) output; (c) input.

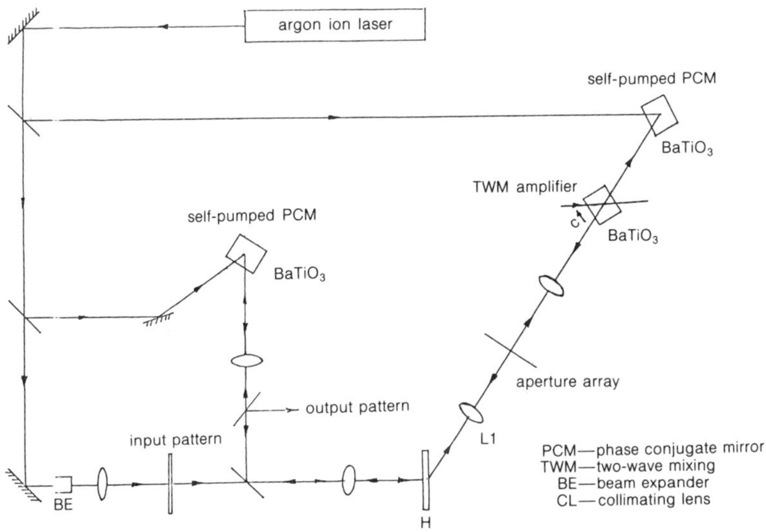

Figure 7.35 Experimental associative memory system.

1987). The nonlinearity allows the system to distinguish between memory patterns that have much higher cross-correlations. In this respect the inner product scheme with nonlinearities in the correlation domain becomes analogous to the so called higher-order neural network models (Psaltis *et al.*, 1988c). These are generalizations of the ordinary network structure and have connection patterns which become tensors instead of simple matrices. The implementation of these systems is more complicated but is essentially based on schemes described so far. A more detailed description of the optical implementation of higher-order networks can be found in the work of Psaltis *et al.* (1988c).

The function of the inner product associative memory can be generalized to the mapping of an arbitrary input pattern onto a different but associated output pattern. This system is called a heteroassociative memory and can be used for classifying input data. The architecture of the original inner product implementation can be transformed into a heteroassociative memory by exchanging the second hologram in the system for one containing the desired output patterns. These patterns are arranged to correspond to particular images on the first hologram. When the input produces a strong correlation with one of the templates stored on the first hologram, its corresponding output pattern will be strongly reconstructed. Thresholding at the output will select the pattern with the strongest reconstruction. The scheme is shown schematically in Figure 7.36. An optically addressed SLM is placed in the

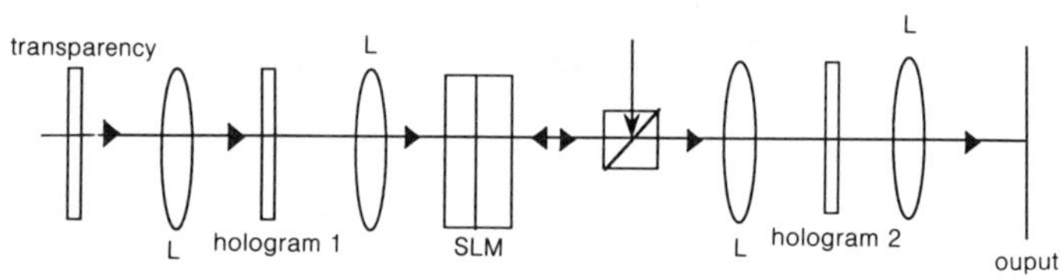

Figure 7.36 Inner product heteroassociate memory.

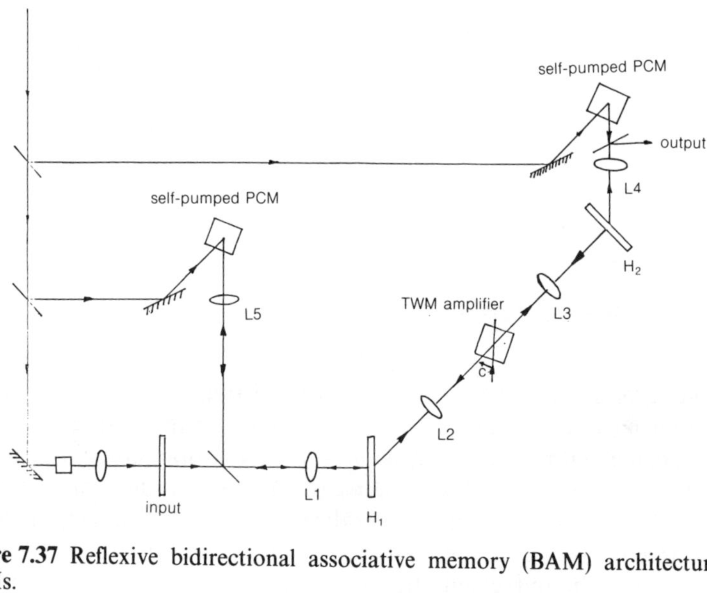

Figure 7.37 Reflexive bidirectional associative memory (BAM) architecture using PCMs.

correlation plane to help discriminate between the correlations by thresholding the light distribution. The resulting signal is then amplified by using a strong read beam to interrogate the SLM and to reconstruct the second hologram. This system can also be implemented by using a joint-transform correlator in the first stage (Song and Yu, 1989). This architecture can no longer be iterated since the input and output patterns are different. Iteration can be re-established if a second system performing the inverse mapping is included in the feedback arm; the input and feedback patterns then become compatible. This structure is identical to Kosko's BAM. One suggestion for implementing this scheme is to use PCMs in a reflexive architecture (Bostel *et al.*, 1989). The scheme is illustrated in Figure 7.37. A photorefractive amplifier is shown in the correlation plane to provide the thresholding and gain required.

7.5. ADAPTIVE SYSTEMS

A learning machine is an automatic system that can be trained to perform a task by using typical examples of the problem in order to learn the rules which underlie its operation. It is hoped that such a system will then be able to generalize these rules to other examples of the task that were not originally included in the training set. Neural networks are learning machines whose structure offers the possibility for direct implementation (Farhat, 1989).

In a neural network the learning procedures generally attempt to modify the connection weights. This is done in order to force the network to generate specific output patterns when particular inputs are present. There are many algorithms which can accomplish this. Some procedures take the form of 'training with a teacher' in which the response of the network is supervised by an external agent, while others are 'self-organizing' in the sense that the network is allowed to formulate its own representation in response to the input.

From the implementational viewpoint, the fact that adaptation is accomplished through connection modification is very attractive. Optical systems are particularly appealing because of the possibilities they provide for reconfigurable interconnections. An optical network can be made adaptive if a spatial light modulator or real-time hologram is used to form the connections.

The most frequently examined learning algorithms are usually supervised rules employing a teaching procedure. This is because they tend to be the simplest to implement. The two procedures that are used most often are the Hebbian rules (Rumelhart and McCelland, 1986; Hebb, 1949; Fisher *et al.*, 1987) and the perceptron algorithms such as back-propagation (Rumelhart and McCelland, 1986; Szu, 1986; Widrow *et al.*, 1988; Rumelhart *et al.*, 1986; Fisher *et al.*, 1987).

7.5.1. Hebbian learning

In the Hebbian procedure, the outer product matrix $(\mathbf{V}\mathbf{U}^{\mathrm{T}})^k$ is generated. **V** is the desired output vector for input **U** in the kth memory. This is added to the $k-1$ memories already recorded by the system. The updating rule for the connection matrix is thus

$$M_{n+1} = M_n + g_1[\mathbf{V}^k(\mathbf{U}^k)^{\mathrm{T}}] \tag{7.19}$$

where M_n is the connection matrix after n pairs of patterns have been presented, and g_1 is a gain function related to the learning rate.

The outer product matrix for two one-dimensional vectors can be formed optically using two 1D SLMs and a suitable anamorphic imaging system (Athale and Lee, 1984). The scheme is shown in Figure 7.38. Light from the first SLM is expanded vertically and focused horizontally onto the second SLM which is mounted perpendicularly to the first. This transmits the light so that the image at the recording plane is the display on the first SLM modulated by the transmittance of the second. If the first SLM displays the vector **V** and the second SLM the vector **U**, the pattern formed in the recording plane is the outer product of the two vectors $\mathbf{V}\mathbf{U}^{\mathrm{T}}$. A 2D recording medium such as an SLM with memory or a photographic plate or photorefractive crystal can be placed in the recording plane to store the pattern. The full connection matrix is slowly built up by superimposing the outer product images for each of the desired memory vectors on the storage device. The device can then be used to recall a specific output vector by using it to

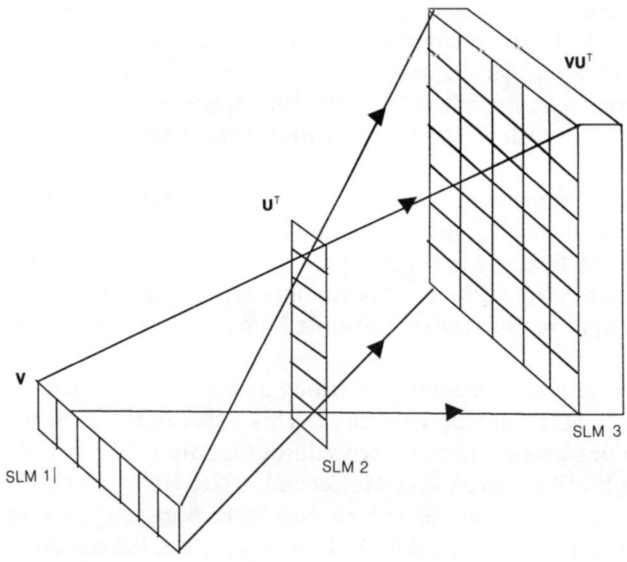

Figure 7.38 Optical generation of outer product matrix.

modulate a beam of light carrying the input pattern. This procedure could be used to update systems implementing associative memories such as the Hopfield model.

7.5.2. The perceptron algorithm

The Hebbian learning rule does not store information in an optimal fashion. This leads, in general, to a relatively low storage capacity or alternatively a restriction on the classes of vectors it can store efficiently. The Widrow–Hoff or least-mean-square (LMS) algorithm (Widrow *et al.*, 1988; Fisher *et al.*, 1987) is a method of storing general classes of vector in an optimal way. The learning rule is of the form

$$\frac{\mathrm{d}M}{\mathrm{d}t} = g_2(\mathbf{V}^k - \mathbf{V})(\mathbf{U}^k)^{\mathrm{T}} \tag{7.20}$$

where g_2 is the gain factor and $\mathbf{V}$ is the actual output from the system. This equation can be more conveniently implemented when it is expressed as a difference equation:

$$M_{n+1} = M_n + g_3\mathbf{V}^k(\mathbf{U}^k)^{\mathrm{T}} - g_3\mathbf{V}(\mathbf{U}^k)^{\mathrm{T}} \tag{7.21}$$

where $g_3 = g_2\,\Delta t$ and Δt is the time interval over which the patterns are held constant. This algorithm can be implemented using an MSLM (Fisher *et al.*, 1987). The MSLM can be used to store the outer product $\mathbf{V}(\mathbf{U}^k)^{\mathrm{T}}$ which can be generated optically as before. The recall operation which extracts the actual output is performed by reflecting the horizontally expanded vector $\mathbf{U}^k$ from the MSLM and summing the resulting light distribution vertically. The term $-g_3\mathbf{V}(\mathbf{U}^k)^{\mathrm{T}}$ is produced by using the system's actual output in place of the memory pattern $\mathbf{V}^k$ to generate the outer product and running the MSLM in its subtractive mode. The whole system is shown schematically in Figure 7.39. The process is controlled by a shutter which blocks the feedback path during the recording of the desired output pattern. When the actual output is being used, the first modulator is left unmodulated. Additional gain can be introduced by including a second optically addressed SLM into the system to enable the light from the input vector modulator to act also as the read-out source in the recall phase. This is a slightly different arrangement to the system described by Fisher *et al.* (1987).

7.5.3. Holographic learning systems

The use of holography has already been mentioned in the context of neural network implementation. It was shown how the introduction of real-time holography using photorefractive materials can simplify the design of some networks by exploiting the phenomena of phase conjugation. Real-time holograms have an even more important role to play in optical neural

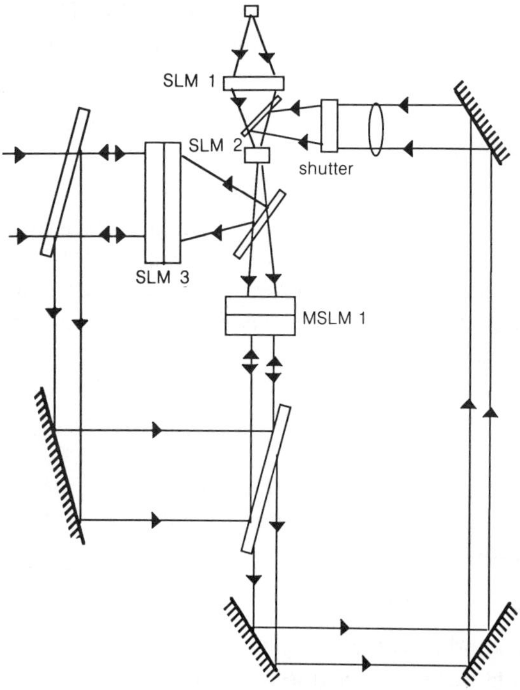

Figure 7.39 Widrow–Hoff learning system. (After Fisher *et al.*, 1987.)

networks, however. If they are used to store the connection patterns, they not only allow two-dimensional networks to be formed but also make the implementation of learning algorithms possible (Wagner and Psaltis, 1987; Psaltis *et al.*, 1988b; Pack *et al.*, 1989; Psaltis *et al.*, 1990a). In its simplest form it is envisioned that a single real-time hologram will perform all the desired mappings for each neuron in the input array. This is an $N \rightarrow N^2$ transformation. If the neurons are arranged in a two-dimensional array with $M(=\sqrt{N})$ neurons per side, then this mapping operation becomes one of $M^2 \rightarrow M^4$. Thus the hologram needs to specify M^4 connections between the input and output planes. However, if it is assumed that an elementary volume specifies a single mapping then only M^3 mappings will be possible using a single hologram. This somewhat arbitrary definition of storage capacity can be explained as follows (Psaltis *et al.*, 1987, 1990a). When a single hologram is used to record all the connection patterns, the information is mixed together and relies on the selectivity of the recording medium to distinguish the different sets of data. This selectivity is related to the dimensions of the recording media through the number of degrees of freedom available to specify mappings. When the hologram has to incorporate complicated phase information, as in the case of two-dimensional connection patterns, the amount of space left to specify connection values is reduced. In other words if A_1 is the area of the input plane and A_2 the area of the output plane, then the maximum number of points that can be specified optically is

$$\frac{A_1 A_2}{\lambda^4}$$

where λ is the wavelength and λ^2 is the area of the smallest possible spot. If a photorefractive crystal has a volume V_{H}, then the total number of samples of wavelength size that can be contained in the crystal is V_{H}/λ^3. Thus in order for the crystal to implement the desired mapping

$$V_{\mathrm{H}} > \frac{A_1 A_2}{\lambda} \tag{7.22}$$

If the dimensions of the input and output planes are the same then, from equation 7.22, only one point in the output plane can be specified by a wavelength scale cube. Thus, for a crystal of volume $N^3\lambda^3$, the number of output points is given by

$$\frac{A^2}{\lambda} = N^3\lambda^3 \tag{7.23}$$

with $A = m\lambda^2$, where m is the total number of output points that can be specified by the hologram. This gives $m = N^{3/2}$ or M^3, which is the maximum number of mappings that can be made by the hologram. In other words, if the dimensions of the input and output planes are to be the same, only $N^{3/2}$ of the N^2 possible points can be used.

In order to prevent cross-talk, each pair of input and output points must use a separate grating in the crystal. This means that in order to form arbitrary connections the $N^{3/2}$ points on the input and output planes must be selected systematically from the N^2 possibilities (Psaltis *et al.*, 1987). The choice of points is arbitrary. For example, the $N \times N$ array could be split up into blocks with each block containing identical numbers of points (the total number of points being $N^{3/2}$) arranged into a set pattern. The positions of the output points are now selected to break the symmetry between input and output. This can be done, for example, by cyclically permuting the pattern within each block. In this way the holograms formed for each input–output contribution are different.

If several holograms are to be stored without needing to be constantly rewritten, the material employed must have a long dark storage time. $LiNbO_3$ crystals are often used for this reason. A system for storing a set of two-dimensional connection patterns within a single-volume hologram using a Hebbian storage prescription is shown in Figure 7.40 (Psaltis *et al.*, 1987). In the recording phase, the connections are established by recording a set

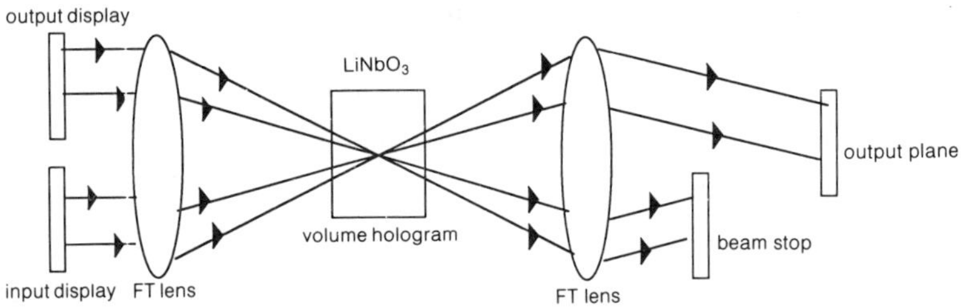

Figure 7.40 Holographic learning system. (After Psaltis *et al.*, 1987.)

of holograms using the input patterns and their desired outputs. These patterns consist of arrangements of points in the allowable positions described above. These patterns are placed in the input plane in a joint-transform arrangement. The complete connection hologram is built up by exposing the crystal to each input–output pair in turn until the full set has been presented. In the replay phase an input pattern is presented and a pattern is generated in the output plane which should be the desired image.

The perceptron algorithm can also be implemented using photorefractive crystals (Pack *et al.*, 1989; Psaltis *et al.*, 1988b). The LMS procedure can be implemented by a simple two-layer network with the connections between the layers specified by a real-time hologram. One possible architecture using this algorithm is shown in Figure 7.41. The system can be used to classify two-dimensional patterns by a one-dimensional output vector (Pack *et al.*, 1989). The dimensionalities of the input and output planes are no longer the same and so sampling the input plane is no longer required. An $LiNbO_3$ hologram is used to store the weights. Negative weights are implemented using electronic subtraction at the 'neuron'. In operation the input pattern is displayed on a monitor and transferred onto a coherent beam via an SLM. This pattern is then imaged into the crystal where the diffracted image is the classification pattern for the input. This is summed in one dimension and imaged onto a detector array. The positive and negative elements are subtracted and the result is thresholded to yield the classification signal. The system is trained to generate specific output classifications by implementing the perceptron learning algorithm.

The set of training patterns is presented to the system one at a time. From random initial conditions (due to background noise and scattering from the crystal) the required connection strengths are slowly built up by measuring the diffracted output, comparing it electronically with the target pattern and displaying the error signal on the one-dimensional array of modulators (LCMs). These are focused into the crystal where the error signal interferes with the input to rewrite the hologram. After each presentation of the error the signal is blocked to allow only the measurement of the diffracted beam. After training, the final connection patterns can be recorded and stored for

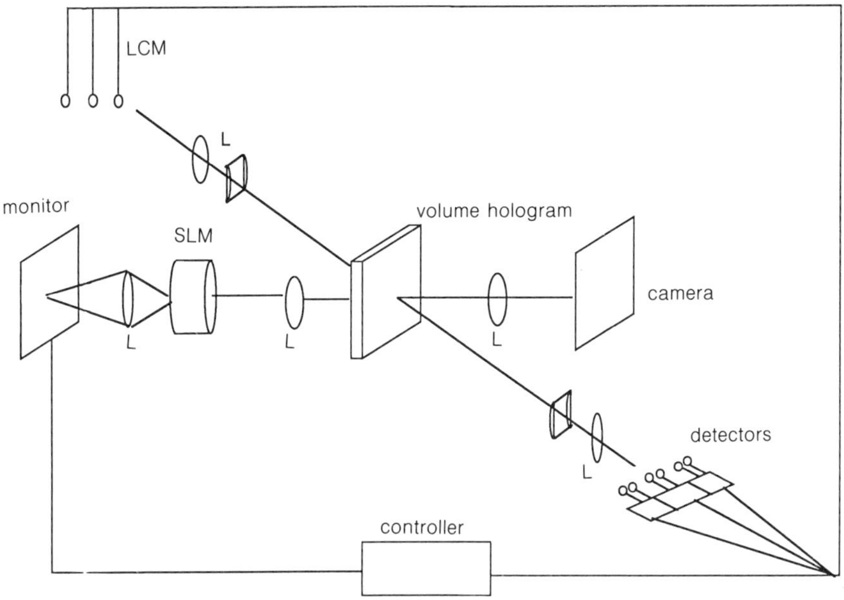

Figure 7.41 Holographic perceptron. (After Pack *et al.*, 1989.)

later use by reconstructing the hologram with each LCM separately. The image can be recorded on an optical disk and reloaded into the system when that set of classifications is required again. This is necessary since photorefractive crystals have a finite decay time which means that with use the holograms will eventually fade away. This system has been experimentally demonstrated using five capital letters as the training patterns (Pack *et al.*, 1989).

The perceptron algorithm can also be implemented in a fashion which is suitable for extension to multilayer systems (Wagner and Psaltis, 1987). A possible architecture is shown in Figure 7.42. The system consists of sets of input and output elements arranged in sparse two-dimensional arrays. The output device must be capable of injecting an optical signal back into the system. The hologram is a photorefractive crystal such as $Bi_{12}SiO_{20}$, $LiNbO_3$, or $BaTiO_3$ which operates a polarization switching diffraction mechanism. In operation light from the input array is directed towards the output with its polarization rotated to the orthogonal state. The undiffracted light retains its original polarization and is reflected back towards the hologram by a phase conjugate mirror. A set of training patterns are presented to the system in a random sequence. The resulting output patterns are thresholded and compared with their desired outputs. An error signal is injected back into the system by the output devices and this records a new hologram in the crystal using the phase conjugate of the input as the reference beam. This

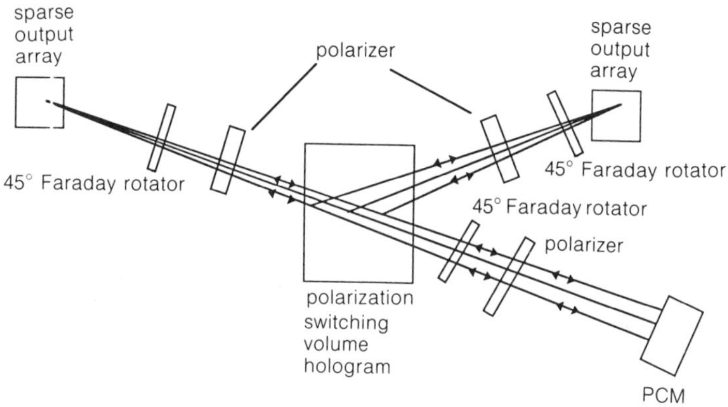

Figure 7.42 Optical back-propagation system. (After Wagner and Psaltis, 1987).

hologram modifies the connectivity by selectively erasing or strengthening the existing holographic grating. The connection strength between two particular devices can be decreased by recording a new hologram which is shifted in phase by π with respect to the old one. This phase shift can be introduced directly on to the error signal or, in cases where there is an external bias field applied to the crystal, by recording the new hologram with the field in the reverse direction. Alternatively the natural decay of the crystal can be exploited to reduce the connection. Connection strengths are increased if the new hologram is in phase with the old one. In this case the new gratings add to the old one, producing a stronger connection. The polarization of the input vector is rotated so that it is perpendicular to the polarization of the output and so the forward and backward travelling signals do not interfere. In this case the polarization of the phase conjugate must also be rotated to ensure that a hologram is still formed between them. This is accomplished by the use of the 45° Faraday rotator which is placed in the path of the undiffracted beam.

At the same time as the error is forming a new hologram, the existing grating is diffracting the error pattern back towards the input array. This makes this particular architecture attractive for multilayer systems as this beam could be used as the error signal for the preceding layers. Unfortunately the phase conjugate reflection also sends some light towards the input and this must be removed if the error pattern is to be used successfully. This may be accomplished by placing a second 45° Faraday rotator in the output arm. The polarization of the error beam will then be in the same plane as the output when it leaves the crystal. The diffracted error signal now possesses the same polarization as the input which is polarized orthogonally to the conjugate. Thus, the backward propagating conjugate reflection may be removed, without affecting the rest of the system, by placing a suitably orientated analyser in the input path. Another analyser can also be included in the output arm in order to eliminate that part of the phase conjugate which is diffracted towards the output array. If a third 45° Faraday rotator is placed in the input arm the polarization of the backward travelling error wave will be shifted by $-45°$. Since the forward propagating input wave is required to have the same polarization at the crystal as the error beam it must originally have been polarized at $+45°$. Thus at the input plane the forward and backward travelling beams will have orthogonal polarization. This is consistent with the output plane and allows the system to be cascaded more easily. Finally, an analyser placed before the phase conjugate mirror will reduce the unwanted reflections by eliminating light of the wrong polarization.

By repeated presentation of the input patterns the system slowly forms a hologram which performs the desired input–output transformation for each of the patterns. When this occurs the error signal will, for all intents and purposes, be zero and the hologram will no longer change. For the algorithm to converge to give the desired result this procedure must occur slowly (Rumelhart and McCelland, 1986; Wagner and Psaltis, 1987; Rumelhart *et al.*, 1986). However, since the response times of ferroelectric photorefractive crystals are slow and a single exposure can be arranged to cause only a small disturbance to existing holograms, it should be possible to find an operating regime in which the learning proceeds at the correct rate. Because new holograms are formed using the phase conjugate beam the system is self-aligning and should learn to correct for some of the aberrations present in the optical components.

The single-layer system can only formulate a transformation hologram if the training patterns are linearly separable (Lippman, 1987). This means that the patterns must fall into categories which are distinguished by boundaries formed from linear functions of the possible neuron values. This restricts the system to fairly simple classifications. The more interesting problems require much more complex decision surfaces. These are generally defined by nonlinear functions of the variables. These functions can be implemented by using several layers which can be trained in a similar way to the above using the back-propagation algorithm. In this procedure, the backward travelling error signals are modulated by the derivative of the transmission function for the feedforward system. This complicates the implementation since suitable nonlinear optical devices have to be found. There have been several suggestions for possible devices capable of implementing these two distinct modes of transmission (Wagner and Psaltis, 1987; Kagan and Friedman, 1989). They range from specially designed liquid crystal devices to biased nonlinear Fabry–Pérot cavities.

A variant of the perceptron algorithm can be applied to the inner product associative memory system. The basic strategy of back propagation can be adopted to update the memory matrix by using it as a means of erasing old

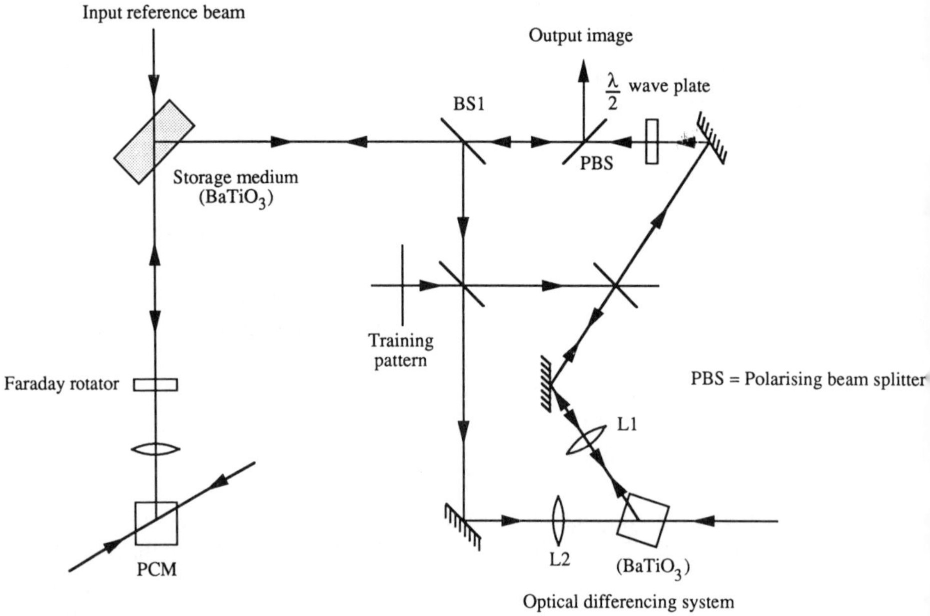

Figure 7.43 Reconfigurable inner product connection matrix using error-driven learning.

memories and replacing them with new ones (Psaltis *et al.*, 1987). A possible optical system for implementing this procedure is suggested in Figure 7.43. The set of memories is stored in a volume hologram and a particular memory is extracted by using a reference beam of the appropriate phase. The output from the system is diffracted into an optical differencing system (Yeh *et al.*, 1988). This consists of a PCM in an interferometer configuration. The output from the hologram is introduced into one arm of the interferometer and the new pattern, which replaces it, into the other. The differencing system generates an error signal which propagates back towards the hologram where it interferes with the phase conjugate of the reference beam. The two beams rewrite the original hologram, gradually erasing the old recording and substituting the new one. The exposure of the hologram must be controlled to ensure that the process does not proceed too quickly otherwise it will tend to destroy other memories in the hologram. Because the hologram is used in a correlator system, the dimensionalities of the input and output images from the hologram are different. The patterns stored in the system therefore do not have to be sampled as drastically as in the other architecture.

7.6. CONCLUSIONS

The network models described in this chapter are very simple. This is partly a reflection of the current level of understanding of the way in which highly interconnected processing units can be made to function, and partly due to the infancy of the necessary optical technology. It is this simple structure, however, coupled with the rich processing behaviour of these systems, which makes them interesting to implement.

Although discussion has concentrated around the Hopfield model and the perceptron, the methods of implementation are quite general and should be applicable, with minor modifications, to most other network models. The Hopfield model and the perceptron are used because they are the simplest and most frequently used examples of neural networks and learning algorithms.

Advances on current implementations will depend on developments in optical technology and progress in the theoretical understanding of the way in which neural systems operate. More specifically, new models are needed which reflect the methods used to construct them. This should make their implementation easier and improve their capabilities. On the other hand, the challenge of implementing more complicated algorithms could also stimulate new developments in optical technology.

The current trend is for computers to become hybrid optoelectronic systems with electronic switching and optical connections. This is particularly relevant with the growing interest in parallel computing with its increased demand on communication. The optical implementation of a neural network forms the basis for far more complex hybrid parallel systems. They therefore provide an ideal test-bed for the ways in which optics can be used to connect these structures. In addition to this, they also open up new avenues for all-optical processing structures which show promise for the future.

The processing and learning capabilities of artificial neural networks are of great interest and could lead to a fundamental re-evaluation of the way we think about computers. The neural processing systems that have been investigated so far represent the elementary stages of the evolution of highly interconnected computer systems. They will inevitably become much more complicated. What the present systems provide are some interesting capabilities in a form that is relatively easy to implement. How they develop is impossible to foresee accurately. In the future, however, a computer may consist of an artificial neural network system working in conjunction with a more conventional computer. The former would perform perceptual tasks such as interpretation, optimization and control, while the latter would carry out the more cumbersome numeric and memory functions. In a large system there could be thousands of neurons and millions of connections, not to mention the connections within its conventional partner. If the current trend for hybrid optoelectronic systems continues, the simple implementations described in this chapter could form the basis of future computer architectures.

ACKNOWLEDGEMENTS

The authors are grateful to H. Imam, C. Stace and A. Kirk without whom the work would not have been possible.

REFERENCES

Athale, R. A. and Lee, J. N. (1984) Optical processing using outer-product concepts, *Proc. IEEE*, **72**, 931–41.

Athale, R. A., Szu, H. H. and Friedlander, C. B. (1986) Optical implementation of associative memory with controlled nonlinearity in the correlation domain, *Opt. Lett.*, **11**, 482–4.

Barhen, J., Toomarian, N. and Protopopescu, V. (1987) Optimisation of the computational load of a hypercube supercomputer onboard a mobile robot, *Appl. Opt.*, **26**, 5007–14.

Bostel, A. J., Stace, C., Swinburn, G. and Hall, T. J. (1989) Optical computing based on neural systems. In *Proc. Int. Symp. on Optics in Computing, SFO. Toulouse, October 17–18, 1989*, pp. 187–94.

Bounds, D. G. (1987) A statistical mechanical study of Boltzmann machines, *J. Phys. A*, **20**, 2133–45.

Bradley, E., Yu, P. K. L. and Johnston, A. R. (1989) System issues relating to laser diode requirements for VLSI holographic optical interconnections, *Opt. Eng.*, **28**, 201–11.

Collins, D. R., Sampsell, J. B., Hornbeck, L. J. *et al.* (1989) Deformable mirror device spatial light modulators and their applicability to optical neural networks, *Appl. Opt.*, **28**, 4900–7.

Collins, S. A., Jr. (1988) Optical computing with spatial light modulators. In *Optical Computing, Proc. 34th Scottish Universities Summer School in Physics, Heriot-Watt University, Edinburgh* (eds B. S. Wherrett and F. A. P. Tooley), Scottish Universities Summer School in Physics, pp. 23–54.

Collings, N., Crossland, W. A., Ayliffe, P. J. *et al.* (1989) Evolutionary development of advanced liquid crystal spatial light modulators, *Appl. Opt.*, **28**, 4740–7.

Cronin-Golomb, M. and Yariv, A. (1986) Phase-conjugate mirrors as thresholding elements for optical associative memories. In *SPIE Proc. Int. Optical Computing Conf.*, pp. 301–3.

Derthick, M. (1984) Variations on the Boltzmann machine learning algorithm, *Carnegie–Mellon University Tech. Rep. CMU-CS-84-120.*

Farhat, N. H. (1987) Opto-electronic analogs of self-programming neural nets: architecture and methodologies for implementing fast stochastic learning by simulated annealing, *Appl. Opt.*, **26**, 5093–103.

Farhat, N. H. (1989) Optoelectronic neural networks and learning machines, *IEEE Circ. Dev. Mag.*, September, 32–41.

Farhat, N. H., Psaltis, D., Prata, A. and Paek, E. (1985) Optical implementation of the Hopfield model, *Appl. Opt.*, **24**, 1469–75.

Feldman, M. R. and Guest, C. C. (1987) Computer generated holographic optical elements for optical interconnection of very large scale integrated circuits, *Appl. Opt.*, **26**, 4377–84.

Fisher, A. D., Lippincott, W. L. and Lee, J. N. (1987) Optical implementations of associative networks with versatile adaptive learning capabilities, *Appl. Opt.*, **26**, 5039–54.

Geman, S. and Geman, D. (1984) Stochastic relaxation, Gibbs distributions, and Bayesian restoration of images, *IEEE Trans. Pat. Anal. Mach. Intel.*, **6** (6), 721–41.

Goodman, J. W., Dias, A. R., Woody, L. M. and Erickson, J. (1979) Application of optical communication technology to optical information processing, *Proc. Soc. Photo-Opt. Instrum. Eng.*, **190**, 485–96.

Guest, C. C. and Tekolste, R. (1987) Designs and devices for optical bidirectional associative memories, *Appl. Opt.*, **26**, 5055–60.

Hall, T. J., Jaura, R., Connors, L. M. and Foote, P. D. (1985) The photorefractive effect—a review, *Prog. Quantum Electron.*, **10**, 77–146.

Hebb, D. O. (1949) *The Organization of Behavior*, Wiley, New York.

Hinton, G. F., Sejnowski, T. J. and Ackley, D. H. (1984) Boltzmann machines: constraint satisfaction networks that learn, *Tech. Rep. CMU-CS-84-119*, Department of Computer Science, Carnegie–Mellon University.

Hopfield, J. J. (1982) Neural networks and physical systems with emergent collective computational abilities, *Proc. Natl. Acad. Sci. USA*, **79**, 2554–8.

Hopfield, J. J. (1984) Neurons with graded response have collective computational properties like those of two-state neurons, *Proc. Natl. Acad. Sci. USA*, **81**, 3088–92.

Hopfield, J. J. and Tank, D. W. (1986) Computing with neural circuits: a model, *Science*, **233**, 625–33.

Huang, T. S. (1971) Digital holography, *Proc. IEEE*, **59**, 1335–46.

Jang, J., Shin, S. and Lee, S. (1988a) Optical implementation of quadratic associative memory with outer-product storage, *Opt. Lett.*, **13**, 693–5.

Jang, J., Jung, S., Lee, S. and Shin, S. (1988b) Optical implementation of the Hopfield model for two-dimensional associative memory, *Opt. Lett.*, **13**, pp. 248–50.

Johnson, K. M. and Moddel, G. (1990) Motivations for using ferroelectric liquid crystal spatial light modulators in neurocomputing, *Appl. Opt.*, **28**, 4888–99.

Kagan, D. and Friedman, H. (1989) Back propagating neurons from bichromic interaction with a three level system, *Appl. Opt.*, **28**, 1697–1700.

Kirkpatrick, S., Gelatt, C. D. and Vecchi, M. P. (1983) Optimisation by simulated annealing, *Science*, **220**, 671–80.

Kohonen, T. (1984) *Self Organization and Associative Memory*, Springer, New York.

Kosko, B. (1988) Bidirectional associative memories, *IEEE. Trans. Sys. Man. Cybern.*, **18**, 49–60.

Lee, L. S., Stoll, H. M. and Tackitt, M. C. (1989) Continuous-time optical neural network associative momory, *Opt. Lett.*, **14**, 162–4.

Lippmann, R. P. (1987) An introduction to computing with neural nets, *IEEE ASSP Mag.*, April, 4–22.

Lu, T., Choi, K., Wu, S., Xu, X. and Yu, F. T. S. (1989) Optical disk based neural network, *Appl. Opt.*, **28**, 4722–4.

Luttrell, S. P. (1985) The implications of Boltzmann-type machines for SAR data processing: a preliminary survey, *RSRE Memo. 3815.*

Marks II, R. J., Atlas, L. E., Oh, S. and Cheung, K. F. (1988) Optical processor architectures for alternating-projection neural networks, *Opt. Lett.*, **13**, 533–5.

McKenzie, D. S. and Sagi, M. A. S. (1986) The random bond Ising model on the Bethe lattice, *J. Phys. A*, **19**, 3883.

Miller, D. A. B. (1988) Quantum well electro-absorptive devices: physics and applications. In *Proc. 34th Scottish Universities Summer School in Physics, Heriot-Watt University, Edinburgh* (eds B. S. Wherrett and F. A. P. Tooley), Scottish Universities Summer School in Physics, pp. 71–94.

Oita, M., Ohta, J., Tai, S. and Kyuma, K. (1990) Optical implementation of large-scale neural networks using a time-division multiplexing technique, *Opt. Lett.*, **15**, 227–9.

Pack, E. G., Wullert II, J. R., and Patel, J. S. (1989) Holographic implementation of a learning machine based on a multicategory perceptron algorithm, *Opt. Lett.*, **14**, 1303–5.

Pack, E. G., Wullert II, J. R., Jain, M. *et al.* (1990) Compact and ultrafast holographic memory using a surface emitting microlaser diode array, *Opt. Lett.*, **15**, 341–3.

Peltier, M. and Micheron, F. (1977) Volume hologram recording and charge transfer process in $Bi_{12}SiO_{20}$ and $Bi_{12}Ge_{20}$ (BSO and BGO), *J. Appl. Phys.*, **48**, 3683–90.

Powell, M., Powles, C. and Bagshaw, J. (1988) The effect of read/write isolation on the resolution of the Marconi spatial light modulator, *Proc. Soc. Photo-Opt. Instrum. Eng.*, **936**, 68–75.

Prager, R. W., Harrison, T. D. and Fallside, F. (1986) Boltzmann machine for speech recognition, *Comput. Speech Lang.* **1**, 3–27.

Psaltis, D. and Farhat, N. H. (1985) Optical information processing based on an associative memory model of neural nets with thresholding and feedback, *Opt. Lett.*, **10**, 98–100.

Psaltis, D. and Hong, J. (1987) Shift-invariant optical associative memories, *Opt. Eng.*, **26**, 10.

Psaltis, D., Yu, Y., Gu, G. and Lee, H. (1987) Optical neural nets implemented with volume holograms. In *Proc. Optical Computing Meet., Lake Tahoe, OSA CA, March 18–20*, pp. 129–32.

Psaltis, D., Sideris, A. and Yamamura, A. A. (1988a) A multilayered neural network controller, *IEEE Control Syst. Mag.*, April, 17–21.

Psaltis, D., Brady, D. and Wagner, K. (1988b) Adaptive optical networks using photorefractive crystals, *Appl. Opt.*, **27**, 1752–9.

Psaltis, D., Park, C. H. and Hong, J. (1988c) Higher order associative memories and their optical implementations, *Neural Networks*, **1**, 149–63.

Psaltis, D., Brady, D., Gu, X. G. and Lin, S. (1990a) Holography in artificial neural networks, *Nature* (*London*), **343**, 325–30.

Psaltis, D., Neifield, M. A., Yamamura, A. and Kobayashi, S. (1990b) Optical memory disks in optical information processing, *Appl. Opt.*, **29**, 2038–57.

Rumelhart, D. E. and McCelland, J. L. (eds) (1986) *Parallel Distributed Processing*, MIT, Cambridge, MA.

Rumelhart, D. E., Hinton, G. E. and Williams, R. J. (1986) Learning representations by back propagating errors, *Nature* (*London*), **323**, 533–5.

Shamir, J., Caulfield, H. J. and Johnson, R. B. (1989) Massive holographic interconnection networks and their limitations, *Appl. Opt.*, **28**, 311–24.

Shariv, I. and Friesem, A. A. (1989) All-optical neural network with inhibitory neurons, *Opt. Lett.*, **14**, 485–7.

Smith, S. D. (1988) Optical circuits. In *Proc. 34th Scottish Universities Summer School in Physics, Herriot–Watt University, Edinburgh* (eds B. S. Wherrett and F. A. P. Tooley), Scottish Universities Summer School in Physics, pp. 95–132.

Soffer, B. H., Dunning, G. J., Owechko, Y. and Marom, E. (1986) Associative holographic memory with feedback using phase-conjugate mirrors, *Opt. Lett.*, **11**, 118–20.

Song, Q. W. and Yu, F. T. S. (1989) Holographic associative memory system using a thresholding microchannel spatial light modulator, *Opt. Eng.*, **28**, 533–6.

Staebler, D. L. and Amodei, J. J. (1972) Coupled-wave analysis of holographic storage in $LiNbO_3$, *J. Appl. Phys.*, **43**, 1042–9.

Szu, H. (1986) Three layers of vector outer product neutral networks for optical pattern recognition, *Proc. Soc. Photo-Opt. Instrum. Eng.*, **634**, 312–30.

Tekolste, R. D. and Guest, C. C. (1987) Optical competitive neural network with Optical Feedback. In *Proc. of IEEE 1st Int. Conf. on Neural Networks, San Diego, CA, June 21–24*, pp. III625–III629.

Ticknor, A. J. and Barrett, H. H. (1987) Optical implementations in Boltzmann machines, *Opt. Eng.*, **26**, 16–21.

Von Lehman, A., Pack, E. G., Carrion, L. C. *et al.* (1990) Optoelectronic chip implementation of a quadratic associative memory, *Opt. Lett.*, **15**, 279–81.

Wagner, K. and Psaltis, D. (1987) Multilayer optical learning networks, *Appl. Opt.*, **26**, 5061–76.

White, H. J. and Wright, W. A. (1988) Holographic implementation of a Hopfield model with discrete weightings, *Appl. Opt.*, **27**, 331–8.

White, H. J., Aldridge, N. B. and Lindsay, I. (1988) Digital and analogue holographic associative memories, *Opt. Eng.*, **27**, 30–7.

Widrow, B., Winter, R. G. and Baxter, R. A. (1988) Layered neural nets for pattern recognition, *IEEE Trans. Acoust. Speech. Sig. Process.*, **36**, 1109–18.

Williams, D., Latham, S. G., Powles, C. M. J. *et al.* (1988) An amorphous silicon/chiral smectic spatial light modulator, *J. Phys. D*, **21**, 5156–9.

Yariv, A. and Kwong, S. (1986) Associative memories based on message bearing optical modes in phase conjugate resonators, *Opt. Lett.*, **11**, 186–8.

Yeh, P., Chang, T. Y. and Beckwith, P. H. (1988) Real-time optical image subtraction using dynamic holographic interference in photorefractive media, *Opt. Lett.*, **13**, 580–8.

8 Third-order nonlinear guided-wave optical devices

Ian Bennion and Martin J. Goodwin

8.1. INTRODUCTION

The efficiency of most nonlinear optical interactions is dependent on the power densities of the interacting light waves and the length over which the interaction is sustained. When the interaction is induced in a bulk sample of a material, high power density is usually achieved by bringing the incident laser beam, or beams, to a focus within the sample. Focusing to smallest spot sizes produces highest power densities, but the more strongly focused the beam, the more rapidly it diverges from the focus. Natural diffraction spreading limits the length over which a given power density can be maintained and thereby limits the achievable nonlinear interaction efficiency. This limitation may be overcome by carrying out the interaction in an optical waveguide. By confining the interacting light waves in a waveguide of small cross-sectional dimensions, typically of the order of the wavelength, very high power densities can be achieved from sources of relatively moderate power and can be maintained over long propagation distances.

The most widely familiar optical waveguide is, perhaps, optical fibre (nonlinear optics in optical fibres is described in Chapter 9) which is now put to extensive practical use, notably in long-distance trunk telecommunications networks. In standard optical fibre, light is constrained within and guided by a circular core region of refractive index higher than that of its concentric circular cladding. A simple ray optics interpretation describes light wave propagation in the fibre as a succession of total internal reflections at the core–cladding interfaces. Many other optical waveguide types are possible, however, and it is primarily with these that we shall be concerned here. A simple thin film of a transparent dielectric material supported on a substrate of lower refractive index, for example, provides one-dimensional optical confinement between the upper and lower film boundaries. Confinement in a second dimension may be achieved by masking a narrow stripe region and selectively removing all of the film except for the stripe. Subject to certain rules, the stripe may be defined in an arbitrarily shaped path along which injected light will be guided. More complex patterns may be produced containing multiple stripes, and in some cases junctions between them, forming passive waveguide devices such as power splitters, combiners, directional couplers, interferometers, resonators, and others.

Techniques and devices based on thin film and stripe optical waveguides have been referred to collectively as **integrated optics** since the late 1960s since when they have been extensively researched for many applications (Miller, 1969; Tien, 1971). From the outset, there have been two independent motivations for integrated optics. Firstly, as we have indicated above, it was recognized that optical waveguides offered the means to circumvent the unattractively high pump powers required for nonlinear optics, particularly second harmonic generation, and likewise the high voltages demanded by bulk electro-optic devices. Secondly, it was perceived that passive and active guided-wave optical devices might be combined into compact monolithic circuits capable of complex functions after the fashion of integrated electronics. In the case of electro-optics, these two paths have converged and single-substrate, low-voltage circuits of considerable complexity performing modulation and switching functions have been demonstrated for many applications in recent years (Bennion, 1989; Bennion and Walker, 1990). Second-order guided-wave nonlinear optics has been pursued with great success in lithium niobate (Sohler *et al.*, 1986; Regener and Sohler, 1988). Major efforts are now devoted to harnessing the potentially large optical second-order nonlinearities of organic materials for integrated optics.

Third-order ($\chi^{(3)}$) nonlinear integrated optics is a more recent extension of the field which seeks to access the potentially subpicosecond response times of the reactive nonlinearity for very fast signal processing devices (Stegeman *et al.*, 1988). The earliest proposals for devices were based on configurations familiar from linear integrated optics, the directional coupler (Jensen, 1982) and the Mach–Zehnder interferometer (Lattes *et al.*, 1983), wherein all-optical functionality was derived from the intensity-dependent refractive index of the waveguide material. These and other device structures are discussed below.

The second half of the 1970s saw an upsurge in interest in optical devices exhibiting bistable behaviour, offering attractive possibilities in optical switching and computing. The majority of bistable devices were configured around a Fabry–Pérot cavity exhibiting a third-order optical nonlinearity. In a theoretical treatment, however, Kaplan discussed an alternative arrangement (Kaplan, 1977): he predicted that the reflectivity of the interface between two dielectric media, one of which has a pronounced $\chi^{(3)}$ optical nonlinearity, would under certain conditions exhibit hysteresis and bistability. With no resonator involved, this effect was seen as being able to operate on input signals with wide spectral linewidths with an ultimate speed limited only by the intrinsic response time of the material nonlinearity. Although subsequent experiments revealed instabilities which largely militated against its usefulness (Smith and Tomlinson, 1984), both the nonlinear interface and the resonant bistable devices motivated investigation of their guided-wave counterparts.

In subsequent sections, our objective is to provide an introduction to third-order nonlinear guided-wave structures with emphasis on device behaviour and minimal recourse to mathematical formulation. In each section, we introduce first the relevant linear waveguide concepts. We have provided extensive references for those who wish to pursue any topic in greater depth.

8.2. $\chi^{(3)}$ INTERACTIONS IN PLANAR WAVEGUIDE STRUCTURES

The simplest planar optical waveguide structure consists of a thin film of a transparent dielectric of refractive index n_f supported on a substrate of refractive index n_s, with $n_f > n_s$. Figure 8.1(a) shows a thin film waveguide surmounted by a thick cladding medium of refractive index n_c, where $n_c < n_f$. It is thoroughly documented (e.g. Adams, 1981) that light injected into this structure may be confined to propagate in discrete modes characterized by propagation constants β_m satisfying the transcendental mode equation

$$k_0 W(n_f^2 - N_m^2)^{1/2} = \tan^{-1} q_{fc}\left(\frac{n_m^2 - n_c^2}{n_f^2 - N_m^2}\right)^{1/2} + \tan^{-1} q_{fs}\left(\frac{N_m^2 - n_s^2}{n_f^2 - N_m^2}\right)^{1/2} + m\pi \tag{8.1}$$

where $N_m = \beta_m/k_0$, $k_0 = 2\pi/\lambda_0$, λ_0 is the free space optical wavelength, m the mode index, W the film thickness, and $q_{fc,fs} = 1$ for TE modes or $q_{fc,fs} = (n_f/n_{c,s})^2$ for TM modes. In this form, equation 8.1 shows particularly clearly the roles of the refractive indices, the film thickness and the mode number m. However, for later convenience, we define the parameters

$$b \equiv \frac{N_m^2 - n_c^2}{n_f^2 - n_c^2} \tag{8.2}$$

$$d \equiv \frac{n_f^2 - n_s^2}{n_f^2 - n_c^2} \tag{8.3}$$

$$V \equiv \frac{k_0 W}{2}(n_f^2 - n_c^2)^{1/2} \tag{8.4}$$

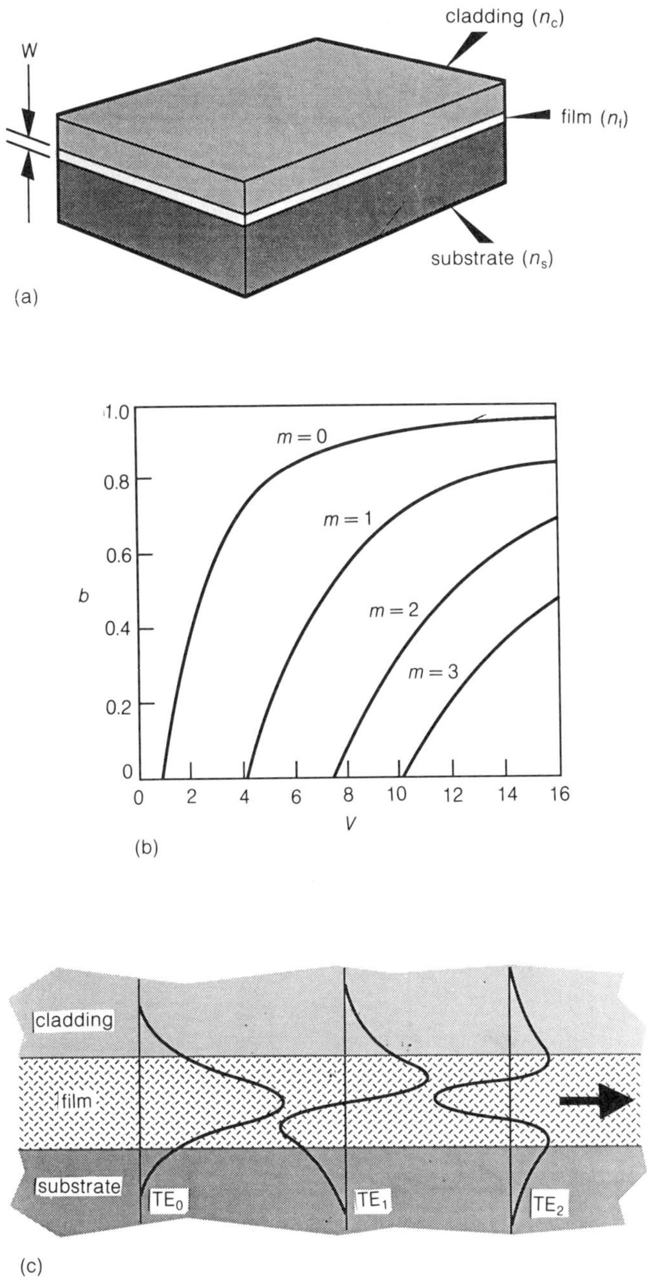

Figure 8.1 (a) A planar optical waveguide comprising a thin film of refractive index n_f and thickness W, supported on a substrate of index n_s, and overlaid by a cladding of index n_c. (b) Plots of b against V for the TE modes of the planar waveguide with $d = 10$. (c) Field profiles of the TE_0, TE_1, and TE_2 modes.

Applying these definitions, and taking tangents on both sides of equation 8.1, leads to the alternative form of the mode equation:

$$\tan 2V(1-b)^{1/2} = \frac{(1-b)^{1/2}[(d+b-1)^{1/2}+b^{1/2}]}{1-b-(d+b-1)^{1/2}b^{1/2}} \tag{8.5}$$

Solutions of equation 8.5 for guided waves of the structure, with a representative value of $d = 10$, are shown in Figure 8.1(b) as plots of b against V. Note that the regime of existence of guided-wave solutions is $n_{c,s} < N_m < n_f$ and that there exists a threshold thickness below which guided waves are not supported. The quantity N_m can be shown to be the functional equivalent in guided-wave optics of the refractive index in bulk optics (Tien, 1971) and, accordingly, is termed the **effective refractive index**. The E field profiles for the first three TE modes of the structure are shown in Figure 8.1(c): in each case the fields are oscillatory in the film and evanescent in the substrate and cladding.

The thin film waveguide is a practically convenient format compatible with established fabrication techniques such as evaporation, sputtering, epitaxial growth, or solution deposition. However, alternative processes also yield waveguides providing one-dimensional optical confinement. By these processes, the waveguide is produced by increasing the refractive index of the substrate in a surface layer. Important and well-known examples are diffusion, best known in the case of titanium-diffused $LiNbO_3$ waveguides, ion exchange in glasses, and ion implantation. The refractive index distribution in waveguides of these types does not exhibit the step-like change at the film–substrate boundary, rather it is smoothly graded from a peak value at or near the substrate surface to the bulk substrate value, as shown in Figure 8.2. These structures are mathematically more cumbersome to deal with than the thin film waveguide; nevertheless, the principles of propagation are fundamentally similar and it is possible to define an effective thickness within which the field solutions are oscillatory and beyond which they are evanescent.

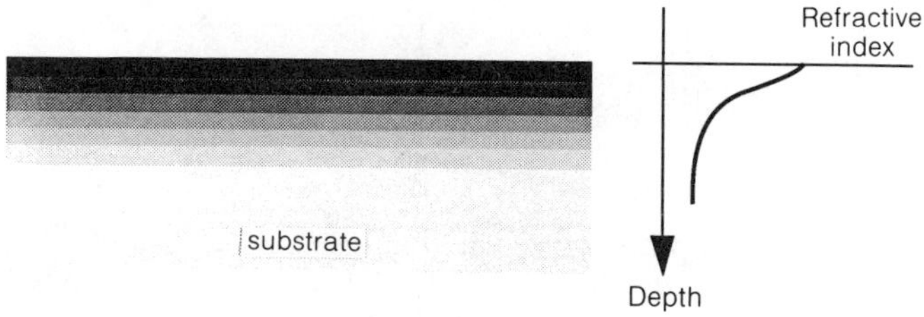

Figure 8.2 A graded-index optical waveguide.

8.2.1. The nonlinear planar waveguide

We are interested here in structures embodying waveguides of any of the types discussed above in which one or more of the film, substrate and cladding exhibit a pronounced $\chi^{(3)}$ nonlinearity. In this case, the structure exhibits a refractive index that is dependent on the local optical intensity and, for an ideal Kerr law medium, is given by

$$n_i(I) = n_{0i} + n_{2i} I \tag{8.6}$$

where n_{0i} is the low intensity value of the refractive index, n_{2i} the nonlinear coefficient, I the intensity, and the subscript i refers to the ith medium. For high speed applications, n_{2i} ideally has its origin in a purely electronic response of the material. For a waveguide mode, the refractive index is spatially averaged over the modal field profile which leads to a **power dependence** of the effective refractive index (Stegeman, 1982) given by

$$N_m = N_{m0} + \Delta N_{m0} P \tag{8.7}$$

where P is the power per unit width (of the unconfined dimension) of the mode in $\mathrm{W\,m^{-1}}$. Since $\beta_m = N_m k_0$, equation 8.7 yields a power-dependent propagation constant $\beta_m = \beta_{m0} + \Delta\beta_{m0} P$.

Any or all of the media constituting the film, substrate and cladding in Figure 8.1(a) may exhibit a value of n_{2i} of sufficient magnitude to produce interesting nonlinear effects. Here we concentrate on the case where only the cladding layer exhibits a nonlinear coefficient n_{2c} of a pronounced magnitude, and we take $n_{2f} = n_{2s} = 0$. This case corresponds to the most experimentally convenient arrangement since it admits the use of established linear waveguide (film–substrate) combinations to which a nonlinear cladding layer may simply be added as an overlayer (Bennion *et al.*, 1984; Vach *et al.*, 1984).

We now introduce the normalized parameter Q describing the nonlinearity in terms of the field strength $\mathbf{E}_{fc}$ at the interface between the film and the cladding:

$$Q \equiv \frac{W^2 k_0^2 n_c^2 n_{2c} |\mathbf{E}_{fc}|^2}{8\mu_0 c} \tag{8.8}$$

where μ_0 is the free space permeability and c the vacuum speed of light. In terms of Q, the intensity-dependent refractive index at the interface, n_c, is given by

$$n_c = n_c(0) + \frac{8Q}{W^2 k_0^2} \tag{8.9}$$

In the presence of this nonlinearity, the mode equation becomes (Robbins, 1983)

$$\tan\left[2V(1-b)^{1/2}\right] = \frac{(1-b)^{1/2}\left[(d+b-1)^{1/2} \pm (b-Q/V^2)^{1/2}\right]}{1-b \mp (d+b-1)^{1/2}(b-Q/V^2)^{1/2}} \tag{8.10}$$

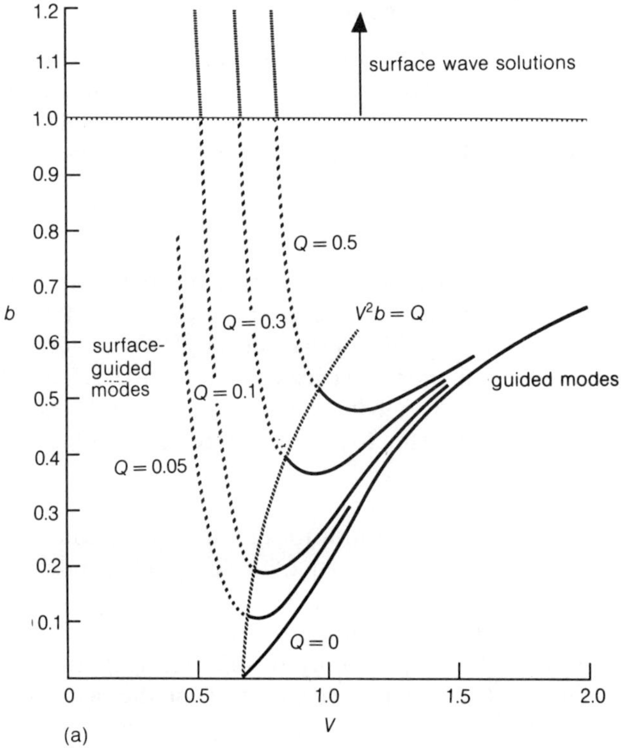

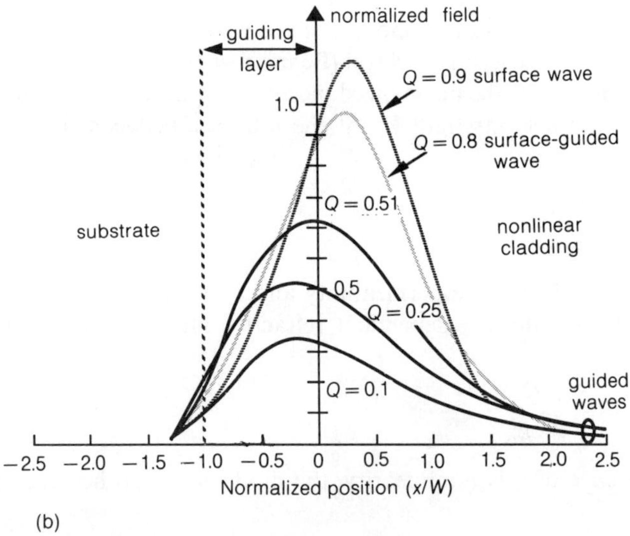

Figure 8.3 (a) b plotted against V for the TE_0 mode of the nonlinear planar waveguide with $d = 10$, for several values of Q. (b) Field profiles for the TE_0 mode with $V = 1$.

The upper signs in equation 8.10 yield solutions wherein the peak of the guided-wave field resides in the film: note that, in this case, in the absence of a nonlinearity $n_{2c} = 0$, $Q = 0$, and equation 8.10 reduces to the linear mode equation 8.5. The lower signs admit solutions only when $Q > 0$ and these solutions correspond to modes for which the field peak is located in the nonlinear cladding, as discussed below.

A plot of b against V for the TE_0 mode of a representative structure with $d = 10$ is given in Figure 8.3(a) for several values of the nonlinear parameter Q. The curve for $Q = 0$ is identical to that plotted in Figure 8.1(b). For $Q > 0$, corresponding to increased power flow down the waveguide, the b–V curves take a form that has no counterpart in the regime of linear guided-wave optics. Corresponding field profiles for $V = 1$ are given in Figure 8.3(b) from which it may be seen that, with increasing Q, the peak of the guided-wave field moves smoothly through the film–cladding interface yielding as it does so mode shapes which cannot exist in linear waveguides. For $Q = 0.51$, the field peak is located on the interface between the linear and the nonlinear media. For greater values of Q, the field peaks in the nonlinear medium and a progression through two distinct mode types may be identified. Firstly, a class of mode is encountered for which the field remains oscillatory within the linear waveguide layer; we refer to this mode type as a **surface guided wave** (see $Q = 0.8$ in Figure 8.3(b)). For modes of this type, $0 \leqslant b \leqslant 1$ and $Q \leqslant bV^2$. The second new mode type is characterized by a field which is evanescent in linear media (see $Q = 0.9$ in Figure 8.3(b)) and is referred to as a surface wave. Surface waves represent solutions of the mode equation

$$\tanh[2V(b-1)^{1/2}] = \frac{-(b-1)^{1/2}[(d+b-1)^{1/2} - (b-Q/V^2)^{1/2}]}{b-1-(d+b-1)^{1/2}(b-Q/V^2)^{1/2}} \tag{8.11}$$

which is obtained directly from equation 8.10 in the limit $(1-b)^{1/2} \to \mathrm{i}(b-1)^{1/2}$. Surface wave solutions exist for $b > 1$ and $Q \leqslant bV^2$. In fact, we do not need the waveguide film at all to propagate modes of this type: surface waves can be supported by the single interface between two dielectric media only one of which has a third-order ($n_2 > 0$) nonlinearity (Maradudin, 1983). It should be noted that only the TE_0 mode can evolve to a nonlinear surface wave. All higher-order modes remain oscillatory in the waveguide film.

Figure 8.4 shows the results of numerical calculations of the guided-wave power, P, as a function of N for the TE_0 and TE_1 modes of a representative planar waveguide with $n_{2c} > 0$ (Stegeman *et al.*, 1984). Note that in each case there is a range of values of P over which N is dual valued leading to the possibility of optical switching and, potentially, bistability under appropriate conditions. Structures of this type have been investigated experimentally. In one experiment (Vach *et al.*, 1984), the central portion of a 1.0 μm film of borosilicate glass ($n_f = 1.61$) supported on a glass substrate ($n_s = 1.52$) was surmounted by the liquid crystal MBBA with $n_c = 1.55$ and a thermal nonlinearity giving $n_{2c}(\text{MBBA}) \approx 10^{-9}\,\text{m}^2\,\text{W}^{-1}$. Light from a CW Ar^+ laser

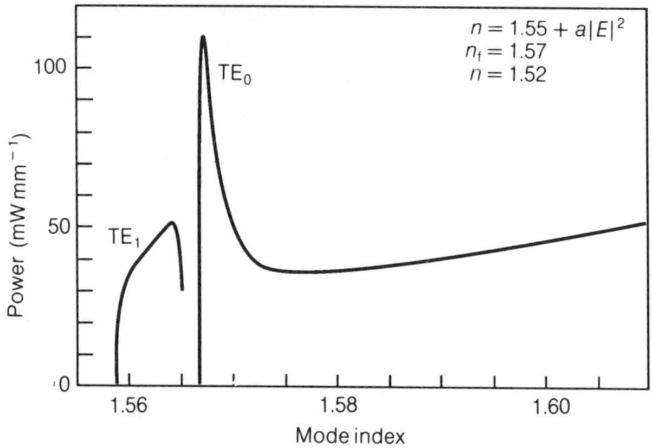

Figure 8.4 Guided-wave power P plotted against the effective index N for the TE$_0$ and TE$_1$ modes of a nonlinear planar waveguide with $n_{2c} > 0$ (Stegeman *et al.*, 1984).

was separately prism-coupled (section 8.2.2) to the TE$_0$ and TE$_1$ modes in an unclad, linear portion of the structure, propagated through the lossy, MBBA-clad region, and coupled out in an unclad region by a second prism. The results of this experiment are shown in Figure 8.5. For the TE$_0$ case, the transmitted power is essentially linear with the input power with evidence of saturation at the highest input power levels. The results for the TE$_1$ mode, however, exhibit pronounced saturation and hysteresis.

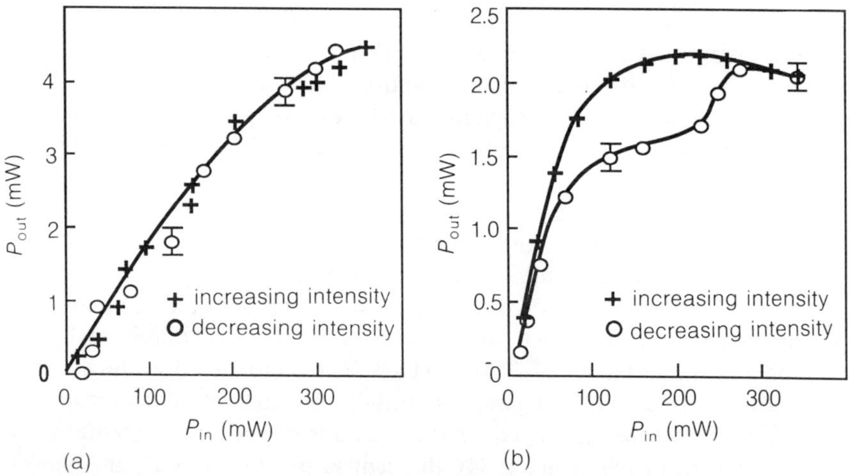

Figure 8.5 Output power vs. input power obtained experimentally for the (a) TE$_0$ and (b) TE$_1$ modes of a glass waveguide with MBBA as a nonlinear cladding (Vach *et al.*, 1984).

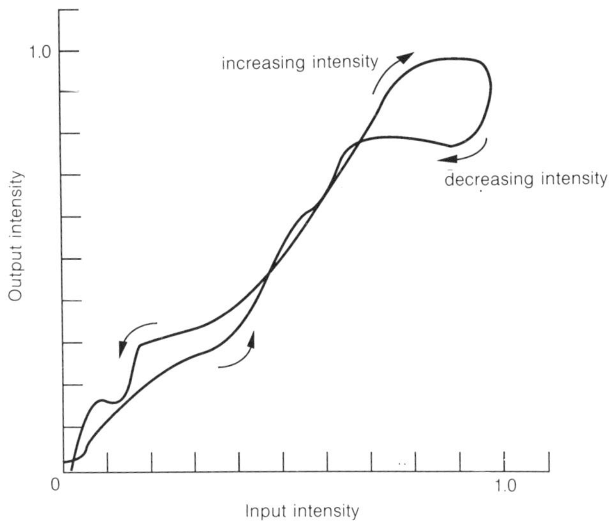

Figure 8.6 Saturation and hysteresis demonstrated in a glass waveguide with nonlinear CS_2 overlayer.

In another experiment (Bennion *et al.*, 1985), the TE_0 mode was excited in a single-mode ion-exchanged glass waveguide, with CS_2 as the nonlinear medium; the electronic nonlinearity in CS_2, with a time constant of ~ 2 ps, is faster by about 11 orders of magnitude than that of MBBA but is also very much weaker with $n_{2c}(CS_2) \approx 3 \times 10^{-18}\,m^2\,W^{-1}$. At low coupled power, the refractive index difference at the nonlinear interface between waveguide and cladding was ~ 0.001. The results of this experiment are shown in Figure 8.6 revealing once again saturation and hysteresis, this time for the TE_0 mode.

The form of the *P–N* curves of Figure 8.4 applies to the waveguides in these experiments albeit with different scales applied to the axes in each case. For the CS_2 experiment, there was no TE_1 mode. Figure 8.7 depicts the transverse boundaries between the three waveguide regions. In the MBBA experiment, the available guided-wave power lay between the Figure 8.4 maxima for the TE_0 and TE_1 modes. As the TE_1 power is increased, the field peak moves progressively out of the film and into the nonlinear medium. This has two effects which together serve to decrease the transmission of the mode across the two transverse boundaries. Firstly, the mode profiles in the linear and nonlinear portions become more and more mismatched, reducing the transmission coefficient at each of the transverse boundaries. Secondly, as more of the light penetrates the lossy MBBA, the fraction absorbed is increased. Thus, the overall transmission coefficient progressively decreases with increased TE_1 power until the maximum depicted in Figure 8.4 is reached, at which point the transmitted power levels off. When the TE_1 power is

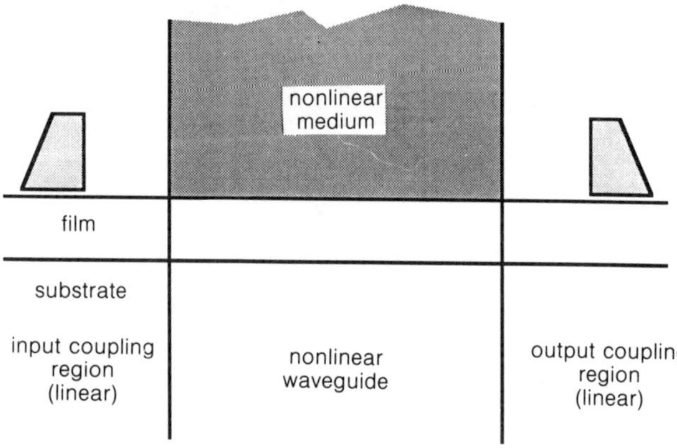

Figure 8.7 The planar waveguide with nonlinear overlayer in the central region flanked by linear coupling regions.

subsequently decreased, both the higher and lower N branches of the TE_1 curve are excited. However, the higher N branch corresponds to higher losses leading to the observed transmission hysteresis. For the TE_0 mode, there is insufficient power available to induce the field peak to shift significantly towards the film–cladding interface, N is almost independent of input power as shown in Figure 8.4, and the transmission is essentially linear with the onset of saturation just evident at the highest input powers.

In the CS_2 experiment, there was sufficient power available from the pulsed Nd:YAG laser utilized to push the TE_0 mode through its Figure 8.4 maximum and hysteresis results. However, the loss mechanism is different since, unlike MBBA, CS_2 is transparent to the optical wavelength used in the experiment. Boardman and Egan (1984) have shown that, in general, the presence of attenuation in the nonlinear cladding will lead to hysteresis. There are two possible loss mechanisms in the CS_2 case: optically imperfect adhesive joints cementing the cell containing the CS_2 to the surface of the waveguide, and β-dependent coupling to the radiation modes of the structure. It is likely that these two mechanisms are not unrelated although a precise theoretical description of radiation mode coupling remains to be worked out.

The increase in N with guided-wave power indicates that the waveguide cut-off condition must also be power dependent. Thus, for a waveguide of thickness less than that required to support the fundamental TE_0 mode at low power, increased input power may adjust the cut-off. We have concentrated here on one possible nonlinear planar waveguide structure and presented our discussion in terms of the mathematically simpler TE modes. Other structures and nonlinear behaviours are possible according to which of the media possess nonlinearities and TM modes can be embraced. Nonlinear switching behaviour may utilize induced birefringence changes in planar waveguides wherein both TE and TM modes are present (Goodwin *et al.*, 1988). Another class of possibilities even exists for structures formed from materials with $n_2 < 0$. Those interested in pursuing these possibilities are referred to the works of Stegeman and Seaton (1985) and Stegeman *et al.* (1988) and the extensive sets of references therein.

8.2.2. Nonlinear prism coupling

The prism–film coupling technique (Tien and Ulrich, 1969), illustrated in Figure 8.8, is a long-established and now standard method in linear integrated optics for coupling an external light wave into a guided mode of a planar waveguide. The prism of refractive index n_p, where $n_p > n_f$, is positioned above the waveguide, separated from it by a narrow gap, and the incident light beam is totally internally reflected from the prism base. When the beam angle, α, is adjusted such that the incident wavevector component in the prism parallel to the plane of the waveguide equals the wavevector of a mode

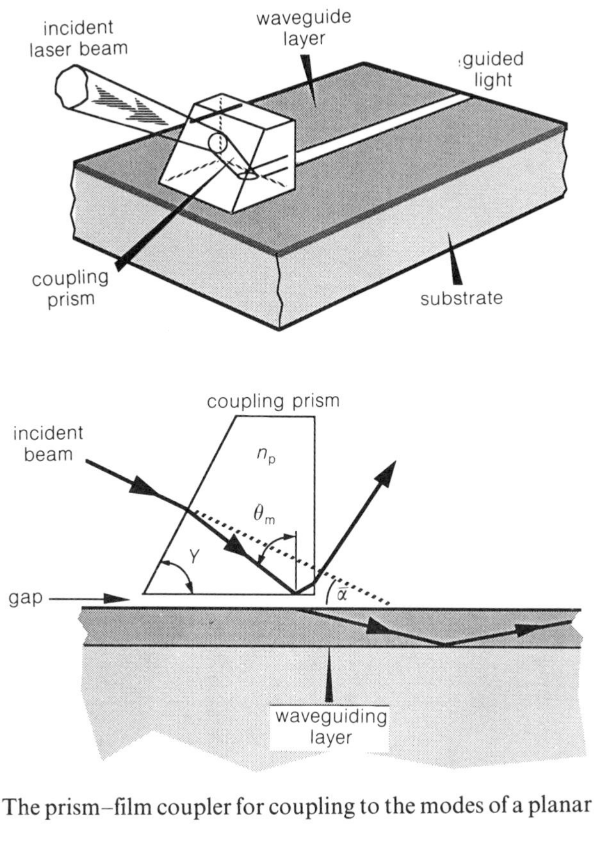

Figure 8.8 The prism–film coupler for coupling to the modes of a planar waveguide.

of the waveguide, that mode is excited by resonant coupling across the gap. This coupling condition translates to

$$N_m = n_p \sin\left\{\gamma + \sin^{-1}\left[\frac{n_g}{n_p}\cos(\alpha_m + \gamma)\right]\right\} \tag{8.12}$$

where γ is the prism angle defined in Figure 8.8, n_g the refractive index of the gap material, and α_m the coupling angle for the mth mode. With a uniform gap, up to ~81% of the incident power can be coupled to a guided mode.

In the presence of a third-order nonlinearity, N_m changes with guided-wave power according to equation 8.7. Thus, when the planar film or substrate, or the medium separating the prism from the film surface, is nonlinear, the phase-matching condition given by equation 8.12 varies with incident light intensity. At a fixed, low power coupling angle, α_m, the coupling efficiency decreases as the guided-wave power is increased, but can be reoptimized by adjustment of α_m (Liao *et al.*, 1985). Figure 8.9 illustrates this behaviour experimentally demonstrated using a nonlinear doped polymer waveguide (Goodwin *et al.*, 1986): the amount of adjustment of α_m may be used, as in this case, to determine the n_2 value of a nonlinear material. If the incident light is pulsed, additional effects are present since the coupling efficiency then varies in time over the pulse profile leading to pulse distortion (Assanto *et al.*, 1988).

Nonlinear prism coupling has been widely investigated experimentally in a wide range of materials. In addition to power-dependent coupling, switching

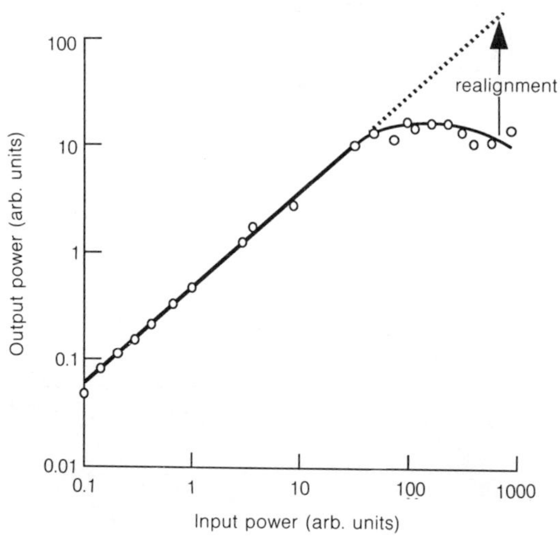

Figure 8.9 Experimentally determined nonlinear response of a prism-coupled doped polymer waveguide.

phenomena (e.g. Fortenberry *et al.*, 1986; Assanto *et al.*, 1987) and optical bistability (Vitrant and Arlot, 1987) have been demonstrated.

8.2.3. Nonlinear grating devices

Waveguide gratings have found many roles in guided-wave optics (Yariv and Nakamura, 1977) with applications embracing input/output couplers, beam splitters and deflectors, filters, distributed feedback resonators, and phase-matching elements for $\chi^{(2)}$ nonlinear interactions. A grating may be formed as a periodic surface corrugation of one of the waveguide boundaries, or as a modulation of the waveguide volume refractive index. A surface relief grating is illustrated in Figure 8.10.

The grating coupler is generally less favoured for input and output beam coupling than the prism coupler discussed in section 8.2.2 owing to its inherently lower efficiency and fabrication inconvenience. On the other hand, it is significantly more compatible with compact device fabrication and is useful in fibre waveguide devices (Bennion *et al.*, 1986). When it is formed in a planar waveguide containing a third-order nonlinearity, the behaviour of the grating coupler is analogous to that described for the prism coupler (Liao *et al.*, 1985). The nonlinear grating coupler has been extensively utilized experimentally.

Rather than coupling the guided modes to the radiation modes of a waveguide, the grating may be designed to couple guided modes to other guided modes within the structure. An efficient interaction occurs only when the Bragg condition is satisfied: in terms of the guided wavevectors,

$$N_d \mathbf{k}_0 = N_i \mathbf{k}_0 + \mathbf{K} \tag{8.13}$$

where N_i, N_d are the effective indices of the incident and diffracted guided waves respectively, and $\mathbf{K}$ is the grating vector, with $|\mathbf{K}| = 2\pi/\Lambda$, and Λ the grating period. When equation 8.13 is not satisfied, the incident wave is passed by the grating without significant deflection. When the waveguide contains a third-order nonlinearity, the Bragg condition becomes power-

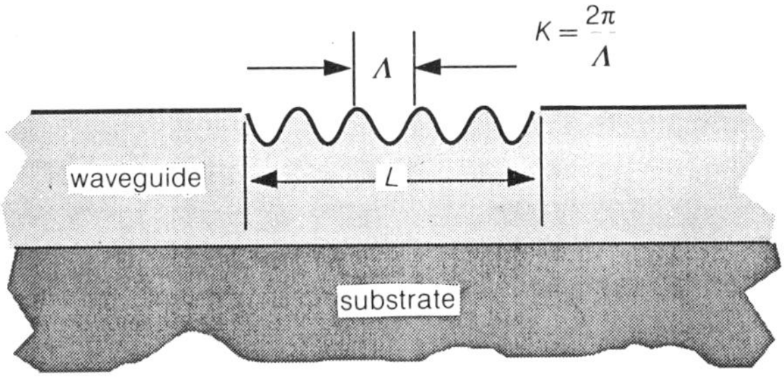

Figure 8.10 A surface relief grating in a planar waveguide.

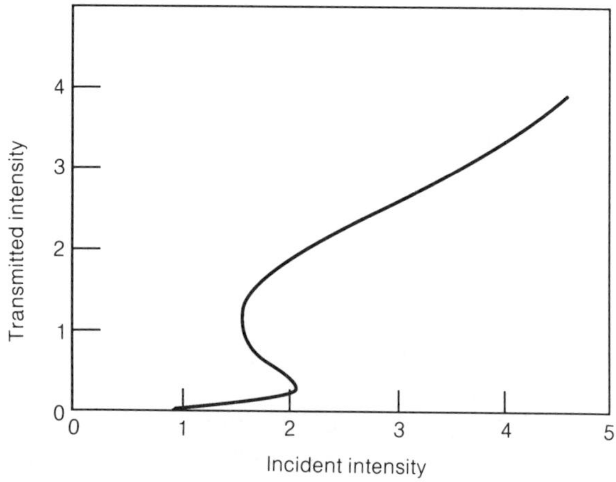

Figure 8.11 The calculated transmission response of a nonlinear DFB grating with $\kappa L = 2$. (After Winful and Stegeman, 1984.)

dependent, leading to a range of possible device functions (Winful and Stegeman, 1984; Seaton *et al.*, 1985).

The distributed feedback (DFB) mechanism provided by a waveguide grating is well known from its application within semiconductor lasers. Nonlinear waveguide structures containing a DFB grating have been modelled theoretically and shown to exhibit optical bistability and switching. More commonly, bistable optical devices have utilized a Fabry–Pérot resonator to provide the optical feedback. The DFB resonator offers a compact alternative that is more compatible with guided-wave optical implementation. Grating interaction efficiencies are governed by the factor κL, where κ is the coupling coefficient, a measure of the strength of the interaction, and L the interaction length. Figure 8.11 shows the calculated transmission response of a nonlinear DFB structure with $\kappa L = 2$ illustrating a single bistable switching region (Winful and Stegeman, 1984). Higher values of κL lead to multiple switching regions, and there is a lower threshold value below which the grating feedback is insufficient to support bistability. Nonlinear DFB bistability in a polydiacetylene film has been demonstrated experimentally by Sasaki *et al.* (1988).

The waveguide grating acts as a bandstop filter, reflecting normally incident light of wavelength λ_B which satisfies the Bragg condition $\lambda_B = 2N\Lambda$. The usual Bragg filter characteristic applies, so that the reflectivity falls continuously to zero for wavelengths on either side of λ_B. The power in a guided mode incident on a nonlinear grating near the Bragg angle changes λ_B, as required for bistability and switching, and so provides for an optically tuned filter. Figure 8.12 shows results obtained by Ehrlich *et al.* (1990), using a thermal nonlinearity in InSb to demonstrate this function. In another mode

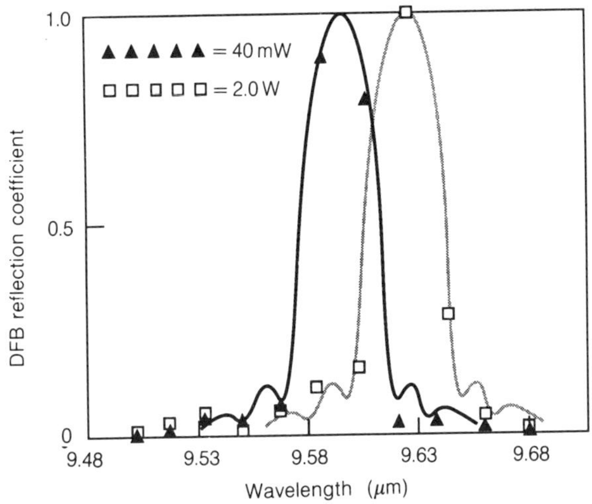

Figure 8.12 All-optical tuning of a waveguide grating filter for incident powers of 40 mW and 2.0 W. (After Ehrlich *et al.*, 1990.)

of operation, a strong control beam incident on the nonlinear grating at an angle other than the Bragg angle may be used to tune the grating to any point on this characteristic for a second, weaker signal beam incident close to the Bragg angle.

The nonlinear grating may be used to achieve power-dependent coupling between the codirectional TE_0 and TE_1 modes of a waveguide (Trillo *et al.*, 1988). In this application, N_i and N_d in equation 8.13 correspond to the effective indices of the two modes, one of which is initially launched into the grating region. At some value of the guided-wave power, equation 8.13 is satisfied and light is converted from one mode to the other by passage through the grating. At a different power, the phase match condition is not satisfied and mode conversion does not occur, and we have the basis for switching action. To utilize this effect to realize an all-optical switch, some means of separating the collinear TE_0 and TE_1 modes is required, a function more readily achieved in the channel waveguide format discussed in section 8.3.

8.3. NONLINEAR CHANNEL WAVEGUIDE DEVICES

The majority of devices used in integrated optics utilize the channel waveguide format and a range of structures has been conceived for modulation and switching functions utilizing the linear electro-optic effect (Bennion, 1989). Generally, these structures form two categories. Devices in the first category utilize voltage-induced changes in the coupling between modes in two adjacent, closely spaced waveguides, or between two modes of different order or polarization state in a single waveguide. In the second category, phase shifts are voltage induced in the well-separated (noncoupled) arms of an interferometer. Electro-optic devices in both categories are now well developed and find practical applications in optical communications and signal processing.

In virtually every case, the linear electro-optic structure has a third-order nonlinear optical counterpart leading to all-optical modulation or switching, wherein the role of the voltage-induced ΔN is played by its guided-wave optical power-dependent equivalent. Following the discussion of section 8.2.1, an optical power P (watts) in a channel waveguide of cross-sectional area A and length L leads to an accumulated nonlinearly induced phase shift given by

$$\Delta\phi(\mathrm{NL}) = k_0\,\Delta N(P/A)L \approx k_0 n_2(P/A)L \tag{8.14}$$

In general, nonlinear waveguide devices of the first category, utilizing mode coupling effects, lead to sharper switching transitions than those based on nonlinear interferometry.

8.3.1. The nonlinear directional coupler

The linear integrated optic directional coupler is illustrated in Figure 8.13. It comprises two identical, initially separated channel waveguides which are brought close together and parallel over a specified length, L, after which they are separated again. L defines the coupling region in which the lateral evanescent fields of the modes in the two waveguides overlap. If light is injected into only one of the two waveguides, e.g. waveguide 1, then the power in each guide evolves as

$$\begin{aligned} P_1(z) &= P_1(0)\cos^2\kappa z \\ P_2(z) &= P_2(0)\sin^2\kappa z \end{aligned} \tag{8.15}$$

where κ is the coupling coefficient which is a function of the overlap of the two modes. Equation 8.15 show that, in the steady state, the power oscillates between the two guides along the propagation length of the coupler. We define the **beat length** of the coupler, L_b, to be the distance over which the

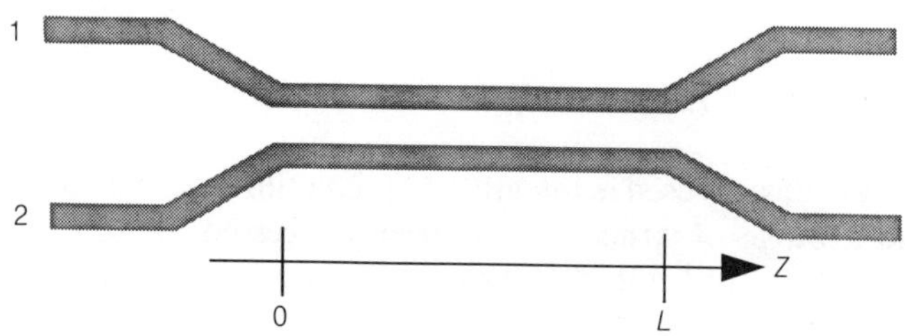

Figure 8.13 Parallel waveguide directional coupler.

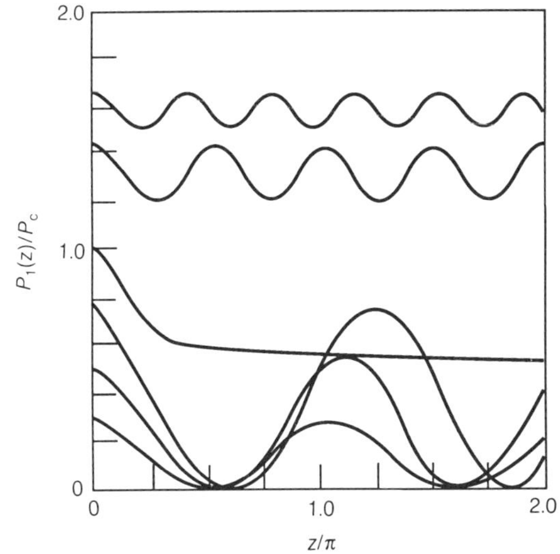

Figure 8.14 The power remaining in the excited waveguide of a nonlinear directional coupler as a function of normalized distance along the coupler, z/π (Jensen 1982).

power initially launched into one guide completes one oscillation: from equations 8.15, $L_b = 2\pi/\kappa$. For a device of length $L = L_b$, all of the input power emerges in the guide in which it was initially launched. We further define the **coupling length** $L_c = L_b/2$. For a device of length $L = L_c$, power launched into one guide emerges entirely in the other guide.

The characteristics of the nonlinear version of this device were first presented by Jensen (1982). Results of his analysis are shown in Figure 8.14 which plots the power remaining in the initially excited waveguide 1, normalized to a critical power, P_c, given by

$$P_c = \frac{4\pi A}{k_0 n_2 L_c} \tag{8.16}$$

with A the waveguide cross-sectional area, as a function of distance along the coupler. For $P_1(0) \ll P_c$, the input power is insufficient to produce any substantial nonlinear effect and the device behaves as predicted by equations 8.15. As $P_1(0)$ is increased, the power dependence of N_1 at the input to waveguide 1 comes into play: $N_1 > N_2$, the coupler is detuned, and power transfer proceeds more slowly than in the linear case. As progressively more power is coupled, however, the extent of the detuning reduces until the point is reached at which there is equal power in each guide and no net detuning. As power continues to build up in the second guide, $N_2 > N_1$, the sense of the detuning reverses and power transfer goes to completion. Thus, for $P_1(0) < P_c$, the effect of the nonlinearity is to increase the coupling length

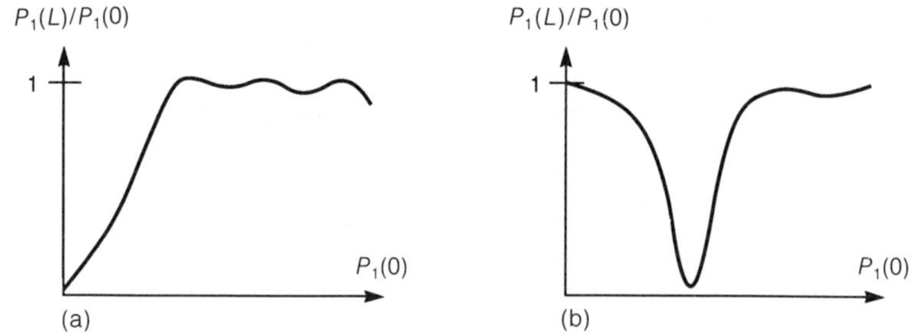

Figure 8.15 Power remaining in the excited waveguide of a nonlinear directional coupler as a function of the input power for (a) $L = L_c$, and (b) $L = L_b = 2L_c$.

L_c. For $P_1(0) = P_c$, coupling is slowed down to the point where the incident power is divided equally between the two waveguides for $z \to \infty$. This is an unstable operating point since any small perturbation of the input power level induces the coupling to proceed to complete transfer of the power to one or other of the outputs. For $P_1(0) > P_c$, power transfer never goes to completion since the magnitude of the induced mismatch in the propagation constants is too large and the equal power condition is never reached. The power oscillates between the guides, the amplitude and the period of the oscillation both decreasing with increased $P_1(0)$.

Figure 8.15(a) shows the power output in waveguide 1 for a coupler of length $L = L_c$ as a function of the power input to the same waveguide. At very low input power, all of the output is in waveguide 2. As the input power is increased, coupling is progressively decreased until at $P_1(0) = 1.25P_c$ the output resides entirely in waveguide 1. As the power is further increased, the induced mismatch between the waveguides is so large that subsequent coupling to waveguide 2 is always small. Thus, the nonlinear coupler demonstrates all-optical switching between its two output ports. The response of a coupler of length $L = L_b$ is shown in Figure 8.15(b). In this case, the input power remains in waveguide 1 except for a narrow range of powers just exceeding P_c where switching occurs.

Experimental demonstration of the operation of a nonlinear directional coupler was first reported by Li Kam Wa *et al.* (1985) for a device of $L = L_c$ produced using strain-induced waveguides in GaAlAs/GaAs multiple quantum well material, and by Gusovskii *et al.* (1985) using dual-core optical fibre. All-optical switching with pulses as short as 10^{-13} s has been demonstrated in a fused quartz, dual-core fibre directional coupler (Friberg *et al.*, 1988).

The nonlinear directional coupler is but one example of an integrated optic device which produces all-optical switching by waveguide mode coupling, in this case between codirectional modes in two waveguides. Other codirectional mode coupling interactions can be used to produce all-optical switching including, for example, grating-coupled $TE_0 \rightarrow TE_1$ conversion in a single-channel waveguide (see section 8.2.3), or coupling between orthogonally polarized modes through nonlinear waveguide birefringence. The DFB grating provides all-optical switching by coupling between contradirectional modes, as discussed in section 8.2.3 for modes of a planar waveguide: it can, of course, be used in a channel waveguide with similar effect. The grating may also be introduced into the coupling region of the two-waveguide directional coupler, providing a contradirectional interaction in that structure.

It is appropriate at this point to emphasize that practical materials rarely exhibit the response of the ideal Kerr-like medium. In particular, enhanced nonlinearities obtained by operation near a material resonance such as, for example, a semiconductor band gap differ from the ideal in quite fundamental respects associated with the promotion of carriers from the valence to conduction bands. Notably, there is a saturation in the achievable index change ΔN_{sat} and a bleachable absorption, as well as delocalization of the nonlinearity as a result of carrier diffusion. A fuller discussion of material properties is given in section 8.5 but for the present we note that these effects modify the experimental behaviour of the devices described in this and subsequent sections.

8.3.2. Nonlinear mode sorter

Figure 8.16 illustrates the waveguide configuration of a device known in linear integrated optics as a BOA (bifurcation optique active) which forms the basis for a class of electro-optic switch (Papuchon *et al.*, 1977). The structure comprises a central waveguide region of length L terminated at either end in identical, symmetric Y-junctions. The central region is wide enough to support two guided modes, i.e. the TE_0 and TE_1 modes. Light injected into one of the input arms excites the two modes approximately equally. Since the two modes propagate with unequal velocities, the total field exhibits a characteristic beat length, $L_b = 2\pi/(N_0 - N_1)k_0$. When the two modes are in phase, the peak in the total field lies to one side of the centre guide, as shown in Figure 8.16; when the modes are out of phase by π, the total field peaks at the other side. The light distribution in the two

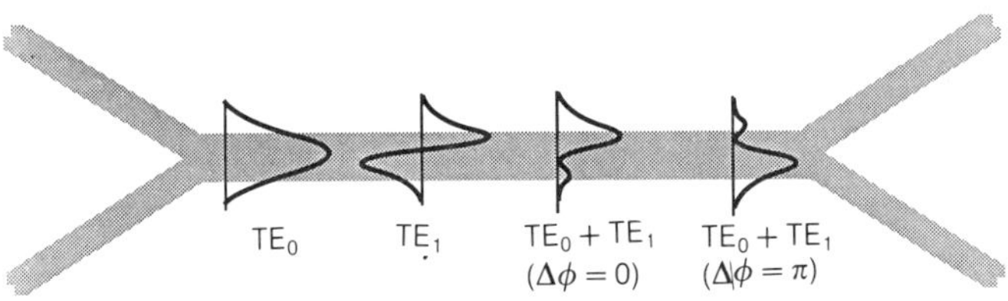

Figure 8.16 The integrated optic BOA (bifurcation optique active) waveguide structure.

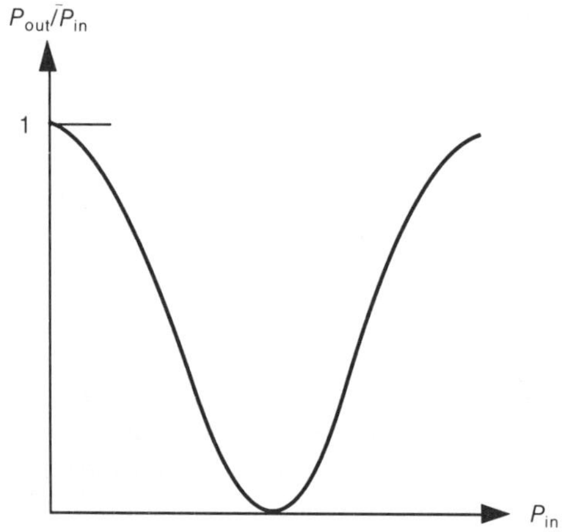

Figure 8.17 Transmission response of the nonlinear mode sorter.

output arms of the structure is dependent on the relative phase between the modes at the branching point. If the phase difference is $2v\pi$ or $(2v+1)\pi$, where $v = 0, 1, 2, \ldots$, light will emerge almost entirely in one or the other arm. In the electro-optic device, voltages are applied to electrodes positioned in the central waveguide region to switch the output between these two states.

In the nonlinear version of this structure, N_0 and N_1 become power dependent and, since their field distributions are dissimilar, the nonlinearly induced $\Delta N_0 \neq \Delta N_1$. The total differential phase shift between the modes is

$$\Delta\phi = \Delta\phi^{(\mathrm{LIN})} + \Delta\phi^{(\mathrm{NL})} = k_0(N_0 - N_1)L + k_0(\Delta N_0 - \Delta N_1)(P/A)L \qquad (8.17)$$

Hence, $\Delta\phi$ changes with P and the interference between the modes may be used to switch the power between the two output arms, with the output characteristic shown in Figure 8.17. It is important to note that this operation is dependent only on accumulating $\Delta\phi = \pi$ along the length of the device, i.e. switching is induced by nonlinear phase shifting rather than by substantial coupling between the modes.

The characteristics of two-mode, single-waveguide devices have been investigated in detail by Silberberg and Stegeman (1987) who have shown that at sufficiently high powers intermode coupling predominates. Figure 8.18 summarizes some results of their analysis when the nonlinear $n_2 > 0$. There is a critical power P_c at which the TE_1 mode is converted entirely to the TE_0 mode. For $P < P_c$, power is exchanged periodically between the modes and the nonlinear phase shift between them grows with propagation distance. For $P \ll P_c$, coupling between the modes is minimal but $\Delta\phi^{(\mathrm{NL})}$ still accumulates and we have the modal interference behaviour described above. For $P > P_c$, intermode coupling is strong and oscillatory power transfer is again observed.

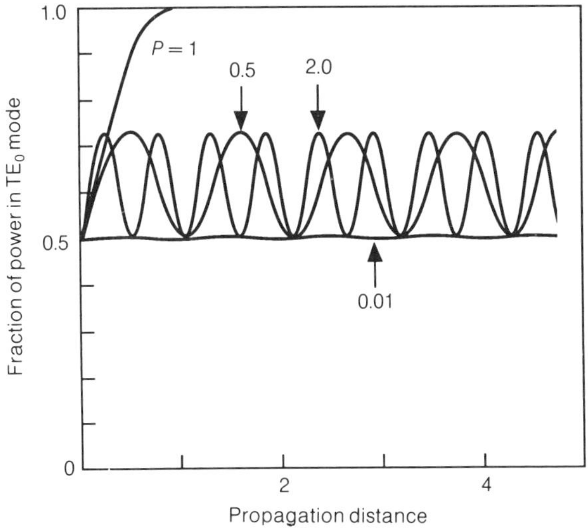

Figure 8.18 Power in the TE_0 mode of a nonlinear mode sorter. Propagation distance is shown in coupling length units. (After Silberberg and Stegeman, 1987.)

8.3.3. Nonlinear Mach-Zehnder interferometer

The Mach–Zehnder interferometer is another device structure widely used as the basis for electro-optic modulation in linear integrated optics. As shown in Figure 8.19(a), its single input waveguide splits at a Y-junction into two widely separated, parallel channels which are subsequently recombined into a single output waveguide by a second Y-junction. Throughout this structure, the waveguides are single moded; in the electro-optic device, electrodes are introduced in the region where the guides are separated. When the two parallel arms are identical, waves arrive at the output waveguide in phase and combine to reconstitute the input signal. With voltages applied, the wave in one arm is phase shifted with respect to that in the other arm. When this phase shift is π, the waves combine in the Y-junction with the antisymmetric field distribution of the first higher-order mode which is not supported by the output waveguide and is lost by radiation. For intermediate phase shifts, two waves sum to a combination of fundamental and first-order mode distributions so that a fraction of the power is guided by the output waveguide and the remainder is lost by radiation. The transfer characteristic of the device is a sinusoidal function of the phase shift as shown in Figure 8.19(b).

In the nonlinear optical Mach–Zehnder interferometer, phase shifts are generated by changes in optical power. However, when the separated arms are identical, equal powers will generate equal phase shifts: no net phase shift will accumulate and there will be no modulation. Whilst this might in principle be circumvented in an all-optical version by using arms of different

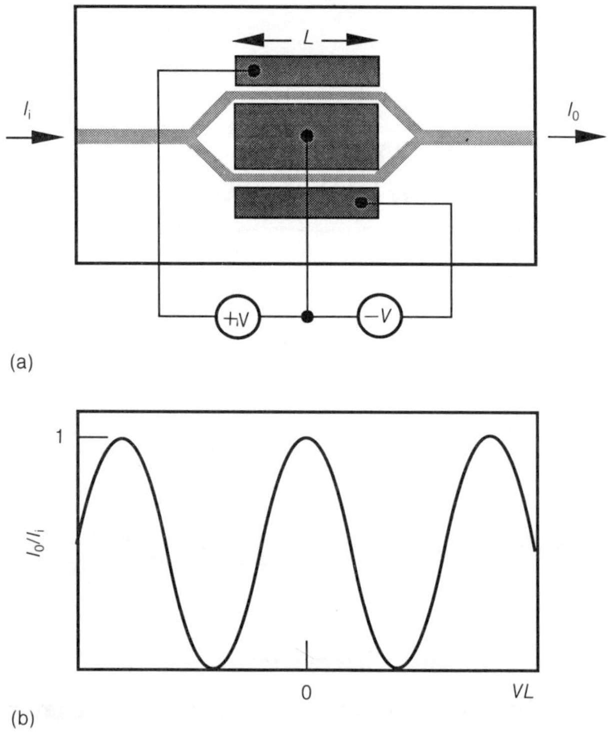

Figure 8.19 (a) The guided-wave electro-optic Mach–Zehnder interferometer and (b) its linear transmission response.

lengths, a more power-efficient approach uses waveguides of different cross-sectional areas or refractive index distributions, within the constraints of single-mode operation (Thylén *et al.*, 1987). Then, the nonlinearly induced phase changes in each arm are dissimilar and a net phase shift can accumulate leading to all-optical modulation and switching. The form of the response of the nonlinear Mach–Zehnder interferometer is identical to that shown in Figure 8.17 for the nonlinear mode sorter. Although its switching characteristic is less sharp than that of the nonlinear directional coupler, the Mach–Zehnder interferometer can be made more power efficient by a factor of ~ 2 (Thylén *et al.*, 1987).

The nonlinear Mach–Zehnder is the basis of the optical logic gate illustrated in Figure 8.20(a) (Lattes *et al.*, 1983). A continuous train of 'clocking' pulses is sent into the structure in the TM_0 mode in the central waveguide where the power divides equally between the two arms of the interferometer. Signal pulses, synchronized to the clock pulse train, enter in the TE_0 mode in either or both of the two outer guides. The polarizer is set to pass only the TM polarization. The logical function of the device is

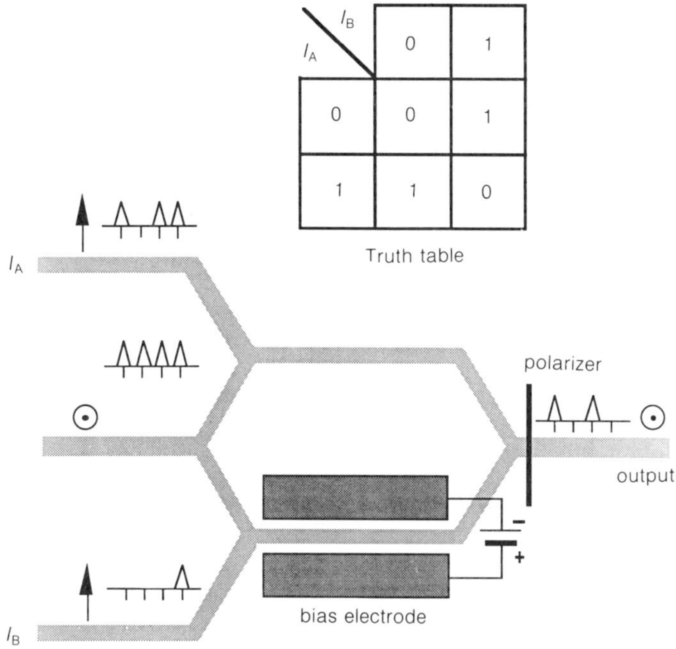

(a)

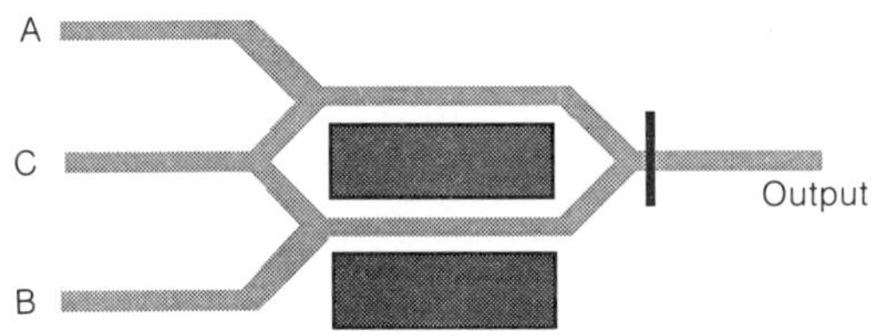

Logic function	Inputs	Phase shift	Comment
Invert	A	0	Pulse stream in B
XOR	A,C	0	Pulse stream in B
XOR	A,C	π	Pulse stream in B
AND	A,B	π	No pulse stream

(b)

Figure 8.20 (a) Optical logic gate using a multiple-input Mach–Zehnder interferometer. (After Lattes *et al.*, 1983.) (b) Alternative logic functions which may be performed by the structure (a).

determined by a bias applied electro-optically. Consider the case when a constant phase shift π is maintained between the interferometer arms. In the absence of any other net phase shift, the interferometer transmission is zero: clock pulses are not transmitted in the absence of signal pulses. If now a signal pulse is introduced in one of the two outer guides and its intensity is adjusted to produce a nonlinear phase shift $\Delta\phi^{(NL)} = \pi$ in one arm of the interferometer, the electro-optic phase shift is thereby cancelled and the polarizer passes a TM pulse. If signal pulses are simultaneously injected in both outer guides, they induce equal nonlinear phase shifts in the interferometer arms, the net phase shift remains at the bias value π, and the polarizer transmission is zero. Thus, the device functions as a logical EXCLUSIVE OR (XOR) gate as shown by the table in Figure 8.20(a). Alternative logic functions may be produced by the same device according to the selection of appropriate values of bias and input ports, as shown in Figure 8.20(b). The multiple-input Mach–Zehnder interferometer is but one example of a logic element which might be constructed based on the third-order optical nonlinearity.

8.4. FURTHER NONLINEAR WAVEGUIDE FUNCTIONS

8.4.1. Degenerate four-wave mixing

Degenerate four-wave mixing (DFWM) is a nonlinear process that occurs when three optical beams interact through the third-order nonlinear susceptibility, $\chi^{(3)}$, of the medium to produce a fourth optical beam. The basic geometry for this interaction is shown in Figure 8.21. Two of the three incident beams (the pump beams) are configured to be counterpropagating, whilst the third beam (the probe beam) is incident on the interaction region from some arbitrary angle. The interaction produces a fourth beam that

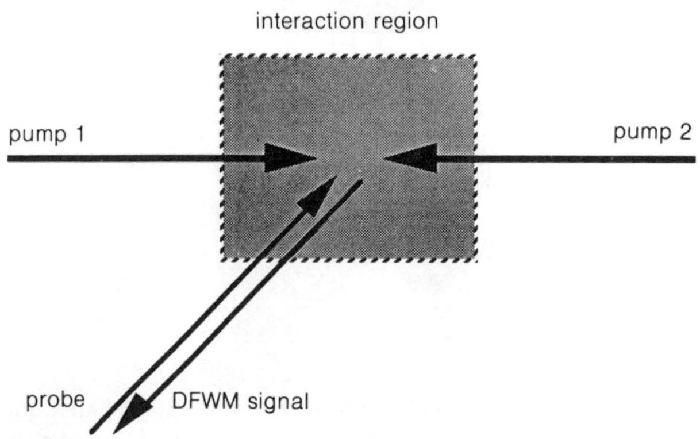

Figure 8.21 Interaction geometry for degenerate four-wave mixing.

counterpropagates along the probe beam. The efficiency of the DFWM process depends on the power in the three incident beams, the size of the interaction region, and the magnitude of the nonlinearity. Additionally, the process is phase matched, so that there is great advantage to be gained by making the interaction length as long as possible for greatest conversion efficiency. DFWM has been used extensively for the characterization of the third-order nonlinear coefficients of materials in bulk samples, though the limited conversion efficiencies achieved have restricted wider applications of the technique. It also permits real-time holography, however, opening the way to many signal processing operations.

The DFWM efficiency can be increased significantly by adopting a waveguide geometry, in which high optical intensities can be maintained over long interaction lengths. This geometry has been analysed by Karaguleff and Stegeman (1984) and the magnitude of the DFWM signal P_{DFWM} shown to be given by

$$P_{DFWM} = |A|^2 (L/H)^2 |G|^2 P_1 P_2 P_3 \tag{8.18}$$

where $|A|^2$ is the nonlinear cross-section for the interaction and is propor-

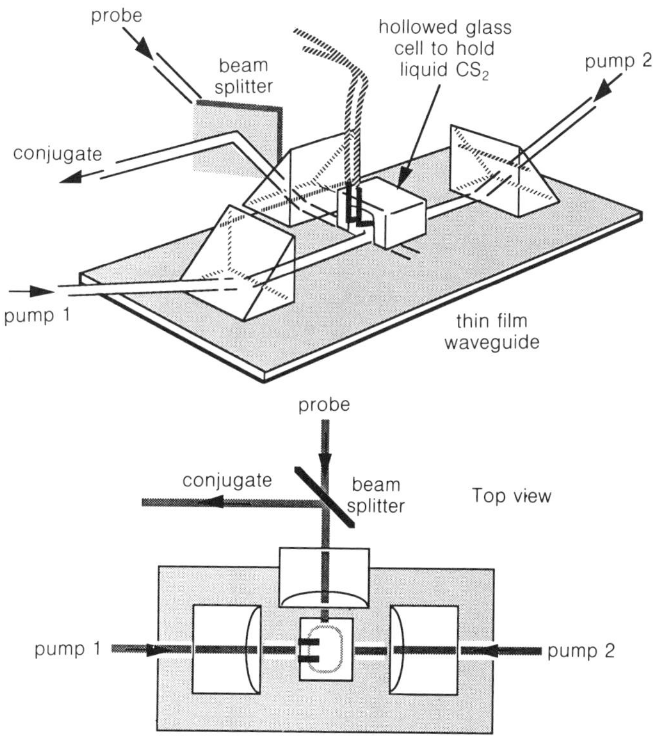

Figure 8.22 Experimental configuration for demonstration of DFWM (Karaguleff *et al.*, 1985).

tional to $|\chi^{(3)}|^2$, $(L/H)^2$ represents the geometrical factor with L and H being the length and width of the interaction region, respectively, P_i the power of the ith incident beam, and $|G|^2$ is a term representing the effect of signal power attenuation through the interaction region.

Guided-wave DFWM has been demonstrated using a CS_2 overlayer on a planar, linear glass waveguide (Karaguleff *et al.*, 1985). The experimental configuration is shown in Figure 8.22. Three prism couplers couple the three input beams simultaneously into the waveguide, where they intersect in the central region over which the CS_2 cell is positioned. The counterpropagating DFWM signal is separated from the probe beam and monitored as a function of incident beam power. Despite relatively low coupling efficiencies, and the penetration of only a small fraction of the incident beam power into the nonlinear medium, a clear DFWM signal was observed with a peak reflectivity of 10^{-9} at $5\,\text{MW}\,\text{cm}^{-2}$ total power in the CS_2.

A second demonstration of waveguide DFWM utilised an ion-exchanged, semiconductor-doped glass nonlinear waveguide in the same experimental geometry as the previous instance (Gabel *et al.*, 1987). In this case, a much larger proportion of the guided light was carried in the nonlinear medium, providing a significant increase in the conversion efficiency. DFWM reflectivities approaching 1% were observed for nanosecond pump pulses of $30\,\text{MW}\,\text{cm}^{-2}$ intensity, rising to 10% with 50 ps pulses.

8.4.2. Nonlinear guided-wave spectroscopy

Nonlinear interactions in waveguide structures can be used to enhance spectroscopic measurements of chemical substances, exploiting the high optical intensities and long interaction lengths thereby accessible. An example of this is provided by coherent anti-Stokes Raman spectroscopy (CARS), in which incident light is used to probe the vibrational spectra of materials. This is achieved through a nonlinear wave mixing process, whereby two incident probe beams of frequencies ω_1 and ω_2 mix to yield a third beam. The conversion efficiency for this is greatly enhanced when the difference frequency, $\omega_1 - \omega_2$, equals one of the molecular vibration frequencies of the material, resulting in resonant scattering of light at frequency $\omega_s = 2\omega_1 - \omega_2$. (A similar scattering process, coherent Stokes Raman spectroscopy (CSRS), utilizes the $\omega_s = 2\omega_2 - \omega_1$ mixing term.) The scattered light can couple into a waveguide mode and, if phase matching conditions are satisfied, grow constructively as it passes through the interaction region to yield a large signal level.

The equation that governs the CARS signal strength, P_{CARS}, has been derived as (Stegeman *et al.*, 1983)

$$P_{CARS} = D^{NL}(L/H)^2 P_1^2 P_2 \tag{8.19}$$

where D^{NL} varies with $|\chi^{(3)}|^2$, and L and H are, again, the length and width of the interaction region. The similarity between equation 8.19 and the

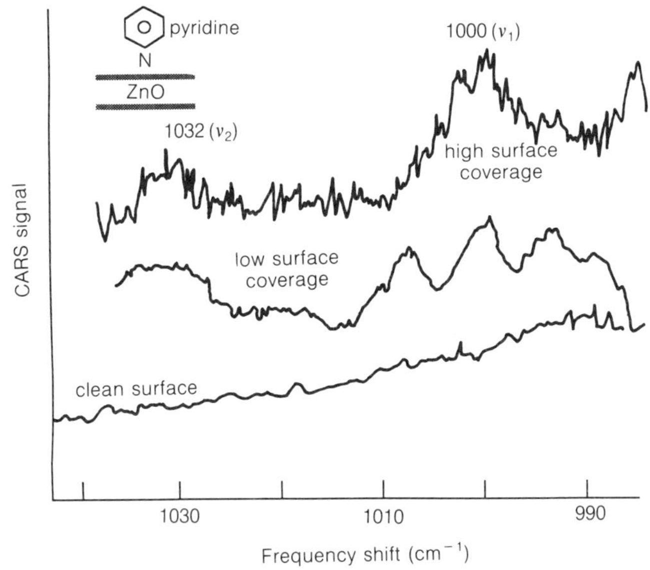

Figure 8.23 Experimental CARS measurement of pyridine monolayer on a ZnO waveguide (Hetherington *et al.*, 1987).

equation governing DFWM (equation 8.18) is indicative of the similarity in the nonlinear wave-mixing processes responsible for the effects. The third-order nonlinear suspectibility takes the form

$$\chi^{(3)} = \chi^{(3)}_{\text{background}} + \sum_p \chi^{(3)}_{rp} \frac{\Gamma_p}{\omega_a - \omega_b - \omega_p - i\Gamma_p} \tag{8.20}$$

where the summation is over the Raman-active vibrational transitions, each with frequency ω_p, susceptibility $\chi^{(3)}_{rp}$, and linewidth Γ_p. As the difference frequency scans each transition frequency, a strong resonant enhancement of $\chi^{(3)}$ and, hence, P_{CARS} is observed. Further enhancement of the process can be achieved through controlling the two input beam modes to produce an interference, which cancels out the contribution to P_{CARS} from the waveguide material itself, leaving only the contribution from the overlayer sample. This technique has been used to achieve very high sensitivities, indicative of a capability for submonolayer spectroscopy. Figure 8.23 shows an example of the results from a CARS measurement using a ZnO waveguide with partial and complete monolayer coverage of pyridine (Hetherington *et al.*, 1987).

8.4.3. Pump–probe interferometry

Interferometry has been used extensively for assessing the nonlinear coefficients of new materials and, when employed with time-resolving pump–probe

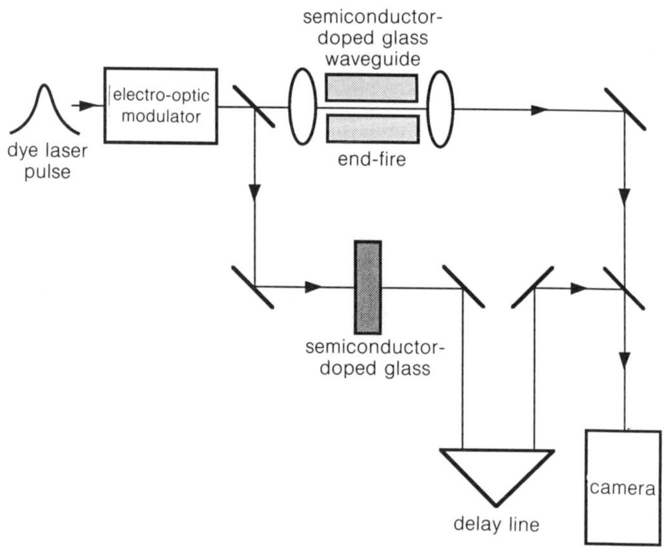

Figure 8.24 Nonlinear pump–probe interferometer arrangement (Finlayson *et al.*, 1989).

techniques, provides a powerful characterization tool. A detailed evaluation of its use for nonlinear waveguide samples has been given by Finlayson *et al.* (1989) whose experiments utilized a semiconductor-doped glass channel waveguide as one arm of a Mach–Zehnder interferometer, as shown in Figure 8.24, excited with a pulsed laser to give time resolution. In the pump–probe arrangement, an initial pump pulse produces a nonlinear excitation in the waveguide which is then interrogated by a subsequent probe pulse. By varying the time delay between the pump and probe pulses, the temporal behaviour of the nonlinear excitation can be followed with resolution determined by the optical pulse width. Using this technique, information on the saturation value of the nonlinear refractive index change can be determined, together with the temporal behaviour of the nonlinearity, including relaxation processes and pulse-narrowing effects. Further, the relative strengths of the electronic and thermal contributions to the overall nonlinearity can be measured.

8.5. NONLINEAR MATERIALS

8.5.1. Waveguide fabrication

For the purposes described herein, a nonlinear material must be compatible with waveguide fabrication processes, usually involving thin film deposition or growth coupled with photolithographic pattern definition and doping. Its refractive index must be uniform in a given sample without macroscopic inhomogeneities. Device interaction lengths and switching powers are influenced by the waveguide losses. In addition to the intrinsic absorption of the material at the wavelength of operation, losses arise from scattering in the waveguide which must be minimized by careful attention to the fabrication processes. Waveguide attenuation of order $1\,\mathrm{dB\,cm^{-1}}$ is achievable with current strongly nonlinear materials.

8.5.2. Nonlinear properties

The dominant requirements on the nonlinear properties of material are concerned with the magnitude and speed of the nonlinearity, and the control of competing nonlinear processes. To be competitive with established electronic devices, all-optical devices should offer a speed advantage, which in practice implies accessing picosecond or subpicosecond response times. The response time embraces both the time taken for the refractive index to change on excitation (turn-on), and the time taken to relax back to its initial value after the excitation has been removed (turn-off). Generally, turn-on occurs extremely rapidly, whilst the turn-off time varies considerably according to the origin of the nonlinearity. The motions and recombination of free carriers and the removal of thermal energy impose limits on the relaxation of nonlinear effects arising from resonant phenomena. The fastest nonlinearities are nonresonant in origin; they occur without breaking chemical bonds or generating charged carrier pairs and derive from the anharmonic motions of bound electrons which yields a nonlinear index change without the absorption of energy. Consequently, the relaxation process is extremely rapid, of order 10^{-14} s. To avoid the presence of additional resonant effects, the excitation wavelength must be chosen to lie beyond any electronic resonances at which bond breaking could occur.

Based on considerations of speed, then, the preferred nonlinearities for device functions are nonresonant in origin. However, in general, such nonlinearities are considerably weaker than those originating in resonant processes. Whilst the search continues to find materials combining a high n_2 with a fast response, significant advances have already been made in a number of material systems permitting the device concepts discussed earlier to be explored. Some compromise is often possible, of course, and enhanced nonlinear response can often be gained by operating at a wavelength closer to an absorption edge at the cost of some reduction in speed.

High incident intensities produce other intensity-dependent effects. Thermally induced refractive index changes are caused by the dissipation of absorbed energy, the magnitude of such changes being dependent on the material absorption coefficient, heat capacity, thermal conductivity and temperature dependence of the linear refractive index. Thermal index changes relax relatively slowly and can interfere with the desired device operation. Note that the nonresonant nonlinear refractive index n_2 is essentially temperature independent. Nonlinear materials must also be optical damage resistant implying, again, a high degree of freedom from physical and chemical defects.

There are further, competing nonlinear processes which must be considered in the evaluation of a material. One of the more important of these is two-photon absorption (TPA), which causes increased absorption in the material at high optical intensity. This can be a critical factor in device realization since a technique frequently adopted to overcome thermal limitations arising from linear absorption is reduction of the average optical power by the use of short, high intensity laser pulses: indeed, this is the preferred mode of operation for most anticipated signal processing applications. Unfortunately, the higher peak intensities result in an increase in the level of TPA, with the net effect that one heat-producing mechanism is simply substituted with another. A careful balance must be achieved between the two processes to yield optimal performance. TPA has been observed and characterized in poly-diacetylenes (Townsend *et al.*, 1989), lead–glass (Mizrahi *et al.*, 1989), and semiconductor-doped glass (Finlayson *et al.*, 1989). In most cases TPA represents a limitation on device operation, although it has also been used to enhance performance (Thakur and Krol, 1990).

We identified in section 8.3 saturation of the nonlinear index change, ΔN_{sat}, which determines the maximum change accumulated nonlinear phase shift achievable for a given device design. Saturation of the induced index change is a material property, associated with effects such as interband absorption in semiconductors and two-level saturable media. Clearly, materials offering a high saturation value are desirable and this is a key parameter which must be evaluated in candidate materials.

8.5.3. Figures of merit

To facilitate comparison of nonlinear materials, a figure of merit may be conceived which reflects the simultaneous requirement for large nonlinearity, n_2, fast response time, τ, and low absorption, α, at the operating wavelength. A conventional figure of merit takes the form

$$M = n_2/\tau\alpha \tag{8.21}$$

A more device-oriented figure of merit, W, has emerged (Stegeman *et al.*, 1988) from consideration of the nonlinear directional coupler which can be more widely applied to other devices that exploit a nonlinearly generated phase shift for their operation:

$$W = \Delta N_{sat}/\alpha\lambda \tag{8.22}$$

Table 8.1 summarizes values of M and W for representative off-resonance, nonlinear materials. It is interesting to note that pure silica glass, because

Table 8.1 Parameters and figures of merit for selected off-resonance nonlinear materials

Material	$n_2\,(\mathrm{m^2\,W^{-1}})$	$\tau\,(\mathrm{s})$	$\alpha\,(\mathrm{cm^{-1}})$	$\Delta\beta_{\mathrm{sat}}$	M	W
PTS	10^{-16}	10^{-14}	10^{2}	10^{-3}	10^{-4}	1
Semiconductor-doped glass	10^{-14}	10^{-11}	1	10^{-5}	10^{-3}	1
Silica	10^{-20}	10^{-14}	10^{-5}	10^{-7}	10^{-1}	100

of its exceptionally low waveguide losses, has the best figures of merit, despite having one of the smallest measured n_2 values. However, from a practical standpoint, its use demands very long device interaction lengths which would add significant internal delay or **latency** to any processor built using silica devices, restricting the choice of compatible processor architectures.

8.5.4. Current materials

No single nonlinear material has yet been identified with an ideal combination of properties for practical nonlinear waveguide devices, and this remains an area with research opportunities. The characteristics of some of the materials systems which have received or continue to receive most attention are briefly reviewed in the following sections.

8.5.4.1. Semiconductors

Typical values for the n_2 coefficients of bulk semiconductor materials lie in the range 10^{-17}–$10^{-18}\,\mathrm{m^2\,W^{-1}}$, rising to $\sim 10^{-16}\,\mathrm{m^2\,W^{-1}}$ for multiple quantum well (MQW) structures in GaAs/GaAlAs (although there is significant variation in the reported values), and a great deal of effort has been devoted to these systems. With semiconductors, even relatively small free carrier densities can cause significant absorption and consequent thermal index changes. This can be reduced by using very short, high-intensity pulses to keep $\Delta n(\text{thermal}) < \Delta n(\text{electronic})$, but this increases considerably the incidence of two-photon absorption which may serve to limit the usefulness of narrow band gap materials in the important near-infrared wavelength region. Larger band-gap materials, such as ZnS, have been used in some experiments and may permit the unwanted absorption processes to be reduced to acceptably low levels.

Improved performance has been obtained when semiconductor material is dispersed in a glass matrix forming a composite. In particular, glasses doped with microcrystallites (100 Å diameter) of CdS_xSe_{1-x} have been widely investigated (Cullen *et al.*, 1986) and found to exhibit large nonlinear coefficients with relatively fast response times (see Table 8.1). The very small size of the crystallites encourages rapid recombination of carriers. High quality waveguides have been produced in these glasses, but the presence of strong competing nonlinear processes has tended to limit their performance in device structrues.

8.5.4.2. Organic materials

Organic polymers and benzene derivatives form a second important class of nonlinear materials. The n_2 value of $2.2 \times 10^{-16}\,\mathrm{m^2\,W^{-1}}$ for the polydiacetylene PTS is among the largest measured to date (Carter *et al.*, 1985). The corresponding off-resonance value of τ, $\sim 10^{-14}$ s, is extremely fast. The origin of this nonlinearity lies in the structure of the polydiacetylenes, with extensive delocalization of bound electrons along a long hydrocarbon chain of alternating single and double bonds. The ability of the materials to form crystalline forms with aligned polymer chains further enhances the nonlinearity. High quality waveguides have been produced in both single-crystal (Thakur and Krol, 1990) and amorphous (Townsend *et al.*, 1989) thin films.

The wider potential significance of organic materials for nonlinear devices resides in the prospect of producing tailor-made materials with enhanced properties by molecular engineering, an active area of research. Although some improvement has been seen in the similar polyacetylene polymers (Kajzar *et al.*, 1987), the long chain structure exhibited by PTS is already a good approximation to the 'ideal' infinitely long, one-dimensional structure with extensively delocalized electrons and is, perhaps, unlikely to be substantially bettered in other materials of this class.

8.5.4.3. Glasses

Optical glasses represent a third significant group of nonlinear materials. The most readily accessible of these is fused silica in the form of optical fibres. As discussed in section 8.5.3, the extremely low waveguide loss achievable in fibres yields a high figure of merit despite a small nonlinear coefficient, and several fibre-based nonlinear devices have been demonstrated. Higher n_2 values, approaching $10^{-18}\,\mathrm{m^2\,W^{-1}}$, have been measured in leaddoped glasses (Friberg and Smith, 1987) and, whilst further data on losses are needed, overall performance improvements can be expected. It is likely, however, that the interaction lengths required for low power operation of high-n_2 fibre devices will remain several metres, with the consequent disadvantage of high latency in devices. Nevertheless, the advantages of low loss and ease of fabrication, and the relative immaturity of the organic materials, tends to make the glasses a preferred material system for applications where high latency can be accommodated.

8.6. SUMMARY

We have introduced the concepts underlying a number of device functions based on third-order nonlinear effects in waveguide structures. Descriptions of operation have tacitly assumed a Kerr law nonlinearity although we have indicated that this ideal behaviour often is not exhibited by real materials. Consequently, detailed modifications to theoretical models of the nonlinearities have had to be developed in many cases. Whilst the number of experimental demonstrations of device functionality has increased steadily in recent years, the quest for improved nonlinear materials continues. The goal of high-speed all-optical logic and signal processing remains, and this field remains one of opportunity.

REFERENCES

Adams, M. J. (1981) *An Introduction to Optical Waveguides*, Wiley, New York.

Assanto, G., Gabel, A., Seaton, C. T., Stegeman, G. I., Ironside, C. N. and Cullen, T. J. (1987) All-optical switching in prism coupling to semiconductor-doped glass waveguides, *Electron Lett.*, **23**, 484–5.

Assanto, G., Fortenberry, R. M., Seaton, C. T. and Stegeman, G. I. (1988) Theory of pulsed excitation of nonlinear distributed prism couplers, *J. Opt. Soc. Am. B*, **5**, 432–42.

Bennion, I. (1989) Applications of nonlinear and electro-optic materials. In *Materials for Non-Linear and Electro-Optics 1989* (ed. M. H. Lyons), Institute of Physics Conference Series 103, Bristol, pp. 1–12.

Bennion, I. and Walker, R. G. (1990) Guided-wave devices and circuits, *Phys. World*, **3**, 47–50.

Bennion, I., Goodwin, M. J., Robbins, D. J. and Stewart, W. J. (1984) An experimental nonlinear optical waveguide device. In *Digital Optical Circuit Technology, Schliersee, Germany, September 1984*, AGARD Conference Proceedings 362, pp. 5.1–5.9.

Bennion, I., Goodwin, M. J. and Stewart, W. J. (1985) Experimental nonlinear optical waveguide device, *Electron. Lett.*, **21**, 41–2.

Bennion, I., Reid, D. C. J., Rowe, C. J. and Stewart, W. J. (1986) High-reflectivity monomode-fibre grating filters, *Electron. Lett.*, **22**, 341–3.

Boardman, A. D. and Egan, P. (1984) Theory of optical hysteresis for TE guided waves, *Philos. Trans. R. Soc. London, Ser. A*, **313**, 363–9.

Carter, G. M., Thakur, M. K., Chen, Y. J. and Hryneiwicz, J. V. (1985) Time and wavelength resolved nonlinear optical spectroscopy of a polydiacetylene in the solid state using picosecond dye laser pulses, *Appl. Phys. Lett.*, **47**, 457–9.

Cullen, T. J., Ironside, C. N., Seaton, C. T. and Stegeman, G. I. (1986) Semiconductor-doped glass ion-exchanged waveguides, *Appl. Phys. Lett.*, **49**, 1403–5.

Ehrlich, J. E., Assanto, G. and Stegeman, G. I. (1990) All-optical tuning of waveguide nonlinear distributed feedback gratings, *Appl. Phys. Lett.*, **56**, 602–4.

Finlayson, N., Banyai, W. C., Seaton, C. T., Stegeman, G. I., O'Neill, M., Cullen, T. J. and Ironside, C. N. (1989) Optical nonlinearities in CdS_xSe_{1-x}-doped glass waveguides, *J. Opt. Soc. Am. B*, **6**, 675–84.

Fortenberry, R. M., Moshrefzadeh, R., Assanto, G., Mai, X., Wright, E. M., Seaton, C. T. and Stegeman, G. I. (1986) Power-dependent coupling and fast switching in distributed coupling to ZnO waveguides, *Appl. Phys. Lett.*, **49**, 687–9.

Friberg, S. R. and Smith, P. W. (1987) Nonlinear optical glasses for ultrafast optical switches, *IEEE J. Quantum Electron.*, **23**, 2089–94.

Friberg, S. R., Weiner, A. M., Silberberg, Y., Sfez, B. G. and Smith, P. W. (1988) Femtosecond switching in a dual-core-fibre nonlinear coupler, *Opt. Lett.*, **13**, 904–6.

Gabel, A., De Long, K. W., Seaton, C. T. and Stegeman, G. I. (1987) Efficient degenerate four-wave mixing in an ion-exchanged semiconductor-doped glass waveguide, *Appl. Phys. Lett.*, **51**, 1682–4.

Goodwin, M. J., Glenn, R. and Bennion, I. (1986) Organic nonlinear optical waveguides formed by solvent-assisted indiffusion, *Electron. Lett.*, **22**, 789–91.

Goodwin, M. J., Edge, C., Trundle, C. and Bennion, I. (1988) Intensity-dependent birefringence in nonlinear organic polymer waveguides, *J. Soc. Am. B*, **5**, 419–24.

Gusovskii, D. D., Dianov, E. M., Maier, A. A., Neustreuev, V. B., Shklovskii, E. I. and Shcherbakov, I. A. (1985) Nonlinear light transfer in tunnel-coupled optical waveguides, *Sov. J. Quantum Electron.*, **15**, 1523–6.

Hetherington III, W. M., Koenig, E. W. and Wijekoon, W. M. K. P. (1987) CARS spectrum of O_2 formed by the trapping of photo-generated electrons on a ZnO surface, *Chem. Phys. Lett.*, **134**, 203–5.

Jensen, S. M. (1982) The nonlinear coherent coupler, *IEEE J. Quantum Electron.*, **18**, 1580–3.

Kajzar, F., Etemad, S., Baker, G. L. and Messier, J. (1987) In *Proc. XV Int. Conf. on Quantum Electronics*, Technical Digest Series, Vol. 21, Optical Society of America, Washington, DC, p. 192.

Kaplan, A. E. (1977) Theory of hysteresis reflection and refraction of light by a boundary of a nonlinear medium, *Sov. Phys. JETP*, **45**, 896–905.

Karaguleff, C. and Stegeman, G. I. (1984) Degenerate four wave mixing with surface guided waves, *IEEE J. Quantum Electron.*, **20**, 716–22.

Karaguleff, C., Stegeman, G. I., Fortenberry, R., Zanoni, R. and Seaton, C. T. (1985) Degenerate four wave mixing in planar CS_2 covered waveguides, *Appl. Phys. Lett.*, **46**, 621–2.

Lattes, A., Haus, H. A., Leonberger, F. J. and Ippen, E. P. (1983) An ultrafast all-optical gate, *IEEE J. Quantum Electron.*, **19**, 1718–23.

Li Kam Wa, P., Sitch, J. E., Mason, N. J., Roberts, J. S. and Robson, P. N. (1985) All-optical multiple-quantum-well waveguide switch, *Electron. Lett.*, **21**, 26–7.

Liao, C., Stegeman, G. I., Seaton, C. T., Shoemaker, R. L., Valera, J. D. and Winful, H. G. (1985) Nonlinear distributed waveguide couplers, *J. Opt. Soc. Am.*, **42**, 590–4.

Maradudin, A. A. (1983) Nonlinear surface electromagnetic waves. In *Optical and Acoustic Waves in Solids* (ed. M. Borissov), World Scientific Publishers, Singapore, p. 72.

Miller, S. E. (1969) Integrated optics: an introduction, *Bell Syst. Tech. J.*, **48**, 2059–71.

Mizrahi, V., De Long, K. W., Stegeman, G. I., Saifi, M. A. and Andrejco, M. J. (1989) Two-photon absorption as a limitation to all-optical switching, *Opt. Lett.*, **14**, 1140–2.

Papuchon, M., Roy, A. and Ostrowsky, D. B. (1977) Electric active bifurcation: BOA, *Appl. Phys. Lett.*, **31**, 266–7.

Regener, R. and Sohler, W. (1988) Efficient second harmonic generation in Ti:LiNbO$_3$ channel waveguide resonators, *J. Opt. Soc. Am. B*, **5**, 267–77.

Robbins, D. J. (1983) TE modes in a slab waveguide bounded by nonlinear media, *Opt. Commun.*, **47**, 309–12.

Sasaki, K., Fujii, K., Tomioka, T. and Kinoshita, T. (1988) All optical bistabilities of polydiacetylene Langmuir–Blodgett film waveguides, *J. Opt. Soc. Am. B*, **5**, 457–61.

Seaton, C. T., Mai, X., Stegeman, G. I. and Winful, H. G. (1985) Nonlinear guided wave applications, *Opt. Eng.*, **24**, 593–9.

Silberberg, Y. and Stegeman, G. I. (1987) Nonlinear coupling of waveguide modes, *Appl. Phys. Lett.*, **50**, 801–3.

Smith, P. W. and Tomlinson, W. J. (1984) Nonlinear optical interfaces: switching behavior, *IEEE J. Quantum Electron.*, **20**, 30–6.

Sohler, W., Hampel, B., Regener, R., Ricken, R., Suche, H. and Volk, R. (1986) Integrated optical parametric devices, *J. Lightwave Technol.*, **4**, 772–7.

Stegeman, G. I. (1982) Guided wave approaches to optical bistability, *IEEE J. Quantum Electron.*, **18**, 1610–19.

Stegeman, G. I. and Seaton, C. T. (1985) Nonlinear integrated optics, *J. Appl. Phys.*, **58**, R57–R78.

Stegeman, G. I., Fortenberry, R., Karaguleff, C., Moshrefzadeh, R., Hetherington III, W. M., Van Wyck, N. E. and Sipe, J. E. (1983) Coherent anti-Stokes Raman scattering in thin-film dielectric waveguides, *Opt. Lett.*, **8**, 295–7.

Stegeman, G. I., Seaton, C. T., Chilwell, J. and Smith, S. D. (1984) Nonlinear waves guided by thin films, *Appl. Phys. Lett.*, **44**, 830–2.

Stegeman, G. I., Wright, E. M., Finlayson, N., Zanoni, R. and Seaton, C. T. (1988) Third order nonlinear integrated optics, *J. Lightwave Technol.*, **6**, 953–70.

Thakur, M. and Krol, M. (1990) Demonstration of all-optical phase modulation in polydiacetylene waveguides, *Appl. Phys. Lett.*, **56**, 1213–15.

Thylén, L., Finlayson, N., Seaton, C. T. and Stegeman, G. I. (1987) All-optical guided-wave Mach–Zehnder switching device, *Appl. Phys. Lett.*, **51**, 1304–6.

Tien, P. K. (1971) Light waves in thin films and integrated optics, *Appl. Opt.*, **10**, 2395.

Tien, P. K. and Ulrich, R. (1969) Theory of the prism–film coupler and thin-film light guides, *J. Opt. Soc. Am.*, **13**, 1325–37.

Townsend, P. D., Jackel, J., Baker, G. L., Shelburne, J. A. and Etemad, S. (1989) Observation of nonlinear optical transmission and switching phenomena in polydiacetylene-based directional couplers, *Appl. Phys. Lett.*, **55**, 1829–31.

Trillo, S., Wabnitz, S. and Stegeman, G. I. (1988) Nonlinear codirectional guided wave mode conversion in grating structures, *J. Lightwave Technol.*, **6**, 971–6.

Vach, H., Seaton, C. T., Stegeman, G. I. and Khoo, I. C. (1984) Observation of intensity-dependent guided waves, *Opt. Lett.*, **9**, 238–40.

Vitrant, G. and Arlot, P. (1987) Demonstration of optical bistability with a nonlinear prism coupler, *J. Appl. Phys.*, **61**, 4744–8.

Winful, H. G. and Stegeman, G. I. (1984) Applications of nonlinear perodic structures in guided wave optics, *Proc. Soc. Photo-Opt. Instrum. Eng.*, **517**, 214–18.

Yariv, A. and Nakamura, M. (1977) Periodic structures in integrated optics, *IEEE J. Quantum Electron.*, **12**, 233.

9 Nonlinearity in optical fibre communications

D. Cotter

9.1. INTRODUCTION

The development of optical fibre telecommunications during the period from the mid-1970s onwards has been one of the great success stories of modern technology. The introduction of optics into telecommunications systems is having an effect just as profound as that due to the introduction of semiconductor electronics in earlier years. With little doubt, the advent of optical fibre communication has also been the most important stimulus for the rapid expansion of effort worldwide in nonlinear optics during the 1980s decade and into the 1990s. Much of this effort has been directed towards the development of new nonlinear materials—low-dimensional semiconductor structures, organic polymers, photorefractive materials, and many others—as discussed elsewhere in this book. However, it was also discovered that the optical fibres development for optical communication can themselves exhibit nonlinear properties.

In general, optical fibres are fabricated from fused silica and other glasses which have nonlinear optical coefficients that are very small in comparison with more conventional 'nonlinear' materials. Despite this, nonlinear effects can have a profound influence on the light propagation in a fibre. Why should this be so? There are two main reasons. First, the tight confinement of the optical field within the core (whose cross-sectional area is perhaps only a few microns square) means that high intensities can be achieved at modest input power. Second, the very low transmission loss of present-day fibres in certain infrared wavelength ranges allows nonlinear effects to become significant during propagation over long path lengths (distances of metres up to tens of kilometres are typical; this may be compared with typical interaction lengths of a few millimetres in other nonlinear optical media in the research laboratory).

During the past decade a wide range of interesting nonlinear effects in fibres have been studied extensively. Some of these have found useful application in a variety of devices such as optical frequency convertors, tunable sources, pulse compressors, sources of ultrashort pulses and high repetition rate pulse trains, optical amplifiers and spectral filters, switching devices and optical logic elements. The list of devices and applications continues to grow. In a more fundamental way, the study of such nonlinear phenomena in fibres is making an important contribution to the rapidly evolving field of nonlinear physics.

Excellent and comprehensive reviews of nonlinear phenomena in optical fibres have appeared recently (Winful, 1986; Marcuse, 1988; Agrawal, 1989). In this chapter we shall consider a more specific topic: the influence of fibre nonlinearities in optical communications systems.

From early days in the development of optical fibre technology, it was recognized that nonlinear optical processes could present ultimate practical limitations on the range and data transmission capacity of communications systems (Smith, 1972; Stolen, 1980). It was noted that effects such as stimulated Raman scattering and stimulated Brillouin scattering could limit the optical power that can be transmitted, whilst self-phase modulation—resulting from the intensity-dependent refractive index—could result in broadening of the transmitted pulses thus limiting the maximum data rate. Despite this early concern, and despite also the very rapid development of optical communications systems during the past decade, nonlinear effects have had a negligible influence on commercial operational systems until now.

However, within the last few years the nonlinear properties of optical fibres have been shown to play an important role in determining the operating limits for several advanced experimental systems, and it seems inevitable that optical nonlinearity will be of increasing significance in the future.

In this chapter it will be assumed that the reader already has some familiarity with the structure and basic linear optical characteristics of optical fibres. These aspects have been reviewed in several texts (for example, Adams, 1981; Snyder and Love, 1983; Neumann, 1988; Miller and Kaminow, 1988).

The losses in silica fibres have now been reduced to levels close to the theoretical limits set by Rayleigh scattering and IR absorption. Major advances have been made in experimental optical communications systems by using erbium-doped fibre amplifiers (see section 9.5.3) to increase the transmission distance that can be achieved (for example, to allow submarine intercontinental communications using only all-optical repeaters) and to increase greatly the number of optical taps or couplers (for example, in a bus distribution network). Semiconductors and other types of solid state lasers capable of launching into a monomode fibre average powers in excess of 100 mW are becoming more widely available. Advanced communications systems will exploit the extraordinary coherence properties of some of these lasers: either single-frequency submegahertz CW operation or ultrashort transform-limited pulsed operation. Thus, a combination of factors such as very low fibre attentuation, long fibre span, higher optical power levels, and the high coherence of advanced laser transmitters means that nonlinear optical effects will certainly pose major systems limitations. In sections 9.2–9.4 the detrimental aspects of the principal nonlinear processes—stimulated Brillouin and Raman scattering, and parametric mixing—will be discussed.

On the other hand, it has now been realized that many of the nonlinear properties of optical fibre might also be utilized to beneficial effect in communications systems, and it is this possibility which is receiving greatest attention currently in many laboratories. Thus the possibilities of using fibres as optical amplifiers are discussed in section 9.5, whilst section 9.6 outlines the use of optical nonlinearity to overcome the limitations of pulse dispersion during transmission in a fibre. In section 9.7 the main conclusions are summarized, and we also indicate future directions: the use of optical fibres in nonlinear devices (such as ultrafast all-optical switches and logic elements) and the development of special fibres with enhanced nonlinear properties.

In this chapter the discussion will be focused almost entirely on nonlinear effects in monomode fibres. Monomode fibres are especially interesting media in which to study nonlinear optical effects; the transverse confinement is greater than for a multimode fibre, and also such waveguides support only one (or two orthogonally polarized) transverse modes of the field. This latter aspect, combined with the influence of 'group-velocity dispersion', is especially significant when we consider effects due to the nonlinear refractive index. In addition to this, monomode fibres are now favoured universally for advanced telecommunications applications and future signal processing devices will rely on monomode technology.

9.2. STIMULATED BRILLOUIN SCATTERING

The nonlinear process in fibres with the largest gain—and hence the lowest threshold in some important instances—is stimulated Brillouin scattering (SBS). SBS can be described as a classical three-wave interaction involving the incident light wave (called the 'pump'), a generated acoustic wave, and a scattered light wave. In the classical picture the pump wave (with frequency ω_p) is scattered by the acoustic wave. This scattered light wave is Doppler shifted to a frequency ω_s which is slightly different from ω_p since the acoustic wave acts as a moving diffraction grating. The scattered light and pump light together form a 'standing wave' (actually the 'standing wave' moves slowly since the two light waves are of slightly different frequencies). This optical 'standing wave' leads, via the process of electrostriction, to an increase in density at the antinodes and a decrease at the nodes. In fact this spatially periodic, moving structure of alternate high and low density coincides with the acoustic wave from which the initial scattering started, so a positive feedback situation exists. The pump light wave produces a scattered wave

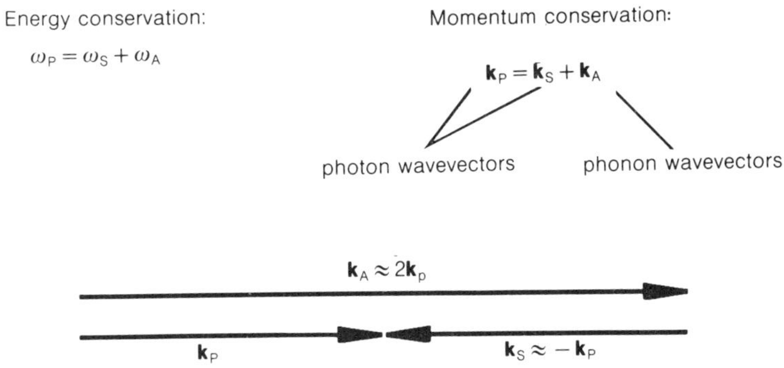

Figure 9.1 Phase matching diagram for stimulated Brillouin scattering.

which, via the 'standing wave' that is formed, then drives the acoustic wave—hence stronger scattering occurs, and so on. This is an example of a 'stimulated' scattering process. It can be initiated by the acoustic waves which are always present in the medium as a result simply of thermal excitation. When the energy conservation and momentum conservation ('phase matching') conditions are considered for the overall process, it is found that the process occurs most strongly for a scattered light wave travelling in the opposite direction to the pump wave. Thus a significant proportion of the optical power travelling in the fibre may be converted into the scattered wave which travels backwards towards the transmitter. The phase matching condition is depicted in Figure 9.1. The scattered wave (known as the 'Stokes wave') is shifted to a lower frequency with respect to the pump, by an amount equal to the acoustic frequency (or 'Stokes shift'):

$$\nu_A = 2nv_A/\lambda \tag{9.1}$$

where v_A is the acoustic velocity, n is the refractive index and λ is the optical wavelength. Taking the values $v_A = 5960\,\text{m}\,\text{s}^{-1}$, $\lambda = 1.55\,\mu\text{m}$, $n = 1.44$ for fused silica gives $\nu_A = 11.1\,\text{GHz}$. Experimentally observed SBS spectra for various optical fibres are shown in Figure 9.2.

SBS was first observed in optical fibre by Ippen and Stolen (1972). They used 5 m and 20 m lengths of fibre having a loss of $1300\,\text{dB}\,\text{km}^{-1}$, and the excitation source was a pulsed narrow band xenon laser operating at 535.5 nm. The threshold for SBS was found to be in the region of 1 W. Later experiments at modern optical communications wavelengths in the infrared region (1.3 and 1.5 μm) using multi-km lengths of low loss fibre have shown that SBS can occur at launched power levels of only a few mW (Cotter, 1983), which is well within the envisaged operating range of communications systems. SBS can be detrimental to such systems in a number of ways: by introducing severe signal attenuation, by causing multiple frequency shifts in some cases, and by introducing a high intensity backward coupling into

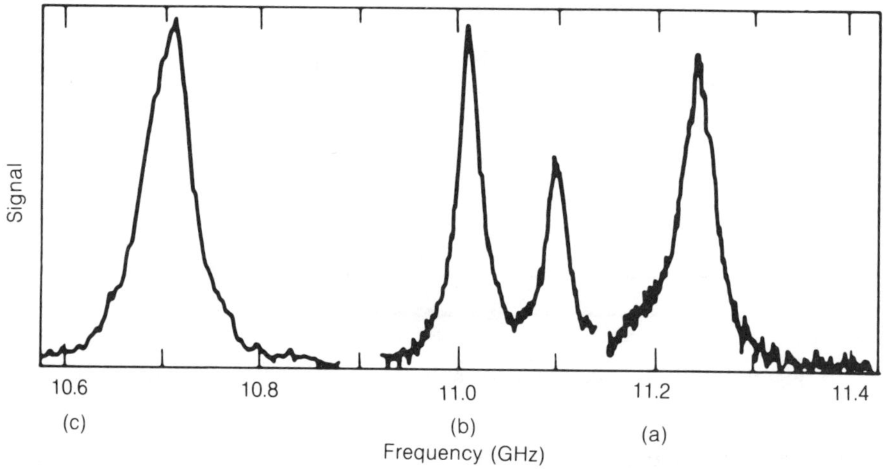

Figure 9.2 Brillouin gain spectra of three fibres at $\lambda = 1.5\,\mu$m: (a) silica core fibre, (b) depressed-cladding fibre, and (c) dispersion-shifted fibre. The horizontal scale refers to the Stokes Brillouin shift; the vertical scale is arbitrary. (After Tkach *et al.* (1986).)

the transmission optics. In some severe instances the intense backward travelling wave may cause fracture and cratering of the fibre input face. It will therefore be necessary to operate at power levels below the threshold for SBS, and this could place a severe limitation on the launch power.

The observed threshold powers for SBS are in close agreement with the theoretical values calculated using the bulk parameters of fused silica. The result of a simple stready state coupled-wave analysis (Smith, 1972; Cotter, 1983) is that the maximum laser power P_L which can be launched into the fibre before SBS becomes detectable is determined by

$$GL_{eff} \simeq 20 \tag{9.2}$$

where G is the SBS exponential gain factor given by

$$G = \frac{2\pi n^7 p_{12}^2 K P_L}{c\lambda^2 \rho_0 v_A \Delta\nu_B S} \quad \text{(units: m}^{-1}\text{)} \tag{9.3}$$

and where ρ_0 is the material density, p_{12} is the longitudinal elasto-optic coefficient, and S is an effective cross-sectional area of the guided mode such that the peak intensity is given by P_L/S. An assumption here is that the laser linewidth is small compared with $\Delta\nu_B$, which is the linewidth for spontaneous Brillouin scattering (Hz, FWHM) at ambient temperature. The factor K is unity for a fibre which maintains polarization, and is equal to one-half otherwise. The effective interaction length L_{eff} is given by

$$L_{eff} = \alpha^{-1}[1 - \exp(-\alpha L)] \tag{9.4}$$

where α is the absorption coefficient (m^{-1}) and L is the fibre length. For long fibre lengths used in communications it is usual for $L \gg \alpha^{-1}$, and hence $L_{\text{eff}} \simeq \alpha^{-1}$. Low loss fibres have longer interaction lengths and thus have lower SBS thresholds. Inserting typical values of parameters appropriate to multi-km lengths of present-day low loss monomode fibre (0.2–0.4 dB km^{-1}, $L_{\text{eff}} \sim 10$–20 km, $p_{12} \sim 0.28$, $\rho_0 \sim 2.2 \times 10^3$ kg m^{-3}, $A = 7.5 \times 10^{-11}$ m^2), together with the values of other parameters listed earlier, gives predicted threshold powers of about 1–5 mW, which are in good agreement with experiment.

It is found also that the SBS gain can be quite different depending on whether the pump bandwidth ($\Delta\nu_{\text{P}}$) is much less or greater than the Brillouin linewidth ($\Delta\nu_{\text{B}}$). In bulk fused silica $\Delta\nu_{\text{B}}$ is found to be 16 MHz at 1.55 μm, and varies as λ^{-2}, although the gain bandwidth in optical fibres can sometimes be much greater (as wide as ~ 100 MHz, as shown in Figure 9.2) as a result of fibre nonuniformity. In the limit when $\Delta\nu_{\text{P}} \gg \Delta\nu_{\text{B}}$, the SBS gain given by equation 9.3 is reduced in the ratio $\Delta\nu_{\text{B}}/\Delta\nu_{\text{P}}$, and thus the threshold pump power is increased by the inverse of this ratio. This explains why SBS was not observed in earlier communications systems which used multimode lasers with spectral widths of ~ 100 GHz or more. SBS does, however, have an important influence in the design of present-day advanced communications systems which use single-frequency lasers with spectral widths in the range 10 kHz–100 MHz. Such lasers are being employed to eliminate bandwidth limitations due to spectral mode hopping and fibre chromatic dispersion, and in coherent transmission systems.

Figure 9.3 depicts a typical experimental arrangement used to observe SBS in the laboratory, and Figure 9.4 shows the result of an experiment in which SBS was observed using a single-longitudinal-mode 1.32 μm Nd:YAG laser. The output power from each end of a 13.6 km monomode fibre cable (0.4 dB km^{-1} attenuation) is shown as a function of input power. At low input power the output power monitored in the backward direction was due only to the Fresnel reflection from the near and distant cleaved end faces of the fibre. However, at input powers exceeding 5 mW, the output power in the backward direction was seen to increase rapidly in a nonlinear fashion, and the efficiency of conversion to the backscattered wave reached 65%. At low input powers the power emitted from the distant end of the fibre was related linearly to the input power, determined by the linear loss of 5.6 dB. However, at input powers exceeding 6 mW, the output power became nonlinear. At input powers greater than 10 mW the output power in the forward direction reached a saturated maximum of about 2 mW. In more recent experiments with low loss fibre (0.2 dB km^{-1}) at 1.5 μm using a colour centre laser as the narrow band source, an SBS threshold as low as $\sim$2mW has been observed.

In many instances it will be necessary to suppress SBS. There are various methods; all use in some way the condition $\Delta\nu_{\text{P}} \gg \Delta\nu_{\text{B}}$ mentioned above. One useful technique is to use short pump pulses which consequently have a large

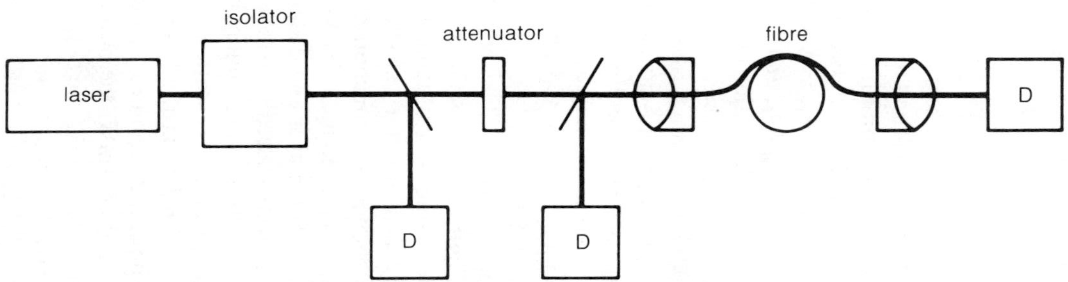

Figure 9.3 Schematic illustration of the experimental set-up used to observe SBS in optical fibres. The devices D are optical power meters or Fabry–Pérot interferometers, and are used to monitor the incident, transmitted and reflected signals.

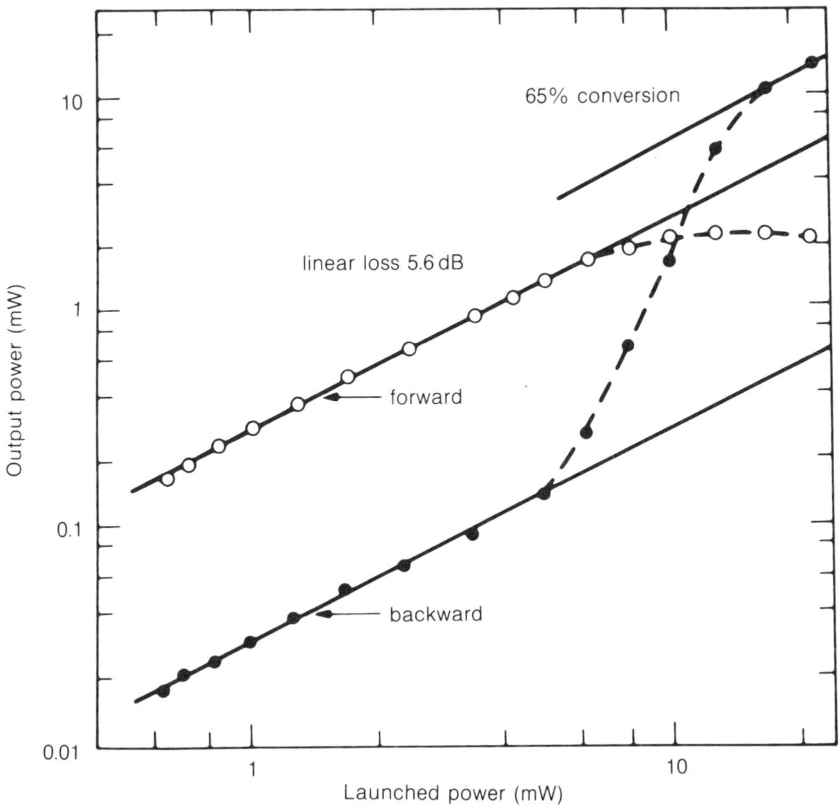

Figure 9.4 Transmitted (forward) and reflected (backward) powers as a function of the input power launched into a 13.6 km long single-mode fibre. (After Cotter (1983).)

bandwidth. Thus, although in the case of a narrow linewidth CW pump the SBS threshold (determined by equation 9.2) is lower than that for stimulated Raman scattering (described below), by using laser pulses of duration 1 ns or less the SBS threshold power can be increased to a level greater than for stimulated Raman scattering, which then becomes the dominant nonlinear scattering process. However, this method of suppressing SBS is unsuitable for those systems in which a narrow band laser is essential, such as telecommunications systems using coherent detection techniques (optical homodyne and heterodyne). Fortunately, in such systems the laser light is modulated before launching into the fibre, so the optical spectrum is substantially broader than the laser linewidth; the condition $\Delta\nu_P \gg \Delta\nu_B$ may then apply. A time-dependent analysis of transient stimulated scattering (Cotter, 1983; Aoki *et al.*, 1988) shows that the gain for SBS can be reduced to a useful degree by using certain digital modulation schemes. For example, whereas pure amplitude-shift keying or frequency-shift keying of the laser output results in only a small increase in the SBS threshold (in terms of the *average* transmitter power), a larger increase in threshold can be achieved by using phase-shift keying of appropriate magnitude, speed and format (Cosentino and Iannone, 1989). This approach has been successfully confirmed in recent experiments (Aoki *et al.*, 1988; Bolle *et al.*, 1989; Sugie *et al.*, 1990). 'Spread-spectrum' modulation techniques have also been proposed to suppress SBS (Hirose *et al.*, 1991).

In multichannel systems using wavelength-divison multiplexing, strong cross-channel interactions due to SBS can occur if the channel spacing is close to the Stokes frequency shift (equation 9.1) (Waarts and Braun, 1985). In low loss fibre at 1.5 μm a cross-talk level of -20 dB can be expected at power levels as low as $\sim 1\,\mu$W, which is a very severe limitation. However, the interaction extends over a frequency range of, at most, only a few hundred MHz centred around the SBS Stokes frequency shift given by equation 9.1, and so this cross talk problem is easily avoided. The very much broader gain bandwidth in stimulated Raman scattering (described below) means that cross talk due to SRS is potentially a much more severe problem.

The application of SBS for optical amplification is discussed in section 9.5.2.

9.3. STIMULATED RAMAN SCATTERING

The peak gain for stimulated Raman scattering (SRS) is 2–3 orders of magnitude lower than for SBS; nevertheless, for the reasons outlined in the previous section, SRS is likely to be the dominant nonlinear process when using a broadband or multimode optical signal.

Similar to SBS, SRS can be viewed as a coupled three-wave process in which the pump wave creates a frequency-down-shifted Stokes wave and a highly damped material excitation wave, which in this case corresponds to the vibrational mode of SiO_2 (Figure 9.5). The Raman gain coefficient for fused silica as derived from the spontaneous Raman spectrum (Stolen and Ippen, 1973) is shown in Figure 9.6. The quantity plotted is the gain coefficient per unit intensity: $g \equiv G/I$ where I is optical intensity (Wm^{-2}) and G is the exponential gain factor for SRS (m^{-1}). The gain is given for a pump wavelength λ of 1.0 μm, and varies as λ^{-1}. At 1 μm a shift of 15 THz corresponds to 50 nm. The different glasses used in standard fibre fabrication have somewhat different peak Raman gain coefficients, bandwidths and frequency shifts, but generally within a factor of 2 of those quoted here for fused silica.

The SRS threshold condition for significant SRS from noise (i.e. in the absence of input radiation in the Stokes frequency band), similar to the expression 9.2 given previously for SBS, is $GL_{\mathrm{eff}} \sim 16$ (Smith, 1972). To estimate the threshold for a multi-km length of low loss fibre at 1.5 μm, we can take the values $L_{\mathrm{eff}} = 10$ km, mode cross-section $= 50\,\mu\mathrm{m}^2$, and $g = 10^{-13}\,\mathrm{mW}^{-1}$ to obtain $P_{\mathrm{L}} = 0.8$ W. Because of the very wide gain bandwidth for SRS, this threshold power would apply equally to a narrow

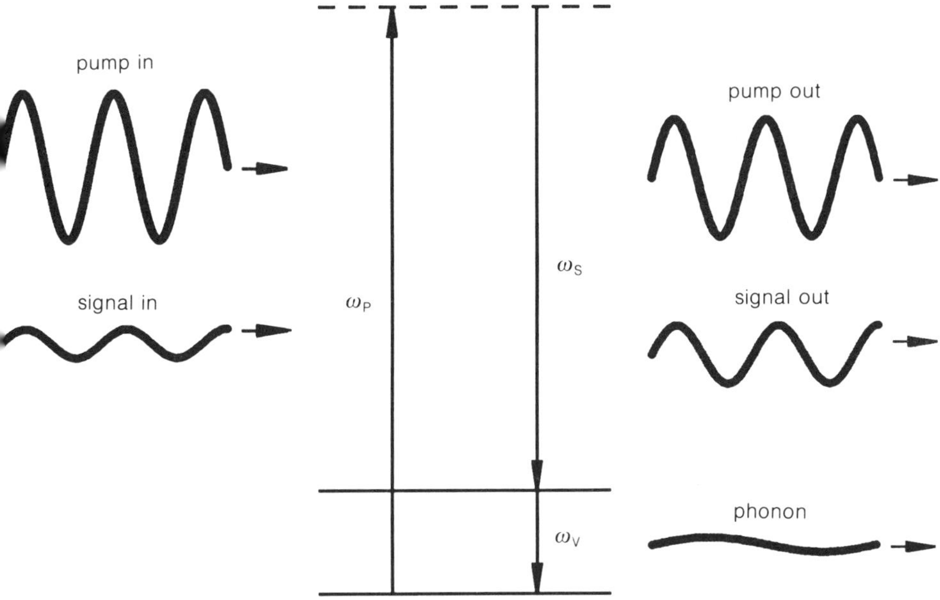

Figure 9.5 Schematic illustration of the process of stimulated Raman scattering. In this process one has simultaneously the absorption of a photon from the pump beam and the emission of a photon at the signal (Stokes) frequency, stimulated by the incident signal beam; the difference in energy is taken up by a high energy phonon (molecular vibration). The emitted photon is in phase with the signal beam. Thus stimulated Raman scattering provides for energy gain at the signal frequency at the expense of the pump.

linewidth CW laser source or one emitting pulses with duration as short as the femtosecond regime. Therefore, in contrast with SBS, pulsing or modulation techniques do not effect the SRS threshold. Nevertheless, the value for the threshold power quoted above, which is typical of that observed experimentally, can be regarded as not too serious a limitation for communications applications in which only one optical channel is present.

The deleterious effects of Raman amplification can appear at much lower powers in wavelength-division multiplexed system, however, where energy can be transferred to a longer wavelength channel; both the depleted and the amplified channels suffer cross talk in this way. The problem of cross talk due to SRS is potentially much more severe than for SBS (described earlier) because the gain extends over a very much wider frequency range (Figure 9.6) and so cross-coupling between adjacent channels is less easily avoided; significant SRS interaction can occur for channel spacings up to about 100 nm. The error rate in a communications system may be significantly degraded when the amplification of one wavelength by another is as little as 1% (−20 dB cross talk). The implies a limit of $GL_{\text{eff}} \simeq 0.01$ to avoid

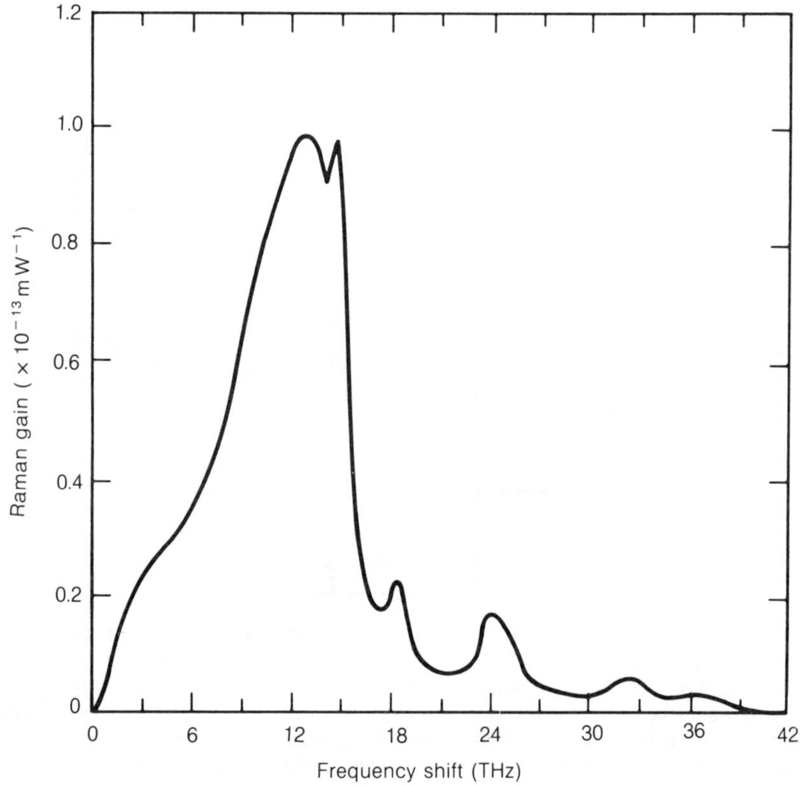

Figure 9.6 Measured Raman gain spectrum for fused silica at a pump wavelength $\lambda_p = 1\ \mu m$. The Raman gain scales inversely with λ_p. (After Stolen (1980); ©1980 IEEE.)

cross talk, rather than the stimulated Raman generation threshold condition $GL_{eff} \simeq 16$, as previously. If we consider the worst-case situation in which the channel separation corresponds to the Raman gain maximum (Figure 9.6) and take the parameter values given in the example above, then significant cross talk can occur at power levels as low as $\sim 1\ mW$.

Before leaving this discussion of SRS, we briefly mention that SRS in fibres provides a very useful source of widely tunable infrared radiation. Using a high power Q-switched mode-locked Nd:YAG laser (1.064 μm) as the pump, for example, the light emerging from the far end of a few tens of metres of monomode fibre comprises a spectral continuum extending from about 1.1 μm to beyond 1.6 μm. This emission is generated by a cascade of SRS orders (the first Stokes wave itself pumps a SRS process to generate the frequency-shifted second Stokes wave, which can pump a third Stokes wave, and so on up to about the fifth Stokes frequency, typically) and the emission is smeared out into a continuous spectrum by a complex sequence of parametric mixing and self-phase modulation processes (Agrawal, 1989). By passing this continuum through a monochromator to select a narrow band of frequencies, the result is a source of pulsed, wavelength-tunable radiation. Many laboratories around the world have used this as a source for measuring and testing optical components, and in particular its use for the measurement of group velocity dispersion in fibres is well known (Cohen and Chinlon Lin, 1978).

The use of SRS for optical amplification is described in section 9.5.1.

9.4. PARAMETRIC FOUR-WAVE MIXING

In the case that several optical frequencies are launched into a fibre simultaneously (for example in a wavelength-division-multiplexed (WDM) system), three waves of frequencies ω_1, ω_2 and ω_3 can interact via the third-order nonlinear susceptibility to generate a fourth wave at frequency ω_4 which obeys the energy conservation relation

$$\omega_4 = \pm\omega_1 \pm\omega_2 \pm\omega_3 \tag{9.5}$$

We will consider the particular case $\omega_1 + \omega_2 = \omega_3 + \omega_4$ because we are here concerned with processes involving frequencies all falling within the same spectral region. This type of frequency mixing could give rise to crosstalk in a WDM communications system. However, the process results in a significant transfer of power to the new frequency ω_4 only if the momentum conservation (phase matching) condition is exactly or nearly satisfied; strict conservation of momentum requires that

$$\Delta k = k_1 + k_2 - k_3 - k_4 = 0 \tag{9.6}$$

where the k_j are propagation constants in the fibre (Stolen and Bjorkholm, 1982).

Parametric mixing can also occur between two input waves (ω_P and ω_3), in which case the energy conservation relation becomes

$$2\omega_P = \omega_3 + \omega_4 \tag{9.7}$$

where $\omega_P\ (=\omega_1 = \omega_2)$ is referred to as the pump, and the lesser of ω_3 and ω_4 is called the Stokes frequency, and the greater is the anti-Stokes. The phase matching condition in this case is

$$\Delta k = 2k_P - k_3 - k_4 = 0. \tag{9.8}$$

Notice however that equations 9.7 and 9.8, which are a degenerate special case of equations 9.5 and 9.6, still describe a third-order nonlinear process and should properly be referred to as four-wave mixing although the term three-wave mixing is often (and confusingly) used in the literature.

In an optical waveguide the propagation constant at a given frequency is the sum of two contributions: the first (Δk_m) is determined by the bulk dispersion of the material, and the second is a contribution depending on the dispersive characteristic of the waveguide itself (Δk_w) (Stolen and Bjorkholm, 1982). The total phase mismatch is then a sum of two parts:

$$\Delta k = \Delta k_m + \Delta k_w \tag{9.9}$$

A good approximation to the contribution Δk_m resulting from material dispersion is

$$\Delta k_m = 2\mu\lambda D(\lambda)\Omega^2 \tag{9.10}$$

where $D(\lambda)$ is the group-velocity dispersion ($D(\lambda) = \lambda^2 d^2n/d\lambda^2$) and $\Omega = |\omega_P - \omega_1|/2\pi c$. This approximation does, however, neglect higher-order dispersion terms and is therefore invalid for pump wavelengths within a few nm of the wavelength (λ_0) for zero group-velocity dispersion ($D = 0$); the latter lies in the range 1250–1600 nm, depending on the waveguide design (section 9.6). In standard monomode fibres the waveguide dispersion Δk_w is quite small. Generally speaking, the material dispersion precludes phase matching in monomode fibre and thus four-wave mixing effects are negligibly weak. There are, however, important exceptions as described below.

Phase matching is possible in monomode polarization-preserving fibre by utilizing the large birefringence (Stolen *et al.*, 1981). For telecommunications applications we are mainly concerned with monomode nonpolarization-maintaining fibre, and in that case there are two situations in which phase matching could be obtained. The first situation is where the pump frequency is slightly longer than λ_0. In that case the small negative Δk_m compensates for Δk_w, and the total Δk given by equation 9.9 is zero. Washio *et al.* (1980) and Chinlon Lin *et al.* (1981) have reported frequency conversion from a 1.32 μm pump frequency to Stokes and anti-Stokes frequency pairs by this process, and the frequency shifts in this case are large—as great as 1600 cm^{-1} (Stokes and anti-Stokes wavelengths as long as 1.67 μm and as short as 1.09 μm respectively). The second situation is where the frequency shift Ω in equation 9.10 is sufficiently small that the coherence length L_{eff} for the parametric interaction ($L_{coh} = 2\pi/\Delta k$) is similar to or greater than the fibre length. For frequency shifts of about 50 GHz or less, the coherence lengths are on the order of kilometres. This is the situation which may arise in systems using wavelength-division multiplexing (WDM) with comb-spectrum sources. Since the gain coefficient for four-wave mixing can be as great as that of SRS (Stolen and Bjorkholm, 1982), this could in certain cases give rise to significant cross talk. This type of cross talk has been investigated by Shibata *et al.* (1986) and Waarts and Braun (1986). Four-wave mixing was observed in a coherent three-channel transmission system operating at 830 nm with a 3.5 km fibre length. With channel separations of ~ 10 GHz the measured parametrically generated power was up to 370 pW for input powers of hundreds of microwatts. The efficiency of generation decreased with carrier frequency separation Ω in the manner predicted by the calculated phase mismatch (equation 9.10). The four-wave mixing efficiency was negligibly small for frequency separations $|\omega_P - \omega_1|/2\pi > 50\,\text{GHz}$ with fibre spans greater than 3.5 km. Waarts and Braun (1986) showed that crosstalk due to four-wave mixing could be the dominant power-limiting mechanism in multichannel WDM systems of the type currently proposed for both trunk and local-loop telecommunications transmission. For example, in a 100-channel monomode fibre link with 5 GHz channel separation and −20 dB acceptable cross talk level, the input power per channel should be lower than 0 to −5 dBm (0 dBm ≡ 1 mW) for a fibre length of 5 km.

9.5. OPTICAL AMPLIFICATION

In the last few years much attention has focused on the use of nonlinear effects in fibres for optical amplification of communications signals. This approach is attractive because it may provide a cost-effective alternative to conventional (optical–electrical–optical) regeneration in transmission repeaters. Here we consider three approaches: fibre Raman amplifiers, fibre Brillouin amplifiers, and fibres doped with rare earth ions which exhibit gain by the process of stimulated emission. In the discussion of the first two of these approaches, it will be demonstrated that the inherent optical nonlinearity of the transmission path in a communications system can be utilized to great advantage.

9.5.1. Fibre Raman amplifiers

Until quite recently the fibre amplification process which received most attention was SRS (Koepf *et al.*, 1982; Desurvire *et al.*, 1983, 1985; Aoki *et al.*, 1982, 1985, 1986; Nakamura *et al.*, 1984; Nakasaka, 1985), although within the last 2–3 years this has been eclipsed somewhat by the advent of doped fibre amplifers (section 9.5.3). The most attractive feature of fibre Raman amplifiers is their very wide bandwidth (multi-THz, as shown in Figure 9.6). This is well suited to the amplification of ultrashort (subpicosecond) pulses (section 9.6). For optical communications, the wide gain bandwidth of fibre Raman amplifiers presents almost no practical limit on the transmission speed of the signal to be amplified, and there is also the possibility of using a single powerful pump to amplify several wavelength channels simultaneously.

Using, for example, a 1.32 μm Nd:YAG laser as the pump, CW signal amplification as high as 20 dB can be obtained in either forward- or backward-pumping configurations with very small spectral distortion of the signal (Aoki *et al.* 1982). However, the signal-to-noise improvement may be somewhat less than this because of spontaneous Raman scattering and low frequency fluctuations in the pump source. Moreover, the effect of these pump fluctuations is greatly increased if SBS is also allowed to occur (Koepf *et al.*, 1982). Hegarty *et al.* (1985), using a KCl:Tl colour-centre laser as the pump source, performed bit error rate measurements of SRS-amplified signals at 1 Gbit s^{-1}, and concluded that a net increase in receiver sensitivity of about 3 dB per 100 mW of pump power could be obtained at signal levels down to −30 dBm for unpolarized light in long monomode fibres at 1.5 µm. For a fibre loss of 0.2 dB km^{-1} this represents an increase in transmission distance of about 15 km per 100 mW of pump power. Olsson and Hegarty (1986) reported the measurements of the noise properties of a Raman optical amplifier. Measurements of CW Raman amplified optical signals from a 1.5 μm single-frequency laser in 32 km of low loss fibre, pumped by a 1.45 μm colour centre laser, revealed that the main noise source was the shot noise generated in the receiver due to the spontaneous Raman radiation. For signal levels down to as low as −50 dBm, no excess noise due to the SRS process was detected. Previously Hegarty *et al.*, (1985) had found that by using narrowband spectral filters before the detection apparatus they could discriminate against spontaneous Raman scattering such that it was not significant for pump powers up to 100 mW and signals levels down to −30 dBm.

An approach used by Aoki *et al.* (1985, 1986) to suppress SBS during SRS amplification was to use a mode-locked pump laser in a backward-pumping configuration. In this case the SBS gain was reduced relative to the case of CW pumping because of the increased spectral width of the pump laser. This allowed signal amplification as high as 14 dB, with a power penalty due to gain fluctuations as low as 1 dB.

In Raman amplification the signal frequencies which can be amplified are determined by the pump frequency and the Raman gain curve for the fibre material (e.g. Figure 9.6). This can pose a restriction because the range of suitable pump lasers capable of providing the necessary power levels is limited (until recently only Nd:YAG, colour centre lasers and Raman-shifted derivatives of these, although very high power semiconductor lasers are now becoming more widely available). Nakamura *et al.* (1984) showed that amplification at the important 1.5 µm wavelength can be obtained by using the second Stokes line (at about 900 cm^{-1} shift frequency) in fused-silica fibre excited by a 1.32 µm Nd:YAG laser. However, in this case, second Stokes amplification can only occur for input pump powers greater than the threshold for stimulated first Stokes scattering from noise; consequently the gain is relatively low (24 dB gain for 9 W input power was obtained) and the signal-to-noise ratio is about an order of magnitude worse than for the corresponding first Stokes amplifier. Another approach is to use an alternative Raman gain medium having a larger vibronic Stokes shift. Molecular deuterium has the required Stokes shift (2988 cm^{-1}) to allow optical gain at 1.56 µm wavelength when pumped with a 1.06 µm Nd:YAG laser. Chraplyvy *et al.* (1983) dissolved deuterium gas in a conventional low loss monomode silica fibre for a few hundred hours at high pressure to achieve a D_2 concentration of $4 \times 10^{20}\,cm^{-3}$. By maintaining this gas-doped fibre at liquid nitrogen temperature, the rate of subsequent out-diffusion is negligible. Using pulses of 130 W peak power from a *Q*-switched Nd:YAG laser to pump 100 m of fibre, the probe signals from a InGaAsP injection laser at 1.56 μm were amplified by 36 dB to an output power of 150 mW. The measured stimulated Raman gain coefficient at 1.56 μm was $7 \times 10^{-14}\,m\,W^{-1}$, about equal to the peak Raman gain of SiO_2 at the same wavelength (Figure 9.6).

As described in section 9.6, SRS has been used to compensate for fibre loss in soliton pulse propagation.

9.5.2. Fibre Brillouin amplifiers

Stimulated Brillouin scattering is another nonlinear amplification mechanism which has been investigated recently for application in advanced telecommunications systems. The various techniques described below provide particularly good examples of how fibre nonlinearity can be harnessed in subtle ways to perform a variety of signal-processing functions.

For many years, SBS amplification was disregarded for optical communications applications because the gain bandwidth of typically 10–20 MHz in bulk silica (see section 9.2) was considered to be much too narrow to amplify effectively the optical carrier and sidebands in operational systems which typically carry data at rates of at least several tens of $Mbit\,s^{-1}$. However, as mentioned in section 9.2, the gain bandwidth in optical fibre can in some cases be several times greater than the value observed in small bulk samples of glass. The feasibility of using SBS for amplification in optical communications, at least for systems operating at modest bit rates, was demonstrated by Olsson and Van der Ziel (1986). They obtained 15 dB of gain in a 37.5 km fibre, cancelling the fibre loss, and giving 5 dB of net gain. The gain occurred over a 150 MHz bandwidth and the pump power was a mere 3.5 mW. Later the same authors developed the technique of broadening the effective gain bandwidth by pump laser frequency sweeping (Olsson and Van der Ziel, 1987), and applied it to the amplification of data-carrying signals at 10 to 90 $Mbit\,s^{-1}$.

A method for exploiting the SBS gain in much wider bandwidth systems was devised by Atkins *et al.* (1986). They showed that SBS can be used to amplify selectively the narrowband carrier component of the received optical signal in a coherent transmission system prior to photodetection. With sufficient carrier gain it should be possible to overcome the inherent noise of the post-detection electronic amplifier and thus to achieve quantum-noise-limited detection. This offers the advantage of increased sensitivity similar to that of coherent optical homodyne detection, but without the formidable engineering problems of phase locking the local oscillator laser in a coherent homodyne receiver. To obtain the necessary Brillouin gain it is necessary to have a pump laser at the receiver to feed energy back down the fibre path

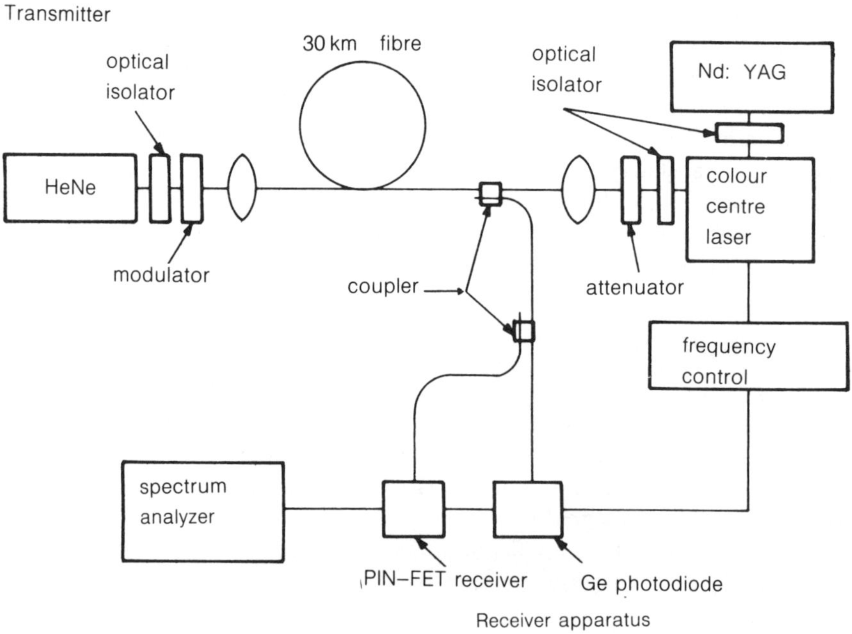

Figure 9.7 Experimental arrangement for selective carrier amplification by stimulated Brillouin scattering in a coherent homodyne transmission system. (After Atkins *et al.*, (1986).)

towards the transmitter. As we have already seen in section 9.2, significant SBS gain can be obtained at mW power levels in long fibre lengths (>10 km) at 1500 nm. The theoretical carrier amplification at a pump power of only 1 mW is 45 dB. The experimental arrangement, shown in Figure 9.7, used a KCl:Tl colour centre laser as the pump and an externally modulated 1520 nm He–Ne laser as the transmitter. The colour centre laser is locked to a frequency some 11 GHz (the Brillouin Stokes shift—equation 9.1) greater than the transmitted carrier. It was shown that carrier amplification by as much as 40 dB could be achieved. In a demonstration of homodyne detection by this technique the transmitted signal was modulated, and with approximately 5 mW of pump power the detected sideband signal was observed to increase by 25 dB (Atkins *et al.*, 1986). In this first published demonstration of the technique, the signal modulation frequency was 80 MHz; in later work this was successfully increased to ~ 1 GHz with similar signal gain. This 'self-homodyne detection' technique using a fibre Brillouin amplifier is likely to offer considerable benefits in terms of signal-to-noise ratio, especially for future systems operating at rates of $\sim 10\,\text{Gbit s}^{-1}$.

An inevitable accompaniment of this narrowband SBS gain is the nonlinear refractive index dispersion (due to the imaginary part of the SBS gain coefficient). This dispersion is manifest as an intensity- and frequency-dependent phase shift of the amplified carrier signal. Cotter *et al.* (1986) showed that this dispersion imposes the most stringent limit on pump and signal frequency stability for self-homodyne detection. For self-homodyne detection of amplitude-shift keyed (ASK) signals, a phase stability of about ± 0.1 rad for the amplified signal carrier may be required, and in a typical case this implies that the pump signal frequency offset should be controlled to within 100 kHz. It was also shown that the nonlinear dispersion can be usefully exploited in those self-homodyne schemes for which specific carrier phase shifts are required. For example, to implement self-homodyne detection of a phase-shift keyed (PSK) signal by amplification of the vestigial carrier, a quadrature phase correction is required. In a typical case, this requires a frequency offset of ~ 1 MHz from the SBS gain line centre, which is well within the SBS 3 dB power gain bandwidth of several MHz. In this case, for a phase stability of ± 0.1 rad, a pump power stability of about $\pm 6\%$ and frequency stability of ± 100 kHz would be required. Although this required degree of frequency stability may appear severe, in practice it is sufficient to lock the pump laser to the optimum frequency using an arrangement similar to that shown in Figure 9.7.

Tkach and Chraplyvy (1989) have shown that the narrow band nature of fibre Brillouin amplifiers can be used to good effect to select channels in densely packed wavelength-division-multiplexed (WDM) systems. Using dispersion-shifted fibres with triangular core dopant profiles, gain bandwidths of 100 MHz were observed, and 25 dB of gain at 1.5 μm was achieved with 14 mW of pump power; 45 Mbit s^{-1} channels as closely spaced as 140 MHz could be successfully demultiplexed (Chraplyvy and Tkach, 1986). The same authors demonstrated optical demodulation of 150 Mbit s^{-1} signals, using the Brillouin gain to amplify only the 'marks' of the signal. In this case a gain of 30 dB was achieved over a 600 MHz bandwidth with 12 mW of pump power, using AlGaAs lasers operating at 0.83 μm (Tkach *et al.*, 1988). When used in a WDM–FSK network, this elegant technique offers considerable advantages of simplicity since the fibre Brillouin amplifier can perform simultaneously the functions of amplification, channel selection and demodulation of the FSK signals.

9.5.3. Rare-earth-doped fibre amplifiers

Another way of overcoming the optical loss in a fibre link is to insert a 'doped fibre amplifier'. This consists of a length of monomode fibre, the core of which is doped with certain rare earth ions during fabrication. When this special fibre is excited optically by a comparatively strong pump source, it provides gain for a weak incident signal in a wavelength range longer than the pump wavelength; the gain occurs via the process of stimulated emission. The wavelength range in which gain is obtained depends, amongst other things, on the particular rare earth dopant that is used. Of particular interest for optical communications applications are fibre amplifiers doped with erbium ions. These can exhibit gain over all or at least a major part of the wavelength range 1.51–1.60 μm, which coincides with the important 'third telecoms window' in which conventional undoped silica fibres exhibit the lowest loss.

It is no exaggeration to say that the recent rapid development of erbium-doped fibre amplifiers is causing a revolutionary change in the design approach for both long-span links and local distribution networks for optical communications. As examples, transoceanic optical cables incorporating fibre amplifiers as all-optical repeaters are being planned currently, and local networks containing fibre amplifiers to allow simultaneous distributions of signals to many thousands of customers are under consideration.

A comprehensive review of erbium-doped fibre amplifiers appears in a book which is the companion to this one. Rare-earth-doped fibre amplifiers are not nonlinear optical devices in the sense implied in this book. Nevertheless, a brief description is included here because these amplifiers will in future have an increasingly significant role in nonlinear optical applications of fibres in telecommunications. For example, by using in-line amplification, the passive losses in a length of optical fibre can be compensated so that the effective length for nonlinear interactions may be greatly increased (to hundreds or thousands of kilometres). Fibre amplifiers may also be used to increase the optical power in a telecommunications system to levels at which nonlinear processes become prominent. In addition, as described in section 9.6, erbium-doped fibre amplifiers provide an attractive alternative to stimulated Raman scattering for the amplification of short pulses and long-distance propagation of solitons.

Figure 9.8 depicts the typical layout of a rare-earth-doped fibre amplifier. A fused fibre coupler is used as a WDM element to combine a low power optical signal and a relatively high power pump source (typically 10–500 mW). The pump and signal wavelengths denoted in the figure are particular examples that can be used with an erbium-doped fibre amplifier, but alternatives are described below. After efficient combination by the WDM coupler, the pump and signal wavelength are passed to a section of doped fibre (which is typically 1–20 m in length). This fibre is usually monomode at the signal wavelength and is fabricated so that typically 1–500 ppm rare earth ions are incorporated into the glass matrix in the fibre core region. The rare earth (lanthanide) elements have the property that they are readily ionized to the trivalent form by removal of one electron from an inner shell (4f) and two from the outermost shell (6s) which is thus emptied completely. The 5s and 5p electronic shells remain filled. The important optical properties of the rare earth ions involve excitations of the remaining electrons in the 4f shell. This configuration means that the optically active electrons are relatively well screened by the 6 filled $5s^2 5p^2$ orbitals from the environment in which the ion finds itself.

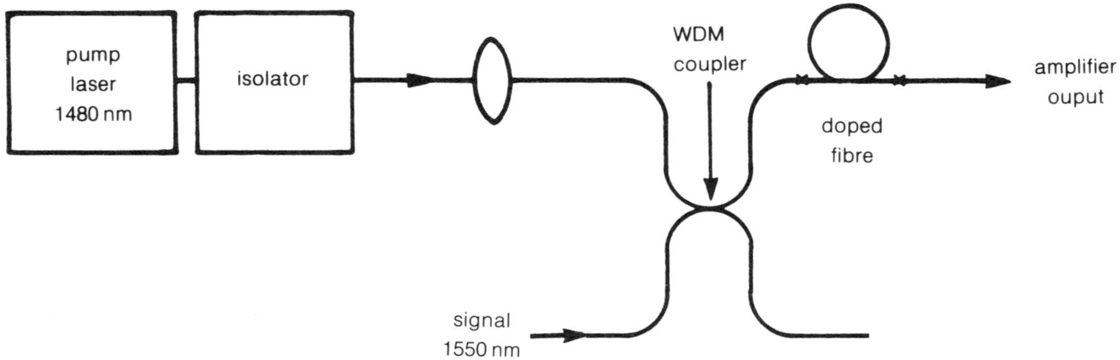

Figure 9.8 Typical layout of an erbium-doped fibre amplifier.

Therefore, compared with other kinds of ionic dopants in glass (transition metals, for example), the electronic transitions of the rare earth ions in glass are similar to those of isolated ions (the transitions of the ions in glass are only slightly shifted in energy and somewhat broadened). Figure 9.9 depicts a part of the energy level diagram of Er^{3+} ions, and shows the most important optical transitions. The optical transition of interest for amplification in the region of 1.55 μm occurs between the $^4I_{13/2}$ and $^4I_{15/2}$ states. This is an example of a three-level laser transition (Siegman, 1986); the lower level ($^4I_{15/2}$) is also the ground state of the ion. Therefore, in common with all three-level lasers, absorption from the ground state to the upper level of the gain transition acts in direct competition with the creation of photons by stimulated emission. Whereas this is a great disadvantage in a conventional laser amplifier which uses a bulk component as the gain medium, in a fibre amplifier high gain can be achieved easily at modest input pump power because of the tight transverse optical confinement (and thus high pump intensity) that can be maintained over a long path length.

Three alternative pump wavelengths for an erbium-doped fibre amplifier

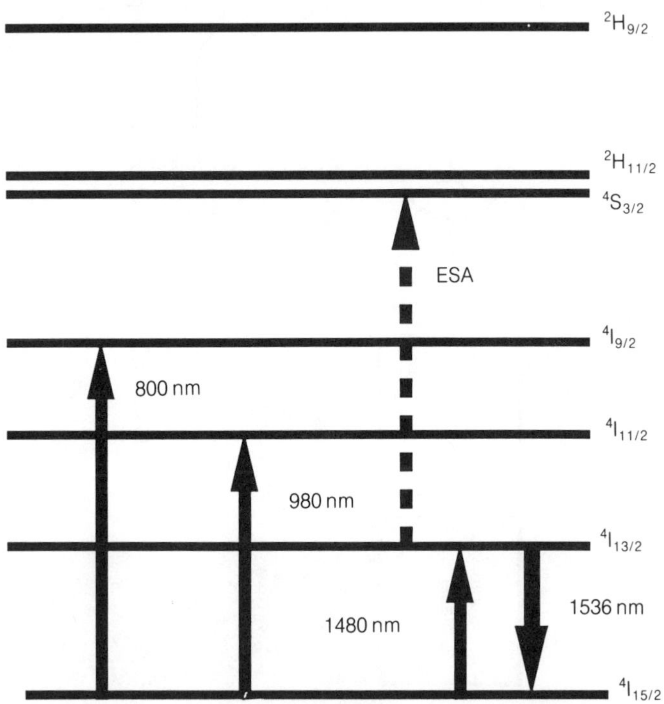

Figure 9.9 Partial energy level diagram of Er^{3+} showing three possible pump wavelengths (800, 980 and 1480 nm) for signal amplification in the region of 1536 nm (diagram not to scale). Pump radiation at 800 nm suffers excited state absorption (ESA).

are depicted in Figure 9.9: 800 nm, 980 nm or 1480 nm. In each case, ions are excited from the ground state to an excited level; the $^4I_{13/2}$ upper level of the gain transition is thus populated directly or via nonradiative decay from the directly excited level. For a typical erbium-doped alumina–silica fibre the $^4I_{13/2}$ level has a spontaneous lifetime which has been measured to be 9.8 ms; this is long enough to allow population inversion and thus optical gain when using a CW source as the pump. Each of the three pump schemes depicted has relative advantages and disadvantages. At first, pumping at 800 nm appeared particularly attractive because of the ready availability of GaAs semiconductor laser sources at that wavelength. However, it was soon discovered (Armitage *et al.*, 1988) that there is an unfortunate coincidence between the pump wavelength and an absorbing transition from the upper laser level; this 'excited state absorption' (ESA) severely reduces the pumping efficiency. In contrast, pumping at 980 nm from the ground state to the $^4I_{11/2}$ level provides efficient amplification because there is no significant ESA. Quantum efficiencies exceeding 80% for the conversion of pump photons at 980 nm to signal photons at ~1550 nm can be achieved (Massicott *et al.*, 1990). Until recently a drawback of the 980 nm pumping scheme for widespread practical application was the scarcity of suitable pump sources. However, during the last two years considerable effort has been devoted to the development of strained-layer semiconductor lasers that can provide the necessary power levels (> 10 mW) for optical pumping of erbium-doped fibre amplifiers at 980 nm, and at the time of writing these are becoming more widely available. The third scheme depicted in Figure 9.9 is pumping at 1480 nm, which lies within the high energy wing of the gain transition itself. Semiconductor lasers that may be used as pump sources at 1480 nm are already widely available. Pump power levels at 980 nm or 1480 nm as low as ~1 mW in a doped fibre can be sufficient to bleach the absorption at ~1550 nm. As shown in Figure 9.10, small signal optical gains in excess of 20 dB can be achieved over a fairly wide spectral range (~1.53–1.56 μm) at pump power levels of a few tens of milliwatts. Small signal gains as high as 45 dB have been reported (Kimura *et al.*, 1989). The preferred choice between 980 nm and 1480 nm pumping is not yet clear; the factors to be considered include differences (albeit small) in the amplifier noise figure, relative complexity of WDM coupler design, pump source reliability and cost. The main advantage of a doped fibre amplifier compared with a Raman amplifer is now obvious, however; the pump power requirement is greatly reduced, allowing the use of a relatively inexpensive semiconductor laser as the pump source.

Figure 9.11 shows the various ways in which an erbium-doped fibre amplifier could be incorporated into a telecommunications system. In Figure 9.11(a) an amplifier is being used to increase the power of a low-level signal. Recent results have demonstrated that erbium-doped silica fibres can be used as very efficient, high gain power amplifiers at 1.55 μm (Massicott

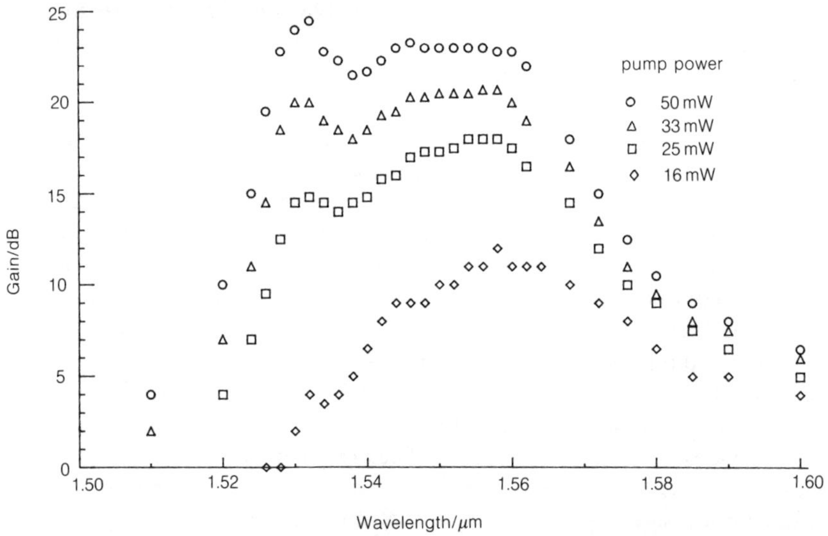

Figure 9.10 Small signal gain spectrum (in dB) for an erbium-doped fibre amplifier pumped at 1.49 μm with the power levels shown. (After Atkins *et al.* (1989).)

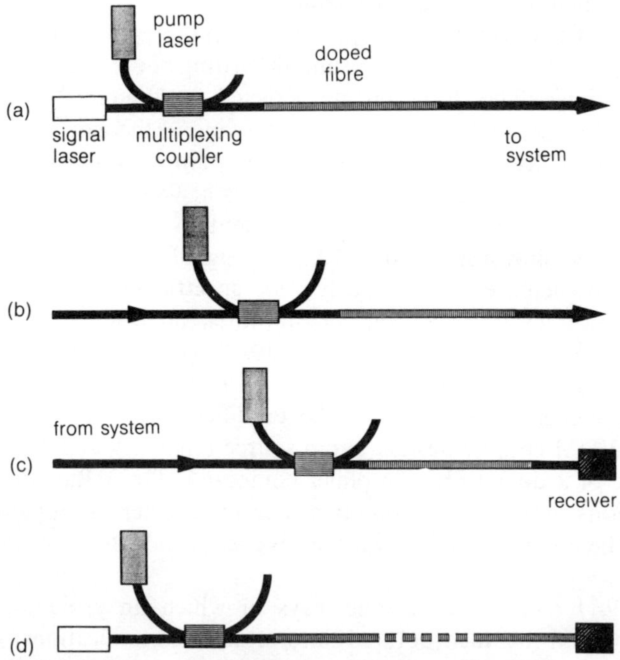

Figure 9.11 Four possible applications of a rare-earth-doped fibre amplifier in a telecommunications system. (a) power amplifier; (b) in-line amplifier; (c) preamplifier; (d) long span amplifier.

et al., 1990). It has been shown, for example, that for a pump wavelength of 1.48 μm, 140 mW output power can be obtained for a signal input of 1 mW at 180 mW pump power. Pumping at 0.98 μm allows output powers greater than 500 mW to be obtained for an input signal of 1 mW. The configurations (b) and (c) shown in Figure 9.11 allow an optical signal in a telecommunications system to be amplified either during transmission or immediately prior to detection. The 'optical repeater' configuration (b) can be used to overcome the effects of loss in a long transmission span, but will not, of course, correct the effects of dispersion. This latter limitation and ways of overcoming it are discussed further in section 9.6. A further possibility, shown in Figure 9.11(d), is to fabricate long lengths of fibre having a low level of erbium (< 1 ppm). Then, instead of having a 'lumped' amplifier as in the configurations (a)–(c), the distributed gain in such a fibre could be adjusted to compensate exactly the optical loss giving a long-span 'transparent' transmission medium. This type of fibre is being developed mainly for application in optical bus distribution networks, but it may also play a useful role in long-distance soliton transmission (section 9.6). As shown in Figure 9.11, co-directional pumping (pump and signal waves travelling in the same direction) is usually preferred since it offers an advantage of lower noise compared with counter-directional pumping because the build-up of spontaneous emission is thus reduced. Finally, by providing mirrors at each end of a doped fibre so as to reflect the signal wavelength, laser oscillation can be obtained easily, and it seems likely that in the future 'fibre lasers' will be increasingly important coherent sources for telecommunications and other applications.

9.6. NONLINEAR REFRACTIVE INDEX

This section is concerned with the influence of the intensity-dependent refractive index on fibre communications systems. As described elsewhere in this book, the refractive index of any medium reveals a dependence on the incident light when sufficiently high intensities are involved. It is usually sufficient to consider only the first nonlinear term, which may be written as

$$n = n_0 + 2n_2\langle \mathbf{E}\cdot\mathbf{E}\rangle \tag{9.11}$$

where the index of refraction n , which has a value n_0 at low intensity, is modified at higher incident intensities by an amount determined by the nonlinear refraction coefficient n_2. Another name for this intensity dependence of the refractive index is the optical Kerr effect. Here $\langle \mathbf{E}\cdot\mathbf{E}\rangle$ denotes the time average of the applied optical field $\mathbf{E}$. For a linearly polarized wave $\langle \mathbf{E}\cdot\mathbf{E}\rangle$ is $|\mathrm{E}|^2/2$ where E is the peak field amplitude, and it is related to the intensity I by $I = \varepsilon_0 cn\langle \mathbf{E}\cdot\mathbf{E}\rangle$. Very often equation 9.11 is written in an alternative form:

$$n = n_0 + n_2^I I \tag{9.12}$$

where $n_2^I = 2n_2/\varepsilon_0 n n_0$. The value of n_2^I (in MKS units) for fused silica is $3 \times 10^{-20}\,\mathrm{m^2\,W^{-1}}$. Significantly higher values of n_2^I are found in certain other glasses (section 9.7). The origin of n_2 in glasses such as silica is mainly the nonlinear response of the electronic motion; this involves a nonresonant virtual electronic excitation which responds very rapidly on a femtosecond time scale.

In a medium that exhibits nonlinear refraction, any spatial and temporal variations of the optical intensity will cause the refractive index to be space and time dependent also. Consider, for example, a beam with a Gaussian radial distribution of intensity as it passes through a bulk nonlinear medium. The induced index change will follow this radial intensity distribution. If the sign of n_2 is positive (as, for example, in the case of fused silica and many other glasses), the refractive index increases with intensity and so the medium acts as a convex lens giving rise to self-focusing; for negative n_2 the effect is self-defocusing. This is one of the most important manifestations of nonlinear refraction in bulk solids. In monomode optical fibres, however, the nonlinearity can be considered to be sufficiently weak ($n_2|E|^2 \ll n_0$) that such transverse effects may be neglected; instead, much more important effects occur as a result of the time variations of the intensity which bring about phase changes in the optical carrier wave (Stolen and Chinlon Lin, 1978). We now consider these temporal effects in more detail.

In the plane-wave limit, the propagation equation for a pulse $E(z,t)$ travelling in the z direction in a medium whose refractive index follows the Kerr law 9.11 can be expressed in the form

$$\mathrm{i}\left(\frac{\partial}{\partial z} + \frac{\mathrm{d}k}{\mathrm{d}\omega}\frac{\partial}{\partial t}\right)E - \frac{1}{2}\frac{\mathrm{d}^2k}{\mathrm{d}\omega^2}\frac{\partial^2}{\mathrm{d}t^2}E = -\frac{\omega n_2}{c}|E|^2 E \tag{9.13}$$

where loss is neglected and the group velocity v_g is $(\mathrm{d}k/\mathrm{d}\omega)^{-1}$. Let us investigate the solutions of this equation. To begin with, we consider the most simple situation in which the second-order dispersion $\mathrm{d}^2k/\mathrm{d}\omega^2$ is negligible. In this regime, the effect of the nonlinear term on the right-hand side of equation 9.13 is known as 'self-phase modulation'. To examine this case, we can further simplify equation 9.13 by eliminating $\partial/\partial t$ with the substitution of a reduced time scale $\xi \equiv t - z/v_g$. A particular value of ξ therefore represents a fixed position in a coordinate frame which travels at the group velocity in the direction of propagation. With E expressed as a function of z and ξ, the propagation equation 9.13 become

$$\frac{\partial E}{\partial z} = \frac{\mathrm{i}\omega n_2}{c}|E|^2 E \tag{9.14}$$

It is straightforward to verify that the solution of equation 9.14 is

$$E(z,\xi) = E(0,\xi)\exp\left(\mathrm{i}\frac{\omega n_2}{c}|E(0,\xi)|^2 z\right) \tag{9.15}$$

This is a pulse whose intensity is unaltered during propagation and which experiences a phase shift ϕ due to self-phase modulation:

$$\phi = \frac{\omega n_2}{c} |E|^2 L \tag{9.16}$$

after a distance L. Since the intensity $|E|^2$ varies throughout the pulse ($|E|^2$ is a function of ξ), the various parts of the pulse undergo different phase shifts, leading to a frequency shift or 'chirp' which is directly proportional to the distance travelled. Suppose, for example, the field $E(z, t)$ is initially in the form of a chirp-free (constant-phase) pulse $E(0, t) = E_0 \operatorname{sech}(t/\tau)$, where τ is a measure of the pulse duration. After such a pulse has travelled a distance L in the medium, the difference between its instantaneous optical frequency (at time $t + L/v_g$) and the carrier frequency ω is obtained by differentiating the phase shift 9.16 with respect to t.

$$\delta\omega = -\frac{d\phi}{dt} = -\frac{\omega n_2}{c} |E_0|^2 L \frac{d}{dt} \operatorname{sech}^2(t/\tau) \tag{9.17}$$

This is illustrated in Figure 9.12. Assuming n_2 is positive, the instantaneous frequencies in the leading half of the pulse are lowered (i.e. there is a red shift), whereas those in the trailing half are raised (blue shifted). The total frequency excursion is $2|\delta\omega_{max}| = 1.54\,\Delta\phi_{max}/\tau$, where $\Delta\phi_{max} = \omega n_2 |E_0|^2 L/c$ is the phase change at the peak of the pulse. This spectral broadening due to self-phase modulation becomes significant when the frequency excursion approaches and exceeds the initial bandwidth of the pulse, which is equal to $1.12/\tau$.

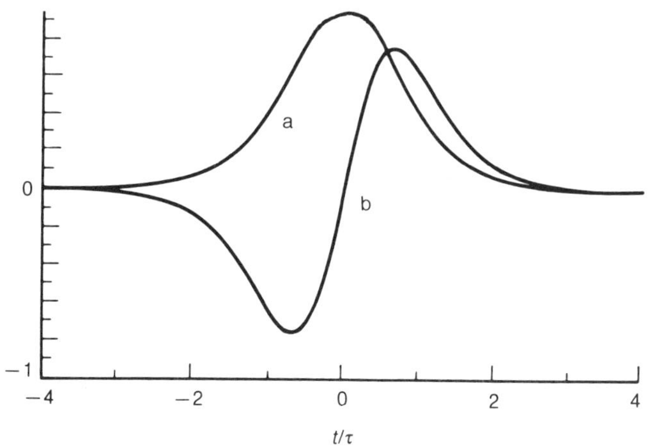

Figure 9.12 The function $\operatorname{sech}^2(t/\tau)$ representing pulse intensity (curve a) and the function $-d[\operatorname{sech}^2(t/\tau)]/dt$ representing the corresponding frequency chirp (curve b) arising from self-phase modulation in a Kerr law nonlinear medium.

We now consider the influence of group-velocity dispersion, which arises when the term $d^2k/d\omega^2$ in equation 9.13 is not zero; in that case the group velocity $v_g \equiv (dk/d\omega)^{-1}$ varies with frequency. (The dispersion term can be written as $d^2k/d\omega^2 = d(v_g^{-1})/d\omega$, where v_g^{-1} is the group delay per unit distance. It is also a common practice, particularly in the optical fibre field, to refer to the group-velocity dispersion in terms of the wavelength dispersion of the group delay, which is $d(v_g^{-1})/d\lambda$. Notice that $dv_g/d\omega$ and $d(v_g^{-1})/d\lambda$ have the same sign, which is the opposite of the sign of $d^2k/d\omega^2$.) The various frequency components within the bandwidth of the pulse travel at different group velocities. It is not difficult to see that, if the medium exhibits a group-velocity dispersion $dv_g/d\omega$ that is positive ($d(v_g^{-1})/d\lambda > 0$), the leading half of the pulse shown in Figure 9.12 is delayed slightly and the trailing half is advanced, leading to pulse narrowing. If, however, $dv_g/d\omega$ is negative ($d(v_g^{-1})/d\lambda < 0$), then we see enhanced pulse broadening. This enhanced effect reinforces the pulse-spreading phenomenon that is well known in the linear optics of dispersive media because any pulse must contain a spread of frequencies determined by its Fourier transform.

Acting in combination with group-velocity dispersion, the nonlinear refractive index can give rise to many interesting and useful pulse-propagation phenomena. As noted in the introduction to this chapter, monomode optical fibres are especially interesting media in which to study these effects; such waveguides support only one transverse mode of the electric field. In the weakly nonlinear regime (which, as mentioned earlier, is a very good approximation for glass fibres), the only significant effect of the nonlinearity is to modify the propagation constant (magnitude of the wavevector) for the guided mode. In other words, despite the transverse distribution of intensity, all parts of the travelling wavefront in a monomode fibre experience the same phase shift. Propagation in a monomode fibre is therefore described rather well by the uniform plane-wave equation 9.13, in which E is now interpreted as the maximum field on the axis.

Figure 9.13 shows the propagation loss of a good quality monomode fibre as a function of wavelength, together with a plot of group-velocity dispersion with wavelength. The material group-velocity dispersion for silica fibres passes through zero at a wavelength close to 1300 nm. The total group-velocity dispersion of a fibre consists mainly of the material group-velocity dispersion plus an additional component, called the waveguide dispersion, which depends on the structure, dimensions and refractive index profile of the fibre. A third, and usually much smaller, component of the total dispersion arises from the fact that the refractive index profile is itself wavelength dependent; this is termed profile dispersion. By fabricating fibres of different designs it is possible to manipulate the waveguide and profile dispersion so that the wavelength λ_0 at which the total group velocity dispersion is zero can be positioned almost anywhere in the range of about 1250–1600 nm. This means that, at any particular wavelength in the spectral range of main interest

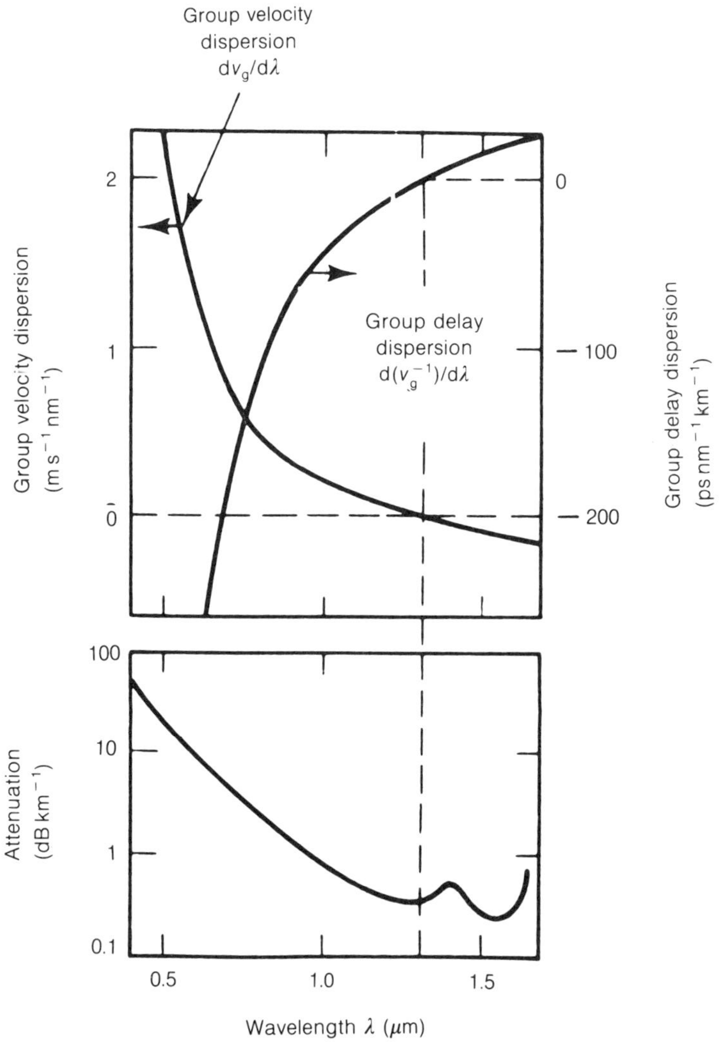

Figure 9.13 Typical loss and group velocity dispersion as functions of wavelength for good quality monomode fibres based on silica glass. The dispersion characteristic shown is that of a 'standard' fibre having zero group velocity dispersion at a wavelength $\lambda_0 \simeq 1300$ nm. By manipulating the waveguide design, other types of fibre can be fabricated with λ_0 positioned anywhere in the range approximately 1250–1600 nm.

for present-day optical communication, it is possible to manufacture fibres having positive, negative or zero group-velocity dispersion.

We now return to consider the wave equation 9.13 which, as already mentioned, provides a good description of the propagation of the pulse envelope in a monomode optical fibre. When the second-order dispersion term is retained but higher-order terms are neglected, equation 9.13 may be written in the form

$$\mathrm{i}\frac{\partial u}{\partial \zeta} + |u|^2 u = \pm \frac{1}{2}\frac{\partial^2 u}{\partial s^2} \tag{9.18}$$

where the parameters are expressed in the normalized forms

$$\begin{aligned} u &= \tau(n_2\omega/c\,|\mathrm{d}^2k/\mathrm{d}\omega^2|^2)^{1/2}E \\ s &= (t - z/v_{\mathrm{g}})/\tau \\ \zeta &= |\mathrm{d}^2k/\mathrm{d}\omega^2|\,z/\tau^2 \end{aligned} \tag{9.19}$$

Here s is a measure of the position in a moving coordinate frame which travels at the group velocity v_{g}, and τ is any convenient time scale (usually taken to be the initial pulse duration). The dispersion term which appears in the right-hand side of equation 9.18 takes the sign of $\mathrm{d}^2k/\mathrm{d}\omega^2$ (which is the opposite of the sign of $\mathrm{d}(v_{\mathrm{g}}^{-1})/\mathrm{d}\lambda$). In this normalized form, equation 9.18 is the celebrated nonlinear Schrödinger equation. The solutions of this seemingly simple equation display extraordinary and interesting properties, many of which have been observed experimentally in optical fibres.

The type of propagation behaviour that occurs depends critically on whether $\mathrm{d}(v_{\mathrm{g}}^{-1})/\mathrm{d}\lambda$ is positive or negative (i.e. for silica fibres, whether the wavelength λ is respectively greater or less than λ_0—see Figure 9.13). For $\mathrm{d}(v_{\mathrm{g}}^{-1})/\mathrm{d}\lambda > 0$ pulse narrowing can occur, as described earlier, and it also becomes possible to observe pulses that propagate either without any change of shape or have shapes that vary periodically with distance. Such extraordinary pulses exist as solitary wave solutions (solitons) of the nonlinear wave equation. The propagation behaviour of solitons is fundamentally different from that of low intensity 'linear' pulses which, by virtue of their finite bandwidth and group-velocity dispersion, must ultimately increase in width during propagation.

Solitary wave solutions of the nonlinear Schrödinger equation have been found both analytically and numerically (Zakharov and Shabat, 1973; Dodd *et al.*, 1982; Blow and Doran, 1987; Agrawal, 1989). In the regime where $\mathrm{d}(v_{\mathrm{g}}^{-1})/\mathrm{d}\lambda > 0$, an exact soliton is obtained when the pulse $u(\zeta, s)$ has the initial form

$$u(\zeta, s) = A \operatorname{sech} s \tag{9.20}$$

and $A = N$, where N is an integer number ($N \geqslant 1$). The amplitude A takes the normalized units defined by equations 9.19. Of course, there is no constraint; we are free to choose any initial boundary condition that we wish. To begin with, however, we focus attention on this particular case of equation 9.20 with $A = N$. (We consider the situation when the initial pulse is different from this later.) The solutions of the wave equation show that the whole energy of such an input pulse propagates subsequently as a single entity with the properties of a soliton. The soliton with $N = 1$ (often called the 'fundamental' soliton) propagates as $u(\zeta, s) = \operatorname{sech} s \exp(-i\zeta/2)$. In other words, apart from a rotating phase, the fundamental solition is a pulse which propagates without change. A simple intuitive explanation for this behaviour is that the frequency chirp generated by self-phase modulation acts together with the group-velocity dispersion to compensate exactly for the 'linear' pulse broadening that the pulse would suffer in the absence of nonlinearity. This picture in which self-phase modulation and group-velocity dispersion are considered as separately identified processes is, however, vastly oversimplified and conceals a wealth of structure and subtleties that can be seen in the mathematical solutions of the nonlinear Schrödinger equation. The fundamental soliton is the only one which propagates without change of shape. Higher-order solitons (integers $N > 1$) undergo oscillations in shape with a period of $\pi/2$ on the distance scale ζ. The pulse changes become more complicated in structure for increasing N value; for $N = 2$ the pulse simply compresses during propagation for $\zeta = 0$ to $\pi/4$, and then re-expands to its original form at $\zeta = \pi/2$ without developing further structure. The evolution is symmetric about $\zeta = \pi/4$ for all solitons. To illustrate the periodic evolution of solitons, Figure 9.14 shows the propagation of an $N = 4$ soliton. Also visible is the periodic compression of the pulse which, in this case, is at its narrowest at $\zeta \simeq 3\pi/40$, $17\pi/40$, $23\pi/40$, $37\pi/40$.

Such behaviour has now been observed experimentally on many occasions.

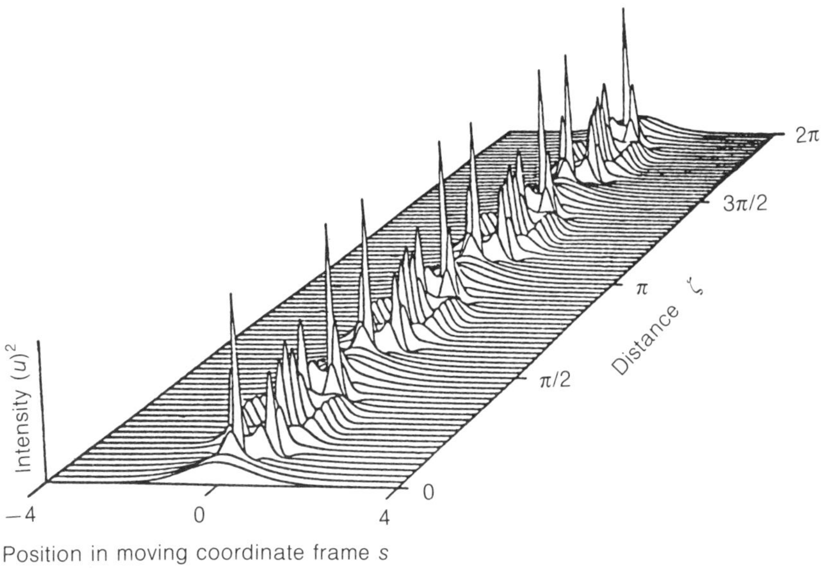

Figure 9.14 Computer-generated solutions of the nonlinear Schrödinger equation showing the evolution of an $N = 4$ soliton as it propagates over a distance equal to four soliton periods. (After Blow and Doran (1987).

The first observation was by Mollenauer *et al.* (1980); soliton pulses were observed in a 700 m fibre using mode-locked pulses from a KCl :Tl colour centre laser at 1.5 μm (Figure 9.15). The peak power levels required for soliton formation are proportional to the group-velocity dispersion and inversely proportional to the square of the pulse duration; in the experiment of Mollenauer *et al.* in which the group-delay dispersion was 15 ps $\mathrm{nm}^{-1}\,\mathrm{km}^{-1}$ and the pulse duration was 7 ps, the peak power required for a fundamental soliton was about 1 W, in good agreement with theory as we now show. The transformations 9.19 allow the parameters in the nonlinear wave equation 9.18 to be expressed in practical units. Taking the values appropriate to the experiment of Mollenauer *et al.* (1980), a pulse having a duration (full width at half-maximum intensity) of 7 ps ($\tau = 4$ ps) and peak power 1 W would generate a fundamental soliton in a standard monomode fibre at $\lambda = 1.5$ μm ($n_0 = 1.45$, core area = 100 $\mu\mathrm{m}^2$, $\mathrm{d}^2k/\mathrm{d}\omega^2 = -(\lambda^2/2\pi c)\,\mathrm{d}(v_g^{-1})/\mathrm{d}\lambda = -1.8 \times 10^{-26}\,\mathrm{s}^2\,\mathrm{m}^{-1}$ from Figure 9.13). The soliton period $\zeta = \pi/2$ corresponds to

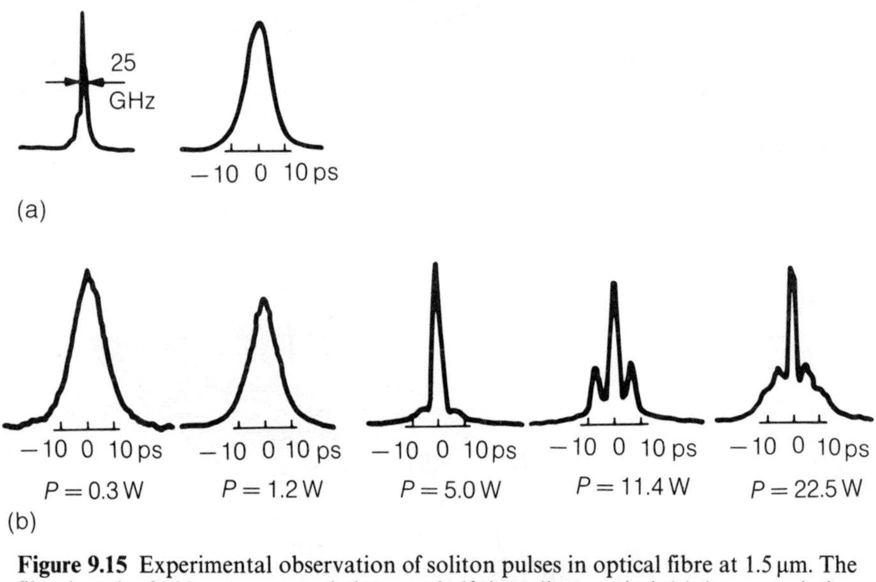

Figure 9.15 Experimental observation of soliton pulses in optical fibre at 1.5 μm. The fibre length of 700 m corresponded to one-half the soliton period. (a) Autocorrelation shape and frequency spectrum of laser pulses launched into the fibre. (b) Autocorrelation shapes of pulses emerging from the fibre for five input peak powers. $P = 0.3$ W: negligible nonlinear effect; only dispersive broadening is seen. $P = 1.2$ W: return of pulse to input width; this corresponds to the fundamental ($N = 1$) solution. P = 5 W: pulse narrowed to minimum width; this corresponds to half-period behaviour of the $N = 2$ soliton. $P = 11.4$ W: first well-resolved pulse splitting; note that the three-fold splitting in autocorrelation corresponds to a two-fold splitting of the pulse itself. This curve and the one at $P = 22.5$ W correspond to the $N = 3$ and $N = 4$ solitons respectively. Note that the powers scale essentially as N^2, as required by theory. (After Mollenauer *et al.*, (1980).)

a distance $\pi\tau^2/2|\mathrm{d}^2k/\mathrm{d}\omega^2|$, which in this case is 1.4 km. The 700 m fibre length used in the experiment would thus correspond to half a soliton period, which is in good agreement with the observations.

It is found, both theoretically and experimentally, that fundamental solitons ($A = 1$) behave as robust entities that are stable with respect to small perturbations of the propagation equation, such as loss and higher-order dispersion. Such solitons can pass through each other and emerge unscathed apart from a change in phase. In contrast, higher-order solitons are intrinsically unstable; minor perturbations cause them to split up into a number of fundamental solitons. In practice, stable propagation of higher-order solitons over long distances does not occur.

Rarely in an experiment is the initial pulse launched into a fibre precisely of the form 9.20. For example, the pulse envelope may not have precisely a hyperbolic secant shape; its amplitude may not be exactly an integer number (in the normalized units); or it may be chirped in frequency. In this case, provided that the initial pulse amplitude exceeds a particular threshold value, a certain proportion of the pulse energy will contribute to a soliton and the remainder will gradually disperse by the process of group-velocity dispersion during propagation. Not only will the soliton contain less energy than the original pulse, it will also have a different width (either narrower or wider). For an initial pulse of the form $u(0, s) = A \operatorname{sech} s$, a soliton is obtained provided that the amplitude $A > 1/2$. As mentioned previously, the special aspect of pulses with the initial condition $A = N$ with integers $N \geqslant 1$ is that all the energy of the initial pulses propagate as solitons; none of the energy disperses.

Hasegawa and Tappert (1973) were the first to propose the use of solitons for long-distance optical fibre communications to overcome one of the primary limitations of such systems in the linear regime, namely pulse broadening (due to group-velocity dispersion), which limits the achievable bandwidth. A further limitation of such systems, however, is the inevitable attenuation over long transmission paths (see Figure 9.13). The effects of loss in soliton propagation were considered by Blow and Doran (1985) using a numerical analysis of the nonlinear Schrödinger equation modified to incorporate loss as a perturbation. They concluded that, for the typical values of loss that occur in optical fibres in the wavelength region ~ 1.3–$1.6\,\mu$m (Figure 9.13), a soliton will broaden during propagation, but that the asymptotic dispersion at long distances is lower than that of an equivalent pulse in the linear regime. Thus the soliton retains aspects of its special character even in the presence of loss. Since loss causes pulse broadening in soliton systems, periodic amplification has been proposed as a means of both maintaining adequate pulse energy and eliminating broadening. Kodama and Hasegawa (1982) showed theoretically that if a soliton pulse propagating in the lossy fibre could be amplified periodically at distances determined by the group-velocity dispersion, the shape of the soliton would be almost unchanged even after several hundred amplifications. A scheme for achieving this using stimulated Raman amplification within the fibre itself was demonstrated by Mollenauer *et al.* (1985). Using the CW output from a 1.46 μm colour centre laser as a backward-travelling pump wave for Raman amplification (section 9.5.1), the 10 ps pulses at 1.56 μm from a second mode-locked colour centre laser were amplified during transmission through a 10 km fibre. With sufficient Raman gain to compensate exactly for net fibre energy loss, Mollenauer *et al.* demonstrated propagation of fundamental soliton pulses without distortion. For the sorts of pulse widths and bit rates under consideration (perhaps ~ 10 ps and ~ 10 GHz) degradation of the signal to noise ratio due to the spontaneous Raman effect (a necessary accompaniment of Raman gain) is not expected to be significant until the system is thousands of kilometres long. Mollenauer and Smith (1988) demonstrated soliton propagation over an effective path length as great as 6000 km (by repeated circulation around a 42 km long fibre loop) with only a minor increase in the pulse width from its initial value of 50 ps. (By contrast, an equivalent input pulse in the linear regime would broaden to ~ 6 ns during propagation over such a distance.)

A simple limit on the bandwidth of a communications system using soliton pulses is that they should be separated in time by at least the maximum pulse width which occurs during their evolution. This is, however, an optimistic limit since it neglects the interaction between neighbouring pulses (Blow and Doran, 1983; Gordon, 1983). The combined field of two adjacent solitons creates an attractive potential between the pulses. The resulting interaction forces the pulses to oscillate around their centre of mass during propagation; they thus exchange places periodically with total coalescence when the pulses meet twice during every period of oscillation. In order to keep the distance of coalescence longer than the system length, consecutive pulses must be spaced sufficiently well apart. Blow and Doran (1983) concluded that for future soliton communication over distances of several hundred kilometres pulse separations on the order of ten times the pulse width (the exact value depending on the pulse width and fibre dispersion) will be necessary, thus reducing the potential bandwidth by about an order of magnitude. This prediction of a short-range interaction between solitons was confirmed by experiments using fibre paths of a few hundred metres (Mitschke and Mollenauer, 1987). In later experiments, however, Smith and Mollenauer (1989) observed an unexpected long-range interaction between solitons during propagation over fibre lengths of thousands of kilometres; alternating attraction and repulsion was observed between 50 ps wide soliton pulses with initial separations as great as 4 ns (i.e. 80 times the pulse width). The origin of this effect is still unclear; explanations that have been suggested are interactions caused by dispersion-wave radiation (Smith and Mollenauer, 1989) and an electrostriction mechanism (Dianov *et al.*, 1990). In more recent experiments, Mollenauer *et al.* (1990) reported an experimental demonstration of soliton transmission over a total effective fibre length of more than 10 000 km. The 50 ps pulses were circulated many times around a 75 km fibre loop which contained three erbium-doped fibre amplifiers spaced 25 km apart (Figure 9.16). An important observation in this experiment was that there were no significant long-range soliton–soliton interactions; Mollenauer *et al.* consider that this is probably connected with the fact that the main fibre used in the loop had low group-velocity dispersion at the 1.53 μm operating wavelength. Indeed, no significant interactions were observed out to 9000 km distance unless the separation between adjacent pulses was less than ~ 4 times the 50 ps pulse duration. Also the jitter in pulse arrival times (which determines the fundamental bandwidth limit in such a system) was consistent with low error rates for data transmission speeds up to $\sim 4\,\mathrm{Gbit\,s^{-1}}$. These observations demonstrate the feasibility of very long-distance soliton communications systems. The first attempts to transmit coded patterns of solitons were reported very recently (Nakazawa *et al.*, 1990).

We now return to the solutions of the nonlinear Schrödinger equation 9.18, and consider the situation when $d(v_g^{-1})/d\lambda < 0$ ($\lambda < \lambda_0$ in Figure 9.13). As

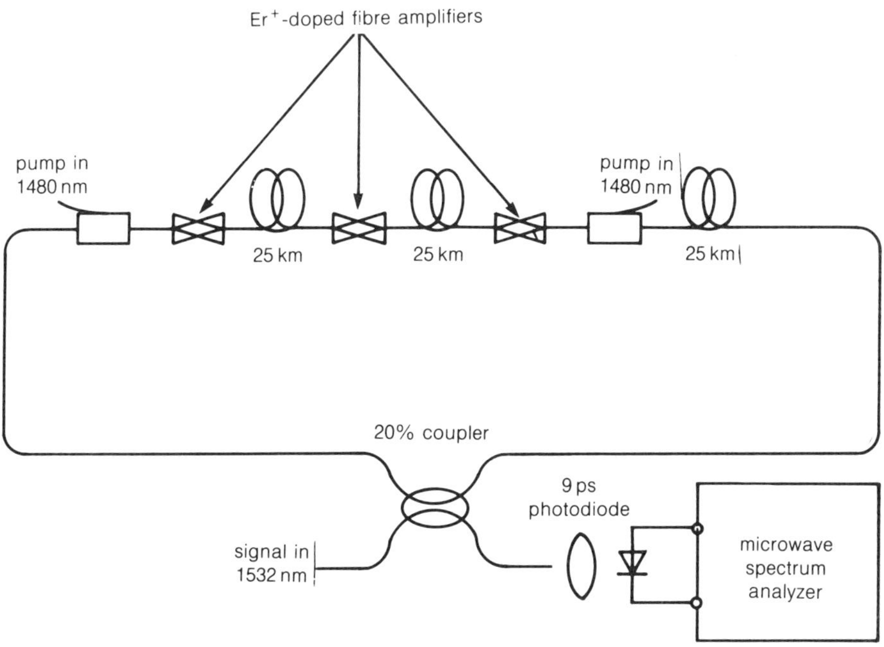

Figure 9.16 Arrangement used to observe transmission of 50 ps solitons in a 75 km recirculating loop of fibre. The detector and microwave spectrum analyzer are used to investigate the pulses after many circuits of the loop, equivalent to transmission over distances of thousands of kilometres. The loop contains three low gain (~ 6.5 dB) erbium-doped fibre amplifiers each 25 km apart which compensate for fibre attenuation. (After Mollenauer *et al.* (1990).)

noted previously, pulses undergo enhanced broadening in this regime (i.e. the combination of nonlinearity and group-velocity dispersion produces a greater degree of pulse broadening for an intense pulse compared with a low intensity one). This is clearly an effect which could limit the transmission bandwidth of an optical communications link. From a numerical analysis of equation 9.18, Tomlinson *et al.* (1984) found an approximate expression for the peak power P which, when launched into a fibre length L, will result in the pulse being broadened to about three times its original width. Converting this into real-world units gives the expression

$$P = \frac{2.0c\tau^2 S}{n_2^I L^2 |\mathrm{d}(v_g^{-1})/\mathrm{d}\lambda| \lambda} \tag{9.21}$$

where τ is the initial FWHM intensity pulse width (assuming smooth unchirped pulses), S is the mode cross-section, and $\mathrm{d}(v_g^{-1})/\mathrm{d}\lambda$ is the group-delay dispersion (i.e. the change in pulse group delay with change in wavelength, normalized to fibre length). This nonlinear broadening may be compared with the corresponding low intensity ('linear') broadening:

$$\frac{\Delta\tau}{\tau} = [1 + (0.44L|\mathrm{d}(v_g^{-1})/\mathrm{d}\lambda|\lambda^2/c\tau^2)^2]^{1/2} - 1 \tag{9.22}$$

Taking the values $\lambda = 1.3\,\mu\mathrm{m}$, $\tau = 100\,\mathrm{ps}$, $S = 50\,\mu\mathrm{m}^2$, $n_2^I = 3.2 \times 10^{-20}\,\mathrm{m}^2\,\mathrm{W}^{-1}$, $L = 10\,\mathrm{km}$ and $\mathrm{d}(v_g^{-1})/\mathrm{d}\lambda = -20\,\mathrm{ps\,nm}^{-1}\,\mathrm{km}^{-1}$ gives $3\times$ nonlinear broadening for $P \simeq 4\,\mathrm{W}$, compared with linear broadening of only $0.001\times$. Nelson *et al.* (1983) observed very large nonlinear broadening in 10–20 km lengths of monomode fibre at 1.06 μm (Figure 9.17), in good agreement with theory. Broading of 150 ps pulses to as much as 2.5 ns was seen at peak pulse powers up to 40 W. Nonlinear broadening of the pulse to twice its original value was observed at peak powers of $\sim 1\,\mathrm{W}$.

On the other hand, this type of nonlinear pulse broading can be utilized for pulse compression. The temporal broadening is accompanied by pronounced spectral broadening with a positive chirp. For example, the almost-rectangular pulse shown in Figure 9.17(c) contains a near-linear frequency sweep which describes essentially all of the pulse energy (Grischkowsky and Balant, 1982). By passing this chirped pulse through a linear despersive delay line with the opposite sign of group-velocity dispersion ($\mathrm{d}(v_g^{-1})/\mathrm{d}\lambda > 0$), such as a pair of suitably aligned diffraction gratings, the pulse can be compressed in time to as short as one-tenth of the initial pulse duration. This technique is now employed widely as a method of optical pulse compression into the femtosecond domain, as described further in the next chapter. For application of the pulse compression technique in the wavelength range of interest in optical communications, it is possible by adjusting the waveguide dispersion to produce fibres with positive and negative dispersion, as described earlier. Thus optical fibres can be used for both the nonlinear pulse broadening and

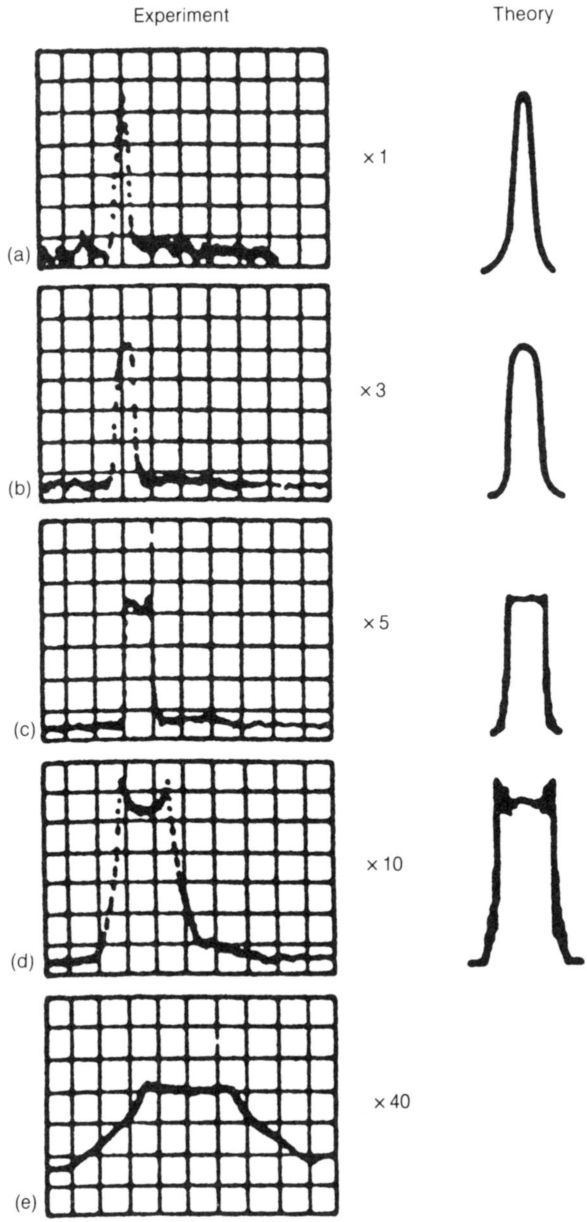

Figure 9.17 Experimentally observed and calculated shapes of 1.06 μm optical pulses (initially 150 ps width) of various peak powers ((a) 0.4 W; (b) 1.5 W; (c) 4.5 W; (d) 6.5 W; (d) 40 W) after transmission through 20 km monomode fibre, showing enhanced pulse broadening. The horizontal scale is 500 ps per division. Vertical scaling factors are shown. (After Nelson *et al.* (1983).)

compression stages, so eliminating the need for diffraction gratings or other dispersive elements (Blow *et al.*, 1983). The pulse compression device is then potentially compact and cheap, no alignment is required between the two elements since they can be fusion jointed, and the throughput and beam quality are good. A device of this type, operating at 1.32 μm, has been demonstrated (Nelson *et al.*, 1985).

Finally, the effects of the intensity-dependent refractive index could influence the performance of communications systems which employ receivers sensitive to changes in carrier phase angle (coherent homodyne or heterodyne detection). There could be significant degradation even in systems (such as those using phase or frequency modulation) in which the optical intensity is nominally constant. Through the nonlinear phase shift 9.16, intensity fluctuations of the transmitted signal (both noise fluctuations and intentional amplitude modulation, if any) are converted into phase noise at the receiver. Chraplyvy *et al.* (1984) considered the effects of this on phase-sensitive detection schemes, and analysed both self-phase modulation (an optical wave acting on itself) and cross-phase modulation in WDM systems (one optical wave modulating a channel at a different wavelength). They concluded that the generated phase noise was tolerable for laser fluctuations not exceeding typically ~ 1 mW rms in angle- (phase- or frequency-) modulated systems, and this is probably not too serious a limitation. However, it was also predicted that for coherent demodulation of intensity-modulated signals (e.g. homodyne detection of amplitude-shift-keyed signals), a peak signal power of only 2 mW in a two-channel WDM system would be sufficient to produce phase noise of amplitude 0.15 rad and therefore a significant noise penalty (>0.5 dB). These quantitative predictions were verified in an experiment in which interference between two channels due to cross-phase modulation was observed (Chraplyvy and Stone, 1984).

9.7. CONCLUSIONS AND FUTURE DIRECTIONS

In this chapter the influence of optical nonlinearity in fibres for advanced telecommunications systems has been considered. Optical nonlinearity can pose significant limitations on the power levels, system range and bandwidth that can be achieved. On the other hand, these nonlinear effects can be put to good use, to compensate for attenuation and group-velocity dispersion in fibres.

At the time of writing, the development of telecommunications networks is advancing at a rapid rate. The recent appearance of doped-fibre amplifiers is having an especially significant influence on the planning of all types of future optical systems. The area of greatest potential growth is in the introduction of optical fibres and optoelectronics into the local network. Passive monomode fibre devices (such as multiway splitters, filters and isolators), together with wavelength-division multiplexing techniques and doped-fibre amplifiers, could allow the construction of a high capacity optical infrastructure reaching large numbers of business and residential users. The optical power levels, spectral channel spacings and other design parameters must be carefully chosen to avoid nonlinear cross talk and the other deleterious effects arising from optical nonlinearity in fibres. At the opposite extreme of the network, soliton transmission systems are at present under serious consideration for intercontinental links. It seems inevitable that fibre nonlinearity will play an increasingly prominent role in many aspects of future development of the global network.

Another area that is rapidly developing currently is the use of fibres as components in devices for ultrafast all-optical switching and logic. These devices make use of the nonlinear refractive index in silica fibre, with its femtosecond response time. Over a number of years a vigorous search has taken place in many laboratories for nonlinear materials well suited to all-optical switching devices; the current conclusion of this search is that optical fibre is the best material currently available for applications in the picosecond and subpicosecond domain (Stegeman and Stolen, 1989). Several different geometries for nonlinear fibre devices have been investigated, including a twin-core fibre coupler (Friberg *et al.*, 1987), a 'rocking-filter' fibre (Trillo *et al.*, 1988), the Mach–Zehnder interferometer (White *et al.*, 1988), and the fibre loop mirror (Doran *et al.*, 1989). Soliton effects can also play an important role in ensuring efficient pulse switching in some of these devices (Blow *et al.*, 1989). Ultrafast all-optical logic gates, based on 'soliton trapping' effects in fibres, have been demonstrated recently (Islam, 1989); these have been shown to be capable of six-fold fan-out and are logically 'complete'.

It is the high speed and low loss of silica-based optical fibres that make them attractive for application in ultrafast switching devices. The disadvantage, however, is that the small nonlinearity leads to a requirement for high optical powers to achieve switching. The switching powers, which vary inversely with effective interaction length, are typically 1 kW for a 1 m effective length using silica fibre (Stegeman and Stolen, 1989). Given that ~ 1 m might be considered a useful length for a device, and that ~ 1–10 W is a suitable operating peak power, glasses with larger nonlinearity are needed. This higher nonlinearity might be obtained at the expense of substantially increased attenuation compared with silica fibre; however, this trade-off may be acceptable because the losses in silica fibre are already at such a low level. Materials under investigation currently include heavy metal oxide glasses (Friberg and Smith, 1987; Hall *et al.*, 1989) and glasses doped with semiconductor microcrystals (Cotter *et al.*, 1989; Cotter and Manning, 1990); in both cases fibres have been fabricated based on these materials. A future prospect is that telecommunications fibre having low loss for pulse transmission will be used in combination with short lengths of high nonlinearity fibre that are configured to perform ultrafast all-optical switching and signal processing.

REFERENCES

Adams, M. J. (1981) *An Introduction to Optical Waveguides*, Wiley, Chichester.

Agrawal, G. P. (1989) *Nonlinear Fiber Optics*, Academic Press, New York.

Aoki, Y., Kishida, S. and Honmou, H. *et al.* (1982) *Electron. Lett.*, **19**, 620.

Aoki, Y., Kishida, S., Washio, K. and Minemura, K. (1985) *Electron. Lett.*, **21**, 191.

Aoki, Y., Kishida, S. and Washio, K. (1986) *Appl. Opt.*, **25**, 1056.

Aoki, Y., Tajima, K. and Mito. I. (1988) *IEEE J. Lightwave Technol.*, **6**, 710.

Armitage, J. A., Atkins, C. G., Wyatt, R. *et al.* (1988) *Proc. CLEO'88*, Optical Society of America, Washington, DC.

Atkins, C. G., Cotter, D., Smith, D. W. and Wyatt, R. (1986) *Electron. Lett.*, **22**, 556.

Atkins, C. G., Massicott, J. F., Armitage, J. R. *et al.* (1989) *Electron. Lett.*, **25**, 910.

Blow, K. J. and Doran, N. J. (1983) *Electron. Lett.*, **19**, 429.

Blow, K. J. and Doran, N. J. (1985) *Opt. Commun.*, **52**, 367.

Blow, K. J. and Doran, N. J. (1987) *IEE Proc. Pt. J (Optoelectron.)*, **134**, 138.

Blow, K. J., Cotter, D. and Doran, N. J. (1983) In *Digest of Topical Meeting on Optical Fiber Communication*, Optical Society of America, Washington, DC.

Blow, K. J., Doran, N. J. and Nayer, B. K. (1989) *Opt. Lett.*, **14**, 754.

Bolle, A., Grosso, G. and Daino, B. (1989) *Electron. Lett.*, **25**, 2.

Chinlon Lin, Reed, W. A. and Pearson, A. D. (1981) *Opt. Lett.*, **6**, 493.

Chraplyvy, A. R. and Stone, J. (1984) *Electron. Lett.*, **20**, 996.

Chraplyvy, A. R. and Tkach, R. W. (1986) *Electron. Lett.*, **22**, 1084.

Chraplyvy, A. R., Stone, J. and Burrus, C. A. (1983) *Opt. Lett.*, **8**, 415.

Chraplyvy, A. R., Marcuse, D. and Henry, P. S. (1984) *IEEE J. Lightware Technol.*, **2**, 6.

Cohen, L. G. and Chinlon Lin. (1978) *IEEE J. Quantum Electron.*, **14**, 855.

Cosentino, A. and Iannone, E. (1989) *Electron. Lett.*, **25**, 1459.

Cotter, D. (1983) *J. Opt. Commun.*, **4**, 10.

Cotter, D. and Manning, R. J. (1990) *Proc. IQEC'90*, Optical Society of America, Washington, DC, Post-deadline Paper QPDP-1, p. 409.

Cotter, D., Smith, D. W., Atkins, C. G. and Wyatt, R. (1986) *Electron. Lett.*, **22**, 671.

Cotter, D., Ironside, C. N., Ainslie, B. J. and Girdlestone, H. P. (1989) *Opt. Lett.*, **14**, 317.

Desurvire, E., Papuchon, M., Pocholle, J. P. *et al.* (1983) *Electron. Lett.*, **19**, 751.

Desurvire, E., Digonnet, M. and Shaw, H. J. (1985) *Opt. Lett.*, **10**, 83.

Dianov, E. M., Luchnikov, A. V., Pilipetskii, A. N. and Starodumov, A. N. (1990) *Opt. Lett.*, **15**, 314.

Dodd, R. K., Eilbeck, J. C., Gibbon, J. D. and Morris, H. C. (1982) *Solitons and Nonlinear Wave Equations*, Academic Press, London.

Doran, N. J., Forrester, D. S. and Nayar, B. K. (1989) *Electron. Lett.*, **25**, 267.

Friberg, S. R. and Smith, P. W. (1987) *IEEE J. Quantum Electron.*, **23**, 2089.

Friberg, S. R., Silberberg, Y., Oliver, M. K. *et al.* (1987) *Appl. Phys. Lett.*, **51**, 1135.

Gordon, J. P. (1983) *Opt. Lett.*, **8**, 596.

Grischkowsky, D. and Balant, A. C. (1982) *Appl. Phys. Lett.*, **41**, 1.

Hall, D. W., Newhouse, M. A., Borrelli, N. F. *et al.* (1989) *Appl. Phys. Lett.*, **54**, 1293.

Hasegawa, A. and Tappert, F. (1973) *Appl. Phys. Lett.*, **23**, 142.

Hegarty, J., Olsson, N. A. and Goldner, L. (1985) *Electron. Lett.*, **21**, 290.

Hirose, A., Takushima, Y. and Okoshi, T. (1991) *Optical Communications*, **12**, 82–5.

Ippen, E. P. and Stolen, R. H. (1972) *Appl. Phys. Lett.*, **21**, 539.

Islam, M. N. (1989) *Opt. Lett.*, **14**, 1257.

Kimura, Y., Suzuki, K. and Nakazawa, M. (1989) *Electron. Lett.*, 1656.

Kodama, Y. and Hasegawa, A. (1982) *Opt. Lett.*, **7**, 339.

Koepf, G. A., Kalen, D. M. and Greene, K. H. (1982) *Electron. Lett.*, **18**, 942.

Marcuse, D. (1988) In *Optical Fiber Telecommunications II* (eds S. E. Miller and I. P. Kaminow), Academic Press, New York.

Massicott, J. F., Wyatt, R., Ainslie, B. J. and Craig-Ryan, S. P. (1990) *Electron. Lett.*, **26**, 1038.

Miller, S. E. and Kaminow, I. P. (eds) (1988) *Optical Fiber Telecommunications II*, Academic Press, New York.

Mitschke, F. M. and Mollenauer, L. F. (1987) *Opt. Lett.*, **12**, 355.

Mollenauer, L. F. and Smith, K. (1988) *Opt. Lett.*, **13**, 675.

Mollenauer, L. F., Stolen, R. H. and Gordon, J. P. (1980) *Phys. Rev. Lett.*, **45**, 1095.

Mollenauer, L. F., Stolen, R. H. and Islam, M. N. (1985) *Opt. Lett.*, **10**, 229.

Mollenauer, L. F., Neubelt, M. J., Evangelides, S. G. *et al.* (1990) *Proc. CLEO'90, Anaheim, CA*, Optical Society of America, Post-deadline Paper CPDP17.

Nakamura, K., Kimura, M., Yoshida, S. *et al.* (1984) *IEEE J. Lightwave Technol.*, **2**, 379.

Nakasaka, M. (1985) *Appl. Phys. Lett.*, **46**, 628.

Nakazawa, M., Suzuki, K., Yamada, E. and Kimura, Y. (1990) *Proc. OFC'90*, Optical Society of America, New York, Post-deadline Paper PD5.

Nelson, B. P., Cotter, D., Blow, K. J. and Doran, N. J. (1983) *Opt. Commun.*, **48**, 292.

Nelson, B. P., Blow, K. J. and Doran, N. J. (1985) *Opt. Lett.*, **10**, 393.

Neumann, E. G. (1988) *Single-mode Fibers—Fundamentals*, Springer Series in Optical Sciences 57, Springer, Berlin.

Olsson, N. A. and Hegarty, J. (1986) *IEEE J. Lightwave Technol.*, **4**, 396.

Olsson, N. A. and Van der Ziel, J. P. (1986) *Appl. Phys. Lett.*, **48**, 1329.

Olsson, N. A. and Van der Ziel, J. P. (1987) *IEEE J. Lightwave Technol.*, **5**, 147.

Shibata, N., Braun, R. P. and Waarts, R. G. (1986) *Electron. Lett.*, **22**, 675.

Siegman, A. E. (1986) *Lasers*, Oxford University Press, Oxford.

Smith, K. and Mollenauer, L. F. (1989) *Opt. Lett.*, **14**, 1284.

Smith, R. G. (1972) *Appl. Opt.*, **11**, 2489.

Snyder, A. W. and Love, J. D. (1983) *Optical Waveguide Theory*, Chapman and Hall, London.

Stegeman, G. I. and Stolen, R. H. (1989) *J. Opt. Soc. Am.*, **6**, 652.

Stolen, R. H. (1980) *Proc. IEEE*, **68**, 1232.

Stolen, R. H. and Bjorkholm, J. E. (1982) *IEEE J. Quantum Electron.*, **18**, 1062.

Stolen, R. H. and Chinlon Lin (1978) *Phys. Rev. A*, **17**, 1448.

Stolen, R. H. and Ippen, E. P. (1973) *Appl. Phys. Lett.*, **22**, 276.

Stolen, R. H., Bosch, M. A. and Chinlon Lin. (1981) *Opt. Lett.*, **6**, 213.

Sugie, T., Imai, T., Ohkawa, N. and Ito, T. (1990) *Proc. ECOC'90 Amsterdam*, Paper MoF 3.2.

Tkach, R. W. and Chraplyvy, A. R. (1989) *Optical and Quantum Electronics*, **21**, S105.

Tkach, R. W., Chraplyvy, A. R. and Derosier, R. M. (1986) *Electron. Lett.*, **22**, 1011.

Tkach, R. W., Chraplyvy, A. R. and Derosier, R. M. (1988) *Electronics. Lett.*, **24**, 260.

Tomlinson, W. J., Stolen, R. H. and Shank, C. V. (1984) *J. Opt. Soc. Am. B.*, **1**, 139.

Trillo, S., Wabnitz, S., Banyai, W. C., Finlayson, N. *et al.* (1988) *Appl. Phys. Lett.*, **53**, 837.

Waarts, R. G. and Braun, R. P. (1985) *Electron. Lett.*, **21**, 1114.

Waarts, R. G. and Braun, R. P. (1986) *Electron. Lett.*, **22**, 873.

Washio, K., Inoue K. and Tanigawa, T. (1980) *Electron. Lett.*, **16**, 331.

White, I. H., Penty, R. V. and Epworth, R. E. (1988) *Electron. Lett.*, **24**, 340.

Winful, H. G. (1986) In *Optical Fiber Transmission* (ed. E. E. Basch), Sams, Indianapolis, IN.

Zakharov, V. E. and Shabat, A. B. (1973) *Sov. Phys. JETP*, **37**, 823.

Ultrashort pulses for nonlinear optical techniques

10

Wilson Sibbett

10.1. INTRODUCTION

The current understanding of ultrafast nonlinear optical techniques relates closely to the generation, characterization and application of light (usually laser) pulses having durations in the picosecond (10^{-12} s) or femtosecond (10^{-15} s) regimes. Although in many instances these ultrashort pulses may have relatively small energies (~ 1 nJ) their peak optical powers can be quite high ($\gtrsim 1$ kW). When tightly focused beams or confined guided-wave geometries are used the available power densities are thus rendered adequate for nonlinear optical processes to be directly exploited and implemented. Where necessary, laser amplifier schemes can be employed such that optical gains of 10^4–10^5 are readily achieved. The nonlinear processes induced at the high peak power levels are relevant to various aspects of pulse generation, manipulation, characterization and to time-domain spectroscopic studies.

Ultrafast nonlinear optical techniques are finding an ever-increasing range of applications in diverse areas of basic physical and biological sciences as well as in technology. Notable amongst these is the developing technology field of nonlinear optics in information processing which represents the main subject matter of this book. A contextual background has been chosen for this chapter on the basis that relatively narrow band gap semiconductor materials or structures (e.g. GaAs/AlGaAs, InGaAsP) and optical waveguides either in fibre form or planar semiconductor geometries are deemed to be of primary importance. For this reason, considerable highlighting has been afforded to sources of near-infrared ultrashort pulses. The overall emphasis here is directed towards the generation and characterization of picosecond and femtosecond laser pulses because of their direct relevance to the field of digital optical processing.

With current advances in research within this topic it can now be recognized that powerful optoelectronic or all-optical techniques will become more appropriate, either when applied separately or in specific combinations, than the slower electronic counterparts. For the all-optical approach the possibilities arising from the exploitability of massive bandwidth and parallelism are especially promising in respect of the processing and transfer of data.

10.2. ULTRASHORT PULSE GENERATION

The generation of ultrashort laser pulses by the technique of mode locking has been an active topic of research since the mid-1960s. Whereas the mode-locking schemes by which picosecond and subpicosecond pulses have been produced have been described in a number of excellent reviews (including Harris (1966), De Maria *et al.* (1967a, b, 1969), P. W. Smith (1970), De Maria (1971), Allen and Jones (1971), P. W. Smith *et al.* (1974), Bradley (1977) and New (1983)) the topic of femtosecond pulse generation has not been so comprehensively reviewed. Therefore in the subsections that follow the treatment of femtosecond pulse sources is dealt with rather more extensively than that for the picosecond regimes.

10.2.1. Basic principles of mode locking

At its most basic a laser oscillator comprises a gain medium within either a standing-wave (Fabry–Pérot cavity) or travelling-wave (ring cavity) resonator as illustrated in Figure 10.1. With the assumption that a fundamental transverse mode (TEM00) beam is produced then the frequency of the *m*th longitudinal (or axial) mode will be given by

$$\nu_m = \frac{mc}{2nL} \text{ or } \nu_m = \frac{mc}{nP} \tag{10.1}$$

where L(or P) is the cavity length (or perimeter), c is the speed of light in vacuum and n is the average refractive index. It follows that the frequency separation, $\delta\nu$, between neighbouring modes will be

$$\delta\nu = \frac{c}{2nL}\left(\text{or } \frac{c}{nP}\right), \text{ i.e. } \delta\omega = \frac{2\pi c}{2nL}\left(\text{or } \frac{2\pi c}{nP}\right) \tag{10.2}$$

When the laser is free running (or unmode-locked) the longitudinal modes have random phases and the output has stochastic or noisy intensity characteristics. The resultant field strength may therefore be expressed as (Herrmann and Wilhelmi, 1987)

$$E(t) = \sum_m \tfrac{1}{2}\hat{E}_m \exp\left[\mathrm{i}\phi_m + \mathrm{i}(\omega_0 + m\,\delta\omega)t\right] + \text{c.c.} \tag{10.3}$$

where m may range from $-N$ to $+N$ for a total of $2N + 1$ longitudinal modes.

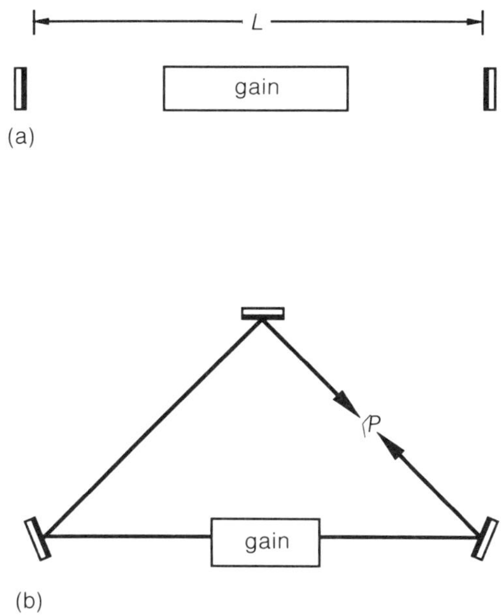

Figure 10.1 Basic configurations for (a) standing-wave (or Fabry–Pérot) and (b) travelling-wave (or ring) resonators.

The corresponding total intensity I will be given by

$$I \sim [E(t)]^2 = \sum_m \tfrac{1}{2}|\hat{E}_m|^2 \tag{10.4}$$

which simply represents the sum of the individual mode intensities.

For the particular case when equal-amplitude, E_0, longitudinal modes are considered to be phase locked then the phase ϕ_m may be regarded as a linear function of m. Supposing that

$$\phi_m - \phi_{m-1} = \theta \tag{10.5}$$

where θ is the constant intermode phase difference, then

$$\phi_m = \phi_0 + m\theta \tag{10.6}$$

Substitution of this form of ϕ_m into equation 10.3 and its subsequent analytical summation will give

$$E(t) = \hat{E}_0 \frac{\sin\left[\dfrac{2N+1}{2}(\delta\omega\, t + \theta)\right]}{\sin\left[\dfrac{\delta\omega\, t + \theta}{2}\right]} \exp[i(\omega_0 t + \phi_0)] + \text{c.c.} \tag{10.7}$$

i.e.

$$E(t) = A(t)\exp\left[\mathrm{i}(\omega_0 t + \phi_0)\right] + \text{c.c.} \tag{10.8}$$

The peak amplitudes of A(t) appear with a periodicity, t, given by

$$t = 2\pi/\delta\omega \equiv 1/\delta\nu \equiv \frac{2nL}{c}(\text{or } nP/c) = T_{\text{RT}} \equiv 1/f_{\text{c}} \tag{10.9}$$

T_{RT} is the round-trip period of the cavity and f_{c} is the so-called cavity frequency. It can also be deduced from equation 10.7 that the pulse duration (FWHM) $\Delta\tau$ is given by

$$\Delta\tau = \frac{2\pi}{2N\,\delta\omega} \equiv \frac{1}{2N\,\delta\nu} \equiv \frac{1}{\Delta\nu_1} \tag{10.10}$$

It must be emphasized that $\Delta\tau$ is related to the inverse of the lasing bandwidth as distinct from the fluorescence or gain bandwidth. Thus, this leads directly to the physical picture of a single discrete pulse circulating within the laser resonator.

It follows from the above rather simplified approach that the shortest pulses will be produced under the conditions where the maximum number of longitudinal modes can be phase locked. (The significance of this will be especially highlighted in section 10.2.6.) A somewhat more rigorous treatment (New, 1983) implies that $\Delta\tau\,\Delta\nu \geqslant K$ where the constant K is pulseshape related. For example, $K = 0.441$ for Gaussian intensity profiles and $K = 0.315$ for sech^2 intensity profiles. Whenever it can be established for a particular pulseshape that the duration – bandwidth product is close to the theoretically predicted value of K then the pulses are described as **bandwidth limited** or **Fourier transform limited**. Some caution must be exercised in respect of this duration – bandwidth product because values that are less than K may arise when some asymmetry is present in the pulseshapes (Diels *et al.* 1985(b); Yoshizawa and Kobayashi, 1984). Ultrashort pulses must therefore be carefully characterized in terms of their specific temporal and spectral features and this complementarity should always be fully recognized.

From equation 10.7 it can also be seen that the peak amplitude will be $(2N+1)E_0$ with a corresponding peak intensity of $(2N+1)^2 I_0$ where I_0 is the assumed equal intensity of the individual longitudinal modes. Thus the peak intensity of a mode-locked pulse exceeds that of the unmode-locked laser by a factor $2N+1$ given by the number of phase-locked longitudinal modes. Clearly, in the circumstance involving a broad-bandwidth gain medium in an extended-cavity laser where a very large number of modes may be phase locked then this intensity enhancement is very significant. For instance, in a KCl:Tl colour centre laser ($\lambda \sim 1.5\,\mu\text{m}$) where coupled-cavity mode-locked (section 10.2.6) bandwidths $\sim 30\,\text{nm}$ are recorded for bandwidth-limited pulses at a repetition period $\sim 12\,\text{ns}$, then approximately

50 000 longitudinal modes can be phase locked. Femtosecond-duration, mode-locked pulses thus directly generated in such a laser oscillator may have relatively high peak powers of several kilowatts. Consequently induced nonlinearities, arising from the optical Kerr effect for example, can play a major role in the kinetics that applies to the evolution of ultrashort (or hypershort) laser pulses.

It may be gathered from the brief consideration outlined above that the goal of locking many longitudinal modes in a 'perfect' phase relationship is not a simple matter. This is generally true because ideal mode locking has been difficult to achieve in relatively narrow bandwidth gain media (e.g. Nd:YAG laser) as well as those having much larger gain bandwidths (e.g. dye, colour centre lasers). Methods by which the optimum mode-locking performance can be obtained for a wide range of laser types has long been a topic of active research and the importance of digital optics has meant that it continues to be vigorously researched.

10.2.2. Mode-locking techniques

Active mode locking and passive mode locking are terms that have gained familiarity and are now widely accepted within the jargon of laser physics literature. Despite this, there is considerable ambiguity associated with the interpretation of passive mode locking in particular. In the present context, therefore, it is probably worthwhile to state how the active and passive mode locking categories are to be distinguished. For **active** mode locking periodic modulation is provided either at the cavity frequency, f_c $(1/T_{RT})$, or at a frequency which is a multiple of f_c by a driving signal that is derived from a source external to the laser. A good example of this is the radio-frequency (RF) electrical signal that is applied to the intracavity acousto-optic modulators that are frequently used with ion (Ar^+, Kr^+) or Nd:YAG lasers. Both acousto-optic and electro-optic modulators find general application in the active mode locking of lasers but an alternative semiconductor device modulator has also been recently demonstrated (Burns and Sibbett, 1990). For all such arrangements a distinction can be made between amplitude modulation (AM mode locking) or phase modulation that leads to the so-called FM mode locking procedures.

Whenever the intracavity flux (or radiation) is responsible for the periodic modulation then the process can be described as **passive**. This general interpretation is used here where, for example, passive coupled-cavity mode locking is mentioned in section 10.2.6. It should be noted, however, that a narrower interpretation is sometimes taken to imply that passive mode locking involves the use of a saturable absorber. In this latter scheme the absorption of the passive medium remains essentially constant at low intensities but it decreases, or **saturates**, at higher intensity levels and so the losses are modified by the intracavity radiation itself. Under appropriate circumstances, the losses can be modulated at the round-trip periodicity of the laser cavity and this condition can be readily achieved without any requirement for external input signals to a modulator. An obvious practical advantage of such a 'passive' method is that it is not necessary to provide a precisely controlled synchronism between the period of a laser resonator and that of an external electrical drive to a modulator. Also, although passive mode locking with this **inherent synchronism** feature has been demonstrated as the technique by which the shortest duration (<20 fs (Finch *et al.*, 1988a and b)) laser pulses can be produced directly, there continues to be a limited supply of suitable saturable absorbers. This has been somewhat alleviated by the identification and use of multiple-quantum-well (MQW) (Silberberg *et al.* 1984; Smith P. W. *et al.*, 1985; Islam *et al.*, 1989a, b; Cesar *et al.*, 1990) and bulk (Soccolich *et al.*, 1990) semiconductor structures which are especially useful in the near-infrared spectral regions for the passive mode locking of semiconductor diode and colour centre lasers.

The limited availability and spectral coverage of suitable saturable absorbing media can be avoided by exploiting other types of optical nonlinearity in passively mode-locked lasers. In particular, the optical Kerr effect in extended waveguides such as silica-based monomode fibres or related carrier-dependent nonlinear phase shifts in semiconductor amplifiers (Miller and Sibbett, 1988; Kean *et al.*, 1989 can be shown to lead to a so-called passive 'coupled-cavity' mode-locking process (section 10.2.6). Although the attendant characteristics such as dispersion will have wavelength dependence the nonlinearity itself can be exploited across a wide range of spectral extent.

As already mentioned in section 10.2.1 the standing-wave or travelling-wave resonators illustrated in Figure 10.1 have longitudinal modes with a frequency separation given by $c/2nL$ or c/nP. In the case of homogeneously broadened laser transitions the initial above-threshold operation will be established for one mode (or more generally because of spatial hole burning and broad gain profiles a number of modes) at the peak region of the gain spectrum (or at a spectral region determined by an intracavity tuning element). A modulation applied at the cavity frequency leads to sideband generation with a frequency coincidence on neighbouring mode frequencies. With inhomogeneously broadened transitions, many longitudinal modes exist above laser threshold and these have independent phases. Modulation at the cavity frequency then serves to phase lock these modes together. The practical techniques by which such mode locking is generally accomplished are now briefly reviewed.

10.2.3. Active mode locking (AM and FM model locking)

In active mode locking an amplitude modulation (AM mode locking) or a phase modulation (FM mode locking) is applied at the cavity frequency or at a multiple of the cavity frequency (harmonic mode locking). Loss modulation is generally produced by an acousto-optic modulator whereas gain modulation may be achieved by synchronous pumping or more directly by a modulated injection current to a semiconductor amplifier chip. Phase modulation, on the other hand, is provided by using an intracavity electro-optic modulator. These procedures are considered in turn below.

10.2.3.1. Loss modulation

An intracavity amplitude modulator may be readily employed within the resonator of CW ion (Ar^+, Kr^+) lasers (Figure 10.2) or Nd:YAG and Nd:YLF lasers. For an RF electrical signal applied at a frequency f_{RF} to the transducer on a silica block or prism then for an appropriate material thickness a standing acoustic wave can be established. The consequent diffraction arising from this **phase grating** leads to a periodic intracavity loss or modulation at a frequency $f_m = 2f_{RF}$. Thus when f_m is chosen to equal the cavity frequency f_c, this amplitude modulation leads to the generation of the sideband frequencies. This can be justified rather simply by considering the field representations in the following manner.

The mode amplitude and frequency at the maximum of the available gain may be assumed to be E_0 and ω_0 respectively. Thus

$$E(t) = E_0 \cos \omega_0 t \tag{10.11}$$

When this mode is modulated at a frequency ω_m and amplitude E_m then

$$E^1(t) = E(t) + E_m \cos \omega_m t \tag{10.12}$$

where $E^1(t)$ is the modulated mode amplitude. This may be rewritten as

$$E^1(t) = E(t)(1 + M \cos \omega_m t) \tag{10.13}$$

where the modulation depth, M, is given by

$$M = \frac{E_m}{E_0} \tag{10.14}$$

Combination of equations 10.11 and 10.13 gives

$$E^1(t) = E_0 \cos \omega_0 t + \frac{ME_0}{2}[\cos(\omega_0 - \omega_m)t + \cos(\omega_0 + \omega_m)t] \tag{10.15}$$

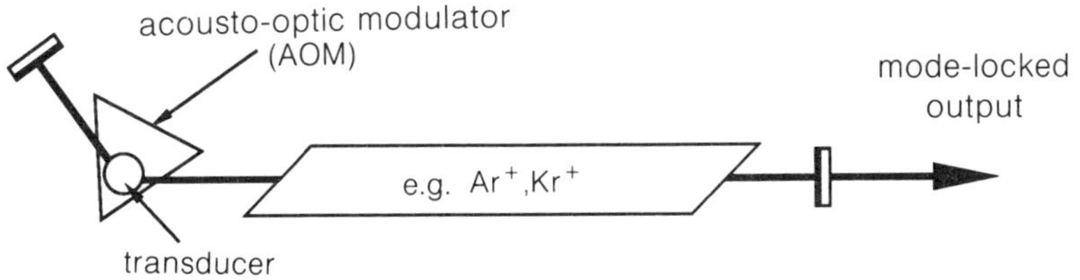

Figure 10.2 Schematic of actively mode-locked argoni on or krypton ion laser.

or

$$E^1(t) = E_0 \cos \omega_0 t + \frac{ME_0}{2} \cos(\omega_0 - \delta\omega)t + \frac{ME_0}{2} \cos(\omega_0 + \delta\omega)t \tag{10.16}$$

where ω_m has been rewritten as $\delta\omega$.

Because the sideband frequencies are coincident with the neighbouring mode frequencies at $\omega_0 \pm \delta\omega$, it can be appreciated that power is transferred in a coherent fashion from the initial mode to an enlarged comb of modes which can be sustained within the accessible gain bandwidth of the active medium. Thus for the previous example (section 10.2.1) where $2N + 1$ modes were assumed then phase locking is achieved over a total bandwidth of $2N\,\delta\omega$ for a total frequency extent from $\omega_0 - N\,\delta\omega$ to $\omega_0 + N\,\delta\omega$.

Because of various sources of noise and phase drift the bandwidth of the mode-locked laser output is usually significantly less than the overall gain bandwidth. As a consequence the durations of pulses from acousto-optically mode-locked CW lasers tend to range from 50 ps (Nd:YLF) through to 70–100 ps (Ar^+, Kr^+, Nd:YAG). For the pulse repetition rates which are typically around 80–100 MHz the average power from these actively mode-locked lasers is usually ~1–2 W for the ion lasers and ~10–20 W for the solid state counterparts. Shorter duration, higher peak power pulses can be produced from such lasers (Nd:YAG in particular) using either the extracavity pulse compression or the coupled-cavity mode-locking techniques described later.

The recent development of titanium–sapphire lasers has made available a new source in the near-infrared (670–1100 nm) spectral region (Moulton, 1986). Application of acousto-optic mode locking to a Ti–sapphire laser pumped by a CW argon ion laser has led to the production of tunable-frequency pulses having durations as short as 1.3 ps with average power levels ~1.5 W at a repetition rate of 82 MHz (Kafka *et al.*, 1990). Subsequent temporal compression enables 50 fs pulses to be generated at an average power of 500 mW which corresponds to an unprecedented peak pulse power of 122 kW (Kafka *et al.* 1990).

Alternatively, picosecond–subpicosecond frequency-tunable pulses can be generated by a synchronous pumping method. This technique is described in the following subsection.

10.2.3.2. Gain modulation

A gain modulation can be provided by optically pumping a **slave** laser by a mode-locked **drive** or **master** laser. For the frequency of this modulation to be matched to the longitudinal mode frequency separation of the slave laser requires that the cavity period of the pump laser should be integrally related to that of the slave laser. In practice, the cavity frequency f_{cp} for the pump laser is usually related to the cavity frequency f_{cs} for the slave laser by

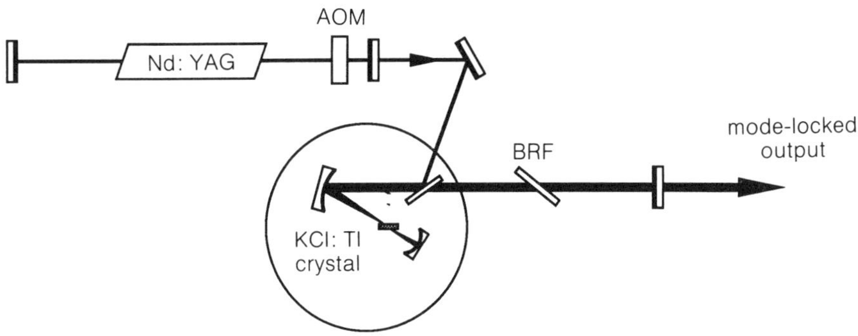

Figure 10.3 Typical layout for a synchronously pumped KCl:Tl colour centre laser. (AOM, acousto-optic modulator; BRF, birefringent tuning filter.)

$f_{cp} = nf_{cs}$, where n is typically 1, 2 or 3. Thus, for an arrangement such as that illustrated in Figure 10.3, a pump Nd:YAG laser pulse provides gain for the pulse that becomes established and circulates within the resonator of the slave colour centre laser. The combination of the modulated periodic gain (or transient gain in a short-lived upper laser state medium such as a dye) and gain saturation, with perhaps additional nonlinear effects (see section 10.2.4) lead to the generation of pulses that have shorter durations than those of the pump pulses. For the ~100 ps pulses from mode-locked Ar^+ or fundamental/frequency-doubled Nd:YAG (YLF) pump lasers, dye laser pulses of picosecond and subpicosecond durations can be produced (Chan and Sari, 1974; Kuhl *et al.*, 1979; Johnson and Simpson, 1983). Similarly with Kr^+ and Nd:YAG pump lasers, picosecond (e.g. Mollenauer, 1985a) and subpicosecond (Langford *et al.*, 1987a) colour centre laser pulses have also been reliably obtained.

As an alternative to the discrete gain medium of a dye jet or colour centre crystal, fibre Raman lasers (Stolen, 1980) have a distributed gain medium which may be exploited in synchronous pumping. Such is the case of the fibre Raman laser which is optically pumped by an actively mode-locked laser source. By appropriate control of group velocity dispersion this can be a source of frequency-tunable picosecond or subpicosecond optical pulses (Kafka *et al.*, 1986).

With a semiconductor diode laser it is convenient to achieve periodic gain by modulating the injection current to the amplifying medium. This may be attempted in the mm wave band at the inherent cavity frequency of the chip (typical length ~250–300 μm, i.e. f_c ~167–139 GHz) but this continues to be technically difficult. While integrated modulator–laser monolithic structures have shown that modulation frequencies ~40 GHz (Tucker *et al.*, 1989) can be used, it is more commonplace to use external laser configurations. By applying high quality antireflection coatings (Ho *et al.*, 1978) and/or suitably

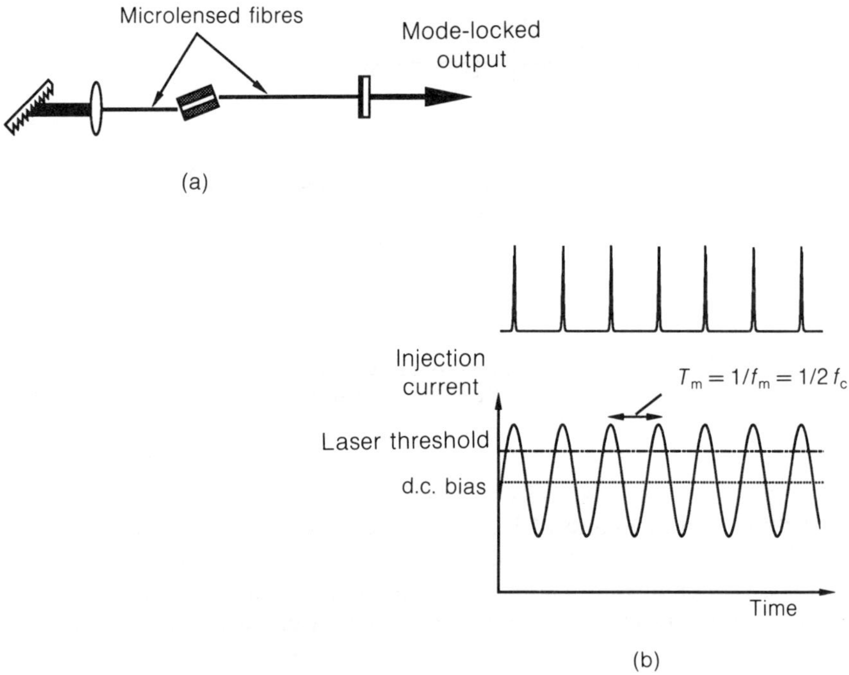

Figure 10.4 (a) External, fibre-based cavity arrangement for an actively mode-locked, frequency-tunable, angled-stripe InGaAsP semiconductor diode laser; (b) typical applied d.c. and RF injection current profiles for such a laser.

angled facet geometries (Bradley *et al.*, 1981); Chen *et al.*, 1982) the resonator role of the semiconductor chip is eliminated and it behaves as a discrete amplifying medium. An external cavity semiconductor laser arrangement such as that illustrated in Figure 10.4(a) for an InGaAsP diode laser can thus be configured. By supplementing a sub-laser-threshold d.c. injection current with an RF current then the laser can periodically reach threshold as indicated in Figure 10.4(b). If the gain medium is located at the centre of the external standing wave resonator then the modulated injection current should have a frequency of $2mf_c$ where m is an integer. (This can be automatically established for a ring resonator configuration.) This ensures that the luminescence that propagates towards, and subsequently returns from, the cavity reflectors experiences gain rather than loss in the semiconductor chip. Also, this choice of arrangement enables the counter-propagating intracavity flux to interfere within the gain medium so that the pulse shortening gain saturation effects can be enhanced. Picosecond duration pulses have been reliably and conveniently produced by this basic type of active mode locking in AlGaAs (Ho *et al.*, 1978; Bradley *et al.*, 1981; Chen *et al.*, 1982) and InGaAsP lasers (Burns *et al.*, 1988).

10.2.3.3. Combined gain–loss modulation (Controlled-amplifier mode locking)

Amplitude modulation has recently been demonstrated for a CW ring erbium fibre laser where an intracavity InGaAsP diode amplifier was used as the modulator (Burns and Sibbett, 1990). The configuration of this actively mode-locked laser is illustrated in Figure 10.5(a) where the erbium fibre was optically pumped either at 530 nm by a frequency-doubled Nd:YAG laser or by a titanium–sapphire laser tuned to 980 nm. An RF sinusoidal injection current supplied to the AR-coated and angled-stripe InGaAsP amplifier means that some of the intracavity radiation suffers loss while some experiences gain depending upon the current density within the semiconductor chip. This excursion from net loss to significant gain can thus produce a

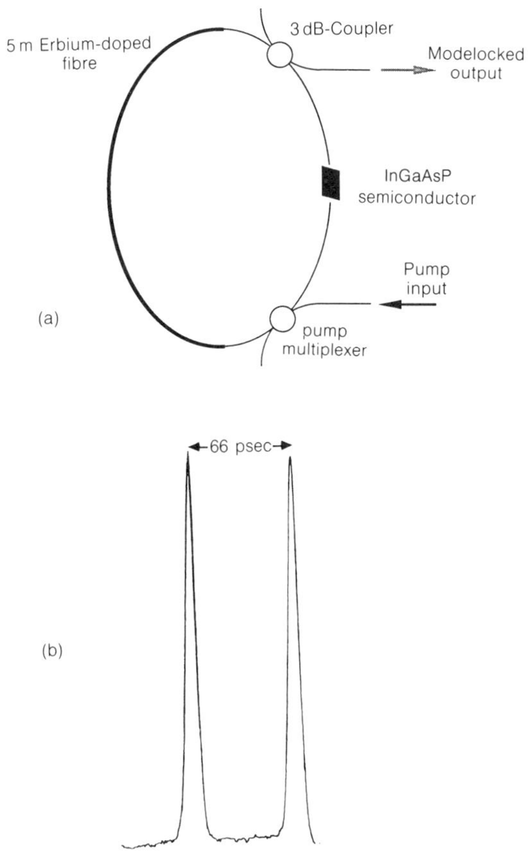

Figure 10.5 (a) Configuration of an erbium fibre ring laser incorporating InGaAsP loss–gain modulator and (b) typical streak intensity profiles of the mode-locked pulses (with calibration delay) from such a laser. The deconvolved pulse duration is 5 ps.

substantial depth of modulation and synchronous mode locking can therefore be achieved by ensuring that the injection current is modulated at a frequency $f_m = mf_c$. In the demonstration of this so-called **controlled-amplifier mode locking** (CAM) scheme $f_c \sim 20$ MHz and $f_f \sim 600$ MHz (i.e. 30th harmonic) and pulse durations around 3–5 ps were obtained (Figure 10.5(b)). With the prospect of the optical pumping of erbium-doped fibre lasers using semiconductor diode laser outputs at 980 nm or 1480 nm then such a mode-locking technique has the attractions that a potentially ruggedized source may simply comprise two semiconductor diode chips together with appropriate lengths of gain and interconnecting monomode fibres.

10.2.3.4. Phase modulation (FM mode locking)

A practical alternative to the techniques involving amplitude modulation is to employ an intracavity phase modulator. In physical terms the mode-locking mechanisms in both cases have a good deal of similarity. It is customary to locate a phase modulator (usually an electro-optic or Pöckels-effect device) close to one of the cavity reflectors. The application of an RF drive signal induces a sinusoidal variation in the phase shift experienced by the intracavity radiation that passes through the modulator (Figure 10.6(a)). As a consequence the carrier frequency of the radiation is shifted by an amount that is determined by the rate of change of the phase (i.e. $d\phi(t)/dt$) hence FM mode locking. During the subsequent cavity periods these frequency shifts are accumulated to the extent that the frequency of the circulating radiation

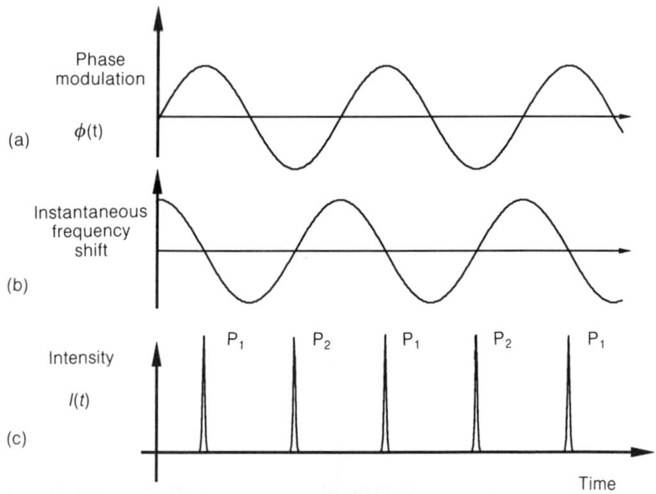

Figure 10.6 Illustration of (a) phase and (b) frequency shifts resulting from applied intracavity phase modulation, and (c) interleaved pulse sequences P_1, P_2 thus produced.

may be progressively displaced from the peak region to the wings of the available gain spectrum. It follows therefore that pulses will be generated at the two extremes of the phase shift during each modulation cycle (i.e. where $dt/d\phi(t) \sim 0$) and so two periodic pulse sequences may be produced as indicated in Figure 10.6(c). Quite often the presence of optical nonlinearity leads to the pulse sequences P_1, P_2 having quite different characteristics such as duration and spectral detail. A distinct feature in the frequency-domain description is that this sinusoidal phase modulation produces a sequence of sidebands (with a Bessel function amplitude series) for each longitudinal mode rather than the two sidebands that arise with amplitude modulation. However, for the relatively small phase shifts arising in practice, only one or two sidebands have sufficient amplitudes to command serious consideration. Thus a coherent power coupling of the modes takes place in a manner similar to that of AM mode locking.

FM mode locking has been successfully applied to a number of laser systems including the recently developed doped-fibre laser systems. In these latter demonstrations both integrated-optical (Geister and Ulrich, 1988) and bulk (Hanna *et al.*, 1989) phase modulator devices have been employed and pulse durations < 20 ps have been obtained (Schlager *et al.*, 1989).

10.2.4. Passive mode locking

As mentioned in section 10.2.2, the concept of passive mode locking is interpreted here as the case where the intracavity radiation induces the periodic modulation that is automatically synchronized to the cavity frequency f_c. In the early days of the mode-locking investigations of gas and solid-state lasers, so-called 'spontaneous' mode locking was reported by a number of researchers (Crowell, 1965; McClure, 1965; Statz and Tang, 1965; Gaddy and Shaefer, 1966). This was justified on the basis that mode competition effects and gain nonlinearities in the active medium could combine to produce some measure of self or spontaneous phase locking of longitudinal modes. However, because of uncertainties arising from the pulse measurement schemes, involving two-photon fluorescence in particular, and because the modelling of the mode-locking process was at a rather primitive stage, the practicality of spontaneous mode locking was not established. By contrast, during the 1980s the role of optical nonlinearities in mode-locked lasers became a major highlight area of research and, interestingly, spontaneous mode locking has recently been unambiguously demonstrated for a femtosecond pulse titanium–sapphire laser (Spence *et al.*, 1990). This topic will not, however, be addressed here.

The most common method for passive mode locking involves the use of saturable absorber media which usually take the form of dye solutions or multiple-quantum-well semiconductor structures. In the case of the frequently employed saturable absorber dye solutions the behaviour can be understood

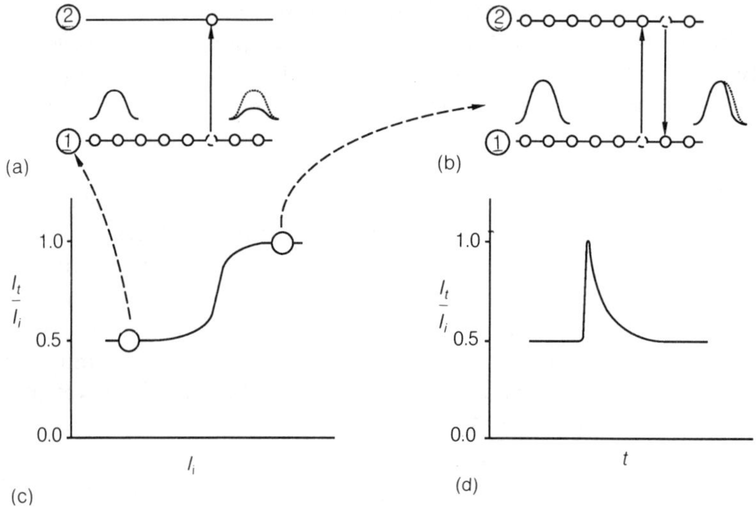

Figure 10.7 Simplified illustration of the behaviour of a saturable absorber dye medium. (a), (c) At low intensities there is net absorption whereas (b), (d) sufficiently high intensity levels the absorption may become saturated. The absorber has a characteristic recovery time as indicated in (d).

by assuming that it can be described in simplified terms to behave as a two-level energy manifold as illustrated in Figures 10.7(a), 10.7(b). At low levels of propagating intensity the predominant lower state population gives rise to net absorption and so the low light level transmission is significantly less than 100% (Figure 10.7(c)). For significantly higher intensities, promotion of molecules from energy state 1 to energy state 2 leads to a reduction in the absorption. Indeed, at a sufficiently high intensity the situation depicted in Figure 10.7(b) may be reached where the population densities in states 1 and 2 may become equal. In other words, the excitation of the dye molecules that constitutes their promotion from energy E_1 to energy E_2 has been due to absorption from the propagating radiation. When the populations in states 1 and 2 become equal then the net absorption is zero and the absorber is said to be **saturated** or **bleached** or **transparent**. The saturable absorber can thus be regarded as an 'optical switch' which opens completely in this transparency condition. Recovery of the absorption will occur with a time that is characteristic of the molecules and the solvent as illustrated in Figure 10.7(d). Thus for a saturable absorber placed within a laser resonator the circulating intracavity flux will influence the transmissivity of this passive medium and a periodic amplitude (or intensity) modulation is thus provided at the cavity frequency f_c. As mentioned earlier an amplitude modulation at the cavity frequency thus applied to a particular longitudinal mode will lead to sideband generation at frequencies that are coincident with those of nearest-neighbour modes and so phase locking is achieved.

The behaviour of saturable absorbers may also be readily described in the time domain. Remembering that the output of intracavity radiation of a free-running laser has a stochastic intensity characteristic then the evolution of discrete ultrashort pulses can be considered. In doing so, it is appropriate to mention at the outset that fast or slow recovery saturable absorbers may be employed. For instance, the saturable absorber dye media used with flashlamp-pumped Nd:glass lasers has a fast recovery time, $\tau_{\text{Recovery}} \sim 10\,\text{ps}$. It is therefore reasonable to suppose that under the appropriate pumping and laser operating conditions selection could be made of the most intense fluctuation or intensity spike in the circulating radiation. In other words, only this most intense feature can saturate the absorber to the point that it receives sufficient preferential gain to evolve at the expense of less intense radiation features. By this means a single discrete pulse may develop and circulate within the resonator while some optical transmission through the output coupler constitutes the passively mode-locked output pulse train or sequence. The typical pulse durations of ~ 2–$10\,\text{ps}$ are comparable with the recovery time of the saturable dye solution and such durations are observed at an early stage of the pulse train (Glenn and Brienza, 1967; Bradley *et al.*, 1969; von der Linde, 1972; Bradley and Sibbett, 1973).

With slow saturable absorbers of the type that are used with passively mode-locked dye lasers, the recovery time of $\sim 1\,\text{ns}$ is not too different from the cavity periods of several nanoseconds. The evolution of discrete ultrashort laser pulses in this case is therefore a combination of the saturation of absorption in the passive medium and the subsequent saturation of the gain in the active medium. The pulse evolution can be readily understood by following the schematic of Figure 10.8. Saturation of the absorption will most probably be accomplished by a family of fluctuations and a burst of radiation 'noise' having a duration $\sim \tau_{\text{Recovery}}$ will be preferentially transmitted. After a substantial number of cavity periods the intensity of this noise burst may be sufficient that the available population inversion in the active medium may be insufficient to provide gain for all of this selected signal. Consequently, the trailing portion of the burst may receive a deficit of amplification. This is the process of so-called **gain saturation**. It can therefore be seen that the establishment of periodic transparency in the saturable absorber leads to a selection of a dominant noise burst while gain saturation is responsible for subsequent temporal shortening of the evolving pulse. Both the saturable absorption and gain influence the pulse-shaping processes that take place. This process continues during many round trips until a steady state condition is achieved. This may be where the increasing bandwidth requirement of the shortening pulse reaches a limit, set by a tuning element for example, or where some additional nonlinearity-related effects (see section 10.2.5.2) may arise to compete with the pulse-shortening mechanisms.

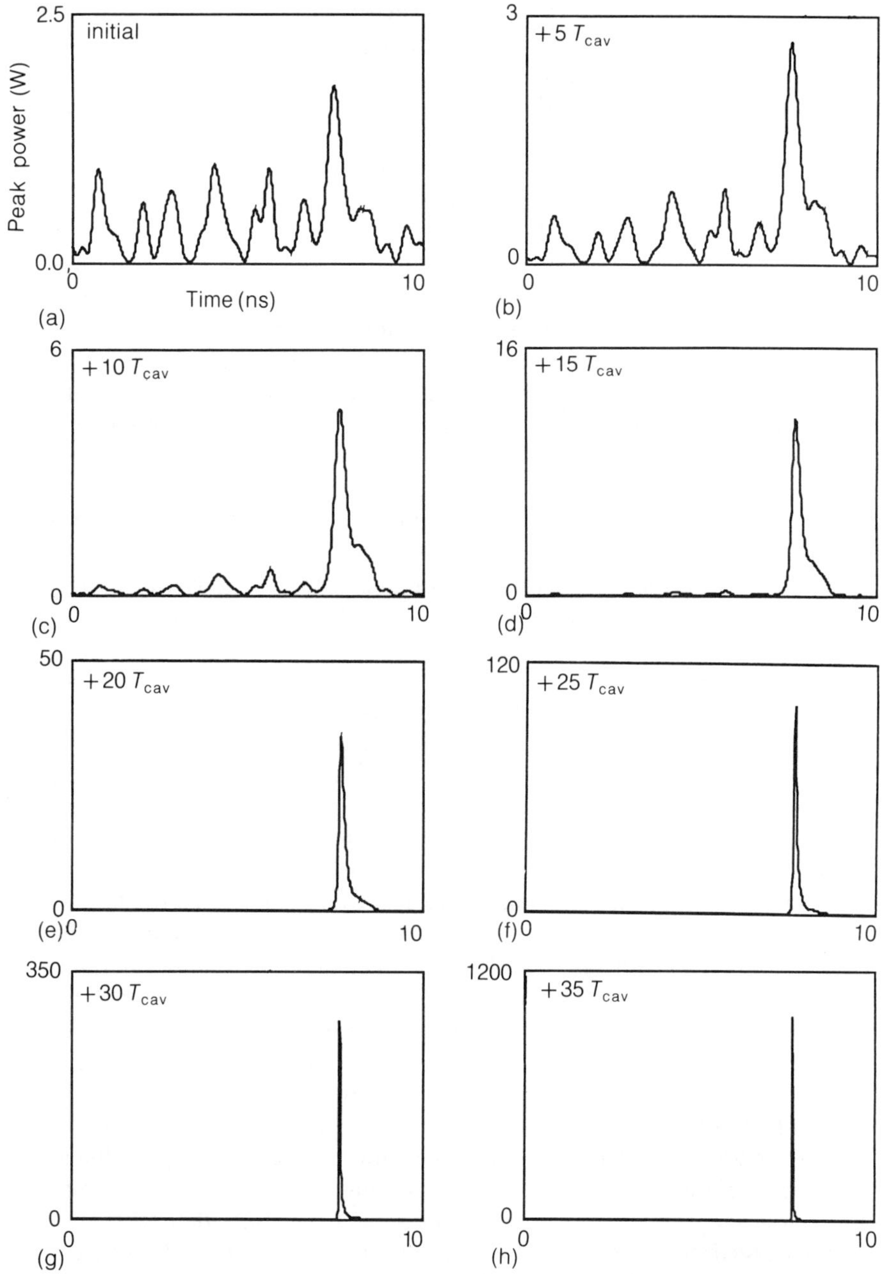

Figure 10.8 Illustration of pulse evolution from noise in a passively mode-locked laser having a slow saturable absorber. (a) Distributed intracavity flux for a ring resonator having a period (T_{cav}) of 10 ns (40 longitudinal modes with random phases assumed). (b)–(h) Selection and subsequent shortening of a dominant noise burst by the combined influences of saturated absorption and gain during 35 cavity periods.

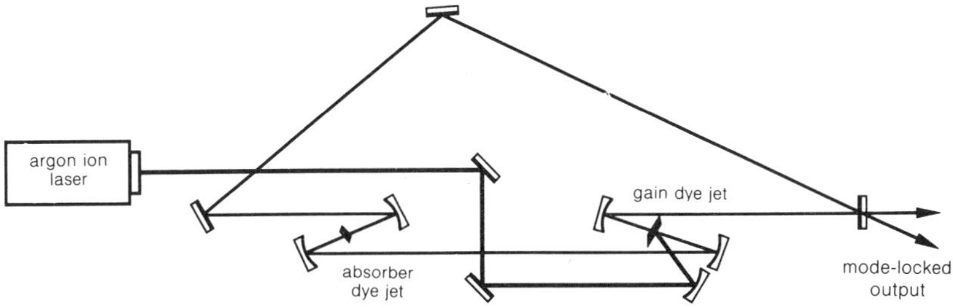

Figure 10.9 Schematic of a colliding-pulse model-locked (CPM) dye laser.

In a flashlamp-pumped dye laser the pulse durations may be ~ 1–5 ps (Arthurs *et al.*, 1972) but in a CW laser subpicosecond (~ 200 fs) pulses can be generated in the type of arrangement shown in Figure 10.9 (Fork *et al.*, 1981; Willson *et al.*, 1982). For this latter CW ring dye laser the propagation of intracavity flux is permitted in both clockwise and counterclockwise directions and the pulse evolution process is initiated when the interacting (or colliding) flux has sufficient intensity to saturate the absorption of the passive medium. It is probably worth noting here that the initial intensity fluctuation patterns, or indeed the subsequent evolving pulses, are unlikely to be indentical because of distinct phase and intensity variations experienced at the reflectors and gain medium by the clockwise and counterclockwise propagating radiation. Also, coherence or interference effects occurring in the saturable absorber will also lead to phase-related behaviour rather than simply intensity- dependent saturation effects in the passive medium. (Additional optical Kerr effects that influence the passive mode locking are dealt with in the next section.) Because of this configuration, the system is usually described as a **colliding-pulse mode-locked** (CPM) dye laser (Fork *et al.*, 1981). A variant on this theme is the antiresonant ring (ARR) arrangement that was proposed originally by Siegman (1981). With this approach one mode-locked beam output is produced but the bidirectional propagation in a 'ring reflection' geometry is retained.

The most common gain medium for the CPM or ARR dye lasers is rhodamine 6G in ethylene glycol solvent and diethyloxadicarbocyanine iodide (DODCI) in ethylene glycol solvent for the saturable absorber medium. With this combination, typical pulse durations are typically ~ 100–200 fs at the optimum operating wavelength around 620–630 nm (Fork *et al.*, 1981); Willson *et al.*, 1982). Subsequently, dye combinations for the blue, green and near-infrared spectral regions were demonstrated first for quasi-CW flashlamp-pumped dye lasers (Sibbett and Taylor, 1983a,b, 1984) and afterwards for CW dye lasers (K. Smith *et al.*, 1985; French and Taylor, 1986). Also in the near-infrared spectral region, the practicality of passive mode locking of colour centre lasers was first demonstrated by Langford *et al.*

(1987b). This has led to the subsequent exploitation of this technique for a range of longer-wavelength colour centre crystal gain media where both bulk and multiple-quantum-well and semiconductor structures have been successfully employed for ultrashort pulse generation as mentioned previously.

The **hybrid** scheme whereby synchronous pumping can be combined with intracavity saturable absorption has also developed. Especially impressive results have been reported for hybrid mode locking in rhodamine 6G (Dawson *et al.*, 1986; Nakazawa *et al.*, 1987) and Styryl 14/9 (Dawson *et al.*, 1987) dye lasers where femtosecond pulses have been produced in both visible and near-infrared spectral ranges.

10.2.5. Exploitable optical nonlinearities in extracavity and intracavity configurations

10.2.5.1. Extracavity optical nonlinearities

During the 1980s the exploitability of optical nonlinearity was recognized and implemented in respect of the further refinement of picosecond and femtosecond pulse laser sources. Most notably, the self-phase-modulation effect (de Martini *et al.*, 1967) especially in monomode optical fibres (Stolen and Lin, 1978) has been frequently highlighted. For example, a typical temporal compressor arrangement is illustrated in Figure 10.10 where the injected Nd:YAG pulses are self-phase-modulated during propagation in the normal dispersion monomode optical fibre. This is the result of the intensity-dependent refractive index $n(t)$ which varies throughout the pulse. This can be expressed as

$$n(t) = n_0 + n_2 I(t) \tag{10.17}$$

where n_0 is the refractive index at low intensity levels. Thus as an ultrashort, high peak power pulse propagates a fibre length L a phase difference arises between the low intensity wings of the pulse profile and the higher intensity regions. Since this is due to the intensity of the pulse itself it is described as

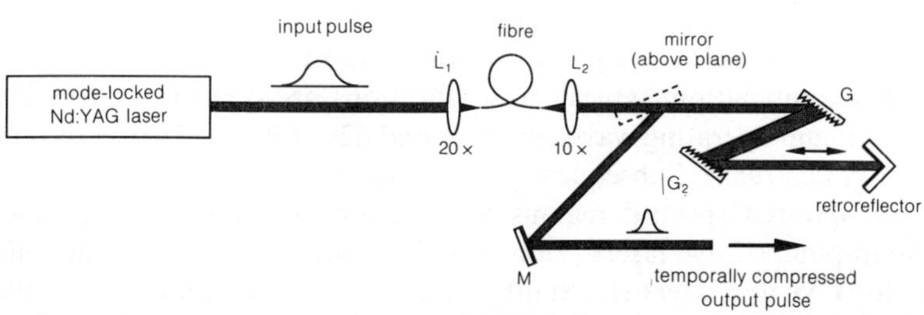

Figure 10.10 Typical layout for a fibre–diffraction grating pair pulse compressor. (From Miller and Sibbett (1988).)

a **self**-phase-modulation (SPM) effect. It gives rise to instantaneous changes in frequency which are related to the rate of change of the phase:

$$d\omega \alpha - \frac{d\phi}{dt} \equiv \frac{d}{dt}\left(\frac{2\pi n_2 L I(t)}{\lambda}\right) \equiv \frac{2\pi n_2 L}{\lambda}\frac{d}{dt}I(t) \tag{10.18}$$

If a prompt (i.e. femtosecond) response is assumed for the optical Kerr effect in silica (Alfano *et al.*, 1972; Owyoung *et al.*, 1972) then the pulse develops a frequency chirp as illustrated in Figure 10.11(c). The SPM-induced spectral broadening implies that frequencies are lowered in the leading edge and raised in the trailing edge of the pulse so that a frequency *up-chirp* is induced. When these self-phase-modulated pulses propagate in a medium having normal group velocity dispersion (GVD) their durations increase and this applies to 1.06 μm Nd:YAG pulses in transit through a silica monomode fibre.

Temporal compression is achieved by passing the spectrally and temporally broadened pulses that exit the fibre through a Treacy diffraction grating pair which provides a negative group velocity dispersion (Treacy, 1968). By this means pulse durations ~ 100 ps from an actively mode-locked Nd:YAG laser are temporally compressed to durations ~ 2–3 ps (e.g. Gomes *et al.*, 1985b)

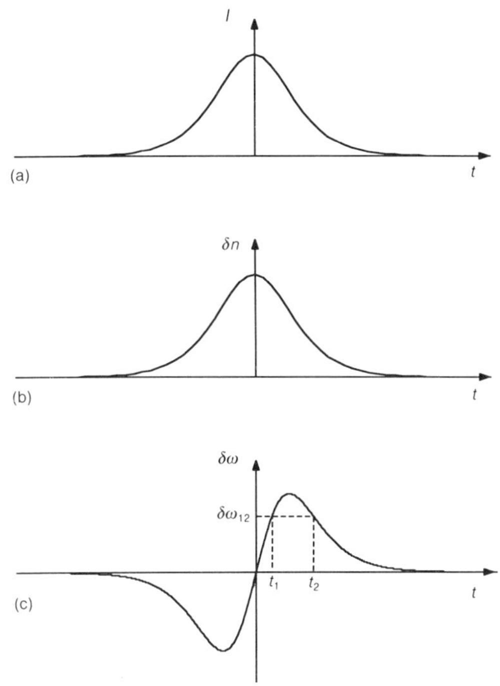

Figure 10.11 Result of (a) an ultrashort pulse propagating in an optical fibre: (b) intensity-induced refractive index changes and (c) self-phase-modulation (SPM) effect.

and cascaded fibre grating pair compressors permit the direct generation of subpicosecond pulses (Gomes *et al.*, 1985b; Zysset *et al.*, 1986). For wavelengths exceeding the first-order dispersion-zero wavelength of the fibre (i.e. inherent negative GVD) then a length of suitable monomode fibre can provide both the frequency chirp and the temporal compression. This has led to the development of all-fibre pulse compressors (Blow *et al.*, 1985). Moreover, the observations (Kuckartz *et al.*, 1988; Golovchenko *et al.*, 1988) that stable and reliable temporal compression is sustained at power levels where significant stimulated Raman scattering exists further confirms the practicality of this technique.

Reference to Figure 10.11(c) shows that the frequency chirp is linear over a substantial central region of the ultrashort pulse profile. This portion of the phase modulation can be temporally compressed by the negative group velocity dispersion imposed by the grating pair but uncompressed wing regions can still remain. This means that this technique can lead to pulses that have some energy in a low intensity, longer-duration pedestal. This can be problematical in particular applications where clean pulses are required and so some care must be exercised in this regard.

The pulse compression scheme has also been spectacularly employed in the femtosecond pulse regime. Amplified pulses from a GVD-compensated, CW ring CPM dye laser can be spectrally broadened in a short length of optical fibre and optically temporally compressed using a diffraction grating pair and prism quadruplet combination. Compressed pulse durations as short as 6 fs around 630 nm (Fork *et al.*, 1987) and ~ 9 fs around 800–850 nm (Becker *et al.*, 1988) have been obtained using this approach.

Enhanced performance characteristics of synchronously mode-locked dye lasers have also resulted from the use of temporally compressed fundamental frequency (Beaud *et al.*, 1986) or frequency-doubled (Johnson and Simpson, 1985) pump Nd:YAG laser pulses. (Post-compressor frequency quadrupling of Nd:YAG laser pulses has also been applied as a means of obtaining picosecond or subpicosecond ultraviolet (266 nm) pulses that find applications such as electron pulse generation in the time-resolved electron microscopy (Halbout *et al.*, 1988)). Indeed, femtosecond pulse generation has been reported over a wide spectral tuning range that reaches into the near-infrared to approximately 1.3 μm (Beaud *et al.*, 1986). Frequency tunability beyond 1.3 μm has been demonstrated by exploiting fibre Raman oscillators (Stolen, 1980). In particular, the synchronously pumped fibre Raman oscillator has been successfully employed as a means of producing subpicosecond near-infrared pulses where the effects of group velocity dispersion were compensated by means of a grating pair configuration (Kafka *et al.*, 1986).

Interestingly, the SPM or related spectral features arising during pulse propagation in an optical fibre may be used to monitor the stability of the output from a mode-locked laser. As the peak power of the pulses is increased the intrafibre spectral broadening increases and fringe detail appears. This arises because it can be seen from Figure 10.11(c) that a particular frequency shift (e.g. $\delta\omega_{12}$) arises at two distinct times t_1, t_2. The fringe visibility of the associated interference will be influenced by the reproducibility of the pulse intensity profile and in particular the rate of change of intensity (i.e. $dI(t)/dt$). Thus the spectra reproduced in Figure 10.12 clearly confirm the pulse stability that can be obtained from the acousto-optically mode-locked Nd:YAG laser.

The frequency chirping associated with self-phase-modulation in an optical fibre has also been exploited in a novel amplification scheme for ultrashort laser pulses. Pulses of 55 ps duration from a CW mode-locked Nd:YLF laser were initially spectrally and temporally broadened in a normal dispersion monomode optical fibre. These pulses were then amplified in a broadband Nd:glass multiple-stage amplifier before being subsequently temporally compressed to become 1 ps duration, 1 TW peak power pulses (Maine *et al.*, 1988). Thus, ultrahigh brightness (10^{18}–10^{19} W cm^{-2}) optical sources are now becoming available in 'table-top' dimensions such that interesting high power processes such as multiphoton ionization and ultrashort XUV–soft X-ray pulse generation can be conveniently investigated in an increasing number of laboratories.

At relatively high peak intensities substantial Raman scattering in the optical fibres may accompany the self-phase-modulation spectral broadening. Under the conditions of negative group velocity dispersion (i.e. anomalous dispersion) soliton Raman pulses may be produced (Gouveia-Neto *et al.*,

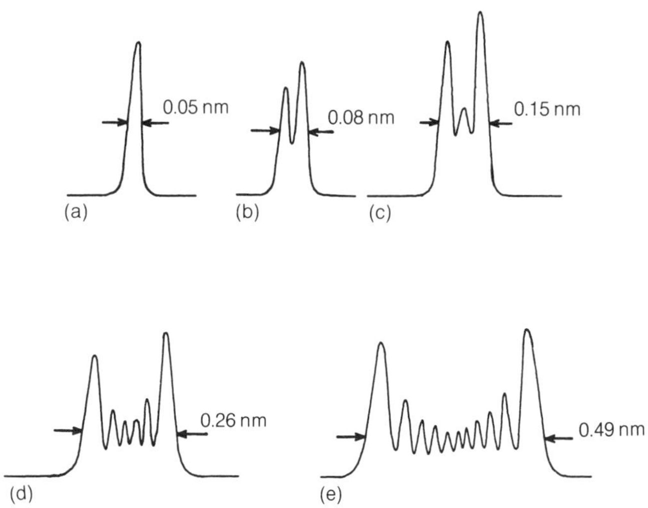

Figure 10.12 Spectral characteristics associated with SPM effects in a 150 m long monomode optical fibre (7 μm core diameter) for pulse peak powers in the 3–53 W range: (a) 3 W; (b) 8 W; (c) 12 W; (d) 25 W; (e) 53 W.

1988). Multisoliton Raman generation in monomode fibres (Vyslouth and Serkin, 1983) has also been demonstrated experimentally (Kafka and Baer, 1987; Gouveia-Neto *et al.*, 1988) as a method by which pulse durations in the femtosecond regime may be achieved. Because the soliton Raman pulse evolves from an initial Raman noise signal then the large observed jitter in the pulse formation is to be expected and this therefore limits the usefulness of such a source. By involving feedback and synchronous amplification in a soliton Raman fibre ring laser, Taylor and co-workers (Gouveia-Neto *et al.*, 1988) have obtained pulses having durations $\sim 130\,\text{fs}$ around $1.4\,\mu\text{m}$. Access to tunability at longer wavelengths extending to $1.7\,\mu\text{m}$ has also been demonstrated for a cascade Raman soliton fibre laser but appropriate dispersion-shifted fibres are required for such arrangements (Gouveia-Neto *et al.*, 1988). These Raman-based schemes offer relative simplicity and convenience, but inherent large phase noise combined with the difficulties in completely suppressing the background signal level represent their primary limitations.

10.2.5.2. Intracavity Optical Nonlinearities

Although subpicosecond pulses can be produced by passively mode-locked CW lasers (e.g. the dye laser, section 10.2.4) it has been found that the control of intracavity self-phase-modulation arising in the dye solvent and due to the off-resonant absorber and gain saturation effects (Valdmanis *et al.*, 1985) gives rise to substantial induced frequency chirps in the intracavity pulses. By incorporating the two pairs of Brewster-angled prisms within the ring resonator as indicated in Figure 10.13 it is practical to provide adjustable GVD to ensure that optimally short duration pulses can be produced. This is because the relative separations of prisms P_1, P_2 and P_3, P_4 constitute a controllable amount of the negative group velocity (or anomalous) dispersion

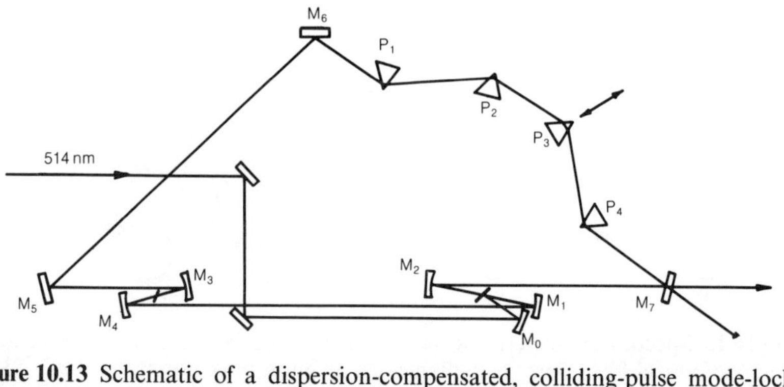

Figure 10.13 Schematic of a dispersion-compensated, colliding-pulse mode-locked (CPM) dye laser. (After Finch *et al.*, (1988a).)

whereas the translation of prism P_3 provides adjustable positive group velocity (or normal) dispersion (Fork *et al.*, 1984).

For the arrangement of Figure 10.13, the radius of curvature of mirrors M_3, M_4 was 5 cm as compared with 10 cm radius for mirrors M_1, M_2. Consequently the focal power density was a maximum in the absorber dye solution. This implied that the optical Kerr effect nonlinearity arose mainly in the solvent of the passive dye but the SPM effect arising from this was supplemented by absorption and gain saturation in the passive and active dyes respectively. Thus, although higher-order (third order in particular) dispersive effects should not be neglected, this fairly simple cavity arrangement can be regarded as a means of compensating for the second-order dispersion that results from the SPM produced by the intracavity radiation (Finch *et al.*, 1988c). This implies that frequency chirping effects that are produced by the relatively high power evolving ultrashort optical pulses in the colliding-pulse mode-locked ring dye laser can be exploited in terms of the increased spectral extent while complementarily the group velocity dispersion can be compensated to ensure that the pulse durations can reach a minimum.

With GVD compensation in a passively mode-locked CW rhodamine 6G dye laser (DODCI as saturable absorber) pulses as short as 27 fs have been directly produced using a six-mirror ring resonator (Valdmanis *et al.*, 1985) and 19 fs when the seven-mirror ring resonator scheme of Figure 10.13 was employed (Finch *et al.*, 1988a). A similar GVD compensation concept has been applied to synchronously pumped, hybridly mode-locked dye lasers where pulses shorter than 100 fs have been produced over a more extended spectral range (Dawson *et al.*, 1986). Because such hypershort laser pulses have inherently high peak powers they can be readily amplified using high repetition rate (~ 5 kHz) copper vapour laser-pumped dye amplifiers (Knox *et al.*, 1984; Knox, 1988) so that extracavity temporal compression can be achieved. With the generation of optical pulses as short as 6 fs around 630 nm (Fork *et al.*, 1987) and 9 fs in the 800–845 nm spectral region (Becker, *et al.*, 1988) new opportunities have been opened up for time-domain spectroscopic studies in the truly femtosecond regime (e.g. Brito-Cruz *et al.*, 1986; Rosker *et al.*, 1986; Mathies *et al.*, 1988).

The technique of GVD-compensated colliding-pulse mode locking has also been exploited in LiF:F_2^+ colour centre laser arrangements so that additional spectral regions could be covered (Langford *et al.*, 1989). With other colour centre gain media (notably KCl:Tl, NaCl:OH^-) femtosecond pulses can be generated in the 1400–1700 nm range using the soliton or coupled-cavity mode locking schemes described in section 10.2.6.

Alternatively, by exploiting intracavity frequency doubling and optical parametric oscillation nonlinear processes in a passively mode-locked laser femtosecond pulses can be made available from the ultraviolet to the near-infrared region. Although demonstrated previously for a synchronously pumped CW dye laser (Sibbett *et al.*, 1980) the current availability of high quality nonlinear crystals (e.g. β-barium borate) has enabled relatively efficient frequency doubling of a CPM Rh6H dye laser to be achieved using ultrathin (~ 55 μm) crystals (Edelstein *et al.*, 1988). Also, by incorporating a KTP nonlinear crystal within a passively mode-locked CPM ring dye laser, optical parametric oscillation has been recently shown to lead to the generation of femtosecond pulses that are tunable over the extended range of 820–920 nm with the potential for tunability over the extended 720–4500 nm region (Edelstein *et al.*, 1989; Wachman *et al.*, 1990).

10.2.6. Coupled-cavity mode locking

When mode-locking schemes were being initially investigated for low gain CW helium–neon laser systems, one approach involved the use of a cavity period matched linear coupled-resonator configuration which incorporated the modulator (Foster *et al.*, 1965; Di Domenico and Czarniewski, 1965). The exploitation of a coupled nonlinear external cavity was first systematically demonstrated by Mollenauer and Stolen in work that led to the development of their **soliton laser** (Mollenauer and Stolen, 1984; Mollenauer, 1985b). Using an anomalously dispersive monomode fibre in an arrangement similar to that shown in Figure 10.14 these researchers exploited the fact that the combined effects of self-phase-modulation and anomalous dispersion could give rise under appropriate experimental conditions to the generation of bright optical solitons (second order observed). The interpretation of the enhancement of the mode-locking performance was that the reinjection of a solitonic shaped and shorted pulse from the control cavity into the master cavity led to the generation of femtosecond duration pulses with excellent stability (Mitschke and Mollenauer, 1987). In subsequent research, both theoretical (Blow and Wood, 1988) and experimental (Kean *et al.*, 1988, 1989; Miller and Sibbett, 1988) studies showed that other types of nonlinear elements in the control cavity would also produce substantial mode locking enhancements. With normal-dispersion fibre the pulses reinjected into the

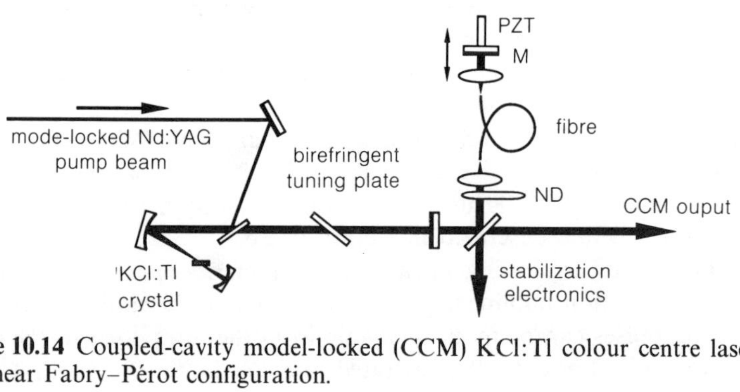

Figure 10.14 Coupled-cavity model-locked (CCM) KCl:Tl colour centre laser in a nonlinear Fabry–Pérot configuration.

master cavity exhibit both spectral and temporal broadening features and bright solitons are not involved. A variety of fibre types, including soliton type, normal dispersion, dispersion shifted, dispersion flattened and erbium doped, have been investigated in the nonlinear external cavity and significant mode locking improvements have been observed over specific ranges of operational parameters for all samples (Sibbett, 1990). It has also been demonstrated that optical nonlinearities associated with saturable absorption–gain and carrier heating in a semiconductor diode amplifier are suitable for the coupled-cavity mode locking of a KCl:Tl, colour centre laser (Sibbett, 1990).

This general mode-locking scheme is variously described as coupled-cavity mode locking (CCM) (Miller and Sibbett, 1988; Kean *et al.*, 1989) or additive-pulse mode locking (APM) (Mark *et al.*, 1989) or interferential mode locking (IML) (Ouellette and Piché, 1988). These designations relate directly to the mechanisms that are understood to dominate the enhanced pulse shaping and shortening processes that arise in such composite-cavity configurations. The coupled-cavity mode locking description will be discussed here in some detail because it can be readily related to the general physical concepts previously outlined in sections 10.2.3 and 10.2.4.

The basic operation of coupled-cavity mode locking can be described in the following terms. Once laser threshold is established in the main cavity then some radiation will be coupled into the external or **control** cavity. Provided the intensity of this radiation in the control cavity is sufficient for a significant nonlinear phase to be induced during propagation in the nonlinear medium, then the feedback signal which experiences gain in the active medium can influence the phase coupling of an increased number of longitudinal modes in the master resonator. This can occur either when an already-formed optical pulse circulates within the main cavity (for example, when active (Yakymyshyn *et al.*, 1989), synchronous (Kean *et al.*, 1988, 1989) or passive (Johnston *et al.*, 1989) modulation is provided) or alternatively when the intracavity radiation takes the form of noise bursts (Sibbett, 1989; Zhu *et al.*, 1989a) or stochastic noise (Goodberlet *et al.*, 1989). Because the main and control cavities are interferometrically matched then the feedback from the control cavity can represent amplitude or phase modulation (or both) that will enhance the coherent coupling of axial modes. It has to be remembered, of course, that this process is subject to major influences such as the gain kinetics of the active medium, group velocity dispersion and bandwidth control so that the shaping and shortening of the pulse (or noise burst) is entirely a compositecavity phenomenon. This is the primary reason for the coupled-cavity mode locking designation and the practical implementation of this scheme in representative experimental configurations is described later.

It is relevant at this point to draw attention to a particularly interesting aspect of the coupling of the longitudinal modes in this type of mode-locking process. The phase-modulated feedback signal from the nonlinear external cavity will experience significant gain in the active medium within the main cavity. The presence of self-phase-modulation spectral broadening will mean that gain is sought at frequencies that progressively lie towards the wings of the gain bandwidth. Under the condition where small signal amplification is being provided in the wings of the gain spectrum while some gain saturation may occur near its peak then the effective gain lineshape may be altered substantially. It is especially significant to consider the case where the coupled-cavity phase-locking process leads to a sufficiently enhanced transfer of power to the modes having frequencies at the extreme of the gain spectrum because the lasing bandwidth may then exceed the conventionally assumed bandwidth parameter.

The implications of this can be directly related to the relationship between pulse duration $\Delta\tau$ and lasing bandwidth $\Delta\nu_l$ given in equation 10.10. As a consequence of efficient coupled-cavity mode locking the effective lasing bandwidth could actually *exceed* the normal fluorescence or gain bandwidth (FWHM) of the active medium. In other words it could be reasonably expected that $\Delta\tau < 1/\Delta\nu_f$ where $\Delta\nu_f$ is the fluorescence bandwidth (FWHM). This could be regarded as particularly advantageous for the relatively narrow bandwidth gain media such as Nd:YAG where a super-mode-locked bandwidth could be produced under the dominating influence of coupled non-linear cavities.

10.2.6.1. Hybrid coupled-cavity mode locking

A typical arrangement for a *hybrid* coupled-cavity laser is illustrated in Figure 10.14. In this example the gain medium is a KCl:Tl colour centre crystal which is synchronously pumped by the 1064 nm output of a CW mode-locked Nd:YAG laser. When synchronously mode locked without any external cavity the pulse durations are in the 10–20 ps range for its characteristic tuning range of 1450–1550 nm. To achieve coupled-cavity mode locking an external cavity which includes a medium that exhibits suitable optical nonlinearity is configured as indicated such that it is interferometrically matched to the main cavity. As mentioned in the previous subsection, an optical fibre is a convenient nonlinear medium. Even for relatively short lengths ($\sim$ 20–50 cm) of monomode silica-based fibres optical power densities are sufficient to produce substantial self-phase-modulation (SPM) spectral broadening. If the fibre has anomalous dispersion (or negative GVD) around the laser wavelength $\sim$ 1500 nm then some temporal compression will occur during pulse propagation in the fibre whereas normal dispersion (or positive GVD) leads to temporal broadening. Provided that a specific soliton generation is not being sought (negative GVD fibre necessary) then enhanced mode locking can be readily achieved for either sign of GVD.

The influence of the coupled nonlinear cavity can be understood by

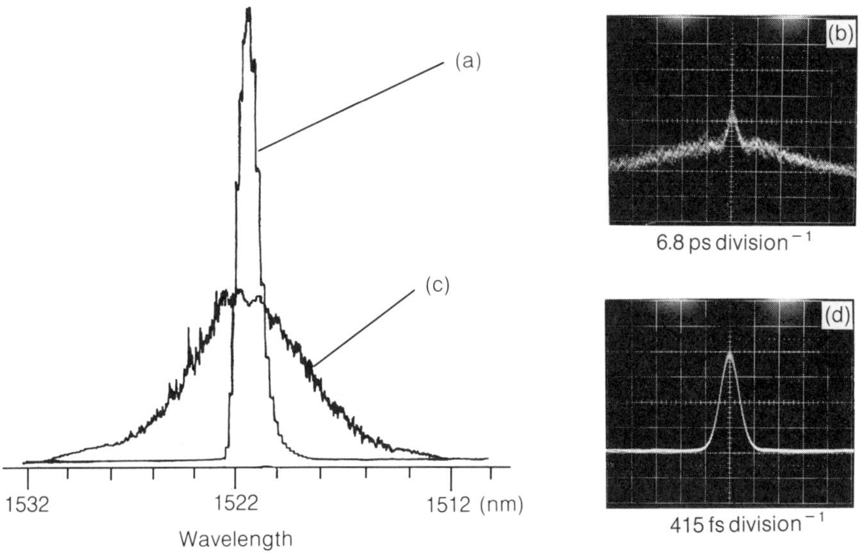

Figure 10.15 Spectral and temporal (SHG autocorrelation) features for the noise bursts produced by a cavity-detuned, synchronously pumped KCl:Tl colour centre laser operating either (a), (b) with the control cavity blocked or (c), (d) with the coupled nonlinear external cavity. (From Zhu *et al.* (1989a).)

referring to Figure 10.15 (Zhu *et al.*, 1989a). The spectrum of Figure 10.15(a) and the SHG autocorrelation trace in Figure 10.15(b) correspond to the noise burst sequence that was produced by the synchronously pumped but deliberately cavity-detuned colour centre laser. The autocorrelation trace for a noise burst exhibits a distinct coherence feature that is superposed on a broader base that is associated with the overall temporal extent of the noise burst. These noise bursts had durations ~ 25 ps and the average output laser power was ~ 70 mW. By coupling ~ 19 mW into the control fibre, the resulting spectral broadening is very evident (Figure 10.15(c)) and dramatic temporal compression (to 360 fs) as shown in Figure 10.15(d) was observed for the composite-cavity laser. In subsequent experimental evaluations involving erbium-doped control-cavity fibres shorter pulse durations ~ 75 fs were achieved at intrafibre powers as low as 3–5 mW (Zhu *et al.*, (1989b). Optical nonlinearity arising from the carrier kinetics in an InGaAsP semiconductor amplifier chip has also been exploited in this type of coupled Fabry–Pérot type arrangement, and pulses as short as 200–300 fs have been reported (Miller and Sibbett, 1988). For this work a d.c. current injection was supplied such that any time-varying amplitude and phase effects were self-induced by the intracavity flux.

Although described above for a KCl:Tl colour centre laser, this CCM scheme has also been demonstrated for other mode-locked colour lasers (Kean *et al.*, 1988, 1989; Yakymyshyn *et al.*, 1989; Johnston *et al.*, 1989), and

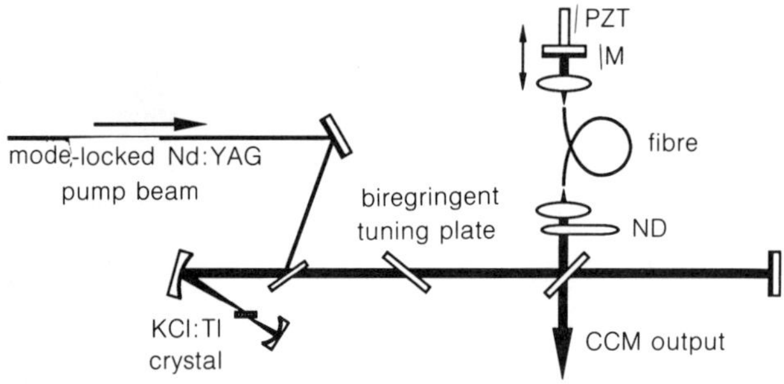

Figure 10.16 Coupled-cavity mode-locked (CCM) KCl:Tl colour centre laser in a nonlinear Michelson configuration.

actively mode-locked titanium–sapphire lasers (French *et al.*, 1989) and is expected to be applicable to other systems such as fibre lasers. By this means near-transform-limited picosecond and femtosecond pulses have been produced where the bandwidths were generally restricted by the gain medium itself or by the intracavity tuning elements that were employed.

An alternative Michelson-type interferometer configuration has also been investigated for pulsed CO_2 lasers (Ouellette and Piché, 1988; Morin and Piché 1989) and CW colour centre lasers (Sibbett, 1990). This scheme which is illustrated in Figure 10.16 can offer a physically more compact geometry and intensity ringing effects that are common to coupled Fabry–Pérot resonators may be avoided. The intensity, interferometric SHG autocorrelation traces and spectra for pulses produced when this configuration was used with a synchronously pumped KCl:Tl laser (Figure 10.17) illustrate the high quality of the enhanced mode locking that can be achieved.

It should be emphasized that the significant influence of the coupled cavity in the CCM lasers derives from an intensity-induced nonlinear effect. This phase-locking enhancement has been shown experimentally to be sufficiently dominant that the mode locking derives principally from the passive optical nonlinearity in the control cavity. Thus even when the gain medium of the main cavity is synchronously pumped by a mode-locked laser, the pulse repetition frequency of the CCM laser in steady state is determined by the colour centre laser cavity and not by the cavity period of the pump laser. This observation was made at an early stage for the soliton laser (Mitschke and Mollenauer, 1986) and it has been subsequently confirmed for the CCM laser arrangement (Finch *et al.*, 1990). Moreover, when it is considered that cavity detuning and the absence of bandwidth restricting filters can be tolerated in CCM lasers (Finch *et al.*, 1990) then this domination by the passive nonlinearity can be more fully appreciated. Thus the modulation simply serves to provide an optical intensity that is appropriate for initiating a control-cavity dynamic phase nonlinearity and the consequent pulse

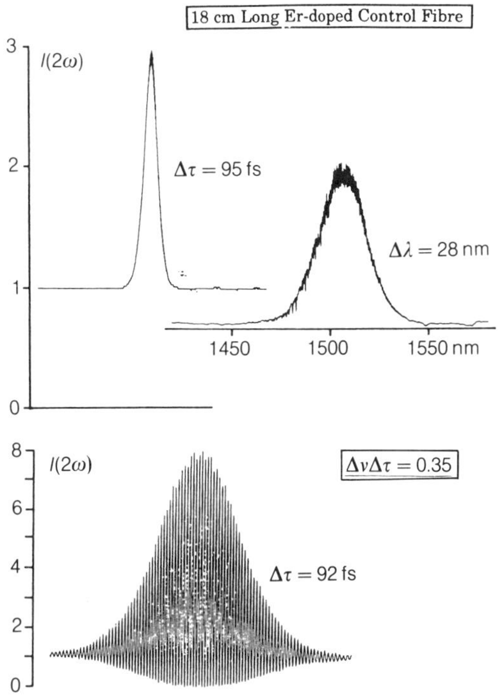

Figure 10.17 Temporal and spectral data relating to pulses produced by the CCM, KCl:Tl laser illustrated in Figure 10.16 (18 cm long erbium-doped control fibre).

shaping and shortening effects are attributable to coherent phase locking of an enlarged comb of longitudinal modes. Obviously, the role of group velocity dispersion is of particular relevance especially where the extended spectra of femtosecond pulses are involved (e.g. Zhu *et al.*, 1989b). It is for these reasons that the mode locking just described can be described as **hybrid coupled-cavity mode locking**. The entirely passive counterpart can also be exploited as outlined in the following subsection.

10.2.6.2. Passive coupled-cavity mode locking

As stated in section 10.2.2 the **passive** designation as used here is taken to mean that the phase locking of longitudinal modes is produced by an effect that is induced solely by the intracavity radiation. In this instance the main-cavity laser is pumped by a source that is not mode locked or modulated in any way. Indeed, the first observation of passive mode locking in a coupled-cavity configuration was reported by Mitschke and Mollenauer (1986) who observed that their soliton laser would continue to operate in the steady state condition whenever the modulator to the Nd:YAG pump laser had been switched off. As pointed out in section 10.4.1 the initiation of the CCM process requires adequate intensity to produce substantial nonlinear phase shifts and in the case of the soliton laser the enhanced mode locking was not self-initiating.

Passive CCM titanium–sapphire laser

The first experimental demonstration of self-starting for passive coupled-cavity mode locking in a CW laser involved the titanium–sapphire (Ti:Al_2O_3) gain medium in an arrangement similar to that illustrated in Figure 10.18 (Goodberlet *et al.*, 1989). A likely explanation for the initiation of the passive CCM process in the Ti:Al_2O_3 laser is that substantial spatial hole burning in the homogeneously broadened gain medium leads to significant mode-beating effects and consequent major intensity fluctuations. (This can be encouraged by siting the Ti:Al_2O_3 rod near the centre of the laser resonator.) These fluctuations have sufficient optical power density to cause spectral broadening in the control-cavity optical fibre by self-phase-modulation. Reinjection of the SPM feedback pulse into the main cavity then gives rise to a coherent phase coupling of many axial modes and the generation of ultrashort laser pulses. The typical duration of pulses produced using the laser arrangement as illustrated is in the 2–5 ps range (Goodberlet *et al.*, 1989; Sibbett, 1990). These pulses also generally show evidence of a strong frequency chirp which can be readily deduced from the loss of fringe visibility in the wings of interferometric autocorrelation traces (Figure 10.19(a)) and from duration–bandwidth products that significantly exceed the Fourier transform limit. These frequency chirps can arise in both the extended optical fibre and the 20 mm long titanium–sapphire rod which was used in this particular laser.

Shorter duration, subpicosecond laser pulses can be obtained by applying the temporal compression scheme described in section 10.2.5. For instance, by using a pair of diffraction gratings, the chirped, picosecond pulses from the passively mode-locked coupled-cavity Ti:Al_2O_3 laser can be temporally compressed to durations ~300 fs. The corresponding interferometric auto-

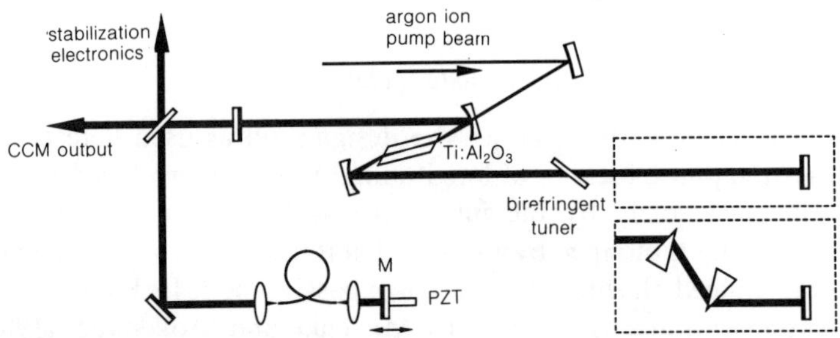

Figure 10.18 Experimental configuration of a passive, coupled-cavity mode-locked (CCM) titanium–sapphire laser (inset with intracavity GVD compensation).

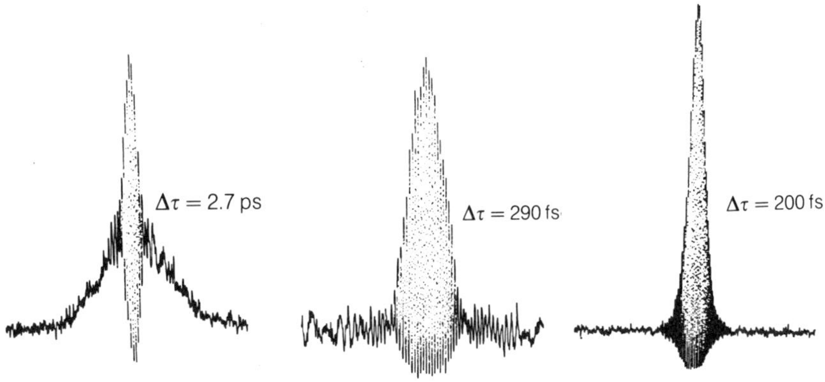

Figure 10.19 Interferometric autocorrelation data for a passive CCM Ti–sapphire laser having (a) no, (b) extracavity, and (c) intracavity dispersion compensation.

correlation trace (Figure 10.19(b)) confirms that the amount of frequency chirp on the pulses has been greatly reduced and near-transform-limited pulse durations can be achieved (Sibbett, 1990).

An alternative scheme involves the incorporation of a pair of Brewster-angled prisms within the master cavity as illustrated in the inset of Figure 10.18. Thus by adjustment of the prism separation and the optical path through the flint glass prisms the intracavity GVD can be suitably controlled. By this means **chirp-compensated** pulses have been generated (Figure 10.19(c)) with durations ~ 150 fs and bandwidths ~ 1.74 THz (i.e. $\Delta\nu\,\Delta\tau \sim 0.35$).

These examples of near-transform-limited pulse generation from a Ti–sapphire laser do not only represent interesting laser physics in themselves. For the wide tuning range offered by the Ti:sapphire gain medium, such pulses are of direct applicability to the time-domain spectroscopic studies of GaAs/AlGaAs bulk and multiple-quantum-well structures. Moreover, the relatively high pulse peak powers enable efficient frequency doubling to be exploited. Interestingly, the UV–blue region of the frequency-doubled spectrum is well suited for time-resolved studies of the wide band gap II–VI semiconductor (e.g. ZnS, ZnSe) structures. These have particular potential in the development of blue semiconductor-based lasers which are expected to be especially applicable in optical disc technology, colour printing and various related aspects of optical signal processing.

Passive CCM Nd:YAG laser

The mode-locked Nd:YAG (and more recently Nd:YLF) lasers are widely used both as pump sources and as picosecond pulse sources in their own right. As an alternative to the extracavity temporal compression schemes described in section 10.2.5 the technique of **passive coupled-cavity mode**

locking is now being frequently employed for both arc lamp and diode array pumped Nd:YAG lasers (Liu *et al.*, 1990; Goodberlet *et al.*, 1990). In such CCM systems there is no requirement for any active mode locking in the main laser cavity. An obvious advantage of this approach is that pedestal-free picosecond pulses can be readily and efficiently generated.

Coupled nonlinear external-cavity configurations similar to that described above for the titanium–sapphire laser can be used. With a 35 cm length of monomode optical fibre in the external control cavity the pulse durations have been reduced from around 90 ps to 6 ps in an arc lamp pumped Nd:YAG laser (Liu *et al.*, 1990). With a diode array pumped Nd:YAG laser, coupled-cavity mode locking has enabled pulse durations as short as 1.7 ps to be obtained (Goodberlet *et al.*, 1990). This latter result may well be an example of the super-mode-locked bandwidth that was discussed previously in this section because the usual gain–bandwidth limited pulse durations for Nd:YAG are typically estimated to be ~ 3 ps.

For a broader bandwidth host medium for the Nd^{3+} ion shorter CCM pulses are of course expected. This has recently been demonstrated to be the case for a CCM Nd:glass laser which produced stable trains of 380 fs pulses (Spielman *et al.*, 1990).

Passive CCM KCl:Tl colour centre laser

The discussion of coupled-cavity mode locking of the KCl:Tl colour centre lasers outlined previously in this section related to the exploitation of optical nonlinearities in a fibre or a semiconductor amplifier chip in the control cavity. The main-cavity laser was already mode locked (or modulated) such that the role of the nonlinear control cavity was one of enhancing the pulse-shortening process by inducing phase coupling of an enlarged comb of longitudinal modes. It was observed that with a fibre-based nonlinear cavity an entirely passive coupled-cavity mode locking could be neither self-initiated nor sustained in anything other than on a very short-term basis when the main cavity modulation was removed.

Recent work has now shown that self-initiating and self-sustaining passive CCM operation can be achieved in a KCl:Tl colour centre laser when a semiconductor device is employed in the nonlinear control cavity (Sibbett, 1990; Grant *et al.*, 1990). In this work the semiconductor device was an anti-reflection-coated, 350 µm long, InGaAsP double-channel, planar, buried-heterostructure diode chip. At d.c. injection currents below the transparency current of 6 mA at wavelengths around 1.5 µm this device exhibited a net absorption whereas at injection currents exceeding 6 mA the device provided net gain. During propagation of intracavity flux in the semiconductor element dynamic changes in the carrier kinetics lead to refractive index, and consequent intensity-dependent phase changes within these saturable absorption or saturable gain regimes. Thus phase-modulated signals are fed back from the nonlinear control cavity to the main laser cavity and the CCM influence proceeds accordingly.

The significant features of this semiconductor-based optical nonlinearity include its accessibility at low power levels and the low group velocity dispersion as compared with the nonlinear fibre counterpart. This is borne out by the observations that a self-initiating–sustaining passive CCM operation can be demonstrated for an intrachip optical power $\sim 700\,\mu W$ and a feedback power into the main cavity of just $5\,\mu W$. Representative interferometric autocorrelation data obtained under the conditions of either saturable absorption or saturable gain indicated that the pulses produced by this scheme of passive coupled-cavity mode locking were essentially free from excess frequency chirp.

This type of scheme which may be employed under low optical power conditions clearly offers potential for semiconductor and doped-fibre based laser systems. Its passive nature also affords the additional advantage that it can have inherently low phase noise characteristics. This is especially significant for applications of periodic ultrashort pulses in electro-optic sampling techniques for example (Weingarten *et al.*, 1988).

10.2.7. Pulse-shaping techniques

The importance and relevance of the shaping and coding of ultrashort optical signals has been recognized since the early 1970s (Desbois *et al.*, 1973) and whereas much of the initial work involved picosecond duration laser pulses (Agostinelli *et al.*, 1979; Froehly *et al.*, 1983; Heritage *et al.*, 1985a, b; Weiner *et al.*, 1986; Thurston *et al.* 1986; Hauer and Warren, 1987) more recent studies have confirmed that profile shaping and encoding can also be successfully undertaken in the femtosecond regime (Weiner and Heritage, 1987; Weiner *et al.*, 1988a, b). In fact, Heritage and Weiner and their co-workers have evaluated the type of temporally nondispersive pulse-shaping grating pair arrangement that was first described by Froehly *et al.* for the significantly longer (~ 30 ps) Nd:YAG laser pulses (Froehly *et al.*, 1983). The inclusion of specially patterned masks to modify the amplitude and phase of the spatially dispersed spectral components enables spectral, and hence temporal, tailoring of the pulses to be undertaken (Weiner *et al.*, 1988b). By this means arbitrarily shaped ultrashort pulses can be synthesized and this has included picosecond square pulses having rise times ~ 100 fs (Weiner *et al.*, 1988b). It is to be expected that this approach will offer a wide variety of opportunities within digital optical communications– data processing and numerous novel spectroscopic applications in the future.

10.3. ULTRASHORT PULSE CHARACTERIZATIONS

10.3.1. Introduction

In the mid-to-late 1960s period when lasers were successfully mode locked to produce picosecond optical pulses the conventional detection schemes involving photodiode–oscilloscope combinations did not have adequate time resolution to enable pulse durations to be measured. Although novel photodetectors have now been demonstrated to have bandwidths ~110 GHz (Parker *et al.*, 1987) the oscilloscopes that have become available to date do not possess sufficient bandwidths for this sub–5 ps resolution to be realized. These ultrafast photodiodes can, nevertheless, be used in conjunction with Fourier harmonic analysis (Parker *et al.*, 1987), electro-optic sampling (Weingarten *et al.*, 1988) or optoelectronic cross-correlation (Auston and Smith, 1982; Manning and Hill, 1986) techniques to provide data indirectly in a picosecond regime. However, the requirement of subpicosecond and femtosecond temporal resolutions also implies that the oscilloscope–photodiode technique would be completely unsuitable to many of the recently developed mode-locked lasers that were described in section 10.2.

It was realized at an early stage that the relatively high peak optical powers associated with ultrashort laser pulses would be suitable for the exploitation of nonlinear correlation techniques in duration-measurement techniques. With the initial development of flashlamp-pumped, mode-locked lasers that produced finite pulse trains the preferred methods involved two-photon fluorescence (TPF) (Giordmaine *et al.*, 1967; Ippen and Shank 1977) and to a lesser extent second-harmonic generation (SHG) intensity autocorrelation (Maier *et al.*, 1966; Armstrong, 1967). With mode-locked CW lasers the latter technique became the most widely employed and intensity and interferometric SHG autocorrelators are discussed in sections 10.3.2 and 10.3.3.

Although SHG intensity autocorrelation offers the best temporal resolution yet reported (~5 fs), it does not provide a direct measure for the pulse durations nor can the pulse shape be readily discerned from this type of monitor. For the pulse duration to be properly determined, pulse spectra must be simultaneously recorded and a pulse profile deduced that is consistent with both the temporal and the spectral data. When interferometric or fringe-resolved detail is retained in the autocorrelation data (see section 10.3.3) then some pulse shape and frequency chirp information can be extracted. Despite its rather more modest temporal resolutions ($\lesssim 1$ ps) the electron-optical streak camera described in section 10.3.4 does offer a linear response characteristic and simultaneous spectral (or spatial) information can be retained in one spatial dimension. This is of particular relevance in the case of time-domain spectroscopic studies.

In applications where optical sampling is carried out the pulse timing jitter or phase noise characteristics of the mode-locked laser are of primary consideration. Interesting insights into such source performance features can be gained by complementing the time-resolving techniques with spectrum analysis. It may therefore be concluded that a wide range of diagnostics is required if ultrashort pulses are to be properly and comprehensively characterized. These considerations constitute the subject matter of this section.

10.3.2. SHG intensity autocorrelation

In the schematic of Figure 10.20 the arrangement of the components of the autocorrelator is typical of the basic Michelson interferometer schemes that are generally employed. The intensity of each pulse in a periodic mode-locked sequence is bisected by a 50:50 beamsplitter BS and the subcomponents are recombined in a second-harmonic generation crystal (type 1, angle tuned) after being retroreflected at roof-top mirrors M_1, M_2. (Typical crystals are KDP and BBO where thicknesses $\sim 100\,\mu m$ and 1 mm respectively ensure that adequate phase-matching bandwidths are available.) The adjustable location of M_1 allows one subcomponent to be temporally advanced or retarded relative to its counterpart such that the instantaneous intensity incident upon the nonlinear crystal is a function of their overlap. A variable optical delay can be provided by mounting one of the retroreflectors on an electrically driven shaker or loudspeaker cone. By synchronizing these two translations with the timebase of an oscilloscope a real-time display of the SHG autocorrelation trace can be derived from the photomultiplier signal. Since the second-harmonic intensity reaching the photomultiplier tube via the cut-off filter F depends quadratically upon the instantaneous intensity of the fundamental frequency, the record represents a second-order intensity autocorrelation function. This may be expressed as $G_2(\tau)$, where

$$G_2(\tau) = 1 + 2\int_{-\infty}^{+\infty} I(t)I(t-\tau)\,dt \Big/ \int_{-\infty}^{+\infty} |I(t)|^2\,dt \tag{10.19}$$

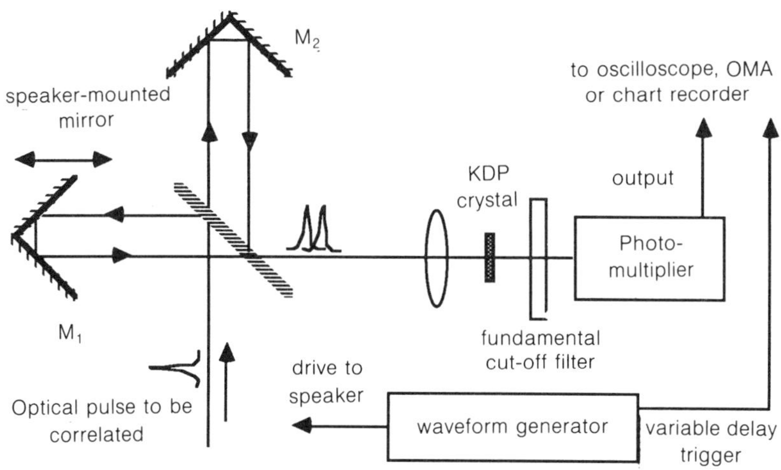

Figure 10.20 Typical layout of a second-harmonic generation (SHG) autocorrelator. (After Finch (1989).)

$I(t)$ represents the intensity profile of the pulse and τ is the optical delay that is imposed by the relative translation of M_1 with respect to M_2.

It can be readily seen from equation 10.19 that the peak-to-background SHG intensity ratio is given by 3:1 and an autocorrelation profile of the type illustrated in Figure 10.21(a) is obtained. The presence of the background

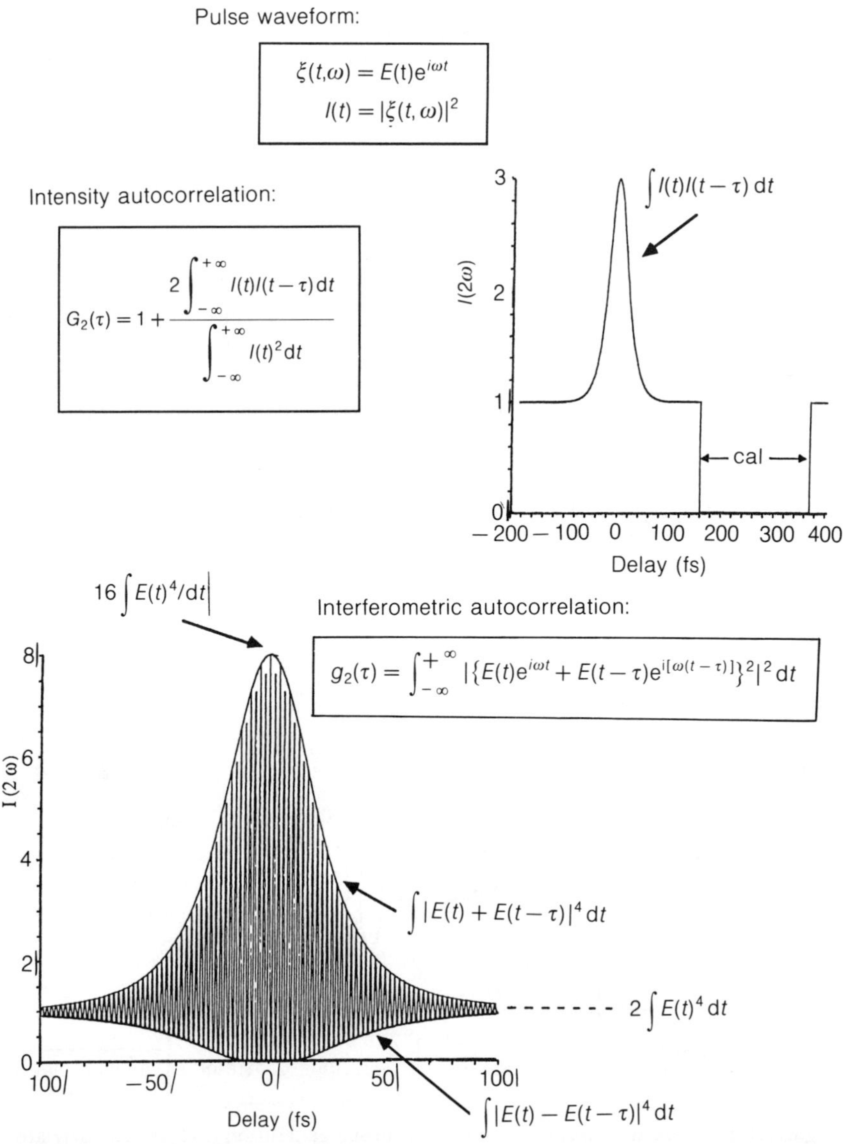

Figure 10.21 Salient characteristics of (a) intensity and (b) interferometric SHG autocorrelation functions relating to femtosecond laser pulses. (After Finch (1989).)

signal offers a useful and convenient check on the alignment of the autocorrelator and of the optimal phase matching in the nonlinear crystal. An alternative **zero-background** SHG autocorrelation trace may be recorded by either having noncollinear propagation of the two component pulses through the nonlinear crystal or selecting a crystal that produces a frequency-doubled signal only when orthogonally polarized pulses are incident (Ippen and Shank, 1977). This is sometimes preferred on the basis that a weak background signal can be more easily discerned but the contrast ratio detail is unfortunately thereby sacrificed.

The duration of the ultrashort laser pulse can be deduced from the recorded autocorrelation duration $\Delta\tau_{ac}$ provided that some prior knowledge exists or an assumption is made regarding the pulse shape. For a specific pulse shape the ratio between the FWHM pulse duration $\Delta\tau_p$ and the autocorrelation FWHM, $\Delta\tau_{ac}$, is a constant. One of the more common intensity profiles that is assumed has a sech^2 description and for this the $\Delta\tau_{ac}/\Delta\tau_p$ ratio is 1.543 and thus the pulse duration can be deduced from the SHG autocorrelation data. The validity of the pulse shape assumption can be cross-checked experimentally by measuring the spectral bandwidth, $\Delta\nu$, of the pulses. For a given intensity profile the duration–bandwidth product $\Delta\tau_p\Delta\nu$ is also a constant and for the example of the sech^2 pulseshape this is 0.315 (cf. the example shown in Figure 10.17).

In section 10.4.2 there was a considerable discussion relating to the exploitation of nonlinear phase shifts that gave rise to frequency chirps on the pulses. An intensity autocorrelation is impervious to such frequency chirp content and relatively insensitive to pulse shape. In other words, the intensity autocorrelation data give a measure of pulse durations irrespective of frequency sweep across the ultrashort pulse. While this can be convenient in respect of pulse duration measurements, a more comprehensive characterization of ultrashort pulses demands additional experimental inputs such as the interferometric autocorrelation data described in the following subsection.

10.3.3. SHG Interferometric Autocorrelation

As its name implies the SHG intensity autocorrelation is the observation of an intensity-based phenomenon as distinct from an amplitude and phase-related characterization of the ultrashort pulses. Consequently, in a time-averaged intensity autocorrelation record the features such as amplitude instability and frequency chirp may be entirely concealed. Additionally, unless particular care is exercised, the calibration of the temporal delay can be prone to inaccuracy arising from imprecise translation of the retroreflector.

These shortcomings can be overcome in a closely related measurement technique which is described as SHG interferometric (or fringe-resolved) autocorrelation (Diels *et al.*, 1978). In a configuration similar to that illustrated in Figure 10.20 the detection system (i.e. the photomultiplier tube, drive electronics and monitor) can be designed such that it possesses a sufficient frequency bandwidth with respect to the rate of the retroreflector translation that interference fringes can be resoved. This means that the amplitude and phase parameters of the optical electric fields are thereby manifested as constructive and destructive interference effects involving the recombined subcomponents as the relative delay is varied.

For a pulse with an electric field envelope described by $E(t)$ and a centre optical frequency of ω_0 then the time-dependent pulse amplitude may be expressed as

$$E(\omega_0, t) = E(t)\exp(\mathrm{i}\omega_0 t) \tag{10.20}$$

where

$$I(t) = |E(t)\exp(\mathrm{i}\omega_0 t)|^2 \tag{10.21}$$

The measured SHG intensity signal is proportional to the correlation function $g_2(\tau)$ given by

$$g_2(\tau) = \int_{-\infty}^{+\infty} |\varepsilon(\omega_0, t) + \varepsilon(\omega_0, t - \tau)|^4 \,\mathrm{d}t \tag{10.22}$$

$$g_2(\tau) \equiv \int_{-\infty}^{+\infty} |E(t)\exp(\mathrm{i}\omega_0 t) + E(t-\tau)\exp[\mathrm{i}\omega_0(t-\tau)]|^4 \,\mathrm{d}t \tag{10.23}$$

This expression for the interferometric autocorrelation has the form shown in Figure 10.21(b) where the peak-to-background ratio is 8:1 as compared with the 3:1 contrast ratio that applies to the intensity autocorrelation. Such an enhancement in contrast ratio has the major practical advantage that this type of pulse measurement technique is much more sensitive to phase and amplitude instabilities in the pulse. Moreover, the fringe detail within the autocorrelation envelope enables frequency chirp effects in the pulses to be readily observed through an associated loss of fringe visibility in the wings of the recorded trace.

It is worth emphasizing at this point that a fringe average of the interferometric autocorrelation is equivalent to the intensity autocorrelation described in the previous subsection. In common therefore with its intensity counterpart, the interferometric autocorrelation has a half-width $\Delta\tau_{ac}$ that can be related to the actual pulse duration $\Delta\tau_p$ (FWHM) by a correction factor that is pulse shape specific. (For example, with a sech^2 intensity profile the ratio $\Delta\tau_{ac}/\Delta\tau_p = 1.897$.)

In calibration terms it is also convenient to note that the delay between two fringe maxima or minima corresponds to one optical period within the pulse. Thus, provided that the central wavelength of the pulse spectrum is known and when the pulses are essentially free from frequency chirp then the fringe spacing can give a self-calibration of the autocorrelation record.

However, the interferometric autocorrelation technique does not directly provide information relating to the profile shape of the correlated pulses. The pulse shape and duration of the pulses can only be determined by applying iterative profile-fitting procedures to the autocorrelation envelope. This can be done for the loci of the fringe maxima and minima and thus the computational demands are somewhat alleviated (Finch, 1989). The envelope integrals can therefore be derived from equation 10.23 by replacing the frequency-dependent terms as indicated in the following equation:

$$g_2(\tau) = \int_{-\infty}^{+\infty} |E(t) \pm E(t-\tau)|^4 \, \mathrm{d}t \tag{10.24}$$

The field addition and subtraction relate to the maxima and minima loci respectively.

The influence of a significant frequency chirp on the ultrashort pulses produced by a mode-locked titanium–sapphire laser is clearly illustrated in the interferometric autocorrelation reproduced in Figure 10.19(a). Interestingly, when the frequency chirp is removed by intracavity compensation of group velocity dispersion (see Figure 10.19(c)), the fringe visibility is restored across the entire autocorrelation and the pulse duration is substantially shorter as indicated.

Despite the information relating to pulse shape and frequency chirp content in interferometric autocorrelation data the measurement is not definite. This is because it is possible to infer rather different combinations of pulse shape and frequency chirp parameters that lead to identical autocorrelation data. Removal of this ambiguity can be achieved by utilizing cross-correlation techniques (Diels *et al.*, 1985b). For example, the pulse can be subjected to a known amount of dispersion in a glass rod placed in one arm of the correlator before being returned with the nondispersed pulse intensity component in an SHG crystal. From this cross-correlator an iterative analysis of the intensity and interferometric data together with a precise knowledge of the dispersive properties of the glass rod should yield an amplitude and phase reconstruction of the original ultrashort pulse. Although nontrivial, this technique represents a powerful characterization method, especially for femtosecond laser pulses.

10.3.4 Electron-optical-chronoscopy

From the description of SHG autocorrelators outlined above it is clear that the interpretation of data from nonlinear measurement techniques can be a rather indirect and involved process. Although the time resolution of electron-optical streak cameras is rather more modest (300 fs at best) (Finch *et al.*, 1988b)) than that of autocorrelators, this alternative linear technique affords direct and unambiguous records of intensity–time profiles not only to discrete laser pulses but also of induced fluorescence/luminescence signals. Additionally, the spectral sensitivities of photocathodes can extend from X-ray to near-infrared wavelengths and so this scheme of electron-optical chronoscopy has a wide applicability with regard to quite general time-domain studies.

Although a variety of streak image tube designs have been reported they all operate on the same basic principle. The operation of a typical streak camera can be readily described with the aid of Figure 10.22(a). Suppose that a discrete optical pulse having a duration of $\Delta\tau_p$ is incident upon a slit as indicated. The transient illumination of the slit is then relayed onto a photocathode where an electron replica of the optical slit image is produced. The photoelectrons are accelerated and focused onto the output phosphor screen by an appropriately designed electrostatic lens arrangement. At the phosphor screen the image information is converted back to optical data. Such an image tube is designed to provide good lensing characteristics whereby the photoelectronic images can be produced with high spatial resolution and minimal geometrical distortion.

In the streak mode the electrons pass between a pair of deflectors to which a linear, time-varying voltage ramp is applied. This deflects or streaks the electrons along a direction which is perpendicular to the input slit length.

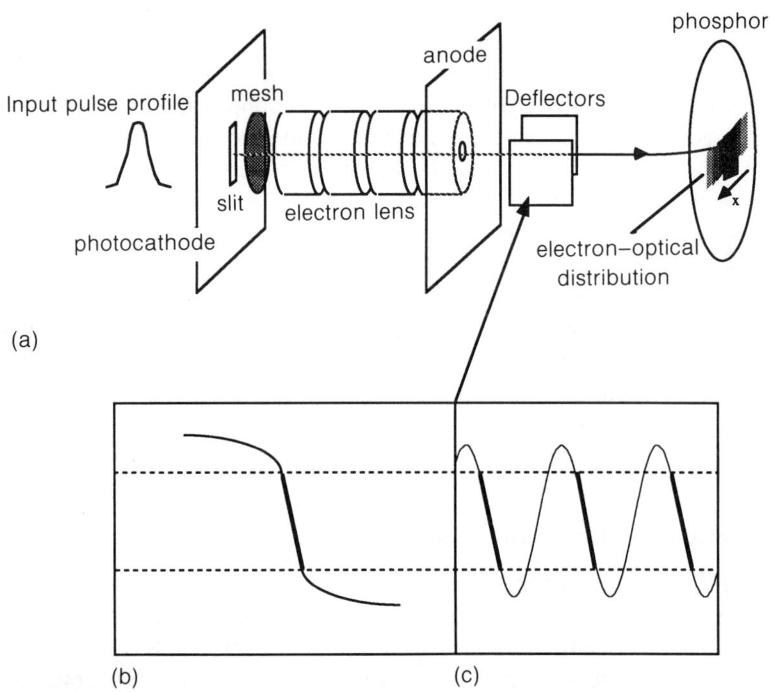

Figure 10.22 Schematic representation of an electron-optical streak image tube (a) with deflection voltage profiles for either (b) single-shot or (c) synchroscan modes of operation.

Thus for a short optical event incident on the input slit, a photoelectron signal level which is linearly proportional to the optical intensity will be produced in the image tube. In the subsequent streaking process at the deflectors the electron distribution is mapped out spatially on the phosphor screen. This is equivalent to saying that one spatial axis of the streak image tube has been converted into a temporal axis or that a time-to-space transformation has taken place. The luminescence from the phosphor screen is directly proportional to the number density of the impacting photoelectrons and thus a recording of the screen intensity as a function of streaked distance provides a linear intensity vs. time representation of the incident optical signal. Importantly, spatial information along the length of the slit is retained such that spatial or spectral data can be time resolved and simultaneously displayed by such a camera.

The dynamic operation of streak cameras can be subdivided into two distinct modes, i.e. the single-shot and the repetitive (synchroscan) operational modes. A single-shot streak record is especially relevant to single-pulse events or to flashlamp-pumped mode-locked lasers where a relatively short train of ultrashort pulses is involved. It is also well suited to applications where relatively low repetition rate cavity-dumped mode-locked CW laser outputs are employed. By contrast, when high repetition rate ($\geqslant 100$ MHz) periodic sequences of ultrashort pulses are available (e.g. mode-locked CW lasers) then the synchroscan operation is more applicable. (For time calibration purposes it is customary to include an optical delay arrangement such as that illustrated in Figure 10.23.) These streak operations are now described in turn.

10.3.4.1. Single-shot streak operation

An initial voltage bias applied to the deflectors effectively switches the streak tube off because photoelectrons are prevented from reaching the recording area of the screen. A suitably time-synchronized application of a single linear range voltage (see Figure 10.22(b)) to the deflectors then enables a temporally isolated event to be displayed on the streak tube screen. Because of space charge and related photocurrent limitations it is necessary to employ either internal or coupled-external image intensification stages so that the streaks have intensity levels that are compatible with image recording techniques.

Typical photoelectron energies in the deflection region are 10–20 keV and streak voltage ranges of several kilovolts which are switched in nanosecond time scales are generally applied. These produce streak velocities (or writing speeds) at the screen which are typically in the 10^7–3×10^8 m s^{-1} range. Thus for screen diameters of 3–4 cm the temporal window of the streak can be as short as 100 ps. For this reason, accurate synchronization must be provided between the application of the deflection voltage ramp and the optical event to be streaked. Suitable ramp generating sources therefore include avalanche transistor chains (Jackson *et al.*, 1986) (jitter < 100 ps) and photoconductive semiconductor (or Auston) switches (Auston, 1975) (jitter ~ 2 ps (Knox and Mourou, 1981)). The latter has the disadvantage that it requires relatively high energy (~ 10–100 μJ) optical pulses to ensure low jitter switching.

10.3.4.2. Synchroscan streak operation

The periodic sequence of identical pulses produced by mode-locked CW lasers has meant that a synchronously repetitive or **synchroscan** streak operation can be exploited (Sibbett, 1973, 1982). In this mode of operation a repetitive linear deflection waveform is precisely synchronized to the periodic sequence of ultrashort pulses to be streaked. For practical implementations of this scheme a high frequency (e.g. ~ 200 MHz) sinusoidal voltage is applied where advantage is taken of the fact that the central sixth of its period and central half of its amplitude are linear to within 5% (cf. Figure 10.22(c)). The frequency of the waveform may be chosen to be equal to or an exact harmonic of the cavity frequency of the mode-locked laser to be monitored. By this means a large number of individual streaks can be superposed on the phosphor screen and thereby an effective signal gain is achieved (Sibbett, 1982).

In operational terms the production of a suitable sinusoidal deflection voltage which is appropriately referenced to the mode-locked laser source is relatively straightforward. This is because an initial low voltage sinusoid can be derived either actively using a tunnel diode oscillator or passively by

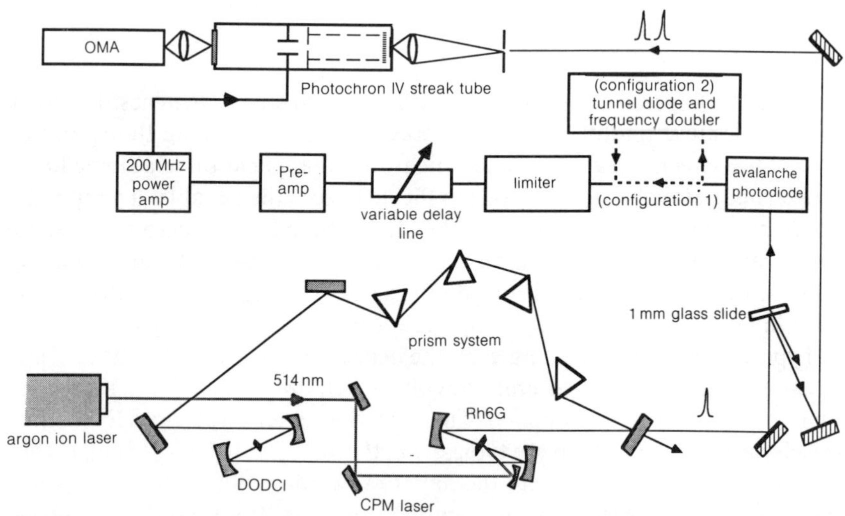

Figure 10.23 Experimental set-up of a synchroscan streak camera used in conjunction with a CPM dye laser. (After Finch *et al.* (1989).)

filtering out a selected Fourier component associated with the pulse sequence (Finch *et al.*, 1989) (Figure 10.23). With subsequent electronic amplification stages the sinusoidal voltage can provide good deflection linearity across the screen of the streak image tube.

This synchroscan streak operation has several notable advantages over the single-shot counterpart. These include (i) compatibility with mode-locked CW lasers, (ii) inherent signal gain and thus no requirement, in most cases, for image intensification, (iii) signal averaging and as a consequence large signal-to-noise ratio, high detection sensitivity and dynamic range, (iv) low photocurrent levels and hence optimal temporal–spatial resolutions and (v) effective real-time monitoring of repetitive events which are conveniently read out by one- or two-dimensional optical multichannel analysers which can be either photoelectronic or photoconductive based. It should also be appreciated, however, that the temporal resolution of a synchroscan streak camera can be limited by accumulated pulse timing jitter in the mode-locked laser output itself. Significantly, this type of streak camera can provide a useful method by which such phase noise in a mode-locked laser can be observed and quantified.

10.3.4.3. Time resolution limitations

The limiting temporal resolution (or instrumental function) of a streak camera is determined by a few key performance parameters. If it is considered that an infinitely short duration optical pulse illuminates the input slit then the spatial extent of the streaked slit image represents an inherent component of the impulse response of the camera. A basic feature of the image tube electron-optics is that of a rather limited spatial resolution (generally expressed in line pairs per millimetre). This implies that there is a minimum spatial extent for a static (or focus-mode) slit image. Under dynamic streaking conditions where higher photocurrents may be involved, a so-called dynamic resolution limit will apply. Thus a streak-determined resolution limit given by the spatial extent of the streaked slit image Δs divided by the streak velocity, v_s, at the phosphor screen. This is sometimes referred to as the technical time resolution limit $\Delta\tau_t$ and may be expressed as

$$\Delta\tau_t = \frac{\Delta s}{v_s} \tag{10.25}$$

For the Photochron II streak tube design, for which the first subpicosecond resolution was reported (Bradley and Sibbett, 1975), $\Delta s \sim 100\,\mu m$ and $v_s \sim 2 \times 10^{10}\,cm\,s^{-1}$ such that $\Delta\tau_t \sim 0.5\,ps$.

To achieve an instrumental function given by the technical resolution limit $\Delta\tau_t$ would require that the photoelectron replica of the ultrashort optical slit image was generated instantaneously and relayed within the streak tube without any time dispersion or distortion. This cannot be the case because of fundamental limitations on the photoemissive process and the subsequent photoelectron kinetics during propagation and deflection in the streak image tube.

The semitransparent photocathodes that are used in streak image tubes are generally of a positive electron affinity type. (Negative electron affinity photocathodes have been investigated but, unless modified, their speed of response is too slow for picosecond–subpicosecond ultrahigh speed camera systems (Phillips *et al.*, 1984).) A minimum photon energy that is equal to the work function ϕ of the photocathode material is required to liberate electrons into the external vacuum environment. This gives rise to the long wavelength, λ_c, limit of sensitivity where $\lambda_c = hc/\phi$. For photon energies in excess of ϕ, the emitted photoelectrons are not monoenergetic because of the fundamental electron scattering and emission process in the cathode itself. (In extreme cases of high photon energies multiple-electron emission can occur but cameras are seldom operated in this nonlinear mode.) This distribution of initial electron energies between zero and $h\nu$-ϕ is a characteristic of particular photocathodes and photocathode types (Sommer, 1968). Such an initial energy distribution (e.g. $\sim 0.5\,\text{eV}$, FWHM distribution for an S20 photocathode illuminated at $\lambda = 600\,\text{nm}$) naturally leads to a distribution in the electron travel times between the cathode and phosphor. This is usually referred to as the photoelectron transit time dispersion, $\Delta\tau_d$.

A convenient means by which the emitted photoelectrons can be made to 'forget' their energy differences involves the application of a high strength electric field (e.g. $> 20\,\text{kV}\,\text{cm}^{-1}$) in the vicinity of the photocathode. This enables the electron to be rapidly accelerated to energies in the keV range thus making their initial energies relatively insignificant (Bradley *et al.*, 1971). For an electric field close to the photocathode of E_{pc} then the transit time speed Δt_d in the streak tube may be expressed as

$$\Delta t_d = \frac{m\,\Delta u}{eE_{pc}} \tag{10.26}$$

m, e are the electron mass and charge respectively and Δu is the half-width of the initial electron velocity distribution. (This simple expression is valid for camera resolutions of several picoseconds but a much more rigorous determination is required for the entire image tube when resolutions $\lesssim 1\,\text{ps}$ are involved (Niu *et al.*, 1982).) The application of such 'extraction' electric fields is made relatively convenient by the introduction of a mesh electrode in close proximity to the photocathode surface (Bradley *et al.*, 1971). Application of an accelerating voltage to this mesh thus provides the field E_{pc} by which $\Delta\tau_d$ can be reduced. However, the transmissivity of this electrode (typically $\sim 50\%$) leads to a loss in signal electrons and some reduction in image quality can arise as a result of the microlensing effects of the mesh apertures.

If Gaussian profiles are assumed for the technical time resolution limit $\Delta\tau_t$ and the transit time dispersion $\Delta\tau_d$ then an instrumental function $\Delta\tau_i$ can be defined for the single-shot streak camera to be

$$\Delta\tau_1 = [(\Delta\tau_t)^2 + (\Delta\tau_d)^2]^{1/2} \qquad (10.27)$$

For the subpicosecond resolution Photochron II streak camera system, $\Delta\tau_t \sim 0.5$ ps and $\Delta\tau_d \sim 0.5$ ps so that $\Delta\tau_i$ was estimated to be 0.7 ps which was in fair agreement with the experimentally measured value (Bradley and Sibbett, 1975). The present state-of-the-art experimental single-shot streak camera has a demonstrated time resolution of 300 fs (Finch *et al.* 1988b).

As already mentioned, the repetitively operating or synchroscan camera system can suffer from a timing jitter between the periodic pulse train and the applied sinusoidal deflection signal. Thus the instrumental function for the synchroscan streak camera can be represented by

$$\Delta\tau_i = [(\Delta\tau_t)^2 + (\Delta\tau_d)^2 + (\Delta\tau_j)^2]^{1/2} \qquad (10.28)$$

where $\Delta\tau_j$ describes the jitter contribution. The best performance reported to date involved a Photochron IV camera (Sibbett *et al.*, 1982) used in conjunction with a mode-locked CW ring dye laser. The instrumental function was deduced to be ~ 0.9 ps from recorded experimental data such as those reproduced in Figure 10.24.

Ongoing developments in this field of research indicate that single-shot

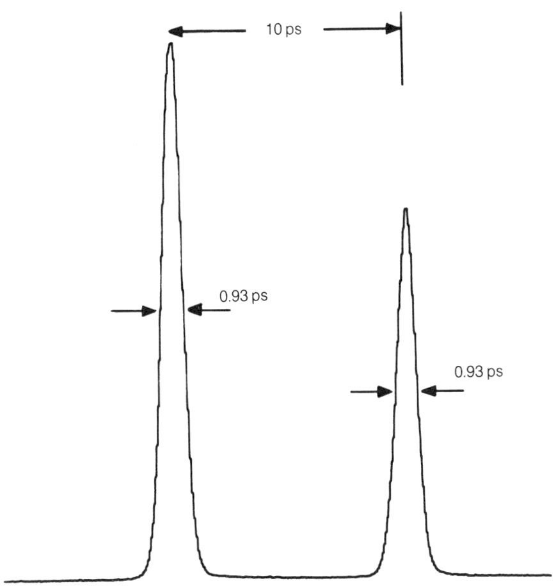

Figure 10.24 Synchroscan camera streak intensity profiles of pulses from a CPM dye laser (After Finch *et al.* (1989).)

cameras having temporal resolutions approaching 100 fs could be realized (Finch *et al.*, 1988b) and synchroscan cameras having real-time instrumental functions ~300 fs appear likely. The latter statement assumes that mode-locked CW lasers can be developed with accumulated jitter of < 300 fs and these are expected to be the all-solid-state or fibre-based systems that are currently being developed and evaluated. Once calibrated, the synchroscan camera can be employed as a monitor by which the long-term relative phase noise in mode-locked CW lasers can be directly quantified. It must be recognized, however, that for femtosecond optical pulses having correspondingly large bandwidths the dispersive effects in relay–coupling optical lenses etc. must be considered in detail. Moreover, the identity of the photo-electron signal to the input optical signal must be seriously examined when substantial temporal scrambling effects can arise in the photocathode and during transit in the streak tube. Furthermore, the detail of the temporal intensity profile of a streak image can only be deduced when the camera resolution is significantly better then the duration of the pulse. Thus for a 1 ps optical pulse duration the camera resolution should be 200 fs or better.

Although the projected improvements in streak camera resolution do not reach the femtosecond regime as such (i.e. < 100 fs), the ability to monitor weak and incoherent optical signals across a wide spectral range is a major asset for electron-optical chronoscopy. This is important not only for time-domain studies of semiconductors and semiconductor structures or devices but it is also useful as a quantitative diagnostic for digital optical data streams at multigigabit to terabit per second transfer rates. The large available dynamic range of intensity, especially for the synchroscan system, is also an important feature where low intensity noise signals may coexist with much higher intensity features.

From an overall perspective it is fair to conclude that the proper characterization of ultrashort pulses involves a combination of diagnostic techniques. When possible, it is best to exploit the unparalleled femtosecond time resolution afforded by the autocorrelation (and cross-correlation) techniques together with the linear, more direct but poorer time resolution associated with electron-optical chronoscopy. These diagnostics can also be usefully complemented with spectrum analysis from which phase noise information can be indirectly inferred (von der Linde, 1986; Rodwell *et al.*, 1989; Finch *et al.*, 1990). It follows, therefore, that unambiguous picosecond and femtosecond pulse characterization requires access to a rather comprehensive array of diagnostics.

10.4. CONCLUSION

It is clear from the above that sources of, and diagnostics for, ultrashort optical pulses now exist in laboratory arrangements. As technologies develop for nonlinear optics it is necessary that these laboratory -based sources and diagnostics be refined to a technologically compatible status for systems implementation in particular. The indications of progress in this direction are now becoming more evident where all-solid-state (e.g. diode array pumped microlasers, doped-fibre lasers etc.) are beginning to replace the less-efficient and much more bulky large-frame lasers of the 1960s–1980s era. With the excellent beam quality and the potential for high peak powers from these modern lasers the prospects for the production of high quality optical pulses for nonlinear optics are very exciting indeed. At sufficiently high peak powers the current ultrashort pulse laser–amplifier combinations having 'table-top' dimensions are finding many novel applications. Within atomic physics multiphoton ionization studies are of particular current interest and these investigations like so many to follow will undoubtedly lead to new insights into the basic science. This and other basic studies in chemistry and biology will likewise be expected to lead initially to a wealth of new avenues within pure and applied science.

At present the sources and diagnostics relating to ultrashort optical pulses have been applied in the main to specific material/device studies in the picosecond and femtosecond regimes. It is to be expected that at least some of the potential speed and bandwidth offered by digital optics will be realized in integrated optical circuitry and subsystem–system architectures. The longer-term prospects would therefore imply that digital optical processing, and ultimately specialist optical computing, will mature into viable and important technologies.

ACKNOWLEDGEMENTS

The contributions of experimental data from my research group colleagues in the J. F. Allen Physics Research Laboratories at St Andrews are gratefully acknowledged. Particular thanks are due to David Crust, Andy Finch, Robert Grant, Colin Johnston, Gordon Kennedy and David Spence who provided some of the figures used in this chapter.

REFERENCES

Agostinelli, J., Harvey, G., Stone, T. and Gabel, C. (1979) *Appl. Opt.* **18**, 2500.
Alfano, R. R., Hope, L. L. and Shapiro, S. L. (1972) *Phys. Rev. A*, **6**, 433.
Allen, L. and Jones, D. G. C. (1971) *Prog. Opt.* **9**, 181.
Armstrong, J. A. (1967) *Appl. Phys. Lett.*, **10**, 16.
Arthurs, E. G, Bradley, D. J. and Roddie, A. G. (1972) *Appl. Phys. Lett.*, **20**, 125.
Auston, D. H. (1975) *Appl. Phys. Lett.*, **26**, 101.
Auston, D. H. and Smith, P. R. (1982) *Appl. Phys. Lett.*, **41**, 599.
Beaud, P., Zysset, B., Schwarzenbach, A. P. and Weber, H. P. (1986) *Opt. Lett.*, **11**, 24.

Becker, P., Fragnito, H. L., Fork, R. L. *et al.* (1988) In *Technical Digest XVI Int. Conf. on Quantum Electronics (IQEC '88, Tokyo)*, Japan Society of Applied Physics, Postdeadline Section, Paper PD-15, p. 36.
Blow, K. J. and Wood, D. (1988) *J. Opt. Soc. Am. B*, **5**, 629.
Blow, K. J., Doran, N. J. and Nelson, B. P. (1985) *Opt. Lett.*, **10**, 393.
Bradley, D. J. (1977) In *Ultrashort Light Pulses* (ed. S. L. Shapiro), Topics in Applied Physics, Vol. 18, Springer, New York, p. 19.
Bradley, D. J. and Sibbett, W. (1973) *Opt. Commun.* **9**, 17.
Bradley, D. J. and Sibbett, W. (1975) *Appl. Phys. Lett.*, **27**, 392.
Bradley, D. J., New, G. H. C. and Caughey, S. J. (1969) *Phys. Lett. A*, **30**, (1969).
Bradley, D. J., Liddy, B. and Sleat, W. E. (1971) *Opt. Commun.*, **2**, 391.
Bradley, D. J., Holbrook, M. B. and Sleat, W. E. (1981) *IEEE J. Quantum Electron.*, **17**, 658.
Brito-Cruz, C. H., Fork, R. L. Knox, W. H. and Shank, C. V. (1986) *Chem. Phys. Lett.*, **132**, 341.
Burns, D. and Sibbett, W. (1990) *Electron. Lett.* **26**, 505.
Burns, D., Crust, D. W., Chang, J. T. K. and Sibbett, W. (1988) *Electron. Lett.*, **24**, 1439.
Cesar, C. J., Islam, M. N., Soccolich, C. E. *et al.*, (1990) *Technical Digest Conf. on Lasers and Electro-Optics (CLEO '90, Anaheim, CA)*, Optical Society of America Paper CFR1, p. 556.
Chan, C. K. and Sari, S. O. (1974) *Appl. Phys. Lett.*, **25**, 403.
Chen, J., Sibbett, W. and Vukusic, J. I. (1982) *Electron Lett.*, **18**, 426.
Crowell, M. H. (1965) *IEEE J. Quantum Electron.*, **1**, 12.
Dawson, M. D., Boggess, T. F., Garvey, D. W. and Smirl A. L. (1986) In *Ultrafast Phenomena V* (eds G. R. Fleming and A. E. Siegman), Springer Series in Chemical Physics 46, Springer, New York, p. 5.
Dawson, M. D., Boggess, T. F. and Smirl, A. L. (1987) *Opt. Lett.*, **12**, 590.
De Maria, A. J. (1971) *Prog. Opt.*, **9**, 33.
De Maria, A. J., Gagosz, R., Heynau, H. et al. (1967a) *J. Appl. Phys.*, **38**, 2693.
De Maria, A. J. (1971) *Prog. Opt.*, **9**, 33.
De Maria, A. J., Gagosz, R., Heynau, H. *et al.* (1967a) *J. Appl. Phys.*, **38**, 2693.
De Maria, A. J., Stetser, D. A. and Glenn, W. H. (1967b) *Science*, **156**, 1557.
De Maria, A. J., Glenn, W. H., Brienza, M. J. and Mack, M. E. (1969) *Proc. IEEE*, **57**, 1.
de Martini, F., Townes, C. H., Gustafson, T. K. and Kelley, P. L. (1967) *Phys. Rev.*, **164**, 312.
Desbois, J., Gires, F. and Tournois, P. (1973) *IEEE J. Quantum Electron.*, **9**, 213.
Di Domenico, Jr, M. and Czarniewski, V. (1965) *Appl. Phys. Lett.*, **6**, 150.
Diels, J.-C., van Stryland, E. W. and Gold, D. (1978) In *Picosecond Phenomena I*, Springer Series in Chemical Physics, Vol. 4, Springer, Berlin, 117.
Diels, J. C., Dietel, W., Fontaine, J. J. *et al.* (1985a) *J. Opt. Soc. Am. B*, **2**, 680.
Diels, J. C., Fontaine, J. J., McMichael, I. C. and Simon, F. (1985b) *Appl. Opt.*, **24**, 1270.
Edelstein, D. C., Wachman, E. S., Cheng, L. K. *et al.* (1988) *Appl. Phys. Lett.*, **52**, 2211.
Edelstein, D. C., Wachman, E. S. and Tang, C. L. (1989) *Appl. Phys. Lett.*, **54**, 1728.
Finch, A. (1989) *PhD Thesis*, University of St Andrews.
Finch, A., Chen, G., Sleat, W. E. and Sibbett, W. (1988a) *J. Mod. Opt.* **35**, 345.
Finch, A., Liu, A., Niu, H., Sibbett, W. *et al.* (1988b) In *Ultrafast Phenomena VI Springer Series in Chemical Physics* 48, (eds T. Yajima, K. Yoshihara, C. B. Harris and S. Shionoya), Springer, New York, p. 159.

Finch, A., Williams, E. and Sibbett, W. (1988c) *Technical Digest XVI Int. Conf. on Quantum Electronics (IQEC'88, Tokyo)*, Paper MP-16, p. 92.

Finch, A., Sleat, W. E. and Sibbett, W. (1989) *Rev. Sci. Instrum.*, **60**, 839.

Finch, A., Kean, P. N., Zhu, X. and Sibbett, W. (1990) *IEEE J. Quantum Electron.*, **26**, 1115.

Fork, R. L., Greene, B. I. and Shank, C. V. (1981) *Appl. Phys. Lett.*, **48**, 671.

Fork, R. L., Martine, O. E. and Gordon, J. P. (1984) *Opt. Lett.*, **9**, 150.

Fork, R. L., Brito-Cruz, C. H., Becker, P. C. and Shank, C. V. (1987) *Opt. Lett.*, **12**, 483.

Foster, L. C., Ewy, M. D. and Crumly, C. B. (1965) *Appl. Phys. Lett.*, **6**, 6.

French, P. M. W. and Taylor, J. R. (1986) In *Ultrafast Phenomena V.* (eds C. R. Fleming and A. E. Siegman), Springer Series in Chemical Physics 46, Springer, New York, p. 11.

French, P. M. W., Williams, J. A. R. and Taylor, J. R. (1989) *Opt. Lett.*, **14**, 48.

Froehly, C., Colonbeau, B. and Vampoille, M. (1983) In *Progress in Optics XX* (ed. E. Wolf), North Holland, Amsterdam.

Gaddy, O. L. and Shaefer, E. M. (1966) *Appl. Phys. Lett.*, **9**, 281.

Geister, G. and Ulrich, R. (1988) *Opt. Commun.*, **68**, 187.

Giordmaine, J. A., Rentzepis, P. M., Shapiro, S. L. and Wecht, K. W. (1967) *Appl. Phys. Lett.*, **11**, 216.

Glenn, W. H. and Brienza, M. J. (1967) *Appl. Phys. Lett.*, **10**, 221.

Golovchenko, E. A., Dianov, E. M., Mamyshev, P. V. and Prokhorov, A. M. (1988) *Opt. Quantum. Electron.*, **20**, 343.

Gomes, A. S. L., Österberg, U., Sibbett, W. and Taylor, J. R. (1985a) *Opt. Commun.*, **54**, 377.

Gomes, A. S. L., Sibbett, W. and Taylor, J. R. (1985b) *Opt. Lett.*, **10**, 338.

Goodberlet, J., Wang, J. and Fujimoto, J. G. (1989) *Opt. Lett.*, **14**, 1123.

Goodberlet, J., Jacobson, J. and Fujimoto, J. G. (1990) In *Technical Digest Conf. on Lasers and Electro-Optics (CLEO'90, Anaheim, CA)*, Optical Society of America Paper CFNS, p. 544.

Gouveia-Neto, A. S., Gomes, A. S. L. and Taylor, J. R. (1988) *IEEE J. Quantum Electron.*, **24**, 332.

Grant, R. S., Kean, P. N. and Sibbett, W. (1990) In *Technical Conf. on Lasers and Electro-Optics Optical Society of America (CLEO'90, Anaheim, CA)*, Paper CTUH 38.

Halbout, J.-M., May, P. and Chiu, G. (1988) *J. Mod. Opt.*, **35**, 1995.

Hanna, D. C., Kazer, A., Phillips, M. W. *et al.* (1989) *Electron. Lett.*, **25**, 95.

Harris, S. E. (1966) *Proc. IEEE.*, **54**, 1401.

Hauer, M. and Warren, W. S. (1987) *Opt. Lett.*, **12**, 398.

Heritage, J. P., Thurston, R. N., Tomlinson, W. J. *et al.* (1985a) *Appl. Phys. Lett.*, **47**, 87,

Heritage, J. P., Weiner, A. M. and Thurston, R. N. (1985b) *Opt. Lett.*, **10**, 609.

Herrmann, J. and Wilhelmi, B. (1987) *Lasers for Ultrashort Light Pulses*, Akademic-Verlag, Berlin.

Ho, P. T., Glasser, L. A., Ippen, E. P. and Haus, H. A. (1978) *Appl. Phys. Lett.*, **33**, 241.

Ippen, E. P. and Shank, C. V. (1977) In *Ultrafast Light Pulses* (ed. S. L. Shapiro), Topics in Applied Physics, Vol. 18, New York, Springer, p. 83.

Islam, M. N., Sunderman, E. R., Bar-Joseph, I. *et al.* (1989a) *Appl. Phys. Lett.*, **54**, 1203.

Islam, M. N., Sunderman, E. R., Soccolich, C. E. *et al.* (1989b) *IEEE J. Quantum Electron.*, **25**, 2454.

Jackson, M. C., Long, R. D., Lee, D. and Freeman, N. J. (1986) *Laser Particle Beams*, **4**, 145.
Johnson, A. M. and Simpson, W. M. (1983) *Opt. Lett.*, **8**, 554.
Johnson, A. M. and Simpson, W. M. (1985) *J. Opt. Soc. Am. B*, **2**, 619.
Johnston, C. I., Spence, D. E., Grant, R. S. and Sibbett, W. (1989) *Opt. Lett. Commun.*, **73**, 370.
Kafka, J. D. and Baer, T. (1987) *Opt. Lett.*, **12**, 181.
Kafka, J. D., Head, D. F. and Baer, T. (1986) In *Ultrafast Phenomena V* (eds G. R. Fleming and A. E. Siegman), Springer Series in Chemical Physics 46, Springer, New York, p. 51.
Kafka, J. D., Watts, M. L., Roach, D. J. *et al.* (1990) In *Technical Digest Conf. on Lasers and Electro-Optics (CLEO'90, Anaheim, CA)*, Optical Society of America, Paper CPDP8, p. 614.
Kean, P. N., Grant, R. S., Zhu, X. *et al.* (1988) In *Technical Digest Conf. on Lasers and Electro-Optics (CLEO'88, Anaheim CA)*, Optical Society of America Paper PD-7.
Kean, P. N., Zhu, X., Crust, D. W. *et al.* (1989) *Opt. Lett.*, **14**, 39.
Knox, W. H. (1988) *IEEE J. Quantum Electron.*, **24**, 388.
Knox, W. and Mourou, G. (1981) *Opt. Commun.*, **32**, 203.
Knox, W. H., Downer, M. C., Fork, R. L. and Shunk, C. V. (1984) *Opt. Lett.*, **9**, 552.
Kuckartz, M., Schulz, R. and Harde, H. (1988) *J. Opt. Soc. Am. B*, **5**, 1353.
Kuhl, J., Klingenberg, H. and von der Linde D. (1979) *Appl. Phys.*, **18**, 279.
Langford, N., Smith, K. and Sibbett, W. (1987a) *Opt. Lett.*, **12**, 817.
Langford, N., Smith, K. and Sibbett, W. (1987b) *Opt. Lett.*, **12**, 903.
Langford, N., Grant, R. S., Johnston, C. I. *et al.* (1989) *Opt. Lett.*, **14**, 45.
Liu, L. Y., Huxley, J. M., Ippen, E. P. and Haus, H. A. (1990) *Opt. Lett.*, **15**, 553.
Maier, M., Kaiser, W. and Giordmaine, J. A. (1966) *Phys. Rev. Lett.*, **17**, 1275.
Maine, P., Strickland, D., Bado, P. *et al.* (1988) *IEEE J. Quantum Electron.*, **24**, 398.
Manning, R. J. and Hill, J. R. (1986) *Opt. Acta*, **33**, 483.
Mark, K., Liu, L. T., Hall, K. L. *et al.* (1989) *Opt. Lett.*, **14**, 48.
Mathies, R. A., Pollard, W. T., Brito-Cruz, C. H. and Shank, C. V. (1988) In *Ultrafast Phenomena VI* (eds T. Yajima, K. Yoshihara, C. B. Harris and S. Shionaya), Springer Series in Chemical Physics 48, Springer, New York, p. 584.
McClure, R. E. (1965) *Appl. Phys. Lett.*, **7**, 148.
Miller, A. and Sibbett, W. (1988) *J. Mod. Opt.*, **35**, 1871.
Mitschke, F. M. and Mollenauer, L. F. (1986) *IEEE J. Quantum Electron.*, **22**, 2242.
Mitschke, F. M. and Mollenauer, L. F. (1987) *Opt Lett.*, **12**, 407.
Mollenauer, L. F. (1985a) In *Handbook of Lasers* (eds M. L. Stitch and M. Bass), Vol. 4, Amsterdam, North-Holland, Chapter 2.
Mollenauer, L. F. (1985b) *Philos. Trans. R. Soc. London, Ser. A*, **315**, 437.
Mollenauer, L. F. and Stolen, R. H. (1984) *Opt. Lett.*, **9**, 13.
Morin, M. and Piché, M. (1989) *Opt. Lett.*, **14**, 1119.
Moulton, P. F. (1986) *J. Soc. Am. B*, **3**, 125.
Nakazawa, M., Nakashima, T., Kubuta, H. and Seikai, S. (1987) *Opt. Lett.*, **12**, 681.
New, G. H. C. (1983) *Rep. Prog. Phys.*, **46**, 877.
Niu, H., Sibbett, W. and Baggs, M. R. (1982) *Rev. Sci. Instrum.*, **53**, 563.
Ouellette, F. and Piché, M. (1988) *Can. J. Phys.*, **66**, 903.
Owyoung, A., Hellwarth, R. W. and George, N. (1972) *Phys. Rev. B*, **5**, 628.

Parker, D. G., Say, P. G., Hansom, A. M. and Sibbett, W. (1987) *Electron. Lett.*, **23**, 527.
Phillips, C. C., Hughes, A. E. and Sibbett, W. (1984) *J. Phys. D*, **17**, 1713.
Rodwell, M. J. W., Bloom, D. M. and Weingarten, K. J. (1989) *IEEE J. Quantum Electron.*, **25**, 817.
Rosker, M. J., Wise, F. W. and Tang, C. L. (1986) *Appl. Phys. Lett.*, **49**, 1726.
Schlager, J. B., Yamabayashi, Y., Franzer, D. L. and Juneau, R. I. (1989) *IEEE Phot. Tech. Lett.*, **1**, 264.
Sibbett, W. (1973) PhD Thesis, Belfast.
Sibbett, W. (1982) *Proc. Soc. Photo-Opt. Instrum. Eng.*, Optical Society of America, **348**, 15.
Sibbett, W. (1989) In *Technical Digest Conf. on Lasers and Electro-Optics (CLEO'89, Baltimore, MD)*, Paper-FQ1, p. 440.
Sibbett, W. (1990) In *Ultrafast Phenomena V* (eds E. P. Ippen and C. B. Harris), Springer, New York.
Sibbett, W. and Taylor, J. R. (1983a) *Opt. Commun.*, **46**, 32.
Sibbett, W. and Taylor, J. R. (1983b) *IEEE J. Quantum Electron.*, **19**, 558.
Sibbett, W. and Taylor, J. R. (1984) *IEEE J, Quantum Electron.*, **20**, 108.
Sibbett, W., Taylor, J. R. and Welford, D. (1980) *Opt. Commun.*, **35**, 283.
Sibbett, W., Niu, H. and Baggs, M. R. (1982) *Rev. Sci. Instrum.*, **53**, 758.
Siegman, A. F. (1981) *Opt. Lett.*, **6**, 334.
Silberberg, Y., Smith, P. W., Eilenberger, D. J. *et al.* (1984) *Opt. Lett.*, **9**, 507.
Smith, K., Langford, N., Sibbett, W. and Taylor, J. R. (1985) *Opt. Lett.*, **10**, 559.
Smith, P. W. (1970) *Proc IEEE*, **58**, 1342.
Smith, P. W., Duguay, M. A. and Ippen, E. P. (1974) *Progress in Quantum Electronics*, Vol. 3, Part 2, Pergamon, Oxford.
Smith, P. W., Silberberg, Y. and Miller, D. A. B. (1985) *J. Opt. Soc. Am. B*, **2**, 1228.
Soccolich, C. E., Islam, M. N., Young, M. G. and Miller, B. I. (1990) *Appl. Phys. Lett.*, **56**, 2177.
Sommer, A. H. (1968) In *Photoemissive Materials*, Wiley, New York.
Spence, D. F., Kean, P. N. and Sibbett, W. (1990) In *Technical Digest Conference on Lasers and Electro-Optics (CLEO '90, Anaheim, CA)*, Optical Society of America, p. 619.
Spielman, C., Krauz, F., Wintner, E. and Schmidt, A. (1990) In *Ultrafast Phenomena VII* (eds E. P. Ippen and C. B. Harris), Springer Series in Chemical Physics, Springer, Heidelberg.
Statz, H. and Tang, C. L. (1965) *J. Appl. Phys.* **30**, 3923.
Stolen, R. H. (1980) *Fiber Integr. Opt.*, **3**, 21.
Stolen, R. H. and Lin, C. (1978) *Phys. Rev. A*, **17**, 1448.
Thurston, R. N., Heritage, J. P., Weiner, A. M. and Tomlinson, W. J. (1986) *IEEE J. Quantum Electron.*, **22**, 682.
Treacy, E. B. (1968) *Phys. Lett. A*, **28**, 34.
Tucker, R. S., Koren, U., Raybon, G., *et al.* (1989) *Electron. Lett.*, **25**, 621.
Valdmanis, J. A., Fork, R. L. and Gordon, J. P. (1985) *Opt. Lett.*, **10**, 131.
von der Linde, D. (1972) *IEEE J. Quantum Electron.*, **8**, 328.
von der Linde, D. (1986) *Appl. Phys. B*, **39**, 201.
Vyslouth, V. A. and Serkin, V. N. (1983) *Pis'ma Zh. Eskp. Teor. Fiz.*, **38** (Part 3), 170.
Wachman, E. S., Edelstein, D. C. and Tang, C. L. (1990) *Opt. Lett.*, **15**, 136.
Weiner, A. M. and Heritage, J. P. (1987) *Rev. Phys. Appl.*, **22**, 1619.

Weiner, A. M., Heritage, J. P. and Thurston, R. M. (1986) *Opt. Lett.*, **11**, 153.
Weiner, A. M., Heritage, J. P. and Saleki, J. A. (1988a) *Opt. Lett.*, **13**, 300.
Weiner, A. M., Heritage, J. P. and Kirsche, E. M. (1988b) *J. Opt. Soc. Am. B*, **5**, 1563.
Weingarten, K. J., Rodwell, M. J. W. and Bloom, D. M. (1988) *IEEE J. Quantum Electron.*, **24**, 198.
Willson, J. P., Sibbett, W. and Sleat, W. E. (1982) *Opt. Commun.*, **42**, 208.
Yakymyshyn, C. P., Pinto, J. F. and Pollock, C. R. (1989) *Opt. Lett.*, **14**, 612.
Yoshizawa, M. and Kobayashi, T. (1984) *IEEE J. Quantum Electron.*, **20**, 797.
Zhu, X., Kean, P. N. and Sibbett, W. (1989a) *IEEE J. Quantum Electron.*, **25**, 2445.
Zhu, X., Kean, P. N. and Sibbett, W. (1989b) *Opt. Lett.*, **14**, 1192.
Zysset, B., Hodel, W., Beaud, P. and Weber, H. P. (1986) *Opt. Lett.*, **11**, 156.

Index